Plant-based Foods and their Implications in Brain Health

Phytoconstituents are a valuable source of new lead compounds due to their structural variety.

The accelerated rate of human ageing brought about by the 21st century's longer life expectancy seems to be one of the major problems facing varied cultures worldwide. Ageing and the onset of chronic diseases appear to be significantly influenced by the type of food consumption pattern that people follow throughout their lives. There is strong evidence that Mediterranean diets are linked to a lower risk of brain disorders. A greater proportion of rural communities in the Mediterranean basin eat a diet high in plant-based foods daily. Stressors like reactive oxygen and reactive nitrogen species have a major negative influence on the brain in addition to the cardiovascular system.

Foods, particularly those derived from plants, are a great source of vitamins, antioxidants, polyphenols, and vital fatty acids, according to a growing body of epidemiological research. Studies indicate the effect of antioxidant vitamins on brain health, including ascorbic acid and α-tocopherol. Polyphenols and other plant-derived bioactivities influence how animal and human brains operate by reducing the amount of oxidative stress and controlling different signal transduction pathways, according to a large body of experimental evidence.

The present volume provides a collation of recent developments in the field with special reference to various types of plant-based phytoconstituents. The book's content is relevant and interesting to a wide range of audiences from the backgrounds of plant sciences, pharmacy, biochemistry, and molecular biology. Apart from being a good reference material source, the book is relevant to plant biotechnologists and students/researchers in the field.

Plant-based Foods and their Implications in Brain Health

Edited by
Soumya Mukherjee, Akula Ramakrishna, and
Atanu Bhattacharjee

CRC Press
Taylor & Francis Group
Boca Raton London New York

CRC Press is an imprint of the
Taylor & Francis Group, an **informa** business

Designed cover image: Shutterstock

First edition published 2025
by CRC Press
2385 NW Executive Center Drive, Suite 320, Boca Raton FL 33431

and by CRC Press
4 Park Square, Milton Park, Abingdon, Oxon, OX14 4RN

CRC Press is an imprint of Taylor & Francis Group, LLC

ISBN: 978-1-032-29231-1 (hbk)
ISBN: 978-1-032-29232-8 (pbk)
ISBN: 978-1-003-30050-2 (ebk)

DOI: 10.1201/9781003300502

Typeset in Times New Roman
by Newgen Publishing UK

Contents

Foreword

I am delighted to introduce and endorse the forthcoming publication of the book titled "**Plant-based Foods and their Implications in Brain Health**." This comprehensive reference book is a monumental step forward in neuroscience, as it assembles the latest information on edible medicinal plants and their potential role in managing the healthy status of our brains. Authored by distinguished experts, this book is invaluable for professionals, researchers, and students in neuroscience, traditional medicine, psychology, and pharmacognosy.

I appreciate the relevance and significance of this compilation. The book is not only about basic information about the brain but also explores various *in-vitro* and *in-vivo* strategies to encounter progressive neuro-degenerative disorders like Alzheimer's disease. It provides in-depth knowledge about the use of marine resources, phytocompounds and their possible role in maintaining the positivity of the brain to keep it disorder and disease-free. Furthermore, it discusses the crucial role of phytomolecules and natural products in Alzheimer's therapy, an area that holds great promise for future treatments. One of the key strengths of this reference book is its coverage of biomarker analysis for Alzheimer's disease. Identifying reliable biomarkers is pivotal in early diagnosis and monitoring, and the insights offered in this book can potentially impact clinical practice and research efforts. This insight is essential for developing targeted interventions and improving our understanding of the disease process in the case of treatment of Alzheimer's patients.

In conclusion, I can say with confidence that this book offers a comprehensive and up-to-date resource for those who are deeply committed to addressing the challenges of brain health and keeping it fit. I wholeheartedly recommend this book to all the professionals, researchers, and students associated with this field and wish the Editor(s) very best for future endeavours.

Prof. (Dr.) S. P. Singh
Vice-Chancellor
The Assam Royal Global University
Guwahati-781035, Assam, India

Preface

Life expectancy in the 21st century has led to increased rates of human ageing, which appears to be one of the biggest challenges in diverse communities worldwide. The nature of food consumption patterns throughout the human life cycle seems to have a major impact on ageing and the development of chronic diseases. Convincing evidence exists in terms of reduced risk of brain disorder to be associated with Mediterranean diets. In the Mediterranean basin, a wider percentage of rural populations consume a daily diet rich in foods of plant origin. Apart from the cardiovascular system, the brain significantly suffers from the impact of stressors, such as reactive oxygen and reactive nitrogen species. The structural diversity of phytoconstituents makes them a valuable source of novel lead compounds for the quest for drugs to improve brain health. In the human brain, the ageing process is typically accompanied by various histological modifications, such as shrinkage of the brain's grey and white matter volume concomitant with other bio-molecular changes. Increasing evidence from epidemiological studies has revealed that foods, especially those of plant origin, are rich in essential fatty acids, polyphenols, vitamins, and antioxidants. Investigations suggest an impact of antioxidant vitamins, such as α-tocopherol and ascorbic acid on brain health. Considerable experimental evidence reveals that polyphenols and other plant-derived bioactivities modulate animal and human brain function by lowering oxidative stress load and by regulating various signal transduction pathways. Omega-3 polyunsaturated fatty acids (n-3 PUFAs) in Mediterranean diets may exert a significant regulatory role in the prevention of brain diseases. PUFAs also seem to be involved in some neuropsychiatric disorders, particularly depression, as well as in brain ageing, dementia, and Alzheimer's disease. PUFAs cannot be synthesized de novo but can be formed from linoleic acid (LA) and α-linolenic acid (ALA). Various oil seed plants like primrose, sunflower, corn, or safflower oil are rich in LA, while hemp, flax oil, as well as nuts and green leafy vegetables such as purslane are main sources of ALA. Polyphenols (flavonoids, coumarins, stilbene) and neurotransmitters (serotonin, melatonin, GABA, and dopamine) that are widely distributed in the leaves, seeds, bark, and flowers of plants have been identified to date and are known to regulate brain health. Non-alcoholic beverages and narcotics also exhibit a concentration dependence on brain health and the efficiency of the nervous system. Thus, in view of the above-mentioned background, this edited volume compiles recent advancements in investigations pertaining to plant-derived food and brain health.

Editors' bio

Soumya Mukherjee has accomplished his PhD in Plant Physiology from the Department of Botany, University of Delhi, India. He has been an ex-faculty of Ramjas College, University of Delhi. Currently, he is employed as an Assistant Professor in the Department of Botany, Jangipur College, University of Kalyani, West Bengal, India. He has worked as a CSIR-research fellow in the area of abiotic stress physiology of plants. He has published both research and review articles in various peer-reviewed international journals (SCOPUS and SCI indexed). He is serving as an associate editor in the editorial board of Plant Signalling and Behaviour (Taylor and Francis) and the Journal of Plant Growth Regulation (Springer). He has been working in the area of salinity stress physiology in sunflowers. He has worked to decipher the role of serotonin and melatonin in imparting salinity tolerance in sunflower and tomato seedlings under salinity stress. He has been working as a Principal Investigator on a DST-SERB funded 'State University Research Excellence Grant' (2023) on the effects of melatonin supplementation in sunflowers for the augmentation of salinity stress tolerance. He has authored an e-learning module in Plant Physiology and Biochemistry published as an effort of the National Mission on Education through ICT (MHRD Project undertaken by the University of Delhi). He has acted as a subject reviewer for various SCOPUS-indexed peer-reviewed international journals. He has published edited volumes in Plant Signaling, and Communication (Springer), Rhizobiology (Springer), Phytomelatonin (Springer), Crop sustainability and IPR (Apple academic press), Strigolactones, Alkamides and Karrikins (CRC press).

Ramakrishna Akula is currently a scientist at Bayer Science and Innovation Pvt.Ltd., Vegetable R&D, Bangalore, India. Dr. Ramakrishna holds a master's degree from Sri Krishna Devaraya University, Anantapur, Andhra Pradesh, India. He obtained his PhD in biochemistry from the University of Mysore, CSIR-CFTRI, Mysuru. He is a Senior Research Fellow of CSIR, New Delhi. He is involved in various research fields such as Plant Biotechnology, Biochemistry, Plant secondary metabolites, Food Science and Technology and Vegetable quality analytics. He has published several research papers in leading international journals, books, reviews and book chapters. He also serves as an editorial board member and reviewer for reputed international journals. Presented over 20 research papers in symposia in India and abroad, as well as a few invited lectures in India. He attended the Fifth International Symposium on Plant Neurobiology held in 2009 in Florence, Italy. He also attended the Technical Community of Monsanto (TCM) held in 2016 in St. Louis, Missouri. He is a member of the Indian Science Congress Association and Society for Biotechnologists (India). He is a fellow of the Society for Applied Biotechnology, India (2012), and recipient of the Global Vegetable Research Excellence Award (2017), Global Technology Recognition, Rapid Recognition Award, and Special recognition from the Monsanto company.

Atanu Bhattacharjee is currently working as a Professor at the School of Pharmacy, The Assam Royal Global University, Assam, India. He is considered a pioneer in the field of Pharmacognosy & Phytochemistry. Dr. Bhattacharjee earned his Ph.D. in Pharmaceutical Sciences from Nitte University, Mangalore, India. His area of research is phytochemistry and neuro-pharmacology. He worked extensively on various *in-silico*, *in-vitro* and *in-vivo* experiments to encounter neuro-degenerative disorders like Alzheimer's disease, etc., with folklore medicine of Indian diaspora and determination of safety profile, standardization of medicinal plants. More than 25 research and review articles published in internationally acclaimed journals are under his credentials. In 2021, he received the prestigious "Young Scientist Award" from the VDGOOD Professional Association, India, for his significant contribution to scientific research. His extensive work on natural products to encounter SARS COV-2 has been recognized globally as he has received a patent from South Africa in 2022 on 'Bioinformatics platform for identification of quercetin derivatives as immunity booster molecules against SARS COV-2'. He also has 5 Indian patents under his credentials. He was the Honourable Academic Editor of Asian Journal of Medicine and Health, SCIENCEDOMAIN International, India till 2021. He is also an editorial board member and reviewer of many internationally reputed journals. Recently, he has edited a book entitled "Computational and Experimental Studies in Alzheimer's Disease. [ISBN: 9781003412069]" Taylor & Francis, CRC Press. USA. 2024;

Contributors

Navidha Aggarwal
MM College of Pharmacy,
Maharishi Markandeshwar (Deemed to be University),
Mullana 133207, Ambala, Haryana, India

Shopnil Akash
Faculty of Allied Health Science,
Department of Pharmacy,
Daffodil International University, Daffodil Smart City,
Ashulia, Savar, Dhaka 1207, Bangladesh

Athanasios Alexiou
Department of Science and Engineering,
Novel Global Community Educational Foundation,
Habersham, NSW, 2770, Australia

Muhammad Asad Ali
Department of Biosciences,
COMSATS University,
Islamabad, Pakistan

Mamoni Banerjee
Bio-Research Laboratory,
Rajendra Mishra School of Engineering Entrepreneurship,
Indian Institute of Technology Kharagpur,
Kharagpur, India

Atanu Bhattacharjee
Royal School of Pharmacy,
The Assam Royal Global University,
Guwahati 781035,
Assam, India

Bedanta Bhattacharjee
Girijananda Chowdhury Institute of Pharmaceutical Science,
Tezpur 784501, Assam, India

Paramita Bhattacharjee
Department of Food Technology and Biochemical Engineering,
Jadavpur University,
Kolkata, India

Arghya Bhattacharya
Department of Pharmacology,
Bengal School of Technology,
Chinsurah R S, Sugandha,
West Bengal 712102, India

Kunal Bhattacharya
Pratiksha Institute of Pharmaceutical Science,
Panikhaiti, Guwahati, Assam-781026, India; and
Royal School of Pharmacy,
The Assam Royal Global University,
Guwahati, Assam 781035, India

Shabana Bibi
Department of Biosciences,
Shifa Tameer-e-Millat University,
Islamabad 44000, Pakistan

Mainak Chakraborty
Department of Pharmaceutical Technology, NSHM Knowledge Campus,
Kolkata – Group of Institutions,
Kolkata 700053, West Bengal, India

Nongmaithem Randhoni Chanu
Faculty of Pharmaceutical Science, Assam Downtown University,
Guwahati, Assam 781026, India

Bharti Chaudhry
Professor, Department of Botany,
Ramjas College, University of Delhi,
Delhi, India

Hitesh Chopra
Department of Biosciences,
Saveetha School of Engineering,
Saveetha Institute of Medical and Technical Sciences,
Chennai-602105, Tamil Nadu, India

Anamika Das
Department of Food Technology and Biochemical Engineering,
Jadavpur University,
Kolkata, India

Dibyajyoti Das
Pratiksha Institute of Pharmaceutical Science, Panikhaiti,
Guwahati, Assam 781026, India

Shaileyee Das
Department of Pharmaceutical Chemistry, Bharat Technology,
Howrah, Affiliated by Maulana Abul Kalam Azad University of Technology,
West Bengal, India

Sanjukta Datta
Department of Food Technology and Biochemical Engineering,
Jadavpur University,
Kolkata, India

Biswajit Devanath
Pratiksha Institute of Pharmaceutical Sciences,
Guwahati, Assam 781026, India

Ganesh Dey
Department of Pharmaceutical Technology,
University of North Bengal, Darjeeling, West Bengal, India

G.P. Dubey
Department of Kriya Sharir,
Faculty of Àyurveda, I.M.S., B.H.U.,
Varanasi, Uttar Pradesh, India

Nilanjan Ghosh
Department of Pharmaceutical Technology,
Jadavpur University,
Kolkata 700032, West Bengal, India

Farah Alam Gill
Department of Biosciences,
Shifa Tameer-e-Millat University,
Islamabad 44000, Pakistan

Simran Giri
Department of Pharmaceutical Technology,
University of North Bengal, Darjeeling, West Bengal, India

Rashid Hussain
Department of Biosciences,
Shifa Tameer-e-Millat University,
Islamabad 44000, Pakistan

Sana Javed
Department of Biosciences,
Shifa Tameer-e-Millat University,
Islamabad 44000, Pakistan

Sajal Kumar Jha
Department of Pharmaceutical Technology,
Adamas University,
Kolkata 700126, West Bengal, India

Pratap Kalita
Pratiksha Institute of Pharmaceutical Sciences,
Guwahati, Assam, India, 781026

Ramen Kalita
Pratiksha Institute of Pharmaceutical Science,
Panikhaiti, Guwahati, Assam 781026, India

Ranjan Dutta Kalita
Department of Biotechnology,
Royal School of Biosciences, The Assam Royal Global University,
Betkuchi, Guwahati 781035, India

Dipanjan Karati
Department of Pharmaceutical Technology,
School of Pharmacy, Techno India University,
Kolkata, West Bengal 700091, India

Ankit Kumar
Agriculture and Food Engineering Department,
Indian Institute of Technology Kharagpur,
Kharagpur, West Bengal 721302, India

Rinki Kumari
ITRC-ECD_ICMR- Head quarter,
New Delhi, India

Robert Lalonde
Laboratory of Stress, Immunity, Pathogens (EA7300),
University of Lorraine, Medical School,
54500 Vandoeuvre- les- Nancy, France

Garima Malik
ITRC-ECD_ ICMR- Head Quarter,
New Delhi, India

Subhash C. Mandal
Pharmacognosy and Phytotherapy Research Laboratory,
Division of Pharmacognosy,
Department of Pharmaceutical Technology,
Jadavpur University,
Kolkata 700032, India

Soumya Mitra
Dr. B.C. Roy College of Pharmacy & Allied Health Sciences,
Durgapur,
West Bengal 713206, India

Swarupananda Mukherjee
Department of Pharmaceutical Technology,
NSHM Knowledge Campus,
Kolkata –Group of Institutions,
124, B. L. Saha Road, Kolkata 700053, West Bengal, India

Nobendu Mukerjee
Center for Global Health Research, Saveetha Medical College and Hospital,
Saveetha Institute of Medical and Technical Sciences,
Chennai, Tamil Nadu, India

Debabrata Nath
Pratiksha Institute of Pharmaceutical Science,
Panikhaiti, Guwahati, Assam 781026, India

Yutika Nath
Serology Division, State Forensic Science Laboratory,
Directorate of Forensic Sciences,
Kahilipara, Guwahati 781019, India

Olakunle J. Onaolapo
Behavioural Neuroscience/Neuropharmacology Unit,
Department of Pharmacology and Therapeutics,
Faculty of Basic Clinical Sciences,
Ladoke Akintola University of Technology,
Ogbomoso, Oyo State, Nigeria

Adejoke Y. Onaolapo
Behavioural Neuroscience/Neurobiology Unit,
Department of Anatomy, Faculty of Basic Medical Sciences,
Ladoke Akintola University of Technology,
Ogbomoso, Oyo State, Nigeria

Prosanta Pal
Department of Pharmaceutical Technology,
University of North Bengal, Darjeeling, West Bengal, India

Pawan Prabhakar
Bio-Research Laboratory,
Rajendra Mishra School of Engineering Entrepreneurship,
Indian Institute of Technology Kharagpur,
Kharagpur; and
Woxsen University,
Hyderabad 502345, India

Surender Prasad
Pratiksha Institute of Pharmaceutical Sciences,
Guwahati, Assam 781026, India

Tasrina Rahman
Pratiksha Institute of Pharmaceutical Sciences,
Guwahati, Assam 781026, India

Sunil Kumar Rai
Department of Cell and Molecular Biology,
Medical University of the Americas, Nevis,
Saint Kitts and Nevis, West Indies KN 1101

Akula Ramakrishna
Bayer Research and Development Centre,
Kallinayaknahalli Village,
Karnataka 561213, India

Souvik Roy
Department of Pharmaceutical
Technology, NSHM Knowledge Campus,
Kolkata – Group of Institutions,
Kolkata 700053, West
Bengal, India

Ram Kumar Sahu
Department of Pharmaceutical Sciences,
Hemvati Nandan Bahuguna Garhwal
University (A Central University), Chauras Campus,
Tehri Garhwal 249161, Uttarakhand, India

Ammara Saleem
Department of Pharmacology,
Faculty of Pharmaceutical Sciences,
Government College University Faisalabad,
Faisalabad 38000, Pakistan

Bapi Ray Sarkar
Department of Pharmaceutical Technology,
University of North Bengal, Darjeeling,
West Bengal, India

Nilanjan Sarkar
Department of Pharmaceutical
Technology, NSHM Knowledge Campus,
Kolkata – Group of Institutions,
Kolkata 700053, West Bengal, India

Satyabrat Sarma
Faculty of Pharmaceutical Sciences,
Assam Down Town University,
Guwahati, Assam 781026, India

Sneh Shalini
ITRC-ECD_ ICMR-Head Quarter,
New Delhi, India

Rahul Kumar Sharma
Amar Shaheed Baba Ajit Singh Jujhar Singh
Memorial College of Pharmacy,
Bela, Ropar, India

Vanshika Sharma
Department of Biotechnology,
Royal School of Biosciences,
The Assam Royal Global University,
Betkuchi, Guwahati 781035, India

Jasmeet Singh
Department of Dravyaguṇa,
Faculty of Àyurveda, I.M.S., B.H.U.,
Varanasi, Uttar Pradesh, India

Nanki Singh
Govt. Ayu. College,
Varanasi, Uttar Pradesh, India

Catherine Strazielle
Laboratory of Stress, Immunity,
Pathogens (EA7300),
University of Lorraine, Medical School,
54500 Vandoeuvre-les-Nancy, France

Jaweria Syed
Department of Rehabilitation Sciences,
Shifa Tameer-e-Millat University,
Islamabad 44000, Pakistan

Shilpa Thakur
Amar Shaheed Baba Ajit Singh Jujhar Singh
Memorial College of Pharmacy,
Bela, Ropar, India

1 Neurobiology
An Overview

Rashid Hussain, Shabana Bibi, Ammara Saleem, Arghya Bhattacharya, Jaweria Syed, Nobendu Mukerjee, and Athanasios Alexiou

1.1 INTRODUCTION

The nervous system (NS) of humans has been a matter of keen study since the era of Hippocrates, from the Greek civilization that is well renowned as the 'Father of Medicine'. He observed that our brain is the organ used to regulate and recognize functions and emotions in the human body [1]. After his efforts, a physician studied the brain and, after dissections, explored the NS, which had ventricles in the cerebral hemisphere and cerebellum, vessels both for sensory and motor inputs [2].

Human physiology is complex, and so is the barrier system protecting the most vital region known as the NS. Basic immune reactions are regulated and halted by their specialized mechanisms [3]. A functional set of cells making tissues and, ultimately, an organ meant to regulate bodily responses based on any intrinsic or extrinsic stimulus constitutes an NS. Topographically, it has two principal divisions, the central and peripheral NS [4]. The brain and spinal cord constitute the central nervous system (CNS) [5]. There are 24 cranial nerves (12 pairs) and 62 spinal nerves (31 pairs) in the peripheral nervous system (PNS). All ganglia are connected with these two types of nerves (cranial and spinal nerves), along with bilateral sympathetic and parasympathetic pelvic nerve chains [4].

The PNS carries sensory stimuli from the environment, whether internal or external, to the CNS and it also has the motor information carrying fibres from CNS to effectors [6]. The somatic and visceral nerve fibres are also included in PNS. The somatic and autonomic systems make up the human NS functional divisions [7]. The autonomic NS is not under voluntary control and has further divisions such as sympathetic, parasympathetic, and enteric divisions. Its afferent and efferent fibres emerge from the brainstem via specific cranial or spinal nerve roots. Its fibres cross the thoracic, abdominal region, pelvis with perineal viscera, and smooth visceral muscle with glands. The submucosal and myenteric plexuses and their related elements make up the enteric NS within the digestive system. This is an autonomous system functionally, which is regulated by parasympathetic and sympathetic NS [6]. The cranial cavity initially has two sets of blood vessels for arterial blood entry: the internal carotid arteries, which are intended for anterior brain circulation, and the vertebral arteries, which provide blood flushing to the posterior region. The circle of Willis connects them together, and they are located within subarachnoid cisterns placed at the base of the brain. This circle has four vessels, which manage the blood perfusion and serve as a protective mechanism against any blood compromise. The cerebrum is supplied by internal carotid arteries. The basil artery supplies the brain stem and cerebellum as a whole through the vertebral arteries [8].

DOI: 10.1201/9781003300502-1

1.2 NERVOUS SYSTEM DEVELOPMENT

It can be challenging to differentiate between the grey and white matter that make up the complex organ that is the brain [5]. The embryonic neural system is essentially just a straight line in the beginning before becoming progressively more sophisticated. In the adult organism, several structures that appear to be near one another are not linked, and the links that do exist may seem random. However, the system has an underlying order that results from the way various components grow. Whenever sperm and egg cells fuse together, a fertilized egg is produced. An organism is produced by the division of a zygote. One of the three germ layers is made up of embryonic cells that develop into various body tissues 16 days after conception. The mucosae of the various body parts, as well as the respiratory and digestive systems, constitute the endoderm. The mesoderm mainly makes up the muscle and connective tissues. Ultimately, the ectoderm forms the NS and the integumentary system (the skin) [9].

A segment of the embryo's ectoderm divides into a specific portion of the neuroectoderm. It is the ancestor of the NS tissue as the child grows. Cells in this area are induced by molecular impulses to develop into neuroepithelium, developing a neural plate. The tissue buckles and folds inward as the cells start to change shape (Figure 1.1). Along the embryo's dorsal surface, a neural groove develops and appears as a line. The term 'neural fold' refers to the ridge-like margin that resides on either side of the neural groove. The neural tube is a tube-like structure that forms right below the ectoderm as the neural folds come together. The neural crest is produced from the ectoderm by the breakdown of the cells in the neural folds, which extend dorsally to the neural tube. The development of the neural crest into the components of the PNS occurs when the neural crest migrates away from the developing, or embryonic, CNS, which will emerge along the neural groove. Melanocytes, craniofacial bone, and cartilage are only a few tissues that develop from the neural crest that are not connected to the central NS [6].

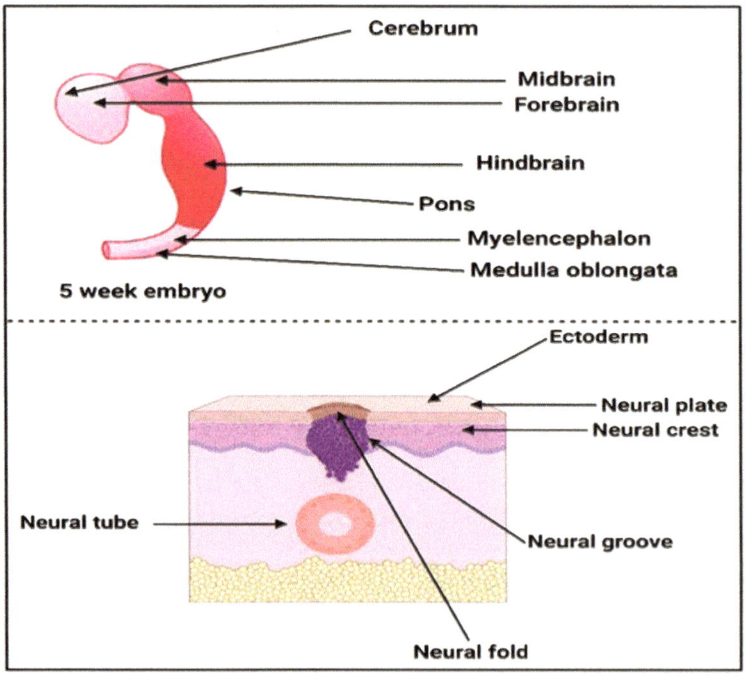

FIGURE 1.1 Graphical view of embryonic development of the brain.

Currently, the primitive NS is a simple, tubular structure. It covers the entire embryo, from the front to the back [10]. The brain begins to form at the front end around 25 days, and the spinal cord forms at the posterior end. By the fourth week of growth, the NS's most basic tissue organization gives way to more intricate structures. A few enlargements occur when the anterior end of the neural tube begins to form the brain, which leads to the formation of pouch chambers. Like a child's balloon animal, the lengthy, plain neural tube begins to take on a new shape. The first step results in the formation of three primary vesicles. Greek words were used for the names of these vesicles, encephalon being the primary root word, which means 'brain' (en- = 'inside'; cephalon = 'head'). Each one's prefix typically reflects where it is along the path of the growing NS. The prosencephalon (pros- = 'in front') is known as the forebrain. The mesencephalon (mes- = 'middle') is known as the midbrain. The rhombencephalon is known as the hindbrain [11]. The vesicles differentiate more as the brain continues to grow (Figure 1.1). Five secondary vesicles are formed from the three original vesicles. Two more vesicles, the telencephalon, and diencephalon, are formed when the prosencephalon enlarges. The cerebrum will develop from the telencephalon [10]. The diencephalon gives birth to various adult structures, including the thalamus and hypothalamus. The eye cup starts to form in the embryo's diencephalon. It grows into the retina, which is the nerve tissue of the eye. This is one of the few cases when CNS components in the embryo end up being peripheral in the adult NS. There are no more precise divisions of the mesencephalon. At the primary vesicle developmental stage, the midbrain is already developed and stays that way. It is surrounded by the remainder of the brain, which makes up the majority of the brain mass. The midbrain only makes up a small percentage of the whole brain [5]. The metencephalon and myelencephalon grow out of the rhombencephalon. The cerebellum develops from the metencephalon, which belongs to the pons in adults. About 10% of the mass of the brain is made up of the cerebellum, which is a crucial component of the brain and is known as the 'small brain' in Latin. The strongest connections between the cerebellum and the rest of the brain are found in the pons, which develop from the same vesicle as the cerebellum. The medulla oblongata and the myelencephalon are mature structures that are alike. The brainstem includes the midbrain, pons, and medulla oblongata [12].

1.2.1 Spinal Cord Development

The cells that make up the neural tube's wall expand and develop into the spinal cord's neurons and glia as the spinal cord grows. Dorsal tissues will be connected to sensory functions, whereas motor activities will be connected to ventral tissues [13]. Cells stretch across the apical and basal sides of the epithelial wall of the pseudo-stratified epithelium that makes up the neural tube. There is originally just one group of undifferentiated stem cells in the neuroepithelium [6]. These cells eventually give birth to the neuronal and glial stem cells, the two major lineages, as well as several sub-lineages. After the last mitotic division, neurons move from the neural tube ventricle to create the mantle zone. Radial glial fibres, which form a scaffold from the ventricular zone to the surface of the developing brain, direct this outward proliferation. As more post-mitotic neuronal cells build up in the neural tube, the neural tube gets thicker, and the CNS, brain, and spinal cord get bigger [7].

1.3 TOUCH AND PAIN

Tiny, myelinated axons and extremely small unmyelinated fibres carry pain impulses from receptors to the spinal cord [14]. Pain is the result of nerve impulses travelling up the spinal cord and being processed by the thalamus, the brainstem, and the cortex. The midbrain is the source of a network of neurons that may inhibit these communications. The information carried by this descending route reaches the spinal cord, where it inhibits the uptake of tissue damage signals by higher brain areas [15]. Touch receptors deliver signals to the spinal cord through sensory nerves, which then synapse or make contact with other nerve cells in the spinal cord and are relayed to the thalamus and sensory cortex [16]. Sensations from the hands and lips occupy a comparatively large portion of the cortex, whereas those from less sensitive portions of the body occupy considerably smaller sections. Touch and pain perception are localized sensations, with certain regions of the body being more sensitive than others [17].

The number and location of receptors play a major role in explaining these divergent reactions. The cornea, for instance, is hundreds of times more responsive to painful sensations than the bottoms of the feet. Nociceptors are the sensory fibres activated by stimuli that cause tissue damage and pain [18]. The reaction to noxious (painful) thermal, mechanical, or chemical stimulus is mediated by chemicals produced by distinct subsets of nociceptor cells. Capsaicin, garlic, and wasabi are just a few examples of plant-based compounds that may cause pain, and they all react to the same molecules in the body [15,19]. Itch is triggered in certain people when certain nociceptors in the skin detect a chemical signal. One such nociceptor is histamine, which is secreted in reactions to insect bites and other allergens. As a result of tissue damage and subsequent inflammation, a wide variety of chemicals are released at the injury site. Increased pain perception is possible because prostaglandins make pain receptors more sensitive to tissue injury [20]. Allodynia is a clinical condition where seemingly harmless stimuli, such as touching burnt skin, may cause severe pain. Prostaglandins play a role in this phenomenon [21]. Changes in the neural system brought on by repeated trauma may increase and extend the 'pain' signal. The end outcome is heightened sensitivity to pain, such that even benign stimuli might cause severe reactions. Many people do not realize that chronic pain is a disorder of the neural system in and of itself rather than just a symptom of another underlying illness [22].

The A-delta and C nerve fibres carry signals of pain and itching to the spinal cord. Since myelinated A-delta fibres insulate the nerve, electrons are routed efficiently and move at a higher velocity, resulting in a pain sensation that is both immediate and distinct. Communication along unmyelinated C fibres is slower because the nerve endings are dispersed over a larger area. It causes general, achy pain that is hard to place. The thalamus perceives signals from the spinal cord and passes them on to the parts of the brain responsible for transforming those signals into perceptual awareness. The skin is a sensory organ that can express a wide range of experiences, from the numbing cold of a breeze to the soothing touch of another's hand on yours to the grating itch of a woollen scarf. Information of various sorts activates distinct receptors, which transduce the stimulation of the skin into electrical nerve impulses. Most painful sensations are the result of nociceptors being activated, which are receptors that only respond to stimuli that might cause permanent damage to nearby tissues. Both chemoreceptors and mechanoreceptors have offshoots that function as nociceptors. A group of receptors called chemo-sensitive nociceptors are activated when cells release chemicals in response to tissue injury or inflammation. High-threshold mechanoreceptive nociceptors are activated by mechanical stimulation that is strong enough to cause tissue injury [23].

As the first sense to develop in the womb, touch is one of the first things a newborn looks for in a loving caregiver. Human orphan research has shown us that touch is also essential for human growth. Babies in Romanian orphanages who were fed properly but not provided any other kind of care had cognitive and neurodevelopmental delays. A soothing touch from a mother is known to lower stress chemicals like cortisol, which may explain why holding a screaming infant might help them relax. Intense stress may cause cell death in the developing brain. As a result, it seems that reducing stress via physical contact, such as hugs and caresses, is crucial not only for the health of children but also for the maturation of the brain in infants [24].

Pain signals are sent from the cerebral cortex to the periaqueductal grey matter, where they activate pathways that modify pain [17]. Endorphins are naturally occurring opioids that have analgesic effects similar to morphine [25]. When you are under intense emotional stress, your body releases adrenaline, which has a dual role as both a painkiller and a stress reliever. By interfering with pain signals as they travel to the spinal cord and brainstem, their release helps regulate and relieve pain. We all have these neural pathways, but how effectively they function and how sensitive they are determine how much pain we experience. It is the reason why some individuals have persistent pain that standard medical care cannot alleviate. There are many different kinds of opioid receptors in the CNS, and endorphins activate all of them. Opioid medications may be sent to the spinal cord before, during, and after surgery to alleviate pain. To alleviate pain without resorting to long-term opioid usage, researchers are exploring the use of electrical stimulation of the spinal cord [4].

Neither pain nor itching are processed by a single region of the brain. Our perception of pain is influenced by a complex web of emotional and sensory factors. Meditative trance states, hypnosis, cognitive behavioural therapy, and regulated cannabis usage are just a few of the effective therapies that aim to address the emotional side of the problem. Chronic pain patients continue to experience pain, although it does not 'hurt' as much as it once did. Brain imaging reveals that cannabis inhibits pain processing in a select few regions, primarily in the limbic system. It is still unclear to researchers exactly how these treatments function [22].

1.4 MEMORY AND LEARNING

We utilize both learning and memory to keep knowledge about our life experiences, which is essential for our survival [26]. We would not be able to develop fundamental motor or communication abilities, the capacity for complicated reasoning, or group relations without learning and memory. Pioneers in neuroscience such as Ramón y Cajal, Hebb, and Marr introduced the concept that memory is coded in the patterns of synaptic connections between neurons [27].

The human brain is the hub of learning, as we have known from the beginning of time. About 100 billion neurons make up the complex network of fibre routes that make up the brain [28]. Since higher-order cognitive processes like memory and reasoning take place in the cerebrum, it is this region of the brain that is most engaged in learning. Vision, listening, speaking, touching, short-term memory, long-term memory, language skills, and reasoning abilities are all specialized functions of the cerebrum that are essential for learning [29].

Short-term memory is a brittle area of the brain that serves as a receiver for the abundance of sensory information we experience every day. Synapses along the nerve fibre convey sensory information through the junctions of neurons and save it briefly in short-term memory. After being processed in short-term memory, the brain's neuronal connections transport these memories to the structural core, where they are compared to pre-existing memories and retained in long-term memory. In reality, some deterioration is visible as information rushes across billions of neurons' axons, which convey messages to the following cell via synapses. Because of this, many of our thoughts are lacking or contain fabricated details that we add to our actual memories to fill the gaps [29].

Neuroscientists think that the strengthening and weakening of connections between brain cells are the primary factors in how learning and memory are formed. Recently, it was demonstrated by the Centre for the Neurobiology of Learning and Memory researchers at the University of California, Irvine. Researchers discovered that when two neurons connect often, they develop a link that makes communication between them easier and more precise. This results in easier memory and more precise thoughts. On the other hand, the transmission was frequently insufficient when two neurons did not connect very regularly, which resulted in either a false memory or nothing at all [30]. Together with the use of positron emission tomography (PET scan) and functional magnetic resonance imaging (MRI), cognitive scientists and neuroscientists have lately made advancements in their understanding of memory processes [31]. The hippocampal formation is made up of the hippocampus, a curving seahorse-shaped organ located underneath the temporal lobes. It aids in learning, memorization, orientation, and spatial awareness [31].

1.4.1 Types of Memory

Neuroscientists and psychologists have divided memory systems into two major categories. Declarative memory is most likely the most well-known memory system. It contains memories of facts and things that actually happened, and it has a conscious component [32], for example, a fact like 'Islamabad is the capital of Pakistan' or an event like 'I visited Islamabad before'. Implicit memory, also known as non-declarative memory, refers to the categories of memory systems that do not have a conscious aspect but are nonetheless quite significant. The memory that is declarative 'knows what'; the memory that is not declarative 'knows how'. Can we improve our memory? Yes, some of the strategies have proved useful for improving our memory.

1. Developing mnemonic devices to remember new information. For instance, creating a pattern or memory for each name in a room might help you recall them all.
2. Solving brain riddles and problems.
3. Forming solid memory associations to assist with memory. Meditation or discussion of recent events may help to strengthen these bonds.
4. Cardiovascular workout to support brain fitness.

1.5 ARTIFICIAL BRAIN AND NEURAL NETWORKS

Neurons are specialized cells that can transmit nerve impulses and respond to stimuli. Dendrites, which are branching projections on neurons, assist in guiding the electrical stimulation that synapses, or connections between brain cells, provide to neighbouring neurons. Previously believed to be axon-specific, dendrites sustain action potentials and release neurotransmitters [33].

1.5.1 Neural Networks

1.5.1.1 Biological Neural Networks

The cortex, brainstem, cerebellum, and other anatomical divisions of the human brain may all be separated anatomically. According to the roles they play and the inherent design of the neural network's architecture, it may be further split into a number of sections and regions [34]. Electrical current can go via the gap junctions that join the postsynaptic and presynaptic cell membranes in an electrical synapse.

1.5.1.2 Artificial Neural Networks

They have been developed to mimic the operations of the brain's neuronal network and architecture. By employing training with examples and algorithms to change the weights of linked neurons, learning is accomplished. Modelling heavily utilizes this process.

1.5.1.3 General Architecture of Artificial Neural Network

An artificial neural network (ANN) is a type of computer software that may be trained to learn how to carry out a given task. The most widely used technique for training neural networks is backpropagation.

1.5.2 Artificial Intelligence (AI)

The study of intelligent behaviour through simulation, modelling, and computational methods is the focus of the discipline of artificial intelligence (AI) in computer science [33]. Expert systems used to identify and resolve engine issues are one example of how AI has been employed to do intelligent tasks.

1.6 NEURODEGENERATIVE DISEASES

Neurodegenerative diseases are defined by the progressive loss of neurons in populations that are particularly vulnerable to metabolic or toxic diseases, as opposed to the selective, static neuronal loss brought on by such conditions [35]. Neurodegenerative diseases are rare and common genetic conditions that affect the CNS. These diseases cause certain groups of neurons and the connections between them to lose function over time [36].

Despite receiving enough training, having an intelligent mind, and having intact sensory abilities, people with dyslexia have trouble spelling and recognizing words accurately or fluently [37]. The point of reading is to understand, which requires both the ability to decode and the ability to understand spoken language [38]. Decoding issues are what characterize dyslexia, but hearing comprehension is often more intact. Despite the fact that those with very low decoding skills (such as young children or those with severe dyslexia) have poor reading comprehension, others with moderate decoding issues can nonetheless sustain acceptable reading comprehension with intact spoken language skills [39].

Reading is a linguistic talent, so it makes sense that it would activate brain regions involved in processing vocal language and certain extra regions involved in processing [40] visual objects and creating visual-linguistic maps (see Figure 1.2). According to functional imaging studies, people with dyslexia have aberrant activations of a widespread left hemisphere language network [41,42]. The left inferior frontal gyrus frequently shows abnormal activity. Studies on structural imaging have shown that this same network has less grey matter. According to new family risk research [43], these grey matter reductions do not just result from reading difficulties but also occur before literacy teaching [44].

Repeated linkage studies that used molecular techniques found that dyslexia is linked to nine risk loci (DYX1 – DYX9, where DYX stands for dyslexia and the number shows the order of discovery) [45,46]. Meaburn et al. (2008), for instance, identified loci that separated a high-reading group from a low-reading group using DNA pooling and more than 100,000 single-nucleotide polymorphisms. The best-replicated dyslexia loci were not among their few significant finds, and each one only partially explained the variance.

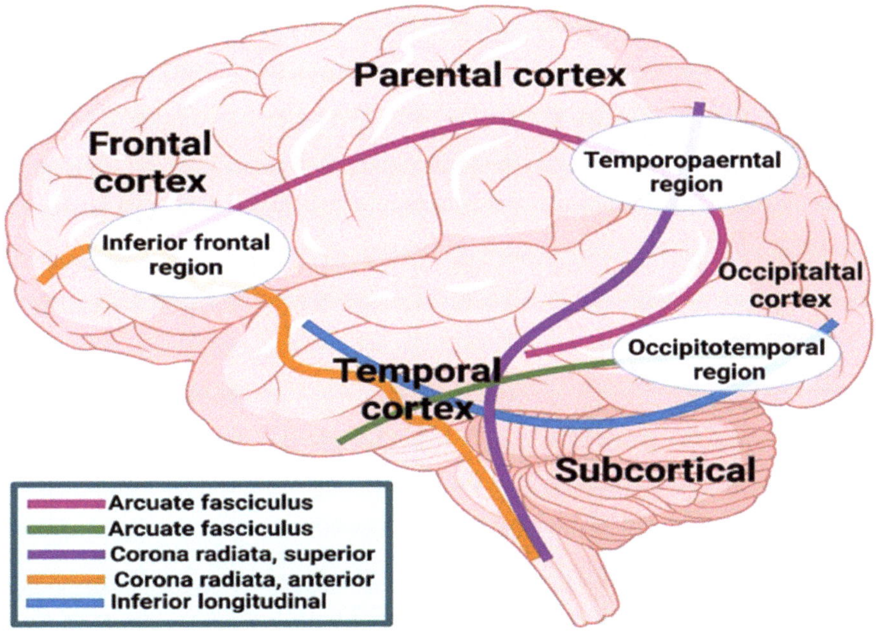

FIGURE 1.2 White matter tracts and possible reading-related grey matter areas are shown schematically as dark grey ovals and colourful lines, respectively.

TABLE 1.1
Candidate genes for dyslexia

Loci	Position	Genes
DYX1	15q21.1	DYXC1
DYX2	6p22.2	KIAA0319
	6p22	DCDC2
DYX3	2p15-16	MRPL19, C20RF3
DYX5	3p12-q13	ROBO1
DYX6	18p11.2	MC5R, DYM, NEDD4L
SPCH1	7q31.1	FOXP2
SLI4	7q35-36	CNTNAP2
SLI1	16q23-24	CMIP, ATP2C2

TABLE 1.2
Summary of neurodegenerative diseases and their mechanism of action in the brain

Disease	Mode of Action	Gene	Reference
Parkinson Disease	Degeneration of dopaminergic neurons in the nigra pars compacta of the mesencephalon.	SNCA, LRRK2, VPS35, PINK1	[48]
Alzheimer's Disorder	It relates to the loss of memory.	APP, PSEN1, PSEN2	[49]
Dementia	High penetrance of autosomal dominant genes is observed.	MAPT, C9ORF72, GRN,	[50]
Ataxia	It affects coordination and speech.	ATXN1/CAG, CACNA1A/CAG	[51]

The prenatal processes of rat neural development have been found to be altered by DYX1C1, DCDC2, KIAA0319, and ROBO1, particularly neurotransmission and the formation of connections when neurons reach their target location (for instance, neurite–axon and dendrite–outgrowth and guidance). Fundamentally, a group or network of alleles that communicate with one another through molecular signalling controls these two prominent brain development mechanisms.

Table 1.1 describes the genes which are associated with dyslexia, and before it, these genes were related to speech impairment/disorder.

Alzheimer's [47] and Parkinson's disease [40,43] are well-researched, while Ataxia and dementia are hotspot diseases. Table 1.2 lists the mode of action and the genes involved in these neurodegenerative diseases.

1.7 DRUGS FOR BRAIN DISORDERS

The effects of addiction on the brain are far-reaching. Stimulants, nicotine, opiates, alcohol, and sedatives all have chemical components that are absorbed into the brain and circulation upon usage. An individual's willpower and self-control might weaken once a drug has entered the brain, leading to reckless behaviour or an insatiable need for a toxic substance [52].

There are a variety of variables, including drug kind and quantity, frequency of use, and addiction severity, that influence how the brain reacts to addiction. Cocaine, as one example, produces a state of exhilaration in its users. Cocaine's psychoactive properties produce this by stimulating the reward and motivation centres of the brain [53]. Many people experience an intense and fleeting rush of dopamine, the neurotransmitter responsible for that happy feeling. This sensation may be quite powerful, leading to a strong urge to keep using it. Delta-9-tetrahydrocannabinol (THC), the psychoactive ingredient in marijuana, attaches to cannabinoid receptors in the brain and induces a calming and relaxing state. High cannabis receptor density areas of the brain are hit particularly hard.

According to the National Institute on Substance Abuse (NIDA), in the United States, marijuana is the abused drug that is most often used. Its usage is highest among teenagers and young adults [54]. The hippocampus is responsible for short-term memory; therefore, marijuana usage may make it more difficult to recall recent events. The cerebellum and basal ganglia, which regulate involuntary muscle movements and aid with coordination, are also affected. Common short-term symptoms of marijuana consumption include impairment in motor abilities, emotional changes, skewed time and sensory perception, diminished memory, and difficulty thinking effectively and solving issues [55].

According to the National Institute on Drug Abuse (NIDA), teens who start using marijuana and do not stop even after they become adults may experience permanent IQ declines. Only around 30% of those who use marijuana do not get addicted; starting before age 18 raises that number by a factor of 4 to 7. There is some evidence that marijuana usage might hasten the loss of neurons in the hippocampus that occurs with ageing, which can exacerbate memory loss. Long-term marijuana usage has been linked to difficulties with coordination, learning, and sleep. Research on the link between cannabis usage and the development of psychosis and mental diseases like schizophrenia is also documented by NIDA. Marijuana usage has also been linked to irregular heartbeat and other cardiac disturbances.

Opioid medicines are CNS depressants and also interfere with the body's normal synthesis of norepinephrine [56]. Opioids dull the sense of pain, make you sleepy, lower your temperature, and slow your cardiovascular, respiratory, and metabolic systems. Abuse of opioids almost always leads to overdose, which often results in life-threatening respiratory depression [57]. The Centers for Disease Control and Prevention (CDC) estimates that 91 people in the United States pass away from an opioid overdose each day, and opioids were implicated in 60% of all overdose fatalities in 2015 [58]. According to NIDA's published research, long-term, chronic heroin use may also cause the degradation of portions of the white matter in the brain, which may have a detrimental effect on stress response, emotion regulation, and decision-making. Long-term risks of persistent opioid drug usage include lung problems and infections of the heart's lining [59].

Stimulant medications are highly addictive because of their effects on dopamine and the limbic reward system. The DEA advises that chronic cocaine use may lead to psychosis and adversely affect CNS functioning. Regular users of meth may experience hallucinations, anxiety, and disorientation due to the same effects on the brain and CNS. Methamphetamine abuse may lead to impairments in learning and memory, motor coordination, and emotional control.

If a person with a heroin addiction suddenly quits taking the drug, they may suffer severe withdrawal symptoms such as cravings, sadness, anxiety, and excessive perspiration. Extensive use of heroin causes significant neural rewiring, which accounts for most of this [60]. Though a complete addiction may not have set in yet, tolerance and dependence may be present. In the long run, the brain becomes used to the chemical overload and the substance's mental consequences. The brain then lowers its synthesis of neurotransmitters, chemical messengers in the brain. Expert care for withdrawal symptoms is recommended since it lowers the likelihood of recurrence and other adverse health outcomes, such as stroke and heart attack.

When a person with an addiction joins a treatment centre, they may get medication and access cutting-edge therapies to help them overcome their problem. After substance abuse treatment, biofeedback is often used to restore balance to the brain. A trained observer can keep tabs on how the brain is doing. The ability to enhance brain function, so mitigating the consequences of addiction and unhealthful urges, is within their reach.

Electroencephalograms (EEGs) are used in biofeedback. Electroencephalograms (EEGs) are primarily used to aid those who have experienced severe brain injuries, but they may also help people with obsessive-compulsive disorder and other brain illnesses. By providing information about one's own body, biofeedback helps people learn to control their own instinctive responses to stress. Meditation, guided visualization, and muscular relaxation are all techniques that may be used in this kind of treatment.

Improved involuntary functions, including heart rate, blood pressure, and muscular contraction, may be achieved by the use of biofeedback in conjunction with cognitive behavioural therapy (CBT) or dialectical behavioural therapy (DBT). There is a kind of biofeedback known as EEQ treatment that focuses on the NS, called neurofeedback. The treatment is a kind of cognitive training. This treatment for addiction involves keeping tabs on mental processes. Reducing tension and anxiety, as well as treating compulsions, are two of its many benefits for patients. Ultimately, with both treatments, the patient receives a reward that encourages the brain to reestablish its previous level of functioning.

1.8 CONCLUSION

The memory and learning process can be enhanced in different ways, like through meditation or cardiovascular workouts. Artificial neural networks have revolutionized the neurobiology field. Machine learning helps in the early diagnosis, prognosis, and development of new therapies for neurodegenerative diseases. To fully integrate machine learning into diagnostic and prognostic neurology practice as well as the development of future medicines, sizable collections of curated datasets and thorough assessments of machine learning methodologies will be required.

REFERENCES

[1] J.M.S. Pearce, *Rev. Neurol. (Paris).* 175 (2019) 119–125. https://doi.org/10.1016/j.neurol.2018.04.011.
[2] F. Acar, S. Naderi, M. Guvencer, U. Türe, M.N. Arda, *Neurosurgery*. 56 (2005) 861–867. https://doi.org/10.1227/01.neu.0000156791.97198.58.
[3] S.K. Ghosh, R.K. Narayan, *Morphologie*. 104 (2020) 267–279. https://doi.org/10.1016/j.morpho.2020.05.005.
[4] M. Zarei-Kheirabadi, A.R. Vaccaro, V. Rahimi-Movaghar, S. Kiani, H. Baharvand, *Stem Cells Dev.* 29 (2020) 266–280. https://doi.org/10.1089/scd.2019.0189.
[5] T. Vanderah, D. Gould, *Nolte's*, Elsevier Health Sciences, 2020.
[6] P.J. Bazira, *Surg. (United Kingdom)*. 39 (2021) 451–462. https://doi.org/10.1016/j.mpsur.2021.06.012.
[7] S. Cunningham, *Nurs. Ski. Control Coord.*, Taylor and Francis Inc., 2021: pp. 5–19. https://doi.org/10.4324/9781351065900-3.
[8] P. Mastorakos, D. McGavern, *Sci. Immunol.* 4 (2019). https://doi.org/10.1126/sciimmunol.aav0492.
[9] C. Molnar, J. Gair, Adaptive immune response. In *Concepts of Biology*, 1st *Canada Edition* (2013).
[10] S. Harzsch, D. Sandeman, J. Chaigneau, *Treatise Zool. Taxon. Biol. Crustac.* 3 (2012) 9–236.
[11] A. de Lahunta, E.N. Glass, M. Kent, *Vet. Clin. Small Anim. Pract.* 46 (2016) 193–216.
[12] D. Djuric, V. Jakovljevic, V. Zivkovic, I. Srejovic, *Can. J. Physiol. Pharmacol.* 96 (2018) 991–1003.
[13] W.A. Alaynick, T.M. Jessell, S.L. Pfaff, *SnapShot: Cell.* 146 (2011) 178.
[14] Myelin, *Membrane | Learn Science at Scitable*, (n.d.).
[15] A.I. Basbaum, D.M. Bautista, G. Scherrer, D. Julius, *Cell.* 139 (2009). https://doi.org/10.1016/j.cell.2009.09.028.
[16] M.F. Yam, Y.C. Loh, C.S. Tan, S.K. Adam, N.A. Manan, R. Basir, *Int. J. Mol. Sci.* 19 (2018). https://doi.org/10.3390/IJMS19082164.
[17] D. and C.I.B. M. Osterweis, A. Kleinman, D. Mechanic, *Anatomy and Physiology of Pain*, (1987).
[18] A.E. Dubin, A. Patapoutian, *Nociceptors:, J. Clin. Invest.* 120 (2010). https://doi.org/10.1172/JCI42843.

[19] R. Vennekens, J. Vriens, B. Nilius, *Curr. Neuropharmacol.* 6 (2008). https://doi.org/10.2174/157015908783769644.
[20] E. Ricciotti, G.A. Fitzgerald, *Arterio Thrombo Vasc. Biol.* 31 (2011), 986–1000. https://doi.org/10.1161/ATVBAHA.110.207449.
[21] J. Sandkühler, *Physiol. Rev.* 89 (2009) 707–758. https://doi.org/10.1152/PHYSREV.00025.2008/ASSET/IMAGES/LARGE/Z9J0020925080004.JPEG.
[22] T.S. Jensen, N.B. Finnerup, *Lancet Neurol.* 13 (2014) 924–935.
[23] C. Chen, Y. Lu, S. Kallakuri, A. Patwardhan, J.M. Cavanaugh, *JBJS.* 88 (2006) 1807–1816.
[24] S.R. Kaler, B.J. Freeman, *J. Child Psychol. Psychiatry.* 35 (1994) 769–781.
[25] Z. Alizadeh, M. Fereidoni, M. Behnam-Rassouli, S. Hosseini, *Iran. J. Neurol.* 13 (2014) 19–27.
[26] S. Tonegawa, K. Nakazawa, M.A. Wilson, *Ser. B Biol. Sci.* 358 (2003) 787–795.
[27] M.K. Johnson, L. Hasher, *Annu. Rev. Psychol.* 38 (1987) 631–668.
[28] Office of Communications and Public Liaison, National Institute of Neurological Disorders and Stroke, National Institutes of Health, M. 20892 Bethesda, Brain Basics: The Life and Death of a Neuron | National Institute of Neurological Disorders and Stroke, (n.d.).
[29] N. Cowan, *Behav. Brain Sci.* 24 (2001) 87–114.
[30] B.J. Baars, N.M. Gage, Academic Press, 2010.
[31] D.A. Sousa, Corwin Press, 2016.
[32] G.J. Wang, N.D. Volkow, L. Chang, E. Miller, M. Sedler, R. Hitzemann, W. Zhu, J. Logan, Y. Ma, J.S. Fowler, *Am. J. Psychiatry.* 161 (2004) 242–248. https://doi.org/10.1176/appi.ajp.161.2.242.
[33] Artificial Neural Networks (ANN) and their Types, (n.d.).
[34] M.C. Nwadiugwu, *Brain. arXiv preprint. arXiv:2101.08635.*
[35] B.N. Dugger, D.W. Dickson, *Cold Spring Harb. Perspect. Biol.* 9 (2017). https://doi.org/10.1101/cshperspect.a028035.
[36] W. Reith, *Radiologe.* 58 (2018) 241–258. https://doi.org/10.1007/s00117-018-0363-y.
[37] F.R. Vellutino, J.M. Fletcher, M.J. Snowling, D.M. Scanlon, S, J. Child Psychol. *Psychiatry Allied Discip.* 45 (2004) 2–40. https://doi.org/10.1046/j.0021-9630.2003.00305.x.
[38] D. Colenbrander, S. Kohnen, K. Smith-Lock, L. Nickels, *Learn. Individ. Differ.* 50 (2016) 210–220. https://doi.org/10.1016/j.lindif.2016.07.021.
[39] S.M. Adlof, T.P. Hogan, *Lang. Speech. Hear. Serv. Sch.* 49 (2018) 762–773. https://doi.org/10.1044/2018_LSHSS-DYSLC-18-0049.
[40] D. Joy, J. Jose, S. Bibi, A. Bandiwadekar, G. Gopan, C. Mariana Gonçalves Lima, T. Bin Emran, F. A Alhumaydhi, H. Ashtekar, C. Adam Conte-Junior, *Bioinorg. Chem. Appl.* 2022 (2022).
[41] J.F. Démonet, M.J. Taylor, Y. Chaix, *The Lancet* 363 (2004) 1451–1460. https://doi.org/10.1016/S0140-6736(04)16106-0.
[42] K. Ryherd, K. Jasinska, J.A. Van Dyke, Y.H. Hung, E. Baron, W.E. Mencl, J. Zevin, N. Landi, *Brain Lang.* 186 (2018) 32–43. https://doi.org/10.1016/j.bandl.2018.08.001.
[43] U. Saleem, A. Shehzad, S. Shah, Z. Raza, M.A. Shah, S. Bibi, Z. Chauhdary, B. Metab. *Brain Dis.* (2021) 1–21.
[44] N.M. Raschle, M. Chang, N. Gaab, *Neuroimage.* 57 (2011) 742–749. https://doi.org/10.1016/j.neuroimage.2010.09.055.
[45] L.M. McGrath, B.F. Pennington, M.A. Shanahan, L.E. Santerre-Lemmon, H.D. Barnard, E.G. Willcutt, J.C. Defries, R.K. Olson, *J. Child Psychol. Psychiatry Allied Discip.* 52 (2011) 547–557. https://doi.org/10.1111/j.1469-7610.2010.02346.x.
[46] S.E. Fisher, J.C. DeFries, *Nat. Rev. Neurosci.* 3 (2002) 767–780. https://doi.org/10.1038/nrn936.
[47] H. Chopra, S. Bibi, I. Singh, M.A. Kamal, F. Islam, F.A. Alhumaydhi, T. Bin Emran, S. Cavalu, *Front. Aging Neurosci.* 14 (2022) 1–17. https://doi.org/10.3389/fnagi.2022.879114.
[48] C.M. Lill, *Mol. Cell. Probes.* 30 (2016) 386–396. https://doi.org/10.1016/j.mcp.2016.11.001.
[49] L.M. Bekris, C.-E. Yu, T.D. Bird, D. Tsuang, Tin: Springer, New York, NY, 2011: pp. 695–755. https://doi.org/10.1007/978-1-4419-7104-3_21.
[50] C.T. Loy, P.R. Schofield, A.M. Turner, J.B.J. Kwok, Genetics of dementia, *Lancet.* 383 (2014) 828–840. https://doi.org/10.1016/S0140-6736(13)60630-3.
[51] T. Ashizawa, G. Xia, *Ataxia, Contin. Lifelong Learn. Neurol.* 22 (2016) 1208–1226. https://doi.org/10.1212/CON.0000000000000362.

[52] Substance Abuse and Mental Health Services Administration, US Department of Health and Human Services, THE NEUROBIOLOGY OF SUBSTANCE USE, MISUSE, AND ADDICTION – Facing Addiction in America, in: Facing Addict. Am. Surg. Gen. Rep. Alcohol, Drugs, Heal., US Department of Health and Human Services, 2016.
[53] E.J. Nestler, *Sci. Pract. Perspect.* 3 (2005) 4–10. https://doi.org/10.1151/spp05314.
[54] J.M. Purcell, T.M. Passley, J.R. Leheste, *Lancet Reg. Heal. – Am.* 14 (2022). https://doi.org/10.1016/j.lana.2022.100325.
[55] S. Zou, U. Kumar, *Int. J. Mol. Sci.* 19 (2018). https://doi.org/10.3390/ijms19030833.
[56] U. Saleem, S. Iman, B. Ahmad, M.A. Shah, S. Bibi, M. Alqarni, M.S. Khan, G.M. Shah, H. Khan, R.H. Alhasani, N.A. Althobaiti, A.E. Albalawi, *Metab. Brain Dis.* (2022). https://doi.org/10.1007/s11011-022-00955-0.
[57] R.A. Rudd, P. Seth, F. David, L. Scholl, I United States, 2010–2015, *MMWR. Morb. Mortal. Wkly. Rep.* 65 (2016) 1445–1452. https://doi.org/10.15585/mmwr.mm655051e1.
[58] C.L. Mattson, L.J. Tanz, K. Quinn, M. Kariisa, P. Patel, N.L. Davis,— *United States, 2013–2019, MMWR. Morb. Mortal. Wkly. Rep.* 70 (2021) 202–207. https://doi.org/10.15585/mmwr.mm7006a4.
[59] S.E. Lakhan, A. Kirchgessner, *Brain Behav.* 2 (2012) 661–677. https://doi.org/10.1002/brb3.78.
[60] T.R. Kosten, T.P. George, *Sci. Pract. Perspect.* 1 (2002) 13–20. https://doi.org/10.1151/spp021113.

2 Plants as Neuroprotective Agents

Role of Plant-Based Natural Products as Potential Brain Health Boosters

Vanshika Sharma, Yutika Nath, and Ranjan Dutta Kalita

2.1 INTRODUCTION

Neurodegenerative diseases (NDDs) are defined as disorders of the central nervous system that result in selective dysfunction and loss of neurons. Neurodegeneration can be defined as a process that is primarily involved in causing both neuropathological conditions and brain ageing. It is known as the brain pathology in the form of cerebrovascular and neurodegenerative disease, which is a leading cause of death globally, with an incidence of about 2/1000 and an 8% total death rate [1]. Several studies have proved that alteration in the physicochemical properties of proteins leads to the accumulation of these proteins in the human brain, which ultimately leads to neuronal degeneration [2]. Neuropsychiatric and neurological disorders like schizophrenia, depression, anxiety, Alzheimer's disease (AD), Parkinson's disease (PD), Huntington's disease (HD), amyotrophic lateral sclerosis (ALS), cerebrovascular impairment, seizure disorders, and head injuries are the key health issues caused due to the neural disorders. Other systemic disorders triggered as a result of neural dysfunction are cardiovascular, respiratory, renal, gastrointestinal, and many others in the 21st century [3].

Several pharmaceutical compounds are available in the market, known for their neuroprotective properties. Drugs that are being used to improve neurofunction generally work by altering the balance of chemicals called neurotransmitters in the brain. This is mainly done by selective enhancement of cerebral blood flow, cerebral oxygen usage metabolic rate, and cerebral glucose metabolic rate in chronically impaired human brain function. It has been reported that several medicines that are derived from medicinal plants have exhibited promising memory-enhancing properties due to the presence of bioactive phytochemical constituents. For instance, in the case of dementia, it is mainly caused by decreased cholinergic activity in the brain. Therefore, an alternative treatment using cholinergic drugs of plant origin is being used. For instance, muscarinic agonists (e.g., arecoline, pilocarpine etc.), nicotinic agonists (e.g., nicotine), and cholinesterase inhibitors (e.g., huperzine) are among the few remedies that can be used for improving memory. Other plant-derived drugs used in dementia are stimulants or nootropics (e.g., piracetam, amphetamine), putative cerebral vasodilators (e.g., ergot alkaloids, and papavarine), calcium channel blockers (e.g., nimodipine) [1].

DOI: 10.1201/9781003300502-2

2.2 HERBS EXHIBITING NEUROPROTECTIVE PROPERTIES

Identification and characterization of new medicinal plants that can cure neurodegenerative diseases and brain injuries arising from stroke have been one of the major and increasing methodical concerns in recent years. Approximately more than 120 traditional medicines are being used for the treatment of central nervous system (CNS) disorders in Asian countries [4]. In the Indian system of medicine, the following medicinal plants have exhibited quite incredible activity in the field of neuropsychopharmacology: *Allium sativum, Bacopa monnierae, Centella asiatica, Celastrus paniculatus, Nicotiana tabaccum, Withania somnifera, Ricinus communis, Salvia officinalis, Ginkgo biloba, Huperiza serrata, Angelica sinensis, Uncaria tomentosa, Hypericum perforatum, Physostigma venosum, Curcuma longa, Terminalia chebula, Crocus sativus, Valeriana wallichii, Glycyrrhiza glabra*, etc. In the Chinese system of medicine, several plants have been used in the treatment of stroke, amongst which *Ledebouriella divaricata, Scutellaria baicalensis, Angelica pubescens, Morus alba, Salvia miltiorrhiza, Uncaria rhynchophylla, and Ligusticum chuanxiong* [5]. Some of the traditional plants that have been known to exhibit neuroprotective properties from the ages are mentioned below:

2.2.1 *GINKGO BILOBA* L.

Ginkgo biloba L. or Ginkgo is a living fossil belonging to the family Ginkgoaceae, which is native to China and Japan and is also known as 'maiden hair tree', 'kew tree', 'ginkyo', etc. It is a large deciduous tree, growing up to a height of about 35 m, with fan-shaped leaves having two lobes and bearing edible seeds. EGb761, an extract from *Ginkgo biloba*, is well known for its beneficial pharmacological effect on neurotransmitter systems and in the treatment of neurological disorders. Various secondary metabolites are found to be in the extract *viz* terpenoids (e.g., bilobalide, ginkgolides, and ginkgolides A, BC, J, and M), and flavonoids (e.g., kaempferol, quercetin, and polymeric flavonoids) [6]. Among these bioactive components, ginkgolides, which serves as a bioactive constituent of EGb761, has been found to interrelate with the cholinergic system, thus exhibiting neuroprotective and regenerative properties of the neural stem cells [7,8]. This herb also improves the circulation of blood into the central nervous system and leads to an increase in the distribution of nutrients and oxygen into the brain. It also helps in the elimination of free radicals generated in the body. Thus, the Ginkgo extract is widely used for the treatment of amnesia and AD, cognitive disorders, and has exhibited therapeutic effects at the biochemical and pharmacological levels [9,10,11].

2.2.2 *PANAX GINSENG*

Panax ginseng C.A. Mey, or Ginseng belongs to the family Araliaceae and is commonly grown in the Northern Hemisphere, mostly in cooler climates of North America and eastern Asia (northeast China, eastern Siberia, Korea, and Bhutan). It is a recurrent, hermaphrodite herb growing up to a height of 80 cm and possessing green-white small flowers and oval-shaped red drupe fruits about 6–7 mm long. This herb plays a potent role in Traditional Chinese Medicine. It is commonly used as a memory tonic and helps to improve memory and learning, especially in elderly people. Saponins are the phytochemical present in this herb, which enhances memory in the scopolamine-induced learning impairment [12].

2.2.3 *BACOPA MONNIERI*

Bacopa monnieri L., Pennel / Brahmi belongs to the family Plantaginaceae, also known as 'water hyssop', is a perennial, non-aromatic herb, which grows up to a height of 30 cm, bearing oblong, succulent, thick leaves in alternative arrangements that are edible in nature. Brahmi is native to India and Australia [13] and is also found in the United States and East Asia. *Bacopa monnieri* is used in the treatment of several ailments like epilepsy, lack of concentration, mental illnesses, sedatives, stroke, anxiety, and many more [14], thereby improving their mental capacity. Studies show that the major bioactive constituents found in this herb are alkaloids such as brahmine, nicotine, and herpestine; steroidal saponins like D-mannitol and hersaponin, bacopasides I-XII, bacopasaponins, bacosides A and B, and monnierin [13,15], which have proven to have some valuable influence on enhancement of memory and cognitive function. *Bacopa monnieri* shows neuroprotective, antioxidant, and hepatoprotective properties [16]. *Bacopa monnieri* has also been documented to exhibit acetylcholinestrase inhibitory activity in patients with AD. Brahmi soothes and relaxes the brain cells and refurbishes them to a normal functioning state in adults [17].

2.2.4 *Terminalia chebula*

Terminalia chebula Retz. / Haritakiis is a deciduous tree belonging to the Combretaceae family growing in parts of South-eastern Asia like India, China, Sri Lanka, Malaysia, and Vietnam. The tree grows up to a height of 30 m, and the leaves are oval in shape with opposite or alternate arrangements bearing yellow flowers emitting an unpleasant odour. The fruits are yellow to orange brown in colour. The ripe fruit of *Terminalia chebula* is treasured because of its beneficial health properties including memory and intellect augmentation. Furthermore, it is known that the consumption of the fruit is efficient in prolonging life, improving eyesight, and delaying ageing [12].

2.2.5 *Convolvulus pluricaulis*

Convolvulus pluricaulis Choisy / Shankhpushpi, also called Bindweed, is a well-reputed 'Medhya herb' belonging to the Convolvulaceae family commonly found in southern India and Burma. This perennial herb bears blue-coloured flowers. Since ancient times, it has been used as a potent memory-improving drug and as a psychostimulant and tranquillizer in traditional Indian medicine [15]. The plant contains miscellaneous phytochemicals like β-sitosterol, vitamin C, silane, decanoic acid, carbohydrates, vitamin E, squalene, linoleic acid, proteins, alkaloids like tropane, flavonoids like kaempferol, glycosides, and many more as bioactive chemicals with assorted beneficial pharmacological effects [19]. It has been observed that bioactive compounds of this herb exhibited anti-ulcerogenic, neuroprotective, antimicrobial including antibacterial, antiviral properties, antioxidant, antianxiety, nootropic, anxiolytic, anticonvulsant, antidepressant, hypotensive, anxiolytic, ameliorative properties [20,21,22].

2.3 EFFECT OF PLANT-DERIVED ANTIOXIDANTS WITH NEUROPROTECTIVE ACTIVITY

Plants possess a wide variety of antioxidants, having the ability to counterattack a wide variety of naturally occurring environmental threats metabolized primarily during aerobic activities. Free radicals are released, which then become oxidized when they come in contact with oxygen-containing substances, thereby damaging the host organism by pairing with the surrounding electrons [23]. The presence of antioxidants in plants can, therefore, prevent the damage caused by free radicals by thwarting electron transfer [24]. Degenerative diseases in nerve cells that pose a threat to neurons and nerve bundles occur due to the damage caused by the generation of oxidative free radicals in the system. Plant-derived antioxidants possess the capability to scavenge free radicals which can help in the reduction of damaged nerve cells caused as a result of oxidative stress, hence maintaining a more active physiological state. Plant metabolites, such as polyphenols, vitamins, alkaloids, etc., play a key role in maintaining the structure and function of neurons, thereby enhancing their healthy state [25].

2.3.1 Polyphenols

Polyphenols are the most abundant antioxidant found in plants and have an exceptional ability to capture oxidative free radicals [26]. Polyphenols can be further classified into flavonoids, phenolic acids, and tannins depending on their physical structure as follows:

2.3.1.1 Flavonoids

Flavonoids have a lower redox potential than oxygen and superoxide radicals, making these antioxidants better suited to bind active oxygen [27]. It has been studied that flavonoids can be readily absorbed into the blood and tissue through the intestine and can effortlessly reach the brain through the blood-brain barrier. Once in the brain, flavonoids and their corresponding metabolites protect the cells from oxidative stress [28]. It has been studied that apigenin, a specific flavonoid, shields the extracellular environment in the nervous system by scavenging free radicals and releasing anti-amyloid proteins [29,30]. The results have also revealed that flavonoids can protect neurons from axonal degradation, myelin rupture *trans*-differentiation, and Schwann cell proliferation through the Krox20 pathway coupled with extracellular signal-regulated kinase-independent processes [29].

Hesperidin, rutin, and rhamno helps in the reduction of free Ca^{2+} located between brain cells when fighting against neurogenic excitotoxicity, weakening the decrease of the mitochondrial membrane potential and thereby increasing caspase-3 activity [31,32,33,34].

2.3.1.2 Phenolic acid

Phenolic acid possesses phenolic hydroxyl groups on the benzene ring and is an allelochemical substance, which helps to regulate the growth state of plants through the process of allelopathy. The antioxidant activity of phenolic acids in any organism is effortlessly affected by environmental influences, which mainly include an oxidative effect under the action of Cu^{2+} [35].

Ferulic acid is the most common phenolic acid which is abundantly found in the leaves and seeds of plants. The coupled structure of the ferulic acid molecule can form a relatively stable phenoxy radical in the presence of ultraviolet light to terminate the free radical chain reaction [36]. Ferulic acid protects neurons from any oxidative stress by constraining the activation of p38 MAPK, caspase-3, and COX-2 and by downregulating the damage caused by the MEK/ERK/p90RSK signalling pathway [35,37].

Another class of phenolic acid, ellagic acid, was found to possess antioxidant activity, which has the ability to cause astrocyte proliferation, glial cell line-derived neurotrophic factor (GDNF) release, and Nrf2 activation [38,39].

Rosmarinic acid has been observed to reduce the amount of reactive oxygen species and malondialdehyde, which in turn condenses cellular oxidative stress, thereby protecting hydrogen peroxide-treated glial cells by phosphorylating protein kinase B (Akt), Ser9 Glycogen synthase kinase-3β (GSK-3β), and Fyn. Rosmarinic acid further regulates the activity of Nrf2 to protect PC-12 cells from oxidative stress [40,41]. induced by amyloid β (Aβ). This substance can also reduce the oxidative stress on SH-SY5Y (human dopaminergic neuron cells) by controlling the upregulation of Bax and the downregulation of Bcl-2, hence reducing mitochondrial membrane potential by reducing reactive oxygen species [42].

2.3.1.3 Tannins

Tannins are broadly distributed in plants, and their antioxidant ability is related to the position and binding mode of the phenolic hydroxyl groups. Different classes of tannins exhibit antioxidant properties based on their substructures [43].

Catechin and its derivatives serve as the main functional components in tea, and many studies have revealed that the constructive effects of catechin on mitochondria-related pathways can prevent neurodegeneration, thereby declining brain function [44]. Catechin protects Aβ- and

6-OHDA-induced neuronal apoptosis by activating the protein kinase C (PKC) pathway and PI3K/AKT, inhibiting the MAPKs pathway [43].

2.3.2 Vitamins

Vitamins in the body have always been regarded as one of the most important nutrients required for the upregulation of the body's immune system. Some vitamins, such as vitamin C, vitamin E, and carotenoids, have strong antioxidant and neuroprotective effects, which are commonly found in plants [45].

Vitamin C, also known as ascorbic acid, has four hydroxyl functional groups, two of which are enol hydroxyl groups, which result in the oxidation and dehydrogenation of vitamin C, making it an extremely reductive and highly effective antioxidant in nature. Vitamins have excellent neuroprotective effects against the oxidative stress induced by metal and non-metals, and precisely, the ability of vitamin C to scavenge free radicals can hereby reduce the degree of oxidative stress and increase the viability of nerve cells in the cerebral cortex and striatum [45].

Vitamin E is fat-soluble in nature as compared to vitamin C. It contains hydroxyl hydrogen on a diacetyl alcohol ring, which produces a strong reduction effect on oxygen free radicals and efficiently inhibits lipid peroxidation. This results in the reduction of oxidative stress, thereby producing free radicals and preventing cognitive decline [46].

Carotenoids are widely found in plant pigments, and their chemical structure bears many conjugated double bonds and electrons that can enrich the structure, thereby maintaining high chemical stability and removing free radicals. Consequently, carotenoids also exhibit obvious effects inhibiting lipid peroxidation, which can effectually prevent brain ageing [46].

In the past ten years, natural carotenoids like astaxanthin [47], crocin [48], and lycopene [49] have been found to show some positive neuroprotective effects. Astaxanthin serves as one of the most effective antioxidants in flavonoids and is widely found in the leaves and fruits of plants. When inside the body, astaxanthin has the ability to cross the blood-brain barrier, thereby wielding antioxidant, anti-inflammatory and neuroplastic effects on the brain [50,51]. Studies have revealed that during oxidative stress and inflammation, crocin was seen to protect cells against neurodegeneration by activating the Akt/GSK, CREB/BDNF, and NF-κB signalling pathways [52,53].

Lastly, lycopene, which is widely present in tomatoes, exhibits high neurotrophic value [54]. It was observed that in a model of Aβ-induced neuronal injury, lycopene did not activate NF-κB, p65, or TLR4 expression [55], but it greatly restored mitochondrial morphological membrane potential and ATP levels [56], thereby inhibiting Bax, mitochondria and the associated decrease in the level of Bcl-2 thereby creating protective effects against inflammation and oxidative stress in nerve cells [57].

2.3.3 Alkaloids

Alkaloids are tagged to be strong antioxidants having a complex structure wherein, the more the nitrogen atoms are exposed in the heterocyclic structure, the easier it is to combine with reactive oxygen and free radicals. There are numerous naturally occurring plant-derived alkaloids that have long been used to develop valuable pharmaceutical substances [58]. Astragalus alkaloids and their derivatives are an important target for biological research mainly because they are most commonly found in food sources such as coffee, tea, and potatoes [59]. The main derivatives of astragalus are methylxanthine, theophylline (1,3-dimethylxanthine), theobromine (3,7-dimethylxanthine), and caffeine (1,3,7-trimethylxanthine) [60]. Caffeine is considered to be a more beneficial derivative of astragalus to humans [61] and can readily excite the central nervous system. Hence, proper caffeine intake can be very beneficial to the nervous system [62]. Caffeine is an antagonist of adenosine A (1) and A(2A) receptors, which are closely related to the regulation of synaptic plasticity and energy metabolism in neurons [63], which is vital for neurodegenerative diseases such as epilepsy [64], causing changes in the level of adenosine that can be quite major in nature. Additionally, memory loss is caused by chronic stress linked to the A2AR receptor, where caffeine acts as an antagonist to prevent and treat such signs [65].

2.4 CONCLUSION

Plants constitute an unlimited source of molecules that are consumed for improving human health. Plants are known to possess secondary bioactive metabolites, a chemical diversity determining the evolutionary success of plants, thereby favouring their adaptation to a changing environment. It is to be expected that due to the presence of different phytochemicals, *in vivo* additive and synergistic effects are enhanced, thus amplifying or reducing their activities. Most of the phytochemicals that have recently been reported to exhibit neuroprotective effects in various experimental models of neurological disorders have an additive activity on brain function and provide neuroprotection. Furthermore, to be a suitable neuroprotective agent, it is very important to cross the blood-brain barrier (BBB) in order to reach the target sites of the CNS and natural products extracted from medicinal plants do fit in this role immensely. Lastly, though the presence of receptors or transporters for polyphenols or other phytochemicals in brain tissues remains to be ascertained, compounds with multiple targets serve as a potential and promising class of therapeutics for the treatment of neural disorders.

REFERENCES

[1] N. Kumar, S. Singh, R. Gupta, Trichosanthes dioica Roxb.: an overview, *Pharmacognosy Reviews* 6(11) (2012) 61. http://doi.org/10.4103/0973-7847.95886.

[2] G. G. Kovacs, Current concepts of neurodegenerative diseases, *Emj Neurology* 1(1) (2014) 10–11.

[3] S. Akhtar, J. Iqbal, Pediatric neurosurgery in Pakistan: Current status and a call-to-action for inclusion in the National Surgical Plan, *Journal of Global Neurosurgery* 2(1) (2022). https://doi.org/10.51437/jgns.v2i1.299.

[4] G. P. Kumar, F. Khanum, Neuroprotective potential of phytochemicals, *Pharmacognosy Reviews* 6(12) (2012) 81. http://doi.org/10.4103/0973-7847.99898.

[5] X. Gong, N. J. Sucher, Stroke therapy in traditional Chinese medicine (TCM): prospects for drug discovery and development, *Phytomedicine* 9(5) (2002) 478–484. https://doi.org/10.1078/09447110260571760.

[6] F. Yuan, R. Yu, Y. Yin, J. Shen, Q. Dong, L. Zhong, L. Song, Structure characterization and antioxidant activity of a novel polysaccharide isolated from Ginkgo biloba, *International Journal of Biological Macromolecules* 46(4) (2010) 436–439. https://doi.org/10.1016/j.ijbiomac.2010.02.002.

[7] C. Wang, Z. Han, Ginkgo biloba extract enhances differentiation and performance of neural stem cells in mouse cochlea, *Cellular and Molecular Neurobiology* 35 (2015) 861–869. https://doi.org/10.1007/s10571-015-0180-z.

[8] C. Ren, Y. Q. Ji, H. Liu, Z. Wang, et al., Effects of Ginkgo biloba extract EGb761 on neural differentiation of stem cells offer new hope for neurological disease treatment, *Neural Regeneration Research* 14(7) (2019) 1152–1157. http://doi.org/ 10.4103/1673-5374.251191.

[9] B. Vellas, N. Coley, P. J. Ousset, et al., Long-term use of standardised Ginkgo biloba extract for the prevention of Alzheimer's disease (GuidAge): a randomised placebo-controlled trial, *The Lancet Neurology* 11(10) (2012) 851–859. https://doi.org/10.1016/S1474-4422(12)70206-5.

[10] M. Jahanshahi, E. Nikmahzar, N. Yadollahi, K. Ramazani, Protective effects of Ginkgo biloba extract (EGB 761) on astrocytes of rat hippocampus after exposure with scopolamine, *Anatomy & Cell Biology* 45(2) (2012) 92. http://doi.org/ https://doi.org/10.5115%2Facb.2012.45.2.92.

[11] C. Zhang, C. Ren, H. Chen, et al., The analog of Ginkgo biloba extract 761 is a protective factor of cognitive impairment induced by chronic fluorosis, *Biological Trace Element Research* 153 (2013) 229–236. https://doi.org/10.1007/s12011-013-9645-4.

[12] S. Halder, U. Anand, S. Nandy, P. Oleksak, et al., Herbal drugs and natural bioactive products as potential therapeutics: A review on pro-cognitives and brain boosters perspectives, *Saudi Pharmaceutical Journal* 29(8) (2021) 879–907. https://doi.org/10.1016/j.jsps.2021.07.003.

[13] S. Aguiar, T. Borowski, Neuropharmacological review of the nootropic herb Bacopa monnieri, *Rejuvenation Research,* 16(4) (2013) 313–326.

[14] A. Srivastava, P. Srivastava, A. Pandey, V. K. Khanna, A. B. Pant, *New Look to Phytomedicine* (2019) 625–655 Academic Press. https://doi.org/10.1089/rej.2013.1431.

[15] X. T. Le, H. T. N. Pham, T. Van Nguyen, K. M. Nguyen, K. Tanaka, H. Fujiwara, K. Matsumoto, Protective effects of Bacopa monnieri on ischemia-induced cognitive deficits in mice: the possible contribution of bacopaside I and underlying mechanism, *Journal of Ethnopharmacology* 164 (2015) 37–45.

[16] M. Rastogi, R. P. Ojha, P. C. Prabu, B. P. Devi, A. Agrawal, G. P. Dubey, Prevention of age-associated neurodegeneration and promotion of healthy brain ageing in female Wistar rats by long term use of bacosides, *Biogerontology* 13 (2012) 183–195.

[17] S. Goswami, N. Kumar, V. Thawani, M. Tiwari, M. Thawani, Effect of *Bacopa monnieri* on Cognitive functions in Alzheimer's disease patients, *International Journal of Collaborative Research on Internal Medicine & Public Health* 3(4) 2011. https://doi.org/10.1007/s10522-011-9367-y.

[18] S. H. Mishra, N. K. Sethiya, Review on ethnomedicinal uses and phytopharmacology of memory boosting herb 'Convolvulus pluricaulis' Choisy. *Australian Journal of Medical Herbalism,* 22(1) (2010) 19–25.

[19] P. Agarwa, B. Sharma, A. Fatima, S. K. Jain, An update on Ayurvedic herb Convolvulus pluricaulis Choisy, *Asian Pacific Journal of Tropical Biomedicine* 4(3) (2014) 245–252. https://doi.org/10.1016/S2221-1691(14)60240-9.

[20] P. Rachitha, K. Krupashree, G. V. Jayashree, et al., Chemical composition, antioxidant potential, macromolecule damage and neuroprotective activity of Convolvulus pluricaulis, *Journal of Traditional and Complementary Medicine* 8(4) (2018) 483–496. https://doi.org/10.1016/j.jtcme.2017.11.002.

[21] V. D. Ravichandra, C. Ramesh, K. A. Sridhar, Hepatoprotective potentials of aqueous extract of Convolvulus pluricaulis against thioacetamide induced liver damage in rats, *Biomedicine & Aging Pathology* 3(3) (2013) 131–135. https://doi.org/10.1016/j.biomag.2013.06.005.

[22] S. Olakkaran, A. Antony, Convolvulus pluricaulis (Shankhapushpi) ameliorates human microtubule-associated protein tau (hMAPτ) induced neurotoxicity in Alzheimer's disease Drosophila model, *Journal of Chemical Neuroanatomy* 95 (2019) 115–122. https://doi.org/10.1016/j.jchemneu.2017.10.002.

[23] K. A. Wojtunik-Kulesza, A. Oniszczuk, T. Oniszczuk, M. Waksmundzka-Hajnos, The influence of common free radicals and antioxidants on development of Alzheimer's Disease, *Biomedicine & Pharmacotherapy* 78 (2016) 39–49. https://doi.org/10.1016/j.biopha.2015.12.024.

[24] Y. J. Taverne, A. Caron, C. Diamond, G. Fournier, T. W. Lyons, Oxidative stress and the early coevolution of life and biospheric oxygen, In *Oxidative Stress* (2020) (pp. 67–85). Academic Press. https://doi.org/10.1016/B978-0-12-818606-0.00005-5

[25] L. A. Del Río, ROS and RNS in plant physiology: an overview, *Journal of Experimental Botany* 66(10) (2015) 2827–2837. https://doi.org/10.1093/jxb/erv099.

[26] T. Jiang, Q. Sun, S. Chen, Oxidative stress: A major pathogenesis and potential therapeutic target of antioxidative agents in Parkinson's disease and Alzheimer's disease, *Progress in Neurobiology*, 147 (2016) 1–19. https://doi.org/10.1016/j.pneurobio.2016.07.005.

[27] G. Duthie, A. Crozier, Plant-derived phenolic antioxidants, *Current Opinion in Lipidology* 11(1) (2000) 43–47.

[28] P. G. Pietta, Flavonoids as antioxidants, *Journal of Natural Products* 63(7) (2000) 1035–1042. https://doi.org/10.1021/np9904509.

[29] L. Zhao, J. L. Wang, R. Liu, X. X. Li, J. F. Li, L. Zhang, Neuroprotective, anti-amyloidogenic and neurotrophic effects of apigenin in an Alzheimer's disease mouse model, *Molecules* 18(8) (2013) 9949–9965. https://doi.org/10.3390/molecules18089949.

[30] U. J. Jung, S. R. Kim, Beneficial effects of flavonoids against Parkinson's disease, *Journal of Medicinal Food* 21(5) (2018) 421–432. https://doi.org/10.1089/jmf.2017.4078.

[31] I. Matias, L. P. Diniz, A. Buosi, G. Neves, J. Stipursky, F. C. A. Gomes, Flavonoid hesperidin induces synapse formation and improves memory performance through the astrocytic TGF-β1, *Frontiers in Aging Neuroscience* 9 (2017) 184. https://doi.org/10.3389/fnagi.2017.00184.

[32] E. S. M. El-Sayed, O. M. Abo-Salem, M. F. Abd-Ellah, G. M. Abd-Alla, Hesperidin, an antioxidant flavonoid, prevents acrylonitrile-induced oxidative stress in rat brain, *Journal of Biochemical and Molecular Toxicology* 22(4) (2008) 268–273. https://doi.org/10.1002/jbt.20237.

[33] S. H. Omar, C. J. Scott, A. S. Hamlin, H. K. Obied, The protective role of plant biophenols in mechanisms of Alzheimer's disease, *Journal of Nutritional Biochemistry* 47 (2017) 1–20. https://doi.org/10.1016/j.jnutbio.2017.02.016.

[34] J. A. Lutz, M. Carter, L. Fields, S. Barron, J. M. Littleton, The dietary flavonoid rhamnetin inhibits both inflammation and excitotoxicity during ethanol withdrawal in rat organotypic hippocampal slice cultures, *Alcoholism: Clinical and Experimental Research* 39(12) (2015) 2345–2353. https://doi.org/10.1111/acer.12896.

[35] X. Cui, Q. Lin, Y. Liang, Plant-derived antioxidants protect the nervous system from aging by inhibiting oxidative stress, *Frontiers in Aging Neuroscience* 12 (2020) 209. https://doi.org/10.3389/fnagi.2020.00209.

[36] C. dos Santos Souza, M. S. Grangeiro, E. P. L. Pereira, et al., Agathisflavone, a flavonoid derived from Poincianella pyramidalis (Tul.), enhances neuronal population and protects against glutamate excitotoxicity, *Neurotoxicology* 65 (2018) 85–97. https://doi.org/10.1016/j.neuro.2018.02.001.

[37] W. C. Lin, Y. F. Peng, C. W. Hou, Ferulic acid protects PC12 neurons against hypoxia by inhibiting the p-MAPKs and COX-2 pathways, *Iranian Journal of Basic Medical Sciences* 18(5) (2015) 478.

[38] G. Q. Wang, X. M. He, G. F. Zhu, D. D. Li, J. S. Shi, F. Zhang, Ellagic acid supports neuron by regulating astroglia Nrf2, *Biotechnology and Applied Biochemistry* 66(5) (2019) 738–743. https://doi.org/10.1002/bab.1791.

[39] Q. S. Liu, R. Deng, S. Li, X. Li, et al., Ellagic acid protects against neuron damage in ischemic stroke through regulating the ratio of Bcl-2/Bax expression, *Applied Physiology, Nutrition, and Metabolism* 42(8) (2017) 855–860. https://doi.org/10.1139/apnm-2016-0651.

[40] F. Firdaus, M. F. Zafeer, E. Anis, M. Ahmad, M. Afzal, Ellagic acid attenuates arsenic induced neuro-inflammation and mitochondrial dysfunction associated apoptosis, *Toxicology Reports* 5 (2018) 411–417. https://doi.org/10.1016/j.toxrep.2018.02.017.

[41] L. P. Gao, H. L. Wei, H. S. Zhao, S. Y. Xiao, R. L. Zheng, Antiapoptotic and antioxidant effects of rosmarinic acid in astrocytes. *Die Pharmazie-An International Journal of Pharmaceutical Sciences* 60(1) (2005) 62–65.

[42] H. J. Lee, H. S. Cho, E. Park, et al., Rosmarinic acid protects human dopaminergic neuronal cells against hydrogen peroxide-induced apoptosis, *Toxicology* 250(2–3) (2008) 109–115. https://doi.org/10.1016/j.tox.2008.06.010.

[43] S. Q. Chen, Z. S. Wang, Y. X. Ma, W. Zhang, J. L. Lu, Y. R. Liang, X. Q. Zheng, Neuroprotective effects and mechanisms of tea bioactive components in neurodegenerative diseases, *Molecules* 23(3) (2018) 512. https://doi.org/10.3390/molecules23030512.

[44] M. Assuncao, J. P. Andrade, Protective action of green tea catechins in neuronal mitochondria during aging, *Frontiers Bioscience* 20 (2015) 247–262.

[45] S. Shohag, S. Akhter, S. Islam, T. Sarker, et al., Perspectives on the molecular mediators of oxidative stress and antioxidant strategies in the context of neuroprotection and neurolongevity: an extensive review, *Oxidative Medicine and Cellular Longevity*, 2022 (2022). https://doi.org/10.1155/2022/7743705.

[46] E. Kesse-Guyot, V. A. Andreeva, V. Ducros, C. Jeandel, C. Julia, S. Hercberg, P. Galan, Carotenoid-rich dietary patterns during midlife and subsequent cognitive function, *British Journal of Nutrition* 111(5) (2014) 915–923. https://doi.org/10.1017/S0007114513003188.

[47] G. Hussein, U. Sankawa, H. Goto, K. Matsumoto, H. Watanabe, Astaxanthin, a carotenoid with potential in human health and nutrition, *Journal of Natural Products* 69(3) (2006) 443–449. https://doi.org/10.1021/np050354.

[48] S. M. B. Asdaq, M. N. Inamdar, Potential of Crocus sativus (saffron) and its constituent, crocin, as hypolipidemic and antioxidant in rats, *Applied Biochemistry and Biotechnology* 162 (2010) 358–372. https://doi.org/10.1007/s12010-009-8740-7.

[49] S. Agarwal, A. V. Rao, Tomato lycopene and its role in human health and chronic diseases, *CMAJ* 163(6) (2000) 739–744.

[50] L. Pan, Y. Zhou, X. F. Li, Q. J. Wan, L. H. Yu, Preventive treatment of astaxanthin provides neuroprotection through suppression of reactive oxygen species and activation of antioxidant defense pathway after stroke in rats, *Brain Research Bulletin* 130 (2017) 211–220 https://doi.org/10.1016/j.brainresbull.2017.01.024.

[51] D. Petri, A. K. Lundebye, Tissue distribution of astaxanthin in rats following exposure to graded levels in the feed, *Comparative Biochemistry and Physiology Part C: Toxicology & Pharmacology* 145(2) (2007) 202–209. https://doi.org/10.1016/j.cbpc.2006.12.008.

[52] L. Zhang, R. Previn, L. Lu, R. F. Liao, Y. Jin, R. K. Wang, Crocin, a natural product attenuates lipopolysaccharide-induced anxiety and depressive-like behaviors through suppressing NF-kB and NLRP3 signaling pathway, *Brain Research Bulletin* 142 (2018) 352–359. https://doi.org/10.1016/j.brainresbull.2018.08.021.

[53] M. Motaghinejad, S. Fatima, M. Karimian, S. Ganji, Protective effects of forced exercise against nicotine-induced anxiety, depression and cognition impairment in rat, *Journal of Basic and Clinical Physiology and Pharmacology* 27(1) (2016) 19–27. https://doi.org/10.1515/jbcpp-2014-0128.

[54] J. Shi, M. L. Maguer, Lycopene in tomatoes: chemical and physical properties affected by food processing, *Critical Reviews in Food Science and Nutrition* 40(1) (2000) 1–42. https://doi.org/10.1080/10408690091189275.

[55] R. Tripathi, Z. Liu, R. Plattner, Enabling tumor growth and progression: recent progress in unraveling the functions of ABL kinases in solid tumor cells, *Current Pharmacology Reports* 4 (2018) 367–379. https://doi.org/10.1007/s40495-018-0149-y.

[56] M. Qu, Z. Jiang, Y. Liao, Z. Song, X. Nan, Lycopene prevents amyloid [beta]-induced mitochondrial oxidative stress and dysfunctions in cultured rat cortical neurons, *Neurochemical Research* 41 (2016) 1354–1364. https://doi.org/10.1007/s11064-016-1837-9.

[57] C. Feng, T. Luo, S. Zhang, K. Liu, Y. Zhang, Y. Luo, P. Ge, Lycopene protects human SH-SY5Y neuroblastoma cells against hydrogen peroxide-induced death via inhibition of oxidative stress and mitochondria-associated apoptotic pathways, *Molecular Medicine Reports* 13(5) (2016) 4205–4214. https://doi.org/10.3892/mmr.2016.5056.

[58] S. Khadem, R. J. Marles, Chromone and flavonoid alkaloids: occurrence and bioactivity, *Molecules* 17(1) (2011)191–206. https://doi.org/10.3390/molecules17010191.

[59] J. P. Monteiro, M. G. Alves, P. F. Oliveira, B. M. Silva, Structure-bioactivity relationships of methylxanthines: trying to make sense of all the promises and the drawbacks, *Molecules* 21(8) (2016) 974. https://doi.org/10.3390/molecules21080974.

[60] G. Li, X. Wang, K. H. Row, Magnetic solid-phase extraction with Fe3O4/molecularly imprinted polymers modified by deep eutectic solvents and ionic liquids for the rapid purification of alkaloid isomers (theobromine and theophylline) from green tea, *Molecules* 22(7) (2017) 1061. https://doi.org/10.3390/molecules22071061.

[61] M. Machnik, S. Kaiser, S. Koppe, et al., Control of methylxanthines in the competition horse: pharmacokinetic/pharmacodynamic studies on caffeine, theobromine and theophylline for the assessment of irrelevant concentrations, *Drug Testing and Analysis* 9(9) (2017) 1372–1384. https://doi.org/10.1002/dta.2097.

[62] S. Cappelletti, P. Daria, G. Sani, M. Aromatario, Caffeine: cognitive and physical performance enhancer or psychoactive drug?, *Current Neuropharmacology* 13(1) (2015) 71–88.

[63] Q. Li, X. Lan, X. Han, J. Wang, Expression of Tmem119/Sall1 and Ccr2/CD69 in FACS-sorted microglia-and monocyte/macrophage-enriched cell populations after intracerebral hemorrhage, *Frontiers in Cellular Neuroscience* 12 (2019) 520. https://doi.org/10.3389/fncel.2018.00520.

[64] S. A. Masino, M. Kawamura Jr, D. N. Ruskin, Adenosine receptors and epilepsy: current evidence and future potential, *International Review of Neurobiology* 119 (2014) 233–255. https://doi.org/10.1016/B978-0-12-801022-8.00011-8.

[65] M. P. Kaster, N. J. Machado, et al., Caffeine acts through neuronal adenosine A2A receptors to prevent mood and memory dysfunction triggered by chronic stress, *Proceedings of the National Academy of Sciences* 112(25) (2015) 7833–7838. https://doi.org/10.1073/pnas.1423088112.

3 *Āyurvedic* Psychotropic Drugs (Medhya Rasayan Drugs) and Their Effect on *Manovaha Srotas*

Jasmeet Singh, Nanki Singh, Sneh Shalini, Sunil Kumar Rai, Garima Malik, Rinki Kumari, and G.P. Dubey*
*Corresponding author

3.1 INTRODUCTION

Health is based on Tridoṣa (primary life forces or biological humours) as per Āyurveda [1]. The five elements (Pancamahābhūta) combine in pairs to constitute the three doṣas-Vāta (ether and air), Pitta (water and fire), and Kapha (water and earth) [1]. The combination of these doṣas inherited at birth indicates an individual's unique constitution called Prakriti [1]. The dynamic balance of tridoṣas creates health [1]. Āyurveda defines mental health as a state of mental, intellectual, and spiritual well-being [1]. The use of plants as a source of medicines and human sustenance has been in vogue since antiquity [2].

India has a rich heritage of plants as medicines; Indian systems of medicines utilize 80 percent of the material derived from plants [2]. The global demand for herbal medicine is not only large but is growing as per the increase in the population [2]. Factors contributing to the growth in demand for traditional medicine include the increase of allopathic medicine in developing countries [2].

3.2 ETYMOLOGY OF *MANA*

The term Mana or Manas is a noun or sometimes used as a verb, which has been derived from the root word Mana or Man, which indicates the sense of Gyāna (knowledge) thinking, indicating, and perceiving one which perceives the Gyāna (knowledge) from Indriya, is called as Manas. In Āyurveda the Bhāva (presence or appearance) and Abhāva (absence or non-appearance) of Gyāna (knowledge) have been attributed as the Lakṣaṇa (Signs) for the Manas [1].

MANOVAHA SROTAS Manovaha srotas are not described in the context of Sroto vigyāna chapters in any of the Samhitā grants, but the description of Manovaha srotas appears in different other chapters related to Mānasika vyādhis (mental disorders). Cakrapāṇī, while commenting on Ca. Ś. in 5/4 states that even though the Manovaha srotas are not mentioned separately anywhere, since the mental faculties are found everywhere in the body except Keśa, Roma, mala, drava, anna, etc., which are unable to perceive the sensation, the Manovaha srotas also deemed to be present throughout the body [2].

3.2.1 Manovaha Srotas Kritya

Manovaha srotas are responsible for all the Sankalpa & Vikalpas (thoughts and ideas) that arise in the mind, plans and precessions, taken by the individual, and sensory and motor activities of organs.

 DOI: 10.1201/9781003300502-3

All these functions are performed by the Manas through the Manovaha srotas only. Manovaha srotas can be assumed as nerve fibres. The nerve fibres that carry impulses from different parts of the body towards the brain are called afferent nerve fibres (Sangyāvaha nāḍīs), and those which carry from the brain to the organs are efferent nerve fibres (Ceṣṭāvaha nāḍīs). The brain consists of sensory and motor areas representing different parts of the body. The areas related to thinking, memory, deglutition, thermo regulation, etc., are also represented in the brain. To state in brief, the whole body functions under the brain alone with the help of Manovaha Srotas [2].

3.2.2 MANO VIṢAYAS

As Mana is Ubhayendriya, it has both objects as well as functions.

"Cintyam – Thinking
Vicāram – Prolongation and expansion of thoughts
Ūhyam – Imagination of past, present and future
Dhyeyam – Deep thinking and concentration on a Thought object
Sankalpam – Planning
yatkincit – All the other things which man can perceive"
(Ca. Śā. 1/20) [1]
ACTIVITIES OF MANA
*"इन्द्रियाभिग्रहः कर्म मनसः स्वस्य निग्रहः|
ऊहो विचारश्च, ततः परं बुद्धिः प्रवर्तते"*

(Ca. Śā. 1/21) [1]

As Manas is ubhayendriya, it has some functions also. Controlling of Indriyas and self-restraint are the functions of Manas. Hypothesis and consideration represent the function of the Mind. Finally, the intellect or Buddhi acts and determines.

Āyurveda has differentiated our sensory and motor organs into gyānendriye (organs of perception) and karmendriye (organs of action). Manas has been given the special status of ubhayendriya (organ of both perception and action). It also says that in our body-mind apparatus, our Mana (mind) controls our body and is more important than the body. The mind is responsible for perception, thinking, understanding, making the right decisions at the right time, etc.

Therefore, in Āyurveda, Ācāryas has given prime importance to Mana (mind). In Bhela-Samhitā, the location of the Manas is somewhat more specific and clear. According to him, the Manas is located at the Śirastalvantargatam (within the skull) and recognized as over and above all the Indriyas. In modern terminology, anatomically, i.e., brain. Ācārya Susṛuta labelled Siras as Uttamānga [1].

3.2.3 CENTRES LOCATED IN MANAS

1. It is the place of Ātmā (soul).
2. The centre of Cetanā (consciousness).
3. The seat of Manas (consciousness).
4. Storehouse of Buddhi, Medhā (intellect).
5. The seat of Citta (consciousness).
6. Storehouse of Smriti (memory).
7. The seat of Antahakaraṇa Catuṣṭhaya (Manas, Buddhi, Ahankāra, and Citta).
8. The seat of Pancendriya (special senses).
9. Centre of Prāṇa or Jivana (life).
10. Regulator of Nidrā (sleep).
11. Seat of Satva, Raja, and Tama (emotion and passion).

12. It is the Snāyu Kendra (nerve centre).
13. Seat of Gyānam (knowledge).
14. It controls Kamanā (desire), Dveṣa (malice), Sukha (happiness), Dukkha (sorrow), and various otherworldly human instincts.

The seeds of Āyurvedic description of Vāyu regarding its correlation with the heart and brain are available in Atharva-veda. In Atharva-veda, the location of the Pavana (Vāta) has been described as Murdhā and Hṛadaya, and it is also mentioned that Murdhā & Śira are both stitched to each other. That is why Vāta, having its top position in the human body, initiates or inspires all the allotted activities to every organ of the body [1][2].

3.3 STRUCTURAL QUALITIES OF MANAS

In Āyurvedic kinds of literature, the Manas have been taken to be Aṇu (atomic) in structure and only one in number. It means Manas has two qualities/properties, which are as follows:

1. Ekatva (onlyness) – only one in number (in a body).
2. Aṇutva – Manas has a dimension of Aṇu (atomic) in structure.

"अणुत्वमथ चैकत्वं द्वौ गुणौ मनसः स्मृतौ"
(Ca. Śā. 1/19) [1]

3.3.1 Swarūpa of Manas

Manas is without any rupa (Nirākāra). It is also nirvikāra. As it is, Anthaindriya swarūpa could be understood through lakṣaṇas. Puruṣa is exposed to vedanās like sukha, dukkha, vicāra, krodha, kāma, etc. only through Manas. Caraka says that we cannot learn without coordination between the soul (Ātmā), Manas, sense organs (Indriyas), and subjects (Indriyārtha).

In Āyurveda, Manas is a process in which knowledge is retained a long time. Āyurveda has accepted the inherent relationship between body and mind (structure and function). The heart and brain are both considered to be the seat of Manas. They are interconnected, but their functions are independent [2]. Manas is the coordinator of all the sense organs. Manas is an instrument that is responsible for the phenomenon of thinking. The presence of Manas is the precondition for knowledge and action. Vāta governs Manas, and all the senses are governed by Manas [3].

3.4 SENSUAL PERCEPTION

Sensual perception is an experience resulting from sense-object contact (Indriya Indiyārtha samyoga), which is referred to as Manas (mind-sense contact), through which it is interpreted and projected to the soul (soul and mind contact) [1]. The sense object is received by the sense organ along with the Manas. Then the Manas analyzes it in the form of merits or demerits and forwards it to Buddhi (intellect), which produces decisive knowledge by which one proceeds to speak or to do something with full knowledge.

The intellect is recognized by the sense organ from which it emanates [4]. That emanated through Manas is known as Manobhāva. Manas undergo modification by the different modes of experiences such as happiness, suffering, passion, anger, and so on. Mana traverses through the channels of the sense organs to grasp the object and assume its shape. This stage of the mind assuming the form of the object is called Vritti-Manovritti [5] [6].

3.4.1 Mano Doṣas

Raja and Tama are the doṣas of Manas, which have a relation with tridoṣas. In vāta rajo guṇa predominates; In pitta satva guṇa, kapha tamo guṇa predominates [2].

3.4.2 Qualities of Mana

The 3 GUṆAS are:

1. Sattva
2. Rajas
3. Tamas

An impure Manas (with Raja and Tama), gets involved with the external worldly things and thus gets affected by Sukha (pleasure) and Dukha (pain/sorrow) [7]. The Manas that is pure with Satvaguṇa alone and concentrates in Ātmā only with all other perceptive Dvāra (doors) closed illuminates like a lamp situated in a wind-free lamp case. One does not get affected by ego, gets away from all attachments, and does not hold anything but rather renounces all; this is called Satyā-Buddhi (absolute true knowledge) [8]. This is the knowledge of eternal, unchangeable, blissful Brahma-Gyāna [9][10].

3.4.3 English Equivalent of Component & Functions

Manas – Thought centre -Thinking, awareness, reasoning, doubting.
Ahankāra – Ego-self-awareness and pride.
Buddhi – Intellect decision, Intellectual.
Citta – Memory centre, Memory.

3.4.4 Characters of an Individual Derived in the Fetus from *Manas*

1. *Bhakti*	Desire
2. *Śiṣṭatā*	Conduct
3. *Śauca*	Purity
4. *Dveṣa*	Hatred
5. *Moha*	Attachment
6. *Tyāga*	Detachment
7. *Mātsarya*	Strong desire or Malice
8. *Śaurya*	Velour
9. *Bhaya*	Fear
10. *Krodha*	Anger
11. *Smṛiti*	Memory
12. *Utasāha*	Enthusiasm
13. *Tandrā*	Drowsiness
14. *Tikṣṇatā*	Sharpness
15. *Mṛadutā*	Softness
16. *Gāmbhirya*	Seriousness
17. *Anāvasthatattva*	Unsatbility

3.5 TYPES OF DISEASES AS PER *ĀYURVEDA*

As per Caraka Samhitā; ***Śārīrika & Mānsika***

Śārīrika

The diseases which reside and are related to the human body by *Nija* or *Āgantuja* caused by the vitiation of *doṣas,* i.e. *Vāta, Pitta,* and *Kapha* [11].

Mānsika

The diseases that are only related to the human mind & thoughts by *Nija* or *Āgantuja* caused by vitiation of Mānsika *doṣas,* i.e. *Satva, Raja,* and *Tama*[12].

(a) *Kevala* (Pure/Only) *Mānasika* e.g. *Kāma, Krodha* etc.
(b) *Anubandhaja* e. g. *Unmāda, Apasmāra*
(c) *Ubhyātamaka*

3.5.1 *MANOVIKĀRAS*

Based on the classical descriptions, *Manovikāras* can be defined as a group of clinical conditions affecting individuals with *Alpasatva* (weak psyche) and characterized by behavioural changes in them due to: (as per above) [13].

A) Discordance of *Manodoṣas* or both *Manodoṣas* and *Śarīra doṣa.*
B) Vitiation of *Manovaha srotamāsi* (channels carrying psychological impulses).
C) Impairment of the function of *Manas* viz., *Indrayābhigraha* (perception and motor control), *Manonigraha* (mental control).

3.5.2 CAUSES OF *MANOVIKĀRAS*

1. Improper diet.
2. Actions committed in this birth.
3. Bad deeds done in the last birth.
4. Physical illnesses and overindulgence.
5. Severe mental or physical stress.
6. Unfavourable – personal, familial, and social environment, etc.

***Āyurveda* also says that negative feelings are emotional toxins [14]**
If they are not driven out of the body in a stipulated time, they give rise to chronic mental disorders like anxiety, neurosis, depression, etc. If this is further ignored, it turns into permanent mental disorders like ***Unmāda*** (unreasonable and irrational state of mind, for example, hysteria) & ***Apasmāra*** (epilepsy).

3.5.3 MENTAL DISORDERS IN *ĀYURVEDA* [15]

1. *Unmāda* - Insanity
2. *Apasmāra* - Epilepsy
3. *Avasāda* - Depression
4. *Citto-udvega* - Anxiety Neurosis
5. *Manasa Mandatā* - Mental Retardation

6. *Atatvā-abhinevişa* - Obsessive Disorders
7. *Madātyaya* - Intoxication etc.

3.6 GENERAL PRINCIPLES OF MANAGEMENT OF MENTAL DISORDERS IN *ĀYURVEDA* [16]

1. *Daiva vyapāsraya cikitsā* (Chanting of *Mantras*, *Homa,* Spiritual healing, Religious rites etc.).
2. *Yukti vyapāsraya cikitsā* (Medicines i.e. *Śodhana & Śamana* – Psychotropic Drugs & Diet).
3. *Satvāvajaya cikitsā* (Psychotherapy).

3.6.1 Treatment by *Āyurvedic* Psychotropic Drugs Preparations & Plants Based Medicine [17][18]

1. Single herb medicine.
2. Multi herbal medicines, i.e. use of *Taila*, *Ghṛata, Āsava*, and *Ariṣṭa* preparations.
3. Herbo-mineral medicine is prepared from metals like mercury, gold, silver, etc.
4. Patent or proprietary medicine.

3.7 TREATMENT BY PSYCHOTROPIC SINGLE PLANTS DRUGS (*MEDHYA* DRUGS) [19]

Aindrī	*Bacopa monniera*
Maṇḍūkaparṇī	*Centella asiatica*
Śankhapuṣpī	*Convolvulus pluricalis*
Bhangā	***Cannabis sativa***
Ahiphena	*Papaver somniferum*
Dhturā	*Dhatura metal*
Vacā	*Acorus calemus*
Aśvagandhā	*Withenia somnifera*
Jyotiṣmatī	*Celastrus paniculatus*
Jaṭāmansī	*Nardostachys jatamansi*
Tagara	*Valeriana wallichii*
Yaṣṭīmadhu	*Glycyrrhiza glabara*
Sarpgandhā	*Rauwulfia serpntina* etc.

Psychoactive substances are used by humans for several different purposes to achieve a specific end. These uses vary widely between cultures. Some substances may have controlled or illegal uses, while others may have shamanic purposes, and still others are used medicinally [20]. Other examples would be social drinking or sleep aids. Here, some psychotropic and *medhya* drugs are mentioned [21].

3.7.1 *Brāhmī*

Brāhmī is widely used as 'Medhya Rasayana' in Āyurveda. Two species are taken as Brāhmī i.e. Aindrī (Bacopa monniera) and Maṇḍūkaparṇī (Centella asiatica).

3.7.2 *Aindrī (Bacopa monniera)*

Bacopa monniera has been classified under 'Medhya rasāyana', i.e., medicinal plants rejuvenating intellect and memory [22]. The name is Brāhmī is derived from the word 'Brahma', the mythical 'creator' in the Hindu pantheon. Because the brain is the centre for creative activity, any compound that improves brain health is called Brāhmī.

Bacopa monniera belongs to the family Scrophulariaceae and is a small creeping herb with numerous branches, small oblong leaves, and light purple or small white flowers with four or five petals. The leaves are succulent, relatively thick, and oblanceolate and are arranged oppositely on the stem [23].

Compounds responsible for the pharmacological effects include alkaloids, saponins, and sterols. Detailed investigations first reported the isolation of the alkaloid 'brahmine' [23]. Later, other alkaloids like nicotine and herpestine have also been reported [24,25]. Subsequently, the isolation of D-mannitol, saponin, hersaponin, and potassium salts was reported [25].

The major chemical entity shown to be responsible for neuropharmacological effects and the nootropic action or antiamnestic effect of Brāhmī is bacoside A, assigned as 3-(a-L-arabinopyranosyl)-O-b-D-glucopyranoside-10, 20-dihydroxy-16-ketdammar-24-ene [26].

Sedative and tranquilizing properties by the effect of glycosides are called hersaponins [27]. Cognitive deficits induced by intracerebroventricularly administered colchicines and injection of ibotenic acid into the nucleus basalis magnocellularis [28]. Anxiety demonstrated that Brahmi contains 25 percent bacoside. Bacoside reduces anxiety without inducing amnesia, the side effects associated with lorazepam, but instead has a memory-enhancing effect [29, 30].

The antidepressant potential of Brāhmī has been evaluated in an earlier study wherein it showed significant antidepressant activity in the most commonly used behaviour paradigms in animal models of depression, namely, the forced swim test and learned helplessness test [31]. Brahmi has been indicated as a remedy for epilepsy in Āyurvedic medicine [32].

3.7.2.1 *Maṇḍūkaparṇī (Centella asiatica)*

It has been used as a medicine in the Āyurvedic tradition of India for thousands of years and is listed in the historic Susṛuta Samhitā, Caraka Samhitā, and other Āyurvedic ancient texts. In China, known as Gotukola, it is one of the reported 'miracle elixirs of life' known over 2000 years ago.

Maṇḍūkaparṇī is a perennial herbaceous creeper, having small fan-shaped green leaves with white or light purple-to-pink or white flowers and bears small oval fruit belongs to the family Apiaceae and is found throughout India in moist places up to an altitude of 1800 m in tropical and subtropical countries in swampy areas. Maṇḍūkaparṇī having saponins (also called triterpenoids) include asiaticosides in which a trisaccharide moiety is linked to the aglycone asiatic acid, madecassoside, and madasiatic acid [33]. In this Brahmoside and brahminoside, are postulated to be responsible for CNS and uterorelaxant actions but are yet to be confirmed by clinical studies. Crude extract that contains glycosides isothankuniside and thankuniside showed antifertility action in mice. Maṇḍūkaparṇī has CNS effects described in Indian literature such as stimulatory-nerve tonic, rejuvenator, sedative, tranquilizer, and intelligence-promoting property [34][35]. It increases the cerebral levels of γ-amino butyric acid (GABA), which explains its traditional use as an anxiolytic and anticonvulsant. It is known that GABA and its agonists inhibit the central cholinergic action by affecting the turnover rate of acetylcholine in the rat brain. The isolated steroids from the plant have been used to treat epilepsy [36].

3.7.3 AHIPHENA (*PAPAVER SOMNIFERUM*)

Papaver somniferum, is commonly known as the opium poppy. *Papaver somniferum* is an annual herb growing to about 100 cm (40 in) tall. The plant is strongly glaucous, giving a greyish-green appearance, and the stem and leaves bear a sparse distribution of coarse hairs. The large leaves are lobed, the upper stem leaves clasping the stem. The flowers are up to 30 – 100 mm (1 – 4 in) in diameter, normally with four white, mauve, or red petals, sometimes with dark markings at the base. The fruit is a hairless, rounded capsule topped with 12 – 18 radiating stigmatic rays or fluted cap [37].

Opium has been eaten, drunk, and smoked for centuries; the poppy, from which opium is derived, is not known from the wild; it has been domesticated for its seeds, which are used for oil and food, and for its dried sap, which produces opium, The opium poppy, as its name indicates, is the principal source of opium, the dried opiate drugs are extracted latex produced by the seed pods. A common method of preparing opium was to dissolve it in alcohol, and this tincture (infusion, extraction, solution), later known as laudanum, became a popular medication for centuries. (Medications at that time were not so much administered to heal as they were simply to alleviate pain) [38].

Opium contains a class of naturally occurring alkaloids known as opiates, which include morphine, codeine, thebaine, oripavine, papaverine, and noscapine. The specific epithet *somniferum* means 'sleep-bringing', referring to the sedative properties of some of these opiates [39].

Tincture of opium or laudanum, consisting of opium dissolved in alcohol or a mixture of alcohol and water, is one of many unapproved drugs regulated by the U.S. Food and Drug Administration (FDA). Its marketing and distribution persist because its historical use preceded the Federal Food, Drug & Cosmetic Act of 1938. Tincture of opium B.P., containing 1% w/v of anhydrous morphine, also remains in the British Pharmacopoeia, listed as a Class A substance under the Misuse of Drugs Act 1971 [40].

Morphine is the predominant alkaloid found in the cultivated varieties of opium poppy that are used for opium production. Other varieties produce minimal opium or none at all, such as the latex-free Sujata type. Non-opium cultivars that are planted for drug production feature a high level of thebaine or oripavine. Those are refined into drugs like oxycodone. Raw opium contains about 8–14% morphine by dry weight or more in high-yield cultivars. It may be used directly or chemically modified to produce synthetic opioids such as heroin.

It is a very powerful hypnotic and narcotic, with powerful analgesic properties. It is the most abundant component of opium, ranging from 4–21% by weight. Morphine is still unsurpassed in its ability to deaden pain. It is considered the most powerful naturally occurring analgesic). It is a general CNS (central nervous system) depressant and, in overdose, can lead to death by completely suppressing the respiratory centre in the brain. The study of opium led to the discovery of the first alkaloid, morphine, and the study of morphine has led to the discovery of the brain's painkillers, the endorphins, perhaps the most important advancement in neurochemistry in the past half-century [41].

3.7.4 *Dhaturā* (Datura metel)

Daturā refers to any one of nine species of shrubby herbaceous plants from the Solanaceae family, which produces large, white, or purple trumpet-shaped flowers.

There are many ways to administer the *Daturā* plant, the most common being tea made from any part of the plants, ingestion of seeds, and smoking blends made from leaves or flowers [42].

Daturā can produce hallucinations, which the user is unable to identify as fantasy. A user will not be able to distinguish between reality and fantasy and, therefore, may become confused, agitated, or scared. Most users report a feeling of discomfort and do not enjoy the trip. The peak may result in memory loss and complete loss of motor/communication functions.

All varieties of *Datura* contain psychotropic tropane alkaloids. *D. metel* contains the highest scopolamine content of the *Datura* genus. The entire plant also contains various withanolides [42][43].

The danger of *Daturā,* because of the unknown quantity of tropane alkaloids in each *Daturā* plant, is that a quantity for overdose is going to be hard to predict. To reduce the chance of overdose, start very small and work your way up to larger doses. Remember, if using another plant, the dosage will be different from last time. *Datura metel* is very dangerous and can lead to insanity or even death, so great care must be taken with its consumption.

There is evidence that *Datura metel* seeds have been used in ancient Indian medicine, modern Indian folk medicine, and *Āyurvedic* medical practices. The most common medicinal uses for

Daturā in these systems are for skin conditions, anxiety disorders, and respiratory ailments, along with a litany of other conditions [44].

3.7.5 VACĀ (ACORUS CALEMUS)

Vacā (botanically recognized as *Acorus calamus* Linn.) is among the rare treatments that are mentioned in the *Vedas*. In later days Ācāryas worked well not only with the nervous system and the brain but also with additional parts of the entire body [45].

It has been described for use in Indian and Chinese systems of medicine for hundreds of years to cure diseases, especially central nervous system (CNS) abnormalities. In *Sanskṛita*, Vacā means 'speaking' and is used for clearing the voice and probably for diseases of the nervous system, anti-stuttering drugs, and other mental disorders. It is demonstrated that it has a strong psycho-pharmacological solution, which has a favourable impact on memory as well as the learning process [46].

Vacā is a herb of choice for use in the case of loud eructation. The powder of *vacā* and white sandal makes a very effective face-pack in the treatment of blemishes and pimples. In many Indian homes, customarily, vacā is administered with honey in a minute quantity to infants on the 11th and 21st days of birth. This is also used for Swarṇa *prāsana Samskāra* in South India on *Puṣya nakṣatra* up to the age of 16 years. It is believed that this practice helps the children to be mentally active and vocal.

3.7.6 AŚWAGANDHĀ (WITHENIA SOMNIFERA)

Withania somnifera (family-*solanaceae*) is a highly esteemed medicinal herb in Āyurveda and most popularized as *Aśvagandhā* although other common names include the King of *Āyurveda*, Indian ginseng (not at all related to *Panax ginseng*), and winter cherry.

The herb is classified as *rasāyana* in *Āyurvedic* medicine due to it being a general tonic. In modern terms, it is called an adaptogen for similar reasons and is also classified as *balya* (increases strength) and *vajikara* (aphrodisiac) [47].

The name *Aśvagandhā* comes from the translation 'smell of horse', which is thought to be due to two main reasons; the root itself smells like a horse, and the root is supposed to imbibe you with the strength and virility of a horse. It produces *Kāmaśakti* like the horse in humans and smells like horses. Beyond those uses, it has been traditionally used as an analgesic, astringent, antispasmodic, and immune-stimulant while being used to treat inflammation, cancer, stress, fatigue, diabetes, and cardiovascular complications while its adaptogenic usage is emphasized for persons with stress-related insomnia, debility, and nervous exhaustion. *Aśvagandhā* has also been reported to be an immune-stimulant compound, with particular emphasis on stress-related suppression in immunity [48].

Aśvagandhā is a highly esteemed medicinal plant in traditional Indian medicine for a wide variety of ailments but usually focused on stress, immune support (with regards to stress), anxiety and depression (again regarding stress), and the treatment of cancer and inflammation; historically, it boasts low toxicity when consumed as the root extract alongside food [49].

3.7.7 JYOTIṢMATĪ (CELASTRUS PANICULATUS)

Celastrus paniculatus is a deciduous woody liana vine commonly known as the black oil plant, climbing staff tree, and intellect tree. It stems up to 10 cm (3.9 in) in diameter and 6 m (20 ft) long with a rough, pale brown exfoliating bark covered densely with small, elongated lenticels. The leaves are simple, broad, and oval, obovate or elliptic in shape, with toothed margins. This climbing shrub grows throughout India at elevations up to 1,800 m (5,900 ft) [50].

The seed is an aphrodisiac, bitter, emetic, laxative, and a stimulant. It is used in the treatment of rheumatism, leprosy, gout, fevers, and paralysis. The crushed seeds, a decoction of the seeds, or the fixed oil from the seed are applied as a poultice or taken internally. Credited with stimulant and diaphoretic properties, they are applied to relieve fevers, joint and muscular pains, and paralysis.

The oil from the seed is a powerful stimulant. It is used in the treatment of beriberi. The seed oil is a major component in various products to relieve rheumatic and muscular pain. The oil is an ingredient of many more herbal medicines, claiming to enhance the nervous and mental system. It is also considered to be an analgesic, aphrodisiac, diaphoretic, emetic, emmenagogue, stimulant, and tonic. The leaf sap has been used as an antidote for opium poisoning. The root is used as an antimalarial and antipyretic [52].

The wood is used in the treatment of tuberculosis. The bark is abortifacient. The juice from the bark is used in the treatment of indigestion. The stem bark is used as an antidysenteric and a decoction of the stem as a diuretic in the treatment of kidney disorders. The fruits are considered antiflatulent, a blood tonic, and a relief for fainting [53].

3.7.8 *Tagara (Valeriana wallichii)*

Valeriana jatamansi, formerly known as *Valeriana wallichii,* is a rhizome herb of the genus *Valeriana* and the family *Valerianaceae,* also called Indian Valerian or *Tagar-Ganthoda,* not to be confused with *ganthoda,* the root of Indian long pepper.

It grows in the Northwest Himalayas in places like Astore (Northern Pakistan) and forests in the region. The genus *Valeriana,* with about 200 species, belongs to the family Valerianaceae and has a distribution throughout the world. The Indian Valerian has long been used in *Āyurveda* (*Carak Samhitā* and *Susṛuta Samhitā*) and Unani systems of medicine. It is a herb useful in Āyurvedic medicine and used as an analeptic, antispasmodic, carminative, sedative, stimulant, stomachic, and nervine. It is described for use in sleep disorders, anxiety, and psychological stress, obesity, skin disease, insanity, epilepsy, and snake poisoning. The crude drugs from roots/ rhizomes and Valerian-derived phytomedicines are used as mild sedatives in the pharmaceutical industry. The activity is largely attributed to the presence of valepotriates [54].

3.7.9 *Yaṣṭhīmadhu (Glycyrrhiza glabara)*

Yaṣṭhīmadhu the common name of *Glycyrrhiza glabra,* is a flowering plant of the bean family Fabaceae, from the root of which a sweet, aromatic flavouring can be extracted.

The liquorice plant is an herbaceous perennial legume native to Western Asia, North Africa, and Southern Europe. Botanically, it is not closely related to anise or fennel, which are sources of similar flavouring compounds. Another such source, star anise, is even more distantly related to anise and fennel than liquorice, despite its similar common name [55].

Liquorice is used as a flavouring in candies and tobacco, particularly in some European and West Asian countries. Properties of glycyrrhizin are under preliminary research, such as for hepatitis C or topical treatment of psoriasis, but the low quality of studies as of 2017 prevents conclusions about efficacy and safety. For thousands of years, *G. glabra* has been used for medicinal purposes including indigestion and stomach inflammation, cough suppression, ulcer treatment, and use as a laxative.

Liquorice extracts have been used in herbalism and traditional medicine. Excessive consumption of liquorice (more than 2 mg/kg per day of pure glycyrrhizinic acid, a liquorice component) may result in adverse effects, such as hypokalaemia, increased blood pressure, muscle weakness, and death [56].

3.7.10 Sarpgandhā (Rauvolfia serpentine)

Sarpagandhā (*Rauvolfia*) is an important medicinal plant distributed in the foothills of the Himalayan range, up to the elevation of 1300 – 1400 m. and almost throughout the country. It is an erect evergreen, perennial under-shrub, 75 cm to 1 m. in height [57].

It is the source of the phytochemical reserpine, which has been used in the treatment of systolic hypertension and for mental disorders, i.e. Mania (also called *Pagal ki jadi*). *Rauvolfia serpentine* contains dozens of alkaloids of the indole alkaloid family, including ajmaline, ajmalicine, reserpine, and serpentine, among others [58].

Rauvolfia serpentine may cause adverse effects by interacting with various prescription drugs or via interference with mechanisms of mental depression or peptic ulcer. The reserpine in *R. serpentina* is associated with diverse adverse effects, including Suicidal tendencies, vomiting, diarrhoea, dizziness, headache, anxiety, or hypersensitivity reactions [58].

3.7.11 Common Useful Herbs-Mineral Yogas of Psychotropic Drugs

Brāhmī vaṭī,
Vāta Kulāntaka Rasa,
Smṛiti Sāgara Rasa,
Yogendra rasa,
Mānasamitra vaṭakama
Stresscom
Zzowin etc.

3.8 CONCLUSION

There are increasing numbers of patients who express a preference for the use of remedies they perceive to be natural, and physicians recommend herbal remedies in the selected cases. It is becoming increasingly important for physicians to be familiar with the herbal remedies commonly used in the patient problems they serve.

Although evidence of the efficacy of herbal preparation in treating psychiatric conditions is growing, translating the results of efficacy studies into effective treatment for patients is hampered by the chemical complexity of the products.

Since mental illness is diverse and individual patients are biochemically unique, a larger number of drugs will increase the likelihood of finding a beneficial medication. Hence, in future times, psychiatric patients will probably have medications with improved effectiveness and with fewer side effects.

There is a lack of drug standardization of commonly available *Āyurvedic* preparations and a paucity of well-controlled studies. This reveals that several herbal drugs are available for the treatment of various mental disorders, but there should be a need to explore the efficacy of many of them. For this, a rigorous study of various traditionally but not scientifically proven herbs must be carried out at the pre-clinical and clinical levels.

REFERENCES

[1] Shastri, P. V., & Chaturvedi, G. N. (1981). Charaka Samhita of Agnivesha, Chikitsa Sthana (Vol. 2). Chaukhamba Bharati Academy.
[2] Sharma, H., Chandola, H. M., & Singh, G. (2008). Basisht Āyurveda kā Siddhānt (Fundamental Principles of Āyurveda). Chaukhambha Orientalia.
[3] Larson, G. J. (2001). The Encyclopedia of Indian Philosophies, Vol. 4: Samkhya, A Dualist Tradition in Indian Philosophy. Princeton University Press.

[4] Dash, V. B. (1991). Fundamentals of Ayurvedic Medicine. Concept Publishing Company.
[5] Frawley, D., & Ranade, S. (2001). Ayurveda, Nature's Medicine. Lotus Press.
[6] Pandey, G. (2003). Ayurvedic Healing: Contemporary Maharishi Ayurveda Medicine and Science. Motilal Banarsidass Publisher.
[7] Chopra, A., Doiphode, V. V., & Padhi, M. M. (2012). A Comprehensive Guide to Ayurvedic Principles and Practice. Singing Dragon.
[8] Svoboda, R. E. (1999). Prakriti: Your Ayurvedic Constitution. Lotus Press.
[9] Frawley, D. (1996). Ayurvedic Psychology: The Mind of Your Dosha. Lotus Press.
[10] Lad, V. (1998). Ayurveda: The Science of Self-Healing. Lotus Press.
[11] Chopra, A., Doiphode, V. V., & Padhi, M. M. (2012). A Comprehensive Guide to Ayurvedic Principles and Practice. Singing Dragon.
[12] Svoboda, R. E. (1999). Prakriti: Your Ayurvedic Constitution. Lotus Press.
[13] Frawley, D. (1996). Ayurvedic Psychology: The Mind of Your Dosha. Lotus Press.
[14] Lad, V. (1998). Ayurveda: The Science of Self-Healing. Lotus Press.
[15] Gogte, V. M. (2000). Ayurvedic Pharmacology & Therapeutic Uses of Medicinal Plants. Chaukhamba Publications.
[16] Sharma, H., Chandola, H. M., & Singh, G. (2008). Basisht Āyurveda kā Siddhānt (Fundamental Principles of Āyurveda). Chaukhambha Orientalia.
[17] Frawley, D., & Ranade, S. (2001). Ayurveda, Nature's Medicine. Lotus Press.
[18] Pandey, G. (2003). Ayurvedic Healing: Contemporary Maharishi Ayurveda Medicine and Science. Motilal Banarsidass Publisher.
[19] Chopra, A., Doiphode, V. V., & Padhi, M. M. (2012). A Comprehensive Guide to Ayurvedic Principles and Practice. Singing Dragon.
[20] Lad, V. (1998). Ayurveda: The Science of Self-Healing. Lotus Press.
[21] Gogte, V. M. (2000). Ayurvedic Pharmacology & Therapeutic Uses of Medicinal Plants. Chaukhamba Publications.
[22] Chopra, A., Doiphode, V. V., & Padhi, M. M. (2012). A Comprehensive Guide to Ayurvedic Principles and Practice. Singing Dragon.
[23] Rai, et al. (2003). Chemical investigation of Bacopa monniera Wettst: Part I—Isolation of chemical constituents. *Journal of Ethnopharmacology*, 89(2–3), 239–244.
[24] Rinki et al. (2015). Efficacy of Ayurvedic Plants." In: Taylor & Francis Group, CRC Press, Pages 397–420, Print ISBN: 978-1-4987-3905-4, eBook ISBN: 978-1-4987-3906-1, DOI: 10.1201/9781315369334-29.
[25] Shastri, et al. (1959). [Title not provided]. *Journal of Scientific & Industrial Research*, 18B, 352.
[26] Chatterji, et al. (1965). Chemistry and pharmacology of Bacoside A. *Indian Journal of Pharmacy*, 27, 98.
[27] Malhotra, & Das. (1959). [Title not provided]. *Journal of Scientific & Industrial Research*, 18B, 340.
[28] Bhattacharya, et al. (1999a). Antioxidant activity of active tannoid principles of Emblica officinalis (amla). *Journal of Ethnopharmacology*, 65(2), 83–90.
[29] Shankar, & Singh. (2000). [Title not provided]. *Phytomedicine*, 7(5), 463–469.
[30] Bhattacharya, & Ghosal. (1998). Anti-oxidant effect of Withania somnifera glycowithanolides in chronic footshock stress-induced perturbations of oxidative free radical scavenging enzymes and lipid peroxidation in rat frontal cortex and striatum. *Phytotherapy Research*, 12(3), 221–229.
[31] Sairam, et al. (2002). Evaluation of adaptogenic properties of Bacopa monniera in an experimental model of chronic stress. *Pharmacology Biochemistry and Behavior*, 73(4), 893–900.
[32] Shanmugasundaram, et al. (1991). Glycolytic and gluconeogenic enzymes in the liver and skeletal muscles of protein-deficient rats: effect of Bacopa monniera and bacoside A. *Fitoterapia*, 62(1), 52.
[33] Singh, B., Rastogi, R. P., & Dhawan, B. N. (1998). Centella asiatica: The Nootropic Brahmi. *Annals of Pharmacotherapy*, 32(3), 376–383.
[34] Kumar, S. R. (2006). Evaluation of Bacopa monniera and Centella asiatica oil extracts on Acetic acid induced writhing response in mice. *Indian Journal of Pharmacology*, 38(2), 101–102.
[35] Gupta, Y. K., Veerendra Kumar, M. H., & Srivastava, A. K. (2003). Effect of Centella asiatica on pentylenetetrazole-induced kindling, cognition, and oxidative stress in rats. *Pharmacology Biochemistry and Behavior*, 74(3), 579–585.

[36] Hausen, B. M. (1993). Centella asiatica (Indian pennywort), an effective therapeutic but a weak sensitizer. *Contact Dermatitis*, 29(4), 175–179.
[37] United Nations Office on Drugs and Crime. (2009). "Opium Poppy Cultivation and Heroin Processing in Southeast Asia". United Nations Office on Drugs and Crime. ISBN: 978-92-1-148240-4. Available at: www.unodc.org/documents/data-and-analysis/Studies/opium-poppy-cultivation-heroin-processing-southeast-asia.pdf.
[38] United Nations Office on Drugs and Crime. (2018). "World Drug Report 2018". UN.
[39] Griffiths, P., & Weimer, B. (2010). "The History of Heroin".
[40] United Nations Office on Drugs and Crime. (2005). "Global Illicit Drug Trends 2003". UN.
[41] National Institute on Drug Abuse. (2020). "DrugFacts: Prescription Opioids". NIDA.
[42] Lindequist, U. (1992). Medicinal and Poisonous Plants of the Tropics. Teubner, Stuttgart.
[43] Ratsch, C. (1998). The Encyclopedia of Psychoactive Plants: Ethnopharmacology and Its Applications. Park Street Press.
[44] Sharma, P. V. (2003). Dravyaguna Vijnana. Chaukhambha Bharati Academy.
[45] Frawley, D., & Lad, V. (2001). The Yoga of Herbs: An Ayurvedic Guide to Herbal Medicine. Lotus Press.
[46] Nadkarni, K. M. (1976). Indian Materia Medica. Popular Prakashan.
[47] Singh, N., & Nath, R. (1982). Herbal Drugs and their Metabolites in Biological Systems. Chaukhambha Bharati Academy.
[48] Kulkarni, R. D. (1999). Ayurveda Chikitsa Rahasya.
[49] Sane, R. T. (2007). Ayurvedic Cure for Common Diseases. Chaukhambha Bharati Academy.
[50] Singh, R. H. (2007). A Text Book of Dravyaguna Vijnana. Chaukhambha Bharati Academy.
[51] Dymock, W., Warden, C. J. H., & Hooper, D. (1890). Pharmacographia Indica: A history of the principal drugs of vegetable origin met with in British India. Trench, Trübner & Co.
[52] Khare, C. P. (2007). Indian Medicinal Plants: An Illustrated Dictionary. Springer.
[53] Paranjpe, P. (2001). Indian Medicinal Plants: Forgotten Healers. Orient Blackswan.
[55] Shastri, P. V. (1983). Charaka Samhita of Agnivesha, Sutra Sthana (Vol. 1). Chaukhamba Bharati Academy.
[56] Dash, V. B. (1992). Materia Medica of Ayurveda: Based on Ayurveda Saukhyam of Todarananda. Concept Publishing Company.
[57] Frawley, D. (1996). Yoga and Ayurveda: Self-Healing and Self-Realization. Lotus Press.
[58] Sharma, P. V. (2002). Dravyaguna Vigyan (Vol. 2). Chaukhambha Bharati Academy.
[59] Sushruta. (1985). Sushruta Samhita. Chaukhamba Orientalia.

Special References

Caraka Sanhita (ni.-7, 8), (chi.-9/3, 96,, 10-/3,57,60,63,64)
Susruta sanhita(ut/60,62)
Astang sangrah
Astang hridaya (ut-7)
Sharangdhar sanhita

FIGURE 3.1 Brāhmī.

FIGURE 3.2 *Maṇḍūkaparṇī* (*Centella asiatica*).

FIGURE 3.3 *Śankhapuṣpī* (*Convolvulus pleuricaulis* Chois.).

FIGURE 3.4 *Bhangā* (*Cannabis sativa*).

FIGURE 3.5 *Ahiphena* (*Papaver somniferum*).

FIGURE 3.6 *Dhaturā* (*Datura metel*).

FIGURE 3.7 *Vacā* (*Acorus calemus* Linn.).

FIGURE 3.8 *Aśwagandhā* (*Withenia somnifera*).

FIGURE 3.9 *Jyotiṣmatī* (*Celastrus paniculatus*).

FIGURE 3.10 *Tagara* (*Valeriana wallichii*).

FIGURE 3.11 *Yaṣṭhīmadhu* (*Glycyrrhiza Alabama*).

FIGURE 3.12 *Sarpgandhā* (*Rauvolfia serpentina*).

4 Potential Impact of Indian Herbs and Spices in Prevention of Pathogenesis of Alzheimer's Disease

Paramita Bhattacharjee, Anamika Das, and Sanjukta Datta*

*Corresponding author

4.1 INTRODUCTION

India is a land rich in spices, and the multitudinous uses of these spices have earned them a distinguished position globally. The term 'spice' has been coined from the Latin words 'Species aromatacea', which means 'aromatic spices'. Traditionally used spices are obtained from several plant parts such as roots, stems, barks, buds, pistils, leaves, seeds, etc.[1] The Spices Board of India (under Ministry of Commerce and Industry, Government of India) documents important spices belonging to the families Zingiberaceae (cardamom, turmeric), Piperaceae (pepper), Iridaceae (saffron), Myrtaceae (clove), Fabaceae (fenugreek), Myristicaceae (nutmeg, mace), Solanaceae (chilli), and Lauraceae (cinnamon) to be the commonly used spices in India. According to reports, India currently has an overwhelming record export of 15,31,154 tonnes of spice/spices products (2021–2022) valued at Rs. 30,576 crores (4102.29 million US$)[2].

Spices and herbs not only serve as flavouring agents enhancing food palatability but also house several bioactive components known as 'active principles', which are reportedly known to have a myriad of health-protective properties such as antioxidative, anti-inflammatory, and anti-carcinogenic. These bioactive principles are generally phenolic compounds (flavonoids) and their derivatives[1]; and many of them, such as gingerol, carvacrol, curcumin, thymol, hydroquinone, and linalool to name a few, have demonstrated anti-inflammatory, anti-amyloidogenic, anti-cholinesterase, and antioxidant effects in prevention of pathogenesis of mild to moderate Alzheimer's disease (AD)[3,4,5,6,7,8]. Table 4.1 exhibits important Indian spices/herbs and their bioactive principles, which are reportedly known to reverse AD pathology.

The major pathways of AD pathogenesis have been illustrated in Figure 4.1. AD pathogenesis progresses mainly by two processes- extracellular deposition of beta amyloid (Aβ), which results in the formation of senile plaques (considered to be the hallmark of AD progression), and/or by intracellular accumulation of tau proteins, which lead to neurofibrillary tangles (which in turn accumulates the enzyme acetylcholinesterase). These two effects eventually lead to neurodegeneration through apoptotic cell death of the cholinergic neurons and synapses and interfere with the normal synthesis of acetylcholine, the key neurotransmitter that aids in learning-memory processes[9]. Acetylcholine is rapidly hydrolyzed to acetate and choline in the presence of acetylcholinesterase (5,000 molecules of acetylcholine are hydrolyzed per second), which is one of the most kinetically efficient enzymes[10] present in humans and thus brings about inactivation of acetylcholine[9]. AD is basically an oxidative stress disorder caused by an imbalance between the production of reactive oxygen species (ROS) and the ability of the human biological system to detoxify them, which results in cytotoxicity,

 DOI: 10.1201/9781003300502-4

TABLE 4.1
List of biologically active components of some common Indian spices reportedly known to have potential impact in prevention of AD pathogenesis and their targeted pathways associated with AD progression

Name of the spices	Bioactive components	Targeted pathways of AD pathogenesis
Turmeric (*Curcuma longa L.*)	Curcumin and its derivatives	• Inhibition of Aβ fibrillation and aggregation • Reduction in oxidative stress as an antioxidant • Inhibition of tau protein aggregation
Black pepper (*Piper nigrum L.*)	Piperine	• Inhibition of AChE activity • Inhibition of Aβ oligomerization • Oxidative stress reduction as an antioxidant • Inhibition of tau phosphorylation
Small cardamom (*Elettaria cardamomum L.)*	1,8-cineole and its derivatives	• AChE-inhibitory activity • Inhibition of Aβ fibrillation and aggregation • Neurotoxicity activity by lowering Bax, caspase -3, -8 and -9 and enhancing the B-cell lymphoma 2 (Bcl-2) levels • Neuroinflammatory activity by regulating the NF-kB signalling pathway
Saffron (*Crocus sativus L.*)	Crocin and its derivatives	• AChE-inhibitory activity • Inhibition of Aβ fibrillation and aggregation • Oxidative stress reduction as an antioxidant • Inhibition of tau phosphorylation • Reduction in Superoxide dismutase (SOD) value
Black cumin (*Nigella sativa L.*)	Thymoquinone and its derivatives	• AChE-inhibitory activity • Inhibition of Aβ fibrillation and aggregation • Inhibition of tau protein aggregation • Neuroinflammatory activity by regulating the NF-kB signalling pathway
Bay leaf (*Laurus nobilis L.*)	1,8-Cineole	• Memory enhancement effects through AChE-inhibitory activity • BChE inhibitory activity
Coriander (*Coriandrum sativum L.*)	Linalool	• Cognitive enhancement by inhibition of Aβ fibrillation and aggregation • Reduction in lipid peroxidation

Reviews/articles that are indexed by Scopus and/or Web of Science have been selected for this compilation

viz. oxidative damage[11,12]. Chronic stresses induced by ROS production and accumulation cause impairment of learning-memory processes, which leads to a progression of AD[13,14,15]. Figure 4.1 also depicts the multifunctional roles the drugs need to play in ameliorating AD progression, such as by prevention of Aβ oligomerization, scavenging of ROS, chelating iron molecules that accumulate in an AD brain, inhibiting the activities of acetylcholinesterase and/or displaying anti-tau protein aggregation effects[16,17].

Table 4.2 presents the individual spice compounds, their extracts, and their formulations, along with their modes of action in targeted pathways of AD pathogenesis and their success in preclinical/clinical investigations (further elaborated in Tables 4.3a and 4.3b). Common spices such as saffron, turmeric, black pepper, coriander, and black cumin have shown efficacies in mitigating AD pathogenesis[3,46,57,60,73] by manipulating the reported pathways illustrated in Figure 4.1. A critical assessment of the consolidated information presented in Table 4.2 reveals that most studies have been performed using the spice bioactive principle in raw form i.e., as a pure chemical standard, and not as a component of the spice extract or isolate.

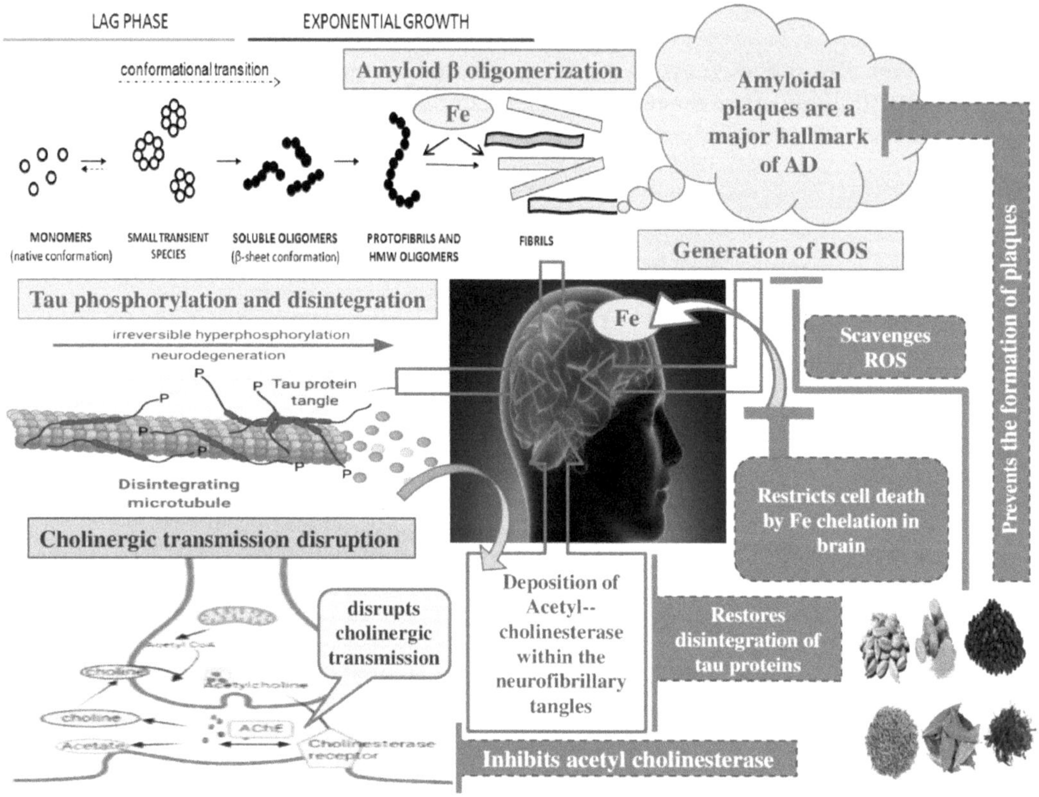

FIGURE 4.1 The major pathways of AD pathogenesis.

Despite the existing volumes of reports, systemic studies on the elucidation of the effects of spice antioxidants on animal models of AD or in human AD patients are rare. In fact, a few studies have mentioned the need to evaluate larger populations to deduce confirmatory outcomes. We have found several studies based on *in vitro* biochemical assays and *ex vivo* experiments using brain cell lines/cultures[58,67]. Turmeric is one of the most researched Indian spices in relation to AD, although when critically assessed, there are obvious indications that more work needs to be conducted to establish curcumin-based natural medicine as an alternative to existing synthetic drugs of AD. The second spice noteworthy in this context is saffron. Studies strongly indicate how its bioactive components (crocin and its derivatives) present in a saffron capsule could act as anti-amyloidogenic agents as effectively as donepezil[63,64]. However, what is lacking is the meticulous execution of any reported study.

This chapter has critically compiled and assessed literature reports that describe methodologies based on strong scientific findings and analyses. Figure 4.2 illustrates a systematic approach for the successful formulation of spiceutical-based medicines, which could be promising futuristic natural alternatives to conventional drugs prescribed for AD.

4.2 CRITICAL APPRAISAL OF THE EXISTING NATURAL DRUGS USED IN THE PREVENTION OF AD PATHOGENESIS

The positive outcomes of the pure biochemicals enumerated in Table 4.2, observed in both *in vitro* pathways and preclinical animal trials, may not reliably predict their success in human clinical trials, since these bioactives, in their pure chemical forms, cannot be consumed by human volunteers.

Although there are few published literature reports on human administration trials using pure chemicals at specific doses, they have not progressed beyond phase-II clinical trials and are, therefore, yet to be Food and Drug Administration (FDA)-approved natural medicines. One study on the administration of a combination of pure curcumin chemicals along with ginkgo leaf extract to human volunteers has shown preliminary success; the compounds responsible for therapeuticity have not been identified in the ginkgo leaf extract[31]. It is rationally opined by us that a synergistic consortium of bioactives, including curcumin and those from ginkgo leaf extract, have possibly contributed to the prevention of AD pathogenesis involving at least one of the pathways elucidated in Figure 4.1. Nonetheless, we also foresee potential danger in the usage of such unidentified mixtures of bioactives since if the bioactives (principally the antioxidants) are not in synergism in the mixture, it would be health-debilitating rather than therapeutic.

These drawbacks could certainly have been circumvented by the usage of the bioactive(s) in the extract form in compliance with the principle of 'food synergy' wherein biomolecules function in synergistic harmony in extracts of its native botanical[110,111,112]. The majority of the extracts mentioned in Table 4.2 are derived by conventional non-selective extraction procedures such as hydrodistillation and/or involving non-GRAS (generally regarded as safe) solvents for solvent extraction, ultrasonication-assisted solvent extraction (UAE), and microwave-assisted solvent extraction (MAE)[23,60]; wherein food synergy in the extracts have not been assessed by the researchers. It is extremely important that the principal bioactive in the extract should be in synergistic harmony with its co-extracted biomolecules, *viz.* present in antioxidant synergy as was present *de novo* in the spice/herb. This synergistic harmony must remain unperturbed during reductionism (extraction) for natural herbal medicines[4,113,114] to be effective alternatives to conventional synthetic drugs prescribed for AD patients.

Moreover, reports on the extracts of the bioactives are substantially incomplete since the ones that have exhibited potencies *in vitro* have not been tested for their efficacies in preclinical trials[67,68,115]; except for one reported work on piperine (related to AD) where investigators have tested a vacuum-dried Soxhlet extract of black pepper suspended in propylene glycol in AD-induced rats[38]. Although piperine was not thermally degraded under the said extraction conditions, the possibility of artefact generation in the extract during prolonged thermal processing cannot be ignored. It is therefore advisable not to employ thermal processing (such as the Soxhlet method of extraction) for obtaining bioactives from spices/herbs since the therapeutic potency of the extract would be compromised due to the presence of undesirable heat labile compounds of potential harm to the human health. Therefore, we advocate green technologies of solvent extraction for the formulation of herbal medicines, which encompass the green principles of use of a safe solvent(s) for both humans and the environment, and involve non-thermal processes for product development[116,117].

These research gaps need to be addressed for natural drug candidates to be successful in the prevention of the pathogenesis of AD. In this context, the green technology of supercritical CO_2 extraction (for extraction of bioactives from spices/herbs) is superior not only to the conventional solvent extraction technologies but also when compared with other non-thermal green technologies of extractions (*viz.* UAE and MAE with GRAS status solvents) since it delivers a 'neat' solventless extract which does not require further down-stream processing for its purification; and also because this high-pressure extraction process can be tuned to make it highly selective for extraction of desirable biomolecules, thereby conserving the natural antioxidant synergy in the extr act[4,113,114,118,119]. Conventional extracts are non-selective and have several undesirable compounds (artefacts) in abundance. The selectivity of the SC-CO_2 extraction process can be realized through vigorous engineering exercises (such as statistical optimization of process variables of extraction, namely, pressure to modify density, viscosity, and diffusivity of SC-CO_2 for enhancing solubility of the desirable molecules in the same, and concomitantly for enhancing penetrability of SC-CO_2 solvent into the sample matrix to ensure its greater accessibility to the extractible molecules; and

TABLE 4.2
Spice compounds, their extracts and formulations: Mode of action in targeted pathways of AD pathogenesis

						Formulations	
Spices	**Raw/ Extract**	**Method of extraction**	**Extracted compositions**	**Yield**	**AD pathways evaluated**	**Native extract**	**Macro/micro-encapsulate**
Turmeric	Extract	USC-CO_2	Curcumin	1.69 % w/w	--	--	--
		EE (α-amylase treatment)	Oleoresin, Curcumin, Volatile oil	10.35 ± 0. 30% w/w 3.92 ± 0.33% w/w 2.72 ± 0.27% w/w	--	--	--
		EE (Glucoamylase treatment)	Oleoresin, Curcumin, Volatile oil	11.15 ± 0.34% w/w 4.10 ± 0.21% w/w 2.46 ± 0.16% w/w			
		SFE (SC-CO_2)	Oleoresin, Curcumin, Total volatiles	35.37 ± 3.43 mg/g 187.09 ± 5.42 µg/g 7.62 ± 0.73 mg/g	--	--	--
		EILE (solvent: ethanol)	Oleoresin, Curcumin	4.41 % w/w 2.82 % w/w	--	--	--
		EILE (solvent: Acetone)	Oleoresin, Curcumin	5.56 % w/w 3.11 % w/w			
		EILE (solvent: DPCARB)	Oleoresin, Curcumin	6.99 % w/w 5.73 % w/w			
		--	--	--	C; D; E	--	Curcumin C3 Complex® (an orange-yellow crystalline powdered plant extract Encapsulate contained 95% curcuminoids with 70% to 80% Curcumin)
		--	--	--	C; D	--	Longvida (containing approximately 80 mg curcumin in a solid-lipid formulation)

Nano-vehicle	Nasal formulations	Characteristics of the extract or the formulations and their pharmaceutical/ pharmacokinetic profiles	Preclinical/ Clinical study	Remarks	References
--	--	--	--	--	3,18
--	--	--	--	--	19
--	--	--	--	--	20
--	--	--	---	--	21
--	--	EPR spectrum for validation of antioxidant potency: N.R. SE value: N.R LD_{50}: ADMET profile: N.R Release kinetics profile: N.R *In silico* study findings: N. R	**Doses:** 2g/day,4g/day **Duration:** 24-48 weeks **Route of administration:** Oral **Subjects:** Randomized, double blind, placebo-controlled groups of human subjects (33 patients with mild to moderate AD)	**Target(s):** MMSE scores; side effects; plasma and cerebrospinal fluid biomarkers (Aβ, tau, and F2-isoprostanes); curcumin and metabolites **Major finding(s) reported:** No differences among groups and no serious adverse events	22
--	--	EPR spectrum for validation of antioxidant potency: N.R SE value: N.R LD_{50}: N.R ADMET profile: N.R Release kinetics profile: N.R *In silico* study findings: N. R	**Doses:** 400 or 800 mg/day **Duration:** 1–3 h (acute); 4 weeks (chronic) **Route of administration**: Oral **Subjects:** 60 human subjects: healthy, aged 60-85 years old	**Target(s):** Cognition; mood and anxiety; blood biomarkers, and Aβ oligomerization **Major finding(s) reported:** Cognition and mood improvements	23

(*continued*)

TABLE 4.2 (Continued)
Spice compounds, their extracts and formulations: Mode of action in targeted pathways of AD pathogenesis

						Formulations	
Spices	Raw/ Extract	Method of extraction	Extracted compositions	Yield	AD pathways evaluated	Native extract	Macro/micro-encapsulate
	Raw	--	Curcumin	--	C; D	--	--
					C	--	--
					C	--	--
					C	--	--

Nano-vehicle	Nasal formulations	Characteristics of the extract or the formulations and their pharmaceutical/ pharmacokinetic profiles	Preclinical/ Clinical study	Remarks	References
--	--	EPR spectrum for validation of antioxidant potency: N.R SE value: N.R LD_{50}: N.R ADMET profile: N.R Release kinetics profile: N.R *In silico* study findings: N.R	**Doses:** 160 ppm (0.43 μmol/g); 5000 ppm (13.6 μmol/g) **Duration:** 6 months **Route of administration:** Oral **Subjects:** Rat model (Tg2576)	**Target(s):** Inhibitory effect on Aβ aggregation and oxidative stress reduction **Major finding(s) reported:** Significant decrease in insoluble and soluble *Aβ* plaque burdens with administration of the first dose but no significant change with the latter dose; lowering of oxidized proteins and pro-inflammatory cytokine (IL-1*β*) in the brain at both doses	24
--	--	EPR spectrum for validation of antioxidant potency: N.R SE value: N.R LD_{50}: N.R ADMET profile: N.R Release kinetics profile: N.R *In silico* study findings: N. R	**Doses**: 500 ppm (1.36 μmol/g);2000 ppm (5.43 μmol/g) **Duration:** 2 months **Route of administration:** Oral **Subjects:** Sprague Dawley rats	**Target(s):** Inhibitory effect on Aβ oligomerization **Major finding(s) reported:** Reduction in *Aβ* deposition; prevention of Aβ-infusion-induced spatial memory deficits	25
--	--	EPR spectrum for validation of antioxidant potency: N.R SE value: N.R LD_{50}: N.R ADMET profile: N.R Release kinetics profile: N.R *In silico* study findings: N. R	**Doses:** 500 ppm (1.36 μmol/g) **Duration:** 5 months **Route of administration:** Oral **Subjects:** Rat model (Tg2576)	**Target(s):** Inhibitory effect on Aβ oligomerization **Major finding(s) reported:** Reduction in amyloid (*Aβ)* plaque levels	26
--	--	EPR spectrum for validation of antioxidant potency: N.R SE value: N.R LD_{50}: N.R ADMET profile: N.R Release kinetics profile: N.R *In silico* study findings: N. R	**Doses:** 7.5 mg/kg/day (0.02 mmol/kg/day) **Duration:** 7 days **Route of administration:** intravenously in a tail vein **Subjects:** Rat model (APPswe / PS1dE9)	**Target(s):** Inhibitory effect on Aβ oligomerization; enhancement of BBB penetrability **Major finding(s) reported:** Curcumin reached brain tissue and cleared existing amyloid (*Aβ)* plaques	27

(*continued*)

TABLE 4.2 (Continued)
Spice compounds, their extracts and formulations: Mode of action in targeted pathways of AD pathogenesis

						Formulations	
Spices	Raw/ Extract	Method of extraction	Extracted compositions	Yield	AD pathways evaluated	Native extract	Macro/micro-encapsulate
					C	--	--
					E	--	--
					C	--	--
					C	--	Capsule or powdered extract

Nano-vehicle	Nasal formulations	Characteristics of the extract or the formulations and their pharmaceutical/ pharmacokinetic profiles	Preclinical/ Clinical study	Remarks	References
--	--	EPR spectrum for validation of antioxidant potency: N.R SE value: N.R LD_{50}: N.R ADMET profile: N.R Release kinetics profile: N.R *In silico* study findings: N. R	**Doses:** 500 ppm (1.36 μmol/g) **Duration:** 4 months **Route of administration:** Oral **Subjects:** Rat model (Tg2576)	**Target(s):** Inhibitory effect on Aβ oligomerization **Major finding(s) reported:** Reduction in A*β* plaque burden and dissociation of insoluble A*β* plaques	28
--	--	EPR spectrum for validation of antioxidant potency: N.R SE value: N.R LD_{50}: N.R ADMET profile: N.R Release kinetics profile: N.R *In silico* study findings: N. R	**Doses:** 500 ppm (1.36 μmol/g) **Duration:** 4 months **Route of administration:** Oral **Subjects:** Rat model (3xTg-AD)	**Target(s):** Inhibitory effect on tau protein phosphorylation **Major finding(s) reported:** Reduction in phosphorylated tau protein aggregates in buffer-extracted hippocampal membrane pellet fractions; performance improved in maze test	29
--	--	EPR spectrum for validation of antioxidant potency: N.R SE value: N.R LD_{50}: N.R ADMET profile: N.R Release kinetics profile: N.R *In silico* study findings: N. R	**Doses:** 5000 ppm (13.6 μmol/g) **Duration:** 10 months **Route of administration:** Oral **Subjects:** Rat model (Tg2576)	**Target(s):** Inhibitory effect on Aβ oligomerization **Major finding(s) reported:** No significant reduction found in A*β* deposition but A11 antibody-marked A*β* oligomers (important class of oligomers associated with AD pathogenesis) decreased in brain	3, 30
--	--	EPR spectrum for validation of antioxidant potency: N.R SE value: N.R LD_{50}: N.R ADMET profile: N.R Release kinetics profile: N.R *In silico* study findings: N. R	**Doses:** 1 g/day, 4 g/day powdered extract or one capsule/day with 120 mg/day standardized ginkgo leaf extract **Duration:** 6 months **Route of administration**: Oral **Subjects:** Randomized, double-blind, placebo control study on human model (36 subjects with possible or probable AD)	**Target(s):** MMSE scores; blood biomarkers (Aβ); plasma curcumin levels; bioavailability study in brain and blood plasma **Major finding(s) reported:** No difference in MMSE scores or blood biomarkers among groups	31

(*continued*)

TABLE 4.2 (Continued)
Spice compounds, their extracts and formulations: Mode of action in targeted pathways of AD pathogenesis

						Formulations	
Spices	Raw/ Extract	Method of extraction	Extracted compositions	Yield	AD pathways evaluated	Native extract	Macro/micro-encapsulate
					--	--	Curcumin micelles
					--	--	Poly (d,l-lactide-co-glycolide)-b-poly (ethylene glycol)-b-poly(d,l-lactide-co-glycolide) (PLGA-PEG-PLGA) micelles

Nano-vehicle	Nasal formulations	Characteristics of the extract or the formulations and their pharmaceutical/ pharmacokinetic profiles	Preclinical/ Clinical study	Remarks	References
--	--	EPR spectrum for validation of antioxidant potency: N.R SE value: N.R LD_{50}: N.R ADMET profile: N.R Release kinetics profile: N.R *In silico* study findings: N.R	**For the short-time bioavailability study:** **Doses:** 1.75 μL curcumin micelles solution/ g body weight (equivalent to 120 μg curcumin/g body weight) **Duration:** 45 min **Route of administration:** Oral **For the long-term feeding study**: **Doses:** 3.5 μL curcumin micelles solution/g body weight (equivalent to 120 μg curcumin/g body weight) **Duration:** 21 days **Route of administration:** Oral **For the short-time bioavailability study:** Female NMRI mice (aged between 8-24 months) **For the long-term feeding study**: Female young and aged NMRI mice (3 Months; 18 months)	**Target(s):** *In vivo* bioavailability study in brain and blood plasma **Major finding(s) reported:** Curcuminoid levels in plasma and brain increased around 10- to 45-fold, respectively, *vis-à-vis* the native curcumin	32
--	--	EPR spectrum for validation of antioxidant potency: N.R SE value: N.R LD_{50}: N.R ADMET profile: N.R Release kinetics profile: N.R *In silico* study findings: N.R	**Doses:** N.R **Duration:** N.R **Route of administration:** Intravenous **Subject**: Mice model	**Target(s):** Modification in pharmacokinetics and tissue distribution **Major finding(s) reported:** Plasma levels of curcumin micelles were increased by 1.31-fold compared with curcumin solution; brain distribution of curcumin micelles increased by 6.28-fold; more effective for AD treatment	33

(continued)

TABLE 4.2 (Continued)
Spice compounds, their extracts and formulations: Mode of action in targeted pathways of AD pathogenesis

						Formulations	
Spices	Raw/ Extract	Method of extraction	Extracted compositions	Yield	AD pathways evaluated	Native extract	Macro/micro-encapsulate
					C; D	--	--
					C	--	--
					C	--	--

Nano-vehicle	Nasal formulations	Characteristics of the extract or the formulations and their pharmaceutical/ pharmacokinetic profiles	Preclinical/ Clinical study	Remarks	References
Nanoliposome: nanostructured lipid carrier coated with lactoferrin	--	EPR spectrum for validation of antioxidant potency: N.R SE value: N.R LD_{50}: N.R ADMET profile: N.R Release kinetics profile: N.R *In silico* study findings: N.R	**Doses:** 2 mg/kg/day **Duration:** 3 weeks **Route of administration:** Intravenous **Subject**: Sprague Dawley rats (180-220 g) and ICR mice (18-22 g)	**Target(s):** Inhibitory effect on Aβ aggregation and oxidative stress reduction; brain uptake **Major finding(s) reported:** Inhibited Aβ oligomerization; acted as showed antioxidant potency; may provide a flexible drug delivery platform for brain targeting	34
Nanocurcumin (A novel polymeric nanoparticle formulation of curcumin, polymerized by N-isopropyla-crylamide, vinylpyrrolidone, and acrylic acid)	--	EPR spectrum for validation of antioxidant potency: N.R SE value: N.R LD_{50}: N.R ADMET profile: N.R Release kinetics profile: N.R *In silico* study findings: N.R	***In vivo* doses:** 23 mg/kg body weight **Duration:** 3 months **Route of administration:** Oral **Subject:** *In -vitro* BBB Model and Tg2576 Mice	**Target(s):** Inhibitory effect on Aβ aggregation; *in vitro and in vivo* bioavailability study in brain and blood plasma; BBB permeability **Major finding(s) reported:** Significantly improved memory function *in vivo*	35
--	Thermosensitive curcumin nasal gel; FMeC1 nasal solution	EPR spectrum for validation of antioxidant potency: N.R SE value: N.R LD_{50}: N.R ADMET profile: N.R Release kinetics profile: N.R *In silico* study findings: N.R	**Doses:** Equivalent to a 5 mg/kg injection **Duration:** One hour respiratory exposure **Route of administration:** Intranasal **Subject:** C57BL/6 and 5XFAD mice (8–12 months old)	**Target(s):** Inhibitory effect on Aβ aggregation; *in vivo* bioavailability study in brain and blood plasma; BBB permeability **Major finding(s) reported:** A promising alternative to the field-dominant intravenous administration method of existing AD drugs	36

(*continued*)

TABLE 4.2 (Continued)
Spice compounds, their extracts and formulations: Mode of action in targeted pathways of AD pathogenesis

Spices	Raw/ Extract	Method of extraction	Extracted compositions	Yield	AD pathways evaluated	Formulations: Native extract	Formulations: Macro/micro-encapsulate
Black pepper		Soxhlet and modified Soxhlet extractions (DBSA)	Piperine	4 ±0.01% w/w		--	--
		Soxhlet- apparatus		10.4 % w/w	A, C, D, E	Vacuum-dried, powdered ethanolic extract, suspend in propylene glycol	--
		UAE		5.8 mg/g	--	--	--
		MAE		92.5 ±0.05% w/w	--	--	--
		SFE (SC-CO_2)		91.31% w/w	--	--	--
		Naviglio Extractor		317.7 mg/ g	--	--	--
					A; D; F	--	PIP-loaded micro emulsion
					A; C; D; E; F	--	--

Nano-vehicle	Nasal formulations	Characteristics of the extract or the formulations and their pharmaceutical/ pharmacokinetic profiles	Preclinical/ Clinical study	Remarks	References
--	--	--	--	--	37
--	--	EPR spectrum for validation of antioxidant potency: N.R SE value: N.R LD_{50}: N.R ADMET profile: N.R Release kinetics profile: N.R *In silico* study findings: N.R	**Doses:** 20mg/kg/day, 200mg/kg /day **Duration:** 2 months **Route of administration:** Oral **Subject:** $AlCl_3$-induced AD rat model	**Target(s):** Inhibitory effect on AChE activity; Aβ oligomerization; oxidative stress; tau protein phosphorylation **Major finding(s) reported:** Black pepper extract showed success in major AD preventive pathways; decreased escape latency time and increased memory in AD-induced rat models	38
--	--	--	--	--	39
--	--	--	--	--	40, 41
--	--	--	--	--	42, 43, 44
--	--	--	--	--	45
--	--	EPR spectrum for validation of antioxidant potency: N.R SE value: N.R LD_{50}: N.R ADMET profile: N.R Release kinetics profile: Done *In silico* study findings: N. R	**Doses:** 2.5 mg equivalent/ kg (0.2 ml/0.15ml)/day **Duration:** 3 days **Route of administration:** Oral **Subject:** Adult Wistar rats	**Target(s):** Inhibitory effect on AChE activity; oxidative stress; SOD value; brain target **Major finding(s) reported:** Micro emulsion increased therapeutic efficacy of piperine for AD and enhanced its delivery to the target region of the brain	46
Polysorbate-80-coated piperine solid-lipid nanoparticles (P-80-PIP-SLN)	--	EPR spectrum for validation of antioxidant potency: N.R. SE value: N.R. LD_{50}: N.R. ADMET profile: N.R. Release kinetics profile: N.R. *In silico* study findings: N.R.	**Doses:** 20mg/kg/day **Duration**: 7 days **Route of administration:** Intraperitoneal **Subject:** Albino Wister rat	**Target(s):** Inhibitory effect on AChE activity; Aβ oligomerization; oxidative stress; tau protein phosphorylation; SOD value; brain target **Major finding(s) reported:** Reduced cholinergic degradation; reduced oxidative stress; reduced Aβ plaques and tau-tangles; reduced SOD values; the formulated SLNs shown to be superior than donepezil	47

(continued)

TABLE 4.2 (Continued)
Spice compounds, their extracts and formulations: Mode of action in targeted pathways of AD pathogenesis

						Formulations	
Spices	Raw/ Extract	Method of extraction	Extracted compositions	Yield	AD pathways evaluated	Native extract	Macro/micro-encapsulate
					A; D; F	--	--
					A; D; F	--	--
					--	--	--
					--	--	--

Nano-vehicle	Nasal formulations	Characteristics of the extract or the formulations and their pharmaceutical/ pharmacokinetic profiles	Preclinical/ Clinical study	Remarks	References
Piperine-loaded chitosan nanoparticles (CS-NPs)	Intranasal formulation using Piperine-loaded chitosan nanoparticles	EPR spectrum for validation of antioxidant potency: N.R SE value: N.R LD_{50}: N.R ADMET profile: N.R Release kinetics profile: Done *In silico* study findings: N.R	**Doses:** 0.1 mL equivalent to 0.045 mg/kg/day; 0.1 mL equivalent to 0.250 mg/kg/day, pH 5.5 (drop-wise) **Duration:** 21 days **Route of administration:** intranasal **Subject:** Adult male Wistar rat model	**Target(s):** Inhibitory effect on AChE activity; oxidative stress; SOD value **Major finding(s) reported:** CS-NPs considered as effective, safe, and non-invasive piperine delivery agent; efficient as standard drug donepezil with additional advantages of dual mechanism (AChE inhibition and antioxidant effect)	48
Piperine-loaded tween integrated monoolein cubosomes	--	EPR spectrum for validation of antioxidant potency: N.R SE value: N.R LD_{50}: ADMET profile: N.R Release kinetics profile: Done *In silico* study findings: N.R	**Doses:** 2.5 mg equivalent/kg **Duration:** 21 days **Route of administration:** Oral **Subject:** Adult male Wistar rat model	**Target(s):** Inhibitory effect on AChE activity; oxidative stress; SOD value **Major finding(s) reported:** Formulation considered as safe and possesses promising *in vitro* and *in vivo* characteristics for effective chronic treatment of AD	49
Piperine-encapsulated PEG-ylated Liposomes (PL)	--	EPR spectrum for validation of antioxidant potency: N.R SE value: N.R LD_{50}: N.R ADMET profile: N.R Release kinetics profile: N.R *In silico* study findings: N.R	**Doses:** 5 mg/kg body weight/ day **Duration:** 14 days **Route of administration:** Oral **Subject:** Male Wistar rat model	**Target(s):** Effective delivery to brain; cognition **Major finding(s) reported:** Improved cognition and showed antidepressant effects; piperine intake was slower in brain	50
--	Piperine-encapsulated PL nasal formulation	EPR spectrum for validation of antioxidant potency: N.R SE value: N.R LD_{50}: N.R ADMET profile: N.R Release kinetics profile: N.R *In silico* study findings: N.R	**Doses:** 7.2 µg/day **Duration:** 14 days **Route of administration:** Intranasal **Subject:** Male Wistar rat	**Target(s):** Effective delivery to brain; cognition **Major finding(s) reported:** Improved cognition and showed antidepressant effects; faster delivery to brain in greater extent compared to oral administration of PL	50

(*continued*)

TABLE 4.2 (Continued)
Spice compounds, their extracts and formulations: Mode of action in targeted pathways of AD pathogenesis

Spices	Raw/ Extract	Method of extraction	Extracted compositions	Yield	AD pathways evaluated	Formulations: Native extract	Formulations: Macro/micro-encapsulate
Small cardamom	Extract	Hydro distillation by Clevenger-type apparatus	1, 8-Cineole; α-terpineol; limonene; α-terpinyl acetate; α-pinene	73.27% w/w; 4.23% w/w; 4.2% w/w; 3.33% w/w; 2.9% w/w	--	--	--
		UAE followed by hydro distillation	1,8-cineole; ᾱ-terpinyl acetate	26.59% to 39.34% w/w; 22.94% to 40.56% w/w	--	--	--
		SFE (SC-CO_2)	1,8-cineole	31.96 ± 0.16 mg/g	--	--	--
					A; C; D; E	--	--
	Oil	--	1,8 cineole, α-terpinyl acetate, limonene, linalyl acetate, linalool	--	A; C; D	Cardamom oil	--
Saffron	Extract	Solvent extraction (water and methanol 1:1 v/v)	Crocin	--	C	Extracted powder suspended in water	--

Nano-vehicle	Nasal formulations	Characteristics of the extract or the formulations and their pharmaceutical/ pharmacokinetic profiles	Preclinical/ Clinical study	Remarks	References
--	--	--	--	--	51
--	--	--	--	--	52
--	--	--	--	--	53
--	PEG-ylated Nanoliposomal formulation	EPR spectrum for validation of antioxidant potency: Done SE value: Done LD_{50}: Done ADMET profile: Done Release kinetics profile: Done *In silico* study findings: Done	--	**Target(s):** Inhibitory effect on AChE activity; Aβ oligomerization; oxidative stress; tau protein phosphorylation **Major finding(s) reported:** Inhibited AChE activity; reduced Aβ oligomerization and tau protein phosphorylation; reduced oxidative stress (*in vitro*)	4, 54, 55, 56
--	--	EPR spectrum for validation of antioxidant potency: N.R SE value: N.R LD_{50}: N.R ADMET profile: N.R Release kinetics profile: N.R *In silico* study findings: N.R	**Doses:** 100 and 200 mg/kg /day **Duration:** 42 days **Route of administration:** Oral **Subject:** Male albino Wistar rat	**Target(s):** Inhibitory effect on AChE activity; Aβ oligomerization; oxidative stress **Major finding(s) reported:** Inhibited AChE activity and reduced oxidative stress; may act as a neuroprotector to treat AD	57
--	--	EPR spectrum for validation of antioxidant potency: N.R SE value: N.R LD_{50}: N.R ADMET profile: N.R Release kinetics profile: *In silico* study findings: N.R	**Concentration:** 15 μg/ml, 15.4 μM (*in vitro* assays)	**Target(s):** *In vitro* assays (amyloid fibril formation of Aβ1-40 (1 mg/mL) **Major finding(s) reported:** Crocin inhibited amyloid fibrillation *in vitro*	58

(*continued*)

TABLE 4.2 (Continued)
Spice compounds, their extracts and formulations: Mode of action in targeted pathways of AD pathogenesis

Spices	Raw/ Extract	Method of extraction	Extracted compositions	Yield	AD pathways evaluated	Formulations: Native extract	Formulations: Macro/micro-encapsulate
				--	C	Extracted powder	--
		Methanolic extract		--	C	Extracted powder	--
				--	D	Pure saffron powder in physiological saline	--
		Hydro-alcoholioc extract using a cold-maceration method		33.11 ± 3.25%	C; D	Extracted dried powder	--
		Ethanolic extract		--	D	Extracted dry powder	--

Nano-vehicle	Nasal formulations	Characteristics of the extract or the formulations and their pharmaceutical/ pharmacokinetic profiles	Preclinical/ Clinical study	Remarks	References
--	--	EPR spectrum for validation of antioxidant potency: N.R SE value: N.R LD_{50}: N.R ADMET profile: N.R Release kinetics profile: N.R *In silico* study findings: N.R	**Concentration:** 15.4 μM (*in vitro* assays)	**Target(s):** *In vitro* assays (amyloid fibril formation of Aβ1-40 (1 mg/mL) **Major finding(s) reported:** Inhibited Aβ amyloid formation and disrupted amyloid aggregates	59
--	--	EPR spectrum for validation of antioxidant potency: N.R SE value: N.R LD_{50}: N.R ADMET profile: N.R Release kinetics profile: N.R *In silico* study findings: N.R	**Dose:** 60 mg/kg bw **Duration:** 1 week **Route of Administration:** Intraperitoneal **Subject:** Aged mice model	**Target(s):** Effect on Aβ fibrillogenesis **Major finding(s) reported:** Saffron stigma constituents inhibited aggregation and deposition of amyloids in the human brain	60
--	--	EPR spectrum for validation of antioxidant potency: N.R SE value: N.R LD_{50}: N.R ADMET profile: N.R Release kinetics profile: N.R *In silico* study findings: N.R	**Doses:** 30 mg/kg bw **Duration:** 21 days **Route of administration:** Intraperitoneal **Subject:** Male Wistar rats (chronic stress induced)	**Target(s):** Chronic stress-induced spatial learning and memory impairments and oxidative stress in the hippocampus **Major finding(s) reported:** Improved learning and memory and repair oxidative stress damage to the hippocampus induced by chronic stress prominent in AD	61
--	--	EPR spectrum for validation of antioxidant potency: N.R SE value: N.R LD_{50}: N.R ADMET profile: Reported Release kinetics profile: N.R *In silico* study findings: N.R	**Doses:** 50 mg/kg/day **Duration:** 1 month **Route of administration:** Oral **Subjects:** C57BL/6 wild-type mice model	**Target(s):** BBB function and integrity, Aβ related pathology **Major finding(s) reported:** Effectively decreased Aβ brain deposits and reduced the total Aβ levels in 5XFAD rat-brain hippocampus	62
--	--	EPR spectrum for validation of antioxidant potency: N.R SE value: N.R LD_{50}: N.R ADMET profile: N.R Release kinetics profile: N.R *In silico* study findings: N.R	**Doses:** 5 μg/rat/; 10 μg/rat **Duration:** 7 days **Route of administration:** Intrahippocampal administration **Subjects:** Adult male Wistar rats	**Target(s):** Learning and memory of EB (ethidium bromide)-demyelinated rats and their oxidative status in the hippocampus **Major finding(s) reported:** Saffron alleviated oxidative stress and improved cognitive deficits in AD patients	6

(*continued*)

TABLE 4.2 (Continued)
Spice compounds, their extracts and formulations: Mode of action in targeted pathways of AD pathogenesis

Spices	Raw/ Extract	Method of extraction	Extracted compositions	Yield	AD pathways evaluated	Formulations: Native extract	Formulations: Macro/micro-encapsulate
		Ethanolic extraction (80%)		--	C	--	Capsule [containing dried ethanolic saffron extract (15 mg), lactose (filler), magnesium stearate (lubricant), and sodium starch glycolate (disintegrant)]
				--	C	--	--
				--	--	--	Saffrotin capsules Impiran, Tehran, Iran, 15 mg (containing 15 mg dried extract of saffron)
		Homogenate extraction method		22.76 %	--	--	--
		Solvent extraction		--	E	Crocin stock (2 mg/ml) was prepared from its powder that was dissolved in piperazine-N, N′-bis 2-ethanesulfonic acid (PIPES) buffer (pH 6.8)	--

Nano-vehicle	Nasal formulations	Characteristics of the extract or the formulations and their pharmaceutical/ pharmacokinetic profiles	Preclinical/ Clinical study	Remarks	References
--	--	EPR spectrum for validation of antioxidant potency: N.R SE value: N.R LD_{50}: N.R ADMET profile: N.R Release kinetics profile: N.R *In silico* study findings: N.R	**Doses:** 30mg/day (15 mg twice a day) **Duration:** 22 weeks **Route of administration:** Oral **Subjects:** Human	**Target(s):** Double-blind study of parallel groups of patients with mild to moderate AD through ADAS-cog and CDR-SB measures **Major finding(s) reported:** Showed anti-amyloidogenic activity significantly comparable to donepezil	63
--	--	EPR spectrum for validation of antioxidant potency: N.R SE value: N.R LD_{50}: N.R ADMET profile: N.R Release kinetics profile: N.R *In silico* study findings: N.R	**Doses:** 30mg/day (15 mg twice a day) **Duration:** 16 weeks **Route of administration:** Oral **Subjects:** Human	**Target(s):** Psychometric measures including the MMSE, ADAS-cog, clinical dementia rating scale-sums of boxes (CDR-SB) **Major finding(s) reported:** Saffron showed effective results in mild to moderate AD	64
--	--	EPR spectrum for validation of antioxidant potency: N.R SE value: N.R LD_{50}: N.R ADMET profile: N.R Release kinetics profile: N.R *In silico* study findings: N.R	**Doses:** 30mg/day (15 mg twice a day) **Duration:** 12 months **Route of administration:** Oral **Subjects:** Human	**Target(s):** Parallel-group, double-blind randomized clinical trial on outdoor patients with dementia, MMSE score **Major finding(s) reported:** Reduced cognitive decline in patients with moderate to severe AD comparable to memantine	65
--	--	--	--	--	66
--	--	EPR spectrum for validation of antioxidant potency: N.R SE value: N.R LD_{50}: N.R ADMET profile: N.R Release kinetics profile: N.R *In silico* study findings: N.R	**Concentration:** 0.2 μg/ml to 600 μg/ml of crocin (*in vitro* assays)	**Target(s):** Inhibitory effect on aggregation of recombinant human tau protein 1N/4R isoform using biochemical methods and cell culture **Major finding(s) reported:** Crocin inhibited tau aggregation with IC_{50} of 100 μg/ml	67

(continued)

TABLE 4.2 (Continued)
Spice compounds, their extracts and formulations: Mode of action in targeted pathways of AD pathogenesis

						Formulations	
Spices	Raw/ Extract	Method of extraction	Extracted compositions	Yield	AD pathways evaluated	Native extract	Macro/micro-encapsulate
		Methanol/water 1:1 v/v		--	A	Methanolic extract	--
		Extraction using 80% ethanol		--	C	Crocin crystals	--
Black Cumin seeds	Extract (essential oil)	Microwave-assisted extraction (micro distillation) and hydro distillation	Thymoquinone	10.8%-24.6 % (MD) and 1.6 %-21.8 % (HD)	--	--	--
		Hydro distillation		37.6 %; 18.4 %-24 %	--	--	--
		Methanolic extraction followed by fractionation in hexane, ethyl acetate and water		--	C	Hexane and water Fractions	--
		Solvent extraction (methanol)		--	--	Oil obtained after methanol extraction	--
		Supercritical fluid (carbon dioxide)		77.2-86.2 %	C	--	--

Nano-vehicle	Nasal formulations	Characteristics of the extract or the formulations and their pharmaceutical/ pharmacokinetic profiles	Preclinical/ Clinical study	Remarks	References
--	--	EPR spectrum for validation of antioxidant potency: N.R SE value: N.R LD_{50}: N.R ADMET profile: N.R Release kinetics profile: N.R *In silico* study findings: Reported	**Concentration:** Various concentrations used for *in vitro* assays and docking studies	**Target(s):** Inhibitory effect on AChE **Major finding(s) reported:** Extract showed moderate AChE-inhibitory activity (up to 30%)	68
--	--	EPR spectrum for validation of antioxidant potency: N.R SE value: N.R LD_{50}: N.R ADMET profile: N.R Release kinetics profile: N.R *In silico* study findings: N.R	**Doses:** 50, 25 and 12.5 mg/ kg/day **Duration:** 2 weeks **Route of administration:** Intraperitoneal **Subjects:** Male Wistar rat	**Target(s):** *In vivo* inhibitory on Aβ aggregation and hyper-phosphorylated Tau-tangles **Major finding(s) reported:** Crocin showed reduction of tau phosphorylation and Aβ concentration	69
--	--	--	--	--	70
--	--	--	--	--	71
--	--	EPR spectrum for validation of antioxidant potency: N.R SE value: N.R LD_{50}: N.R ADMET profile: N.R Release kinetics profile: N.R *In silico* study findings: N.R	**Concentration:** 1,10 & 100 µg/ml *In vitro* rat (Sprague Dawley) cerebellar granule (CGNs) neuron cell culture (*in vitro* assay)	**Target(s):** Aβ toxicity in *in vitro* rat CGNs cell culture **Major finding(s) reported:** Showed effective protection against Aβ 1-40-induced CGNs toxicity	7
--	--	EPR spectrum for validation of antioxidant potency: N.R SE value: N.R LD_{50}: N.R ADMET profile: N.R Release kinetics profile: N.R *In silico* study findings: N.R	**Doses:** 1 ml/kg **Duration:** 80 days **Route of administration:** Oral (gavage) **Subjects:** Male Sprague Dawley rats	**Target(s):** Effect on spatial recognition **Major finding(s) reported:** Helped to improve spatial cognition in rats challenged with chronic cerebral hypoperfusion	72
Nano-emulsions	--	EPR spectrum for validation of antioxidant potency: N.R SE value: N.R LD_{50}: N.R ADMET profile: N.R Release kinetics profile: N.R *In silico* study findings: N.R	**Doses:** 1 mg/kg bw **Duration:** 3 months **Route of administration:** Oral (gavage) **Subjects:** 2-month-old Sprague Dawley rats	**Target(s):** Expression of Aβ and other genes in HFCD rat brains **Major finding(s) reported:** Nanoemulsions effectively regulated Aβ42 levels in rat brain	73, 74

(*continued*)

TABLE 4.2 (Continued)
Spice compounds, their extracts and formulations: Mode of action in targeted pathways of AD pathogenesis

Spices	Raw/ Extract	Method of extraction	Extracted compositions	Yield	AD pathways evaluated	Formulations: Native extract	Formulations: Macro/micro-encapsulate
	Raw	--		--	C	--	--
		--		--	C	--	--
		--		--	D	--	--
		--		--	A	--	--
Bay leaf	Extract	Solvent extraction (UAE in methanol followed by liquid-liquid extraction obtaining three fractions: $CHCl_3$, EtOAc, and MeOH)	1,8-cineole	69.358%w/w	C	--	--

Nano-vehicle	Nasal formulations	Characteristics of the extract or the formulations and their pharmaceutical/ pharmacokinetic profiles	Preclinical/ Clinical study	Remarks	References
--	--	EPR spectrum for validation of antioxidant potency: N.R SE value: N.R LD_{50}: N.R ADMET profile: N.R Release kinetics profile: N.R In silico study findings: N.R	**Concentration:** 0.1 and 1 µM (*in vitro* assays)	**Target(s):** Neuroprotective effects on neurotoxicity Induced by Aβ 1–40 **Major finding(s) reported:** TQ pretreatment repressed amyloid beta induced apoptosis of cultured cerebellar granule neurons (CGNs)	75
--	--	EPR spectrum for validation of antioxidant potency: N.R SE value: N.R LD_{50}: N.R ADMET profile: N.R Release kinetics profile: N.R *In silico* study findings: N.R	**Concentration:** 0.78–400 lM (*in vitro* assays)	**Target(s):** Neuroprotective effects of TQ on neurotoxicity Induced by Aβ 1–40 **Major finding(s) reported:** TQ reduced Aß induced neurotoxicity *in vitro* model of undifferentiated pheochromocytoma rat cell line, PC-12, by inhibiting oxidative stress	76
--	--	EPR spectrum for validation of antioxidant potency: N.R SE value: N.R LD_{50}: N.R ADMET profile: N.R Release kinetics profile: N.R *In silico* study findings: N.R	**Concentration:** 2.5-40 µM (*in vitro* assays)	**Target(s):** Antioxidant effects on microglial cells **Major finding(s) reported:** Could effectively delay neurodegeneration induced by oxidative stress in microglial cells	77
--	--	EPR spectrum for validation of antioxidant potency: N.R SE value: N.R LD_{50}: N.R ADMET profile: N.R Release kinetics profile: N.R *In silico* study findings: N. R	**Concentration:** 5 and 10 mg/kg; p. o. (*in vitro* assays)	**Target(s):** Role on L-Cys induced rats **Major finding(s) reported:** TQ lowered stimulation of AChE enzyme function at the same dose level of donepezil (10 mg/ kg; p. o)	78
--	--	EPR spectrum for validation of antioxidant potency: N.R SE value: N.R LD_{50}: N.R ADMET profile: N.R Release kinetics profile: N.R *In silico* study findings: N.R	**Concentrations:** 25.0, 50.0 and 100.0 µg/ mL (*in vitro* assays)	**Target(s):** *In vitro* assays on SH-5YSY and SK-N-BE (2)-C human neuronal cell lines, as well as on C6 mouse glial cell line **Major finding(s) reported:** Showed anti-amyloidogenic efficacy	79

(continued)

TABLE 4.2 (Continued)
Spice compounds, their extracts and formulations: Mode of action in targeted pathways of AD pathogenesis

Spices	Raw/ Extract	Method of extraction	Extracted compositions	Yield	AD pathways evaluated	Formulations: Native extract	Formulations: Macro/micro-encapsulate
		Hydrodistillation/ steam distillation to obtain EO. Also extracted with ethanol		--	A	Extracted oil as well as ethanolic extract	--
		Aqueous extract		--	A; B	Aqueous extract	--
		Incense		--	A	Bay leaf smoke	--
Coriander	Extract (volatile oil)	Hydro distillation	Linalool	76.33%-87.53 %	--	Extracted oil	--
		Ethanolic extract (seeds)		--	--	Dried extract	--
				--	A	Dried extract	--

Nano-vehicle	Nasal formulations	Characteristics of the extract or the formulations and their pharmaceutical/ pharmacokinetic profiles	Preclinical/ Clinical study	Remarks	References
--	--	EPR spectrum for validation of antioxidant potency: N.R SE value: N.R LD_{50}: N.R ADMET profile: N.R Release kinetics profile: N.R *In silico* study findings: N.R	**Concentrations:** 0.5 and 1 mg/ml (*in vitro* assays)	**Target(s):** *In vitro* biochemical assays **Major finding(s) reported:** Showed good AChE-inhibitory effects	80
--	--	EPR spectrum for validation of antioxidant potency: N.R SE value: N.R LD_{50}: N.R ADMET profile: N.R Release kinetics profile: N.R In silico study findings: N.R	**Concentration:** 0.3 μg/ml (*in vitro* assays)	**Target(s):** *In vitro* study using brain homogenate of male Wister rats **Major finding(s) reported:** Aqueous extracts showed remarkably inhibited effects of both AChE and BChE	81
--	--	EPR spectrum for validation of antioxidant potency: N.R SE value: N.R LD_{50}: N.R ADMET profile: N.R Release kinetics profile: N.R *In silico* study findings: N.R	**Doses:** 5 min before each test **Duration:** 23 days **Route of administration:** Olfactory **Subjects:** 3-month-old male Wister rats treated with scopolamine	**Target(s):** Potential effect of bay leaf incense on memory function and brain antioxidant status **Major finding(s) reported:** Tests suggested the efficiency of the smoke in effectively diminishing induced cognitive deficits and AChE activity	82
--	--	--	--	--	83
--	--	EPR spectrum for validation of antioxidant potency: N.R SE value: N.R LD_{50}: >5000 mg ADMET profile: N.R Release kinetics profile: N.R *In silico* study findings: N.R	**Dose:** 100, 200 mg/kg, **Duration:** 15 days **Route of administration:** Oral **Subjects:** Tacrine treated male Wister rats	**Target(s):** Cognitive behaviour assessed by plus maze learning task, biochemical assessments **Major finding(s) reported:** Showed promising role in ameliorating cognition	5, 84
--	--	EPR spectrum for validation of antioxidant potency: N.R SE value: N.R LD_{50}: >5000 mg ADMET profile: N.R Release kinetics profile: N.R *In silico* study findings: N.R	**Dose:** 5, 10 and 15% w/w of diet **Duration:** 45 days **Route of administration:** Oral **Subjects:** Scopolamine and diazepam treated mice model	**Target(s):** Cognitive functions, total serum cholesterol levels and brain cholinesterase activity **Major finding(s) reported:** Had promising memory improving capacities	5, 85

(*continued*)

TABLE 4.2 (Continued)
Spice compounds, their extracts and formulations: Mode of action in targeted pathways of AD pathogenesis

Spices	Raw/ Extract	Method of extraction	Extracted compositions	Yield	AD pathways evaluated	Formulations: Native extract	Formulations: Macro/micro-encapsulate
		Hydro distillation		33.1 %	--	Extracted oil	--
		Hydro distillation (oil extracted from fruits)		--	C	Volatile oil	--

Reviews/articles that are indexed by Scopus and/or Web of Science have been selected for this compilation.
Ultrasound-assisted supercritical carbon dioxide (USC-CO_2) extraction, Ultra-sonication–assisted solvent extraction (UAE), Enzyme-assisted solvent extraction (EE), Microwave-assisted solvent extraction (MAE), Supercritical fluid extraction (SFE),
Supercritical carbon dioxide (SC-CO_2), Enzyme-assisted ionic liquid extraction (EILE), Not reported (N.R.), Thymoquinone (TQ), Poly lactic-co-glycolic acid (PLGA), Essential oils (EO),
Mild cognitive impairment (MCI), AD Assessment Scale-cognitive subscale (ADAS-cog), Clinical dementia rating scale : sums of boxes (CDR-SB), Cerebellar granules (CGNs), 3-(4,5-dimethylthiazol-2-yl)-2,5diphenyltetrazolium bromide

mass transfer modelling of the extraction process) and the extract thus obtained is neat (solvent-free) and therefore safe to be consumed *per* se[53,118,120,121,122,123,124,125,126]. Selectivity in extraction cannot be achieved by conventional or other green extraction technologies, wherein the solubility of the antioxidants is limited by the polarity and/or temperature of the extracting solvent(s). Besides, SC-CO_2 extraction can successfully preserve antioxidant synergy in a cardamom seed extract rich in its bioactive principle 1,8-cineole, wherein it was found to be synergistically present with a consortium of co-extracted antioxidants, namely, α-terpineol, α-terpinyl acetate, linalool, and limonene[54,55]. This antioxidant harmony showed success *in vitro* in the prevention of β-amyloid oligomerization[4]. In fact, the said extract could be a putative multifunctional biotherapeutic since the same also demonstrated efficacies in reducing the effects of oxidative stress, in the prevention of iron-induced cell death in human neuroblastoma SHSY5Y cell line[4], and as an acetylcholinesterase inhibitor[56].

Nano-vehicle	Nasal formulations	Characteristics of the extract or the formulations and their pharmaceutical/ pharmacokinetic profiles	Preclinical/ Clinical study	Remarks	References
--	--	EPR spectrum for validation of antioxidant potency: N.R SE value: N.R LD_{50}: >5000 mg ADMET profile: N.R Release kinetics profile: N.R In silico study findings: N.R	**Concentration:** 50-800 μg/mL *In vitro* study on cortical neurons derived from enzymatically and mechanically dissociated cortex of day 18 embryonic Wistar SPF rat brain (*in vitro* assays)	**Target(s):** Effect on various aspects of CNS **Major finding(s) reported:** Oil containing linalool showed effect on rat SH-SY5Y neuronal cells	5, 86
--	--	EPR spectrum for validation of antioxidant potency: N.R SE value: N.R LD_{50}: >5000 mg ADMET profile: N.R Release kinetics profile: N.R In silico study findings: N.R	**Dose:** 200 μl (1% & 3%) **Duration:** 21 days **Route of administration:** Nasal **Subject:** Male Wister rats	**Target(s):** Chronic inhalation of coriander volatile oil in Aβ (1-42) neurotoxicity induced Wister male rats **Major finding(s) reported:** Effectively restored brain antioxidant status and significantly improved spatial working memory	5, 87

(MTT) assay, Mini-Mental State Examination (MMSE): The best-known and the most often used short screening tool for providing an overall measure of cognitive impairment in clinical, research and community settings. MMSE (Folstein 1975) is a simple pen-and-paper test of cognitive function based on a total possible score of 30 points; it includes tests of orientation, concentration, attention, verbal memory, naming, and visuospatial skills.

Preventive mechanisms: A: Inhibition of Acetylcholinesterase (AChE) activity, B: Inhibition of activities of β-secretase, C: Inhibition of amyloidal plaque (Aβ) formation/oligomerization in brain and Aβ-induced inflammation, D: Oxidative stress reduction as an antioxidant,

E: Inhibition of tau phosphorylation, F: Superoxide dismutase (SOD) value reduction

4.3 LIMITATIONS OF EXISTING AD DRUGS AND SUGGESTIVE ALTERNATIVES

The FDA-approved synthetic drugs such as donepezil, tacrine, rivastigmine, and galanthamine prescribed for AD have shown success in slowing down neuro-degeneration but are known to have negative impacts on human health because they pose adverse side effects such as aggression, depression, gastrointestinal disturbances, hepatotoxicity, and cardiac problems[127]. Moreover, these drugs do not have a long-lasting effect (effective only for 1–3 years) because of their poor penetrability, i.e., their inability to cross the blood-brain-barrier (BBB) and reach the targeted areas, primarily the blood cells in the hippocampal area and/or acetylcholinesterase accumulating neuronal junctions[128,129]. As a result, the huge intake of these chemicals adds to the toxic insult to the cells, resulting in more harm to individuals than arresting the progression of AD[130].

TABLE 4.3a
Role of spices and their respective bioactive components on behavioural tests in *in vivo* animal model

Name of spices	Bioactive components	*In vivo* behavioural tests in animal model	Remarks	References
Turmeric	Curcumin	Y-maze test in 3xTg-AD rat model	Improved performance in Y-maze test	29
		Morris water maze test (MWM) in Sprague Dawley rats	Prevention of Aβ-infusion-induced spatial memory deficits found in Morris water maze test	25
Black pepper	piperine	Force swimming test (FST) in albino Wistar rats	P-80-PIP-SLN results increased immobility time by FST	47
		Morris water maze test in male Wistar rats	CS-NPs resulted in significant improvements in escape latency or retention of memory	48
		Impaired memory evaluated through passive avoidance apparatus in male Wistar rats	PIP-loaded cubs formulations increased retention latency which reflects learning ability and adequate cognitive improvement	49
		Evaluation of cognitive abilities in adult Wistar rats through passive avoidance test	Cognitive performance partially improved after treatment with PIP-loaded micro emulsion formulations	46
		Morris water maze test, forced swimming test and spontaneous motor behaviour assessment were performed in adult male Wistar albino rats	Decreased escape latency and increased retention time in MWM, decrease in immobility assessed by swimming test confirmed the efficacy of PL in improvement of cognition and assured the same as an antidepressant	50
Small cardamom	1,8 cineole, α-terpinyl acetate, limonene, linalyl acetate, linalool	Morris water maze test, elevated plus maze test, passive avoidance test and locomotor activity in $AlCl_3$-treated male Wistar albino rats	Significantly prevented the increase in escape latency, decreased latency time, significant recovery in retention latency and improved locomotor activity	57
Saffron	Crocin and its derivatives	Morris water maze test, memory abilities, and open field test in Al-treated mice	Crocin improved cognition and memory abilities substantially as measured by their coordination of movements and reduced escape time	88
		Water maze tests of learning and memory on chronic stress induced male Wistar rats	The extract prevented chronic stress induced deficits in spatial learning and memory and oxidative stress in the hippocampus	61
		Morris Water Maze test performed in male Wistar rats induced with multiple sclerosis (Intrahippocampal injection of ethidium bromide)	Learning and spatial memory status was improved after administration of saffron extract	6
		Passive avoidance test conducted for learning and memory ability of Al-exposed Balb-c mice	Dried saffron extract had no effect on Al-induced memory deficit for short duration of time (six days)	89

TABLE 4.3a (Continued)
Role of spices and their respective bioactive components on behavioural tests in *in vivo* animal model

Name of spices	Bioactive components	*In vivo* behavioural tests in animal model	Remarks	References
		Passive and active avoidance tests performed on D-galactose and $NaNO_2$ administered (i.p) ovariectomized mice (induced amnesia)	Hydro alcoholic saffron extract could improve learning and memory deficits in amnestic mice	90
		Morris water maze test in male adult Wistar rats	Intraperitoneal administration of an aqueous solution of crocin or water-alcoholic saffron extract improved cognitive deficits induced by chronic cerebral hypoperfusion in rats	91
		Morris water maze test conducted in intracerebroventricular STZ treated male Wistar albino rats	Crocin administration significantly improved memory consolidation	92
		Morris water maze test conducted in male albino Wistar rats	Intraperitoneal and intrahippocampal administration of crocin significantly improved spatial memory conditions	93
		Passive avoidance paradigm and Y-maze task for spatial cognition evaluation in STZ induced male Wistar rats	Crocin treatment significantly prevented cognitive impairments in STZ induced rats	94
		Passive avoidance tests performed in mice	Alcoholic saffron extract demonstrated ameliorative on learning and memory processes	95
		Object recognition test and the step-through passive avoidance task in scopolamine-induced male Wistar rats	Intraperitoneal injection of aqueous saffron extract had ameliorative effect on cognitive memory	96
Black cumin	Thymoquinone and its derivatives	Morris water maze test and passive avoidance tests conducted in juvenile and neonatal stages of Wistar rats	Improvement in learning-memory during juvenile and neonatal growth	97
		Spatial working memory performance on a radial arm maze	Better performance showed in passive avoidance tests by the animals. Thus, the extract prevented scopolamine-induced spatial memory deficits in rats	98
		Spatial memory performance (SMP) was investigated using eight-arm radial arm maze (RAM) in male Sprague Dawley rats	Learning-memory was significantly enhanced and a significant decrease in the mean number of working memory error (WME) was observed	99

(*continued*)

TABLE 4.3a (Continued)
Role of spices and their respective bioactive components on behavioural tests in *in vivo* animal model

Name of spices	Bioactive components	*In vivo* behavioural tests in animal model	Remarks	References
		Morris water maze and Y-maze tests conducted on scopolamine-induced Wister rats	The extract (oil) showed protective effects on cognitive function and cortico-hippocampal neural alterations in male Wistar rats	100
		Morris water maze test (long-term and short-term working memory)	The fixed oils of the seeds showed spatial cognitive preservation in rats with chronic cerebral hypoperfusion	72
		T-maze alternation task and object recognition test on scopolamine-induced male Wistar rats	Reversal of amnesic effect of scopolamine-induced spatial and non-spatial working memory impairment	101
		Behavioural tests (PA, MWM) on LPS induced male Wistar rats	Memory impairments was improved by reducing hippocampal cytokine levels consequent brain tissue oxidative damage	102
		Morris water maze test conducted in female Sprague Dawley rats	TQ had an ameliorative effect on spatial learning	103
		Morris water maze test conducted on nonylphenol induced male Wistar rats	TQ (5 mg/kg) significantly improved spatial memory and learning	104
		Fear learning behaviour in passive avoidance test in an AD rat model	TQ improved learning function	105
Bay leaf	1,8-Cineole	Y-Maze Spontaneous Alternation Test, radial arm Maze and novel object recognition test conducted on scopolamine-induced male Wistar rats	Bay leaf incense positively attenuated scopolamine-induced memory impairments	82
Coriander	Linalool	Y-maze task and radial arm maze task in Wistar rats	The volatile oil showed significant improvement in spatial memory formation	87
		Morris water maze test and elevated plus maze tests in transgenic AD model (3xTg-AD) and non-transgenic (non-Tg) mice	Linalool restored cognitive and emotional functions	106
		Morris water maze test and passive avoidance task performed with male C57BL/6J mice	Linalool administration was successful in improving cognitive impairment	107

Reviews/articles that are indexed by Scopus and/or Web of Science have been selected for this compilation.

TABLE 4.3b
Role of spices and their respective bioactive components in characteristic AD clinical trial

Name of spices	Bioactive compounds	Clinical trial in humans (AD patients)	Remarks	References
Turmeric	Curcumin	• Mini Mental State Examination (MMSE) • Enzyme-linked immunosorbent assay to measure Aβ 40 level in blood serum • Measurement of isoprostanes iPF2ᾱ-III and antioxidants in plasma • Liver and kidney function	**Major finding(s) reported:** Curcumin has role in Aβ disaggregation, anti-inflammation, and antioxidation; also improves cognition; no adverse side effects on health, hence can be recommended for longer trials	31
		Primary outcomes: • Changes in clinical laboratory tests, MMSE • Alzheimer's Disease Assessment Scale –Cognitive Subscale (ADAS-Cog) at 24 weeks **Secondary outcomes:** • Neuropsychiatric Inventory (NPI) • Alzheimer's Disease Cooperative Study –Activities of Daily Living (ADCS-ADL) scale • Plasma levels of Ab1-40 and Aβ1-42; and levels of Aβ1-42, t-tau and p-tau • F2-isoprostanes in cerebrospinal fluid	**Major finding(s) reported:** No differences among groups, and no serious adverse events	22
		Computerized Mental Performance Assessment System: • Immediate word recall task • Simple reaction time task • Two choice reaction time tasks • Rapid visual information processing task • Digit vigilance task • Serial three and serial seven subtraction tasks • Delayed word recall task • Delayed word recognition task • Delayed picture recognition task	**Major finding(s) reported:** Curcumin showed potential in improving important cognitive functions even at a low dose (approximately 80 mg) and delayed detrimental effects of psychological stresses	23
Saffron	Crocin	• Psychometric measures included MMSE • Alzheimer's Disease Assessment Scale-cognitive subscale (ADAS-cog) • Clinical Dementia Rating Scale—Sums of Boxes (CDR-SB)	**Major finding(s) reported:** Saffron capsule [30 mg/day (15 mg twice per day) for a 22-week study was found to be as effective as donepezil in the treatment of mild to moderate AD after 22 weeks]	63
		• Psychometric measures included MMSE • Alzheimer's Disease Assessment Scale-cognitive subscale (ADAS-cog) • Clinical Dementia Rating Scale—Sums of Boxes (CDR-SB)	**Major finding(s) reported:** Saffron capsule [30 mg/day (15 mg twice per day) for a 22-week study was found to be safe and effective treatment of mild to moderate AD after 16 weeks	64

(continued)

TABLE 4.3b (Continued)
Role of spices and their respective bioactive components in characteristic AD clinical trial

Name of spices	Bioactive compounds	Clinical trial in humans (AD patients)	Remarks	References
		• Mini Mental State Examination • Severe Cognitive Impairment Rating Scale (SCIRS) • Functional Assessment Staging (FAST)	**Major finding(s) reported:** Administration of saffron extract (30 mg/day) capsules could be compared to memantine in reducing cognitive decline in patients with moderate to severe AD for a 12-month study	65
		• Psychometric tests- Geriatric Depression Scale, GDS • Functional Rating Scale of Symptoms of Dementia • Neuropsychiatric Inventory, NPI • Montreal Cognitive Assessment, MoCA • MMSE	**Major finding(s) reported:** Saffron improved MMSE, MRI, EEG, and ERP scores	108
Black cumin	Thymoquinone	• Neuropsychological tests- logical memory test, digit span test, Rey-Osterrieth complex figure test, letter cancellation test, trail making test • Cognitive test- stroop test	**Major finding(s) reported:** NS capsule (500 mg) have shown positive modulatory effects on memory of elderly volunteers	109

Reviews/articles that are indexed by Scopus and/or Web of Science have been selected for this compilation.

To circumvent such complications of existing drugs, botanical ingredients (Table 4.1) are increasingly gaining popularity in the treatment of AD. However, the applications of these active ingredients in the treatment of AD pathogenesis have not delivered their desirable effects, mostly due to improper extraction methods or drug delivery systems (*vide supra*). The extracts, when consumed orally, might not exhibit intended efficacies since they have to bypass an elaborate digestive system, and also several carrier proteins and pathways that can interfere with their proper utilization by body cells[131]. Moreover, the solvent extraction methods used in isolating the phytochemicals might show other harmful effects since their extracts house carcinogenic solvent residues[132,133]. We advocate the usage of green solvents for extraction to deliver safe extracts, and especially supercritical CO_2 extraction to obtain a safe-cum-neat-cum-selective extract (with intact antioxidant synergy) for formulation of herbal drugs in ameliorating AD pathogenesis[4].

4.4 EXTRACT CHARACTERIZATION NEEDED FOR SAFE USAGE AS NATURAL DRUG

In the pursuit of formulation of a safe herbal medicine for AD, it is vital to perform- detailed spectroscopic-cum-chromatographic studies for elucidation of the consortium of antioxidants extracted from the spice/herb, evaluation of synergistic effect (SE) value (which must be greater than unity) of the antioxidants' consortium in the spice/herb extract, assessment of LD_{50} values of the pure chemical/extract in animal models, and standard ADMET profile analysis of the extract constituents (Table 4.2).

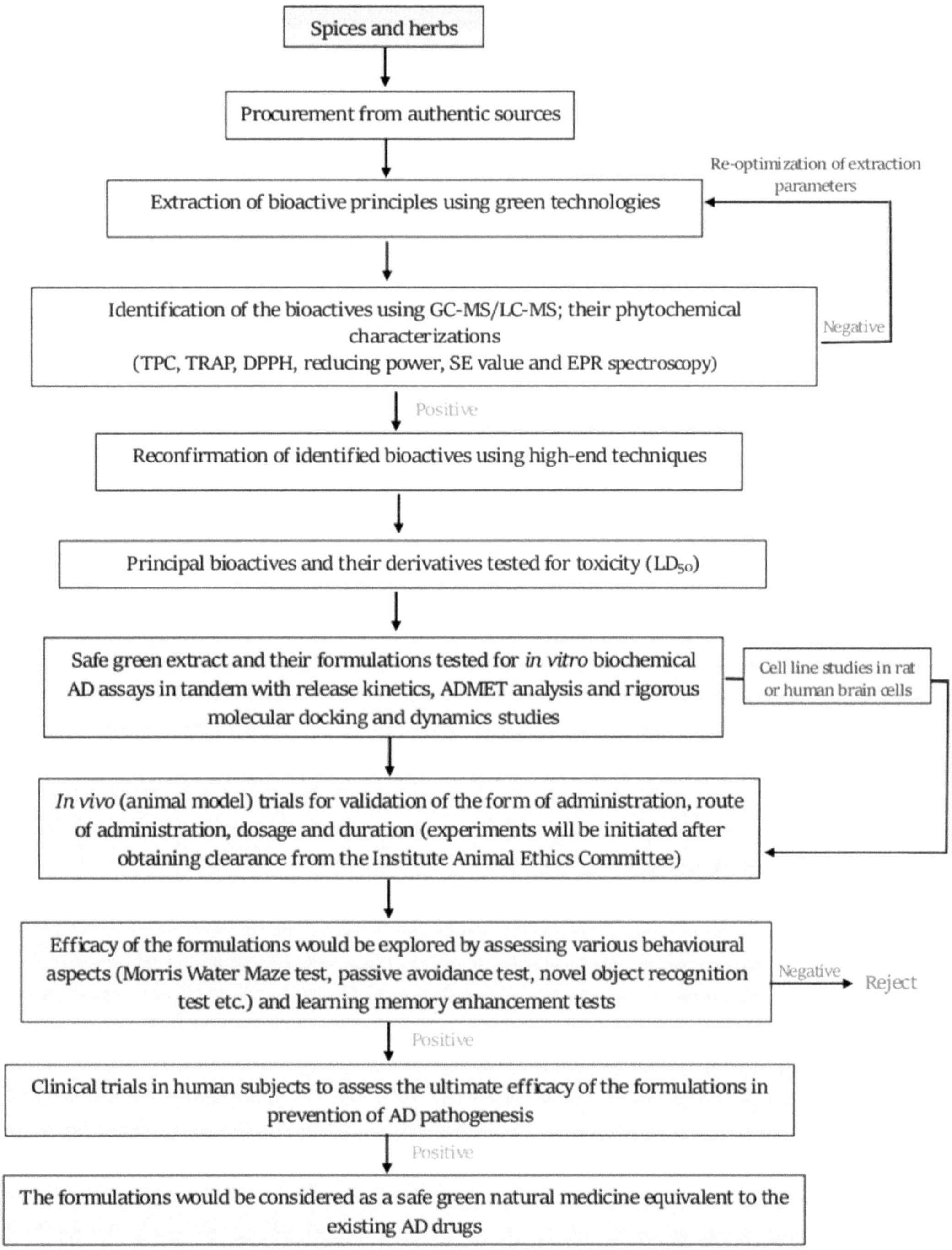

FIGURE 4.2 A systematic approach for the successful formulation of promosing spiceutical-based medicines as futuristic natural alternatives to conventional AD drugs.

4.5 EXTRACT FORMULATIONS AND CHARACTERIZATIONS FOR ENHANCED EFFICACIES

The solvent extracts, including the SC-CO_2 extracts *per se* (Table 4.2), have not been highly successful in demonstrating desirable therapeutic effects in the prevention of AD pathogenesis, and thereby necessitated further processing of the extracts, such as formulating their encapsulates for enhanced bioavailabilities. Macro-encapsulated formulations such as commercial capsules and micelles[32,65]; micro-emulsions such as those of piperine[46], and crocin extracts (from black pepper and saffron, respectively); and nano-formulations such as nanoparticles of commercially available standard piperine[47,48]; nanoliposomes of- curcumin[34], -commercially available standard piperine[50], -1,8-cineole-rich extract of small cardamom seeds[134], when administered orally or intraperitoneally, have demonstrated greater efficacies in the prevention of AD pathogenesis, *vis-à-vis* the native extracts. Among all these, the nano-formulations have exhibited relatively enhanced BBB penetrability and, therefore, have been more effective as anti-cholinergic as well as anti-amyloidogenic agents over their macro/micro counterparts. Moreover, a solitary publication on nanoliposome (of standard piperine powder)-based nasal drops has revealed greater efficacy *vis-à-vis* its nanoliposome even at lower dose-duration[50].

For formulation of extracts into bioavailable forms, it is pertinent that in addition to studies on ADMET profiles, release kinetics profiles of the bioactive(s) from their formulations should also be investigated. It is important to consider the choice of GRAS status excipients for extract formulations.

From Figure 4.1 and Table 4.2, it is evident that most herbal formulations target *in vitro* prevention of amyloid β_{40} and β_{42} oligomerization, while the commercial drugs are primarily acetylcholinesterase inhibitors *in vivo*[135]. For drugs targeting acetylcholinesterase inhibition pathway, it is vital to conduct *in silico* molecular docking and dynamics studies[136,137] in tandem with pharmacokinetic studies. However, the herbal medicines, whose modes of action have been indicated, have not been thoroughly investigated in these specific aspects to enable computational drug design (Table 4.2). Concomitant investigations on routes of administration (oral or intraperitoneal), dose duration, effects on pathways of AD pathogenesis by standard *in vitro* assays, and *in silico* molecular docking-cum-molecular dynamics studies will holistically contribute to ascertaining safety and for assessing suitability for use of pure bioactive compound/extract/formulation as a drug; and its effectiveness for desired action(s).

4.5.1 Cell Line Studies

If the chemical/extract/formulation meets success in *in vitro* pathways of AD pathogenesis, it can be subjected to cell line investigations, using either rat or human brain cells as described in Table 4.2.

4.5.2 Animal Model Studies

If the chemical/extract/formulation meets success in *in vitro* pathways of AD pathogenesis, it can be safely extrapolated to animal studies. Although it has been reportedly established that transgenic mouse models have contributed significantly to understanding the pathophysiology of β-amyloid oligomerization[138], no standardized transgenic AD mouse model exists to date worldwide. Whether animal studies should primarily involve transgenic AD mice for meaningful elucidation of the drug efficacy, especially for drugs targeting *β*-amyloid toxicity, is not yet clearly known. Several authors have used $AlCl_3$-induced Sprague Dawley rats or male Wistar albino rats in their inv estigations[38,46,50]. Many studies have also involved scopolamine or streptozotocin inductions to bring about adverse changes in the rat brain cells[82,85,139,] rendering them amenable to drug intervention studies (Table 4.2). All animal studies reported have used brain cell homogenates and have primarily targeted the amyloid-β oligomerization pathway. However, the pathways of AD are many and complicated, and availability of similar data with animal models is scarce.

Non-invasive animal behaviour assessments (Table 4.3a), post administration of the herbal drugs, have been researched upon by many, wherein mostly simple learning-memory enhancement tests such as Morris water maze test[140], Y-maze test and novel object recognition tests (either through pictures or touch models) find mention[25,49,57,88,94,107]. These widely used visual cognitive tests in animals do not well simulate human behaviour, and thus, the translation of animal to human may be skewed[141]. In assessing neurodegeneration, i.e., impairment of cognition abilities, it is necessary to assess holistically several facets of cognition such as memory, attention, mood and response inhibition, which could be rendered possible by the use of the Bussey-Saskida touchscreen model[142] for conducting human-apposite visuomotor conditional learning, and continuous performance tests with rodents[143,144,145,146]. The use of the Bussey-Saskida model will perhaps enable more meaningful elucidation of strategies for cognitive translation from animal to human. However, reports on the usage of this touchscreen model for the assessment of drug efficacy in AD mouse models are yet to be explicitly reported.

Advanced investigations on AD pathogenesis by Nobel laureate Prof. May-Britt Moser have revealed that the novel object recognition test by animals could be substituted by training animals to link odours of objects with their locations, say, responding to the smell of a banana by navigating left, and chocolate by going right. Her work has added a new dimension to existing research methodologies for drug formulations for AD. Brain odour mapping of the grid cells of the hippocampus and the entorhinal cortex in well-trained animals would enable scientists now to learn more about the causes and progression of AD since, in an AD brain, the grid cells are the first to perish, causing severe spatial memory loss and navigation problems[147]. This could steer a paradigm shift in the classical animal behaviour assessment tests for prospective elucidation of the pathophysiology of AD. Non-invasive investigations on AD pathophysiology must be succeeded by invasive investigations (Table 4.2), such as assessing the neurometabolic flux by analyzing the cerebral metabolic rate of glucose oxidation[148,149].

Of late, global legislation based on the '3R' principle of 'Refining, Reducing and Replacing' is emphasizing the reduction of animal studies and encouraging the adoption of animal-free methods in assessing drug efficacies of AD[150]. Thus, computational or *in silico* studies are increasingly gaining importance and therefore, pathway F in Figure 4.1 needs special attention. To serve as a futuristic robust substitute for animal model studies- biochemical or cell-free assays, and those involving human-derived cell lines, human *ex vivo* tissue (brain tissue, cerebrospinal fluid, blood, saliva, etc.), human stem cells, microfluidic systems, 2D and 3D co-cultures, and organoids; in tandem with *in silico* studies are warranted.

4.5.3 Human Clinical Trials

The ultimate benefit of the entire endeavour of designing a herbal medicine with promising efficacy in the prevention of AD pathogenesis will be in the successful application of the same in a human AD patient. The data obtained from preclinical *in vitro*/*ex vivo*/*in vivo* studies are extrapolated to obtain solutions for human subjects (Table 4.3b). However, reports on human clinical studies involving herbal medicines are not only scanty but less systemized. Very few reports are available that have used the gold standard for testing interventions in people, *viz.* the randomized, placebo-controlled clinical trial with patients aged 60 – 80 years having mild to moderate AD[23]. Although studies on saffron[63,64] have reported on double-blind placebo interventions with patients above the age of 60, the formulation details of the capsules prescribed to the patients are missing. More systematic studies for successful herbal medicines are needed to compete with existing synthetic drugs.

4.6 COMPARATIVE EVALUATION OF THE HERBAL MEDICINES *VIS-À-VIS* EXISTING SYNTHETIC DRUGS

For more meaningful clinical trials, it is suggested to investigate the efficacy of a herbal drug in comparison to a standard synthetic drug of AD, such as donepezil or rivastigmine, by including a positive control cohort who are administered with synthetic drug(s). This must be initiated at the animal trial level before progressing to a human clinical one. Concerning animal trials, there is only one study on a comparative evaluation of the efficacy of curcumin *vis-à-vis* donepezil on the memory function of male Wistar albino rats[151]. There are also few published reports that have included standard AD drugs as positive experimental controls in human clinical trials, such as donepezil, while elucidating the efficacy of a saffron capsule (safranal and crocin) in mild-moderate AD human patients for 22 weeks[63]; and memantine, while enumerating efficacy of a saffron extract in moderate-severe AD human patients for one year[65].

Although investigations by several researchers worldwide impress upon the fact that natural medicines are better than synthetic drugs, natural remedies are still lagging in competition with those by synthetic ones, since they do not meet the criteria highlighted in Figure 4.2.

4.7 FURTHER CONSIDERATIONS

4.7.1 Food-Drug and Drug-Drug Interactions

The ultimate success of clinical trials also would depend to a large extent on *in vivo* food-drug and drug-drug interactions when the drugs are administered to AD patients. Very few systematic published literature are available on drug-drug interactions considering two AD drugs simultaneously administered to patients and/or interactions of AD drugs with other ongoing medications[152,153]. Besides, patients challenged with ailments such as hypertension, diabetes or cardiovascular disorders cannot be prescribed AD drugs without a thorough understanding of their clinical conditions[154,155]. While some drug-drug interactions might have synergistic effects, others can act antagonistically, altering the course and concern of the prescribed medications of AD[156,157].

As far as food-drug interactions are concerned, there is a systematic review of the same by Briguglio *et al.*[158]. Food-drug interactions have been broadly classified into four types: type I (pharmaceutical), type II (pharmacokinetic and pharmaceutical), type III (pharmacokinetic), and type IV (pharmacokinetic). Type I is generally encountered when drugs are administered intraperitoneally since they mix with nutrients taken orally and precipitate consequent to physical incompatibility. Several adverse effects of oral intake of drugs with ionic components of foods are reported in the literature; besides their action being chiefly influenced by pH, they have the propensity to complexation and could be unstable in the gastrointestinal (GI) tract[158]. These facts perhaps explain the limitations of the oral or intravenous routes of drug administration in combating AD and strongly indicate the futuristic use of nasal formulations (Table 4.2) in circumventing AD pathogenesis.

Besides limitations of synthetic drugs used for the treatment of AD, there are also serious concerns about the usage of spiceutical-based natural drugs for AD, such as those based on curcumin and piperine, which are reportedly known to target the enteric and hepatic transporters of the ATP-binding cassette such as MDR1, MRP1 and MRP2. BCRP and the metabolizing enzymes of phase I, namely the cytochrome P450 enzymes CYP1A, CYP2B, and CYP2E, would possibly be targeted by piperine; and CYP2C, CYP2D, and CYP3A by curcumin[158] if these bioactives are administered orally (individually or in combination) as an extract (or its encapsulate) in human volunteers (AD patients with mild to moderate AD) who are on a normal balanced diet. This raises serious concerns about the rampant usage of herbal medicines. Baum *et al.*[31] reported on the success of the co-administration of ginkgo leaf extract along with pure curcumin in human subjects (Table 4.2) who have shown success in MMSE tests (Table 4.3b). The findings of the study cannot be very promising since ginkgo compounds, principally ginkgolic acid, are also known to inhibit enteric and hepatic transporters of the ATP-binding cassette. A study should encompass these adverse aspects prior to recommending the use of curcumin/ginkgo combination in the treatment of AD. It is therefore rationally opined that delivery of bioactives (post elucidation of the antioxidant synergy among them) through the nasal route, bypassing the oral/intravenous route of administration, would be more effective in the treatment of AD. Thus, nasal formulations of bioactives need to be researched.

For a drug to be effective in ameliorating AD pathogenesis, it is imperative to investigate at length the effects of the balanced diet components (fed to the patient) on the pharmacodynamic profiles of the drugs. It is highly complex and challenging to learn about all aspects of food interference(s) on drug action, and it is extremely difficult to mitigate food-food, food-drug, and drug-drug interactions while administering even the best of the drug formulations, *viz.* as nanoliposomes/nanoencapsulates. Moreover, patients with other co-morbidities should also be considered with extra care, and drug-drug interactions can have tremendous health implications if taken carelessly. This raises alarms about the rampant usage of oral drugs for AD treatment. Systematic literature reports on drug-drug interactions are scanty, which is another important aspect that cannot be ignored.

4.8 CONCLUDING REMARKS

AD in itself holds mystery to date as to which path it follows initially or how it propagates in a human. The ramifications of the disease are innumerable; many opine the possibility of cholinergic pathway disruption, while others keep the β-amyloids in the first place. Administration of available synthetic drugs appears to be the only convenient solution for arresting its pathogenesis. However, considering the unhealthy side effects of the same, we envisage the use of green technology in formulating 'green' medicines that can change the future of AD drugs for the betterment of AD patients. Nonetheless, green solutions also need to embrace rigorous studies at molecular levels, as well as properly designed preclinical and clinical studies for herbal medicines to be effective in ameliorating AD pathogenesis.

REFERENCES

1. Peter, K. V. (1st ed). 2012. Introduction. In *Handbook of Herbs and Spices*. Chapter 1. Woodhead Publishing series in Food Science, Technology and Nutrition, Elsevier, Amsterdam. pp. 1–6.
2. Anonymous. 2022a, December, 22. Trade Information and Statistics – Spices Board. Retrieved from www.indianspices.com/box2info.html.
3. Hamaguchi, T., Ono, K., Murase, A. et al. 2009. Phenolic compounds prevent Alzheimer's pathology through different effects on the Amyloid-(beta) aggregation pathway. *American Journal of Pathology* 175: 2557–2565.
4. Paul, K., Ganguly, U., Chakrabarti, S., Bhattacharjee, P. 2020. Is 1, 8-cineole-rich extract of small cardamom seeds more effective in preventing Alzheimer's disease than 1, 8-cineole alone? *NeuroMolecular Medicine* 22: 150–158.
5. Patel, D., Desai, S., Devkar, R. et al. 2012. Acute and sub-chronic toxicological evaluation of hydro-methanolic extract of *Coriandrum Sativum* L. seeds. *EXCLI Journal* 11: 566–575.
6. Ghaffari, S., Hatami, H., Dehghan, G. 2015. Saffron ethanolic extract attenuates oxidative stress, spatial learning, and memory impairments induced by local injection of ethidium bromide. *Research in Pharmaceutical Sciences* 10: 222–232.
7. Ismail, N., Ismail, M., Latiff, L. A. et al. 2008. Black cumin seed (*Nigella Sativa* Linn.) oil and its fractions protect against beta amyloid peptide-induced toxicity in primary cerebellar granule neurons. *Journal of Food Lipids* 15: 519–533.
8. Sharma, N., Tan, M.A., An, S. S. A. 2021. Mechanistic aspects of apiaceae family spices in ameliorating Alzheimer's disease. *Antioxidants* 10: 1571–1589.
9. Pepeu, G., Giovannini, M. G., Braco, L. 2013. Effect of cholinesterase inhibitors on attention. *Chemico-Biological Interactions* 203: 361–364.

10. Potter, P. E., Hadjiconstantinou, M., Meek, J. L. et al. 1984. Measurement of acetylcholine turnover rate in brain: An adjunct to a simple HPLC method for choline and acetylcholine. *Journal of Neurochemistry* 43: 288–290.
11. Sies, H. 1997. Oxidative stress: Oxidants and antioxidants. *Experimental Physiology* 82: 291–295.
12. Storz, G., Imlay, J.A. 1999. Oxidative stress. *Current Opinion in Microbiology* 2: 188–194.
13. Muriach, M., López-Pedrajas, R., Barcia, J.M. et al. 2010. Cocaine causes memory and learning impairments in rats: Involvement of nuclear factor kappa B and oxidative stress, and prevention by topiramate. *Journal of Neurochemistry* 114: 675–684.
14. Palumbo, M.L., Fosser, N.S., Rios, H. et al. 2007. Loss of hippocampal neuronal nitric oxide synthase contributes to the stress-related deficit in learning and memory. *Journal of Neurochemistry* 102: 261–274.
15. Sharma, D.R., Sunkaria, A., Bal, A. et al. 2009. Neurobehavioral impairments, generation of oxidative stress and release of pro-apoptotic factors after chronic exposure to sulphur mustard in mouse brain. *Toxicology and Applied Pharmacology* 240: 208–218.
16. Bartolini, M., Naldi, M., Fiori, J. et al. 2011. Kinetic characterization of amyloid-beta 1–42 aggregation with a multimethodological approach. *Analytical Biochemistry* 414, 215–225.
17. Banerjee, P., Sahoo, A., Anand, S. et al. 2014. Multiple mechanisms of iron-induced amyloid beta-peptide accumulation in SHSY5Y cells: Protective action of negletein. *NeuroMolecular Medicine* 16, 787–798.
18. Chhouk, K., Wahyudiono, W., Kanda, H. et al. 2017. Comparison of conventional and ultrasound assisted supercritical carbon dioxide extraction of curcumin from turmeric (*Curcuma longa* L.). *Engineering Journal* 21: 53–65.
19. Kurmudle, N., Kagliwal, L. D., Bankar, S. B. et al. 2013. Enzyme-assisted extraction for enhanced yields of turmeric oleoresin and its constituents. *Food Bioscience* 3: 36–41.
20. Nagavekar, N., Singhal, R. S. 2019. Supercritical fluid extraction of *Curcuma longa* and *Curcuma amada* oleoresin: Optimization of extraction conditions, extract profiling, and comparison of bioactivities. *Industrial Crops and Products* 134: 134–145.
21. Sahne, F., Mohammadi, M., Najafpour, G. D. et al. 2017. Enzyme-assisted ionic liquid extraction of bioactive compound from turmeric (*Curcuma longa* L.): Isolation, purification and analysis of curcumin. *Industrial Crops and Products* 95: 686–694.
22. Ringman, J. M., Frautschy, S. A., Teng, E. et al. 2012. Oral curcumin for Alzheimer's disease: Tolerability and efficacy in a 24-week randomized, double blind, placebo-controlled study. *Alzheimer's Research and Therapy* 4: 1–8.
23. Cox, K. H., Pipingas, A., Scholey, A. B. 2015. Investigation of the effects of solid lipid curcumin on cognition and mood in a healthy older population. *Journal of Psychopharmacology* 29: 642–651.
24. Lim, G. P., Chu, T., Yang, F. et al. 2001. The curry spice curcumin reduces oxidative damage and amyloid pathology in an Alzheimer transgenic mouse. *Journal of Neuroscience* 21: 8370–8377.
25. Frautschy, S. A., Hu, W., Kim, P. et al. 2001. Phenolic anti-inflammatory antioxidant reversal of Abeta-induced cognitive deficits and neuropathology. *Neurobiology of Aging* 22: 993–1005.
26. Yang, F., Lim, G. P., Begum, A. N. et al. 2005. Curcumin inhibits formation of amyloid beta oligomers and fibrils, binds plaques, and reduces amyloid *in vivo*. *Journal of Biological Chemistry* 280: 5892–5901.
27. Garcia-Alloza, M., Borrelli, L., Rozkalne, A. et al. 2007. Curcumin labels amyloid pathology *in vivo*, disrupts existing plaques, and partially restores distorted neurites in an Alzheimer mouse model. *Journal of Neurochemistry* 102: 1095–1104.
28. Begum, A. N., Jones, M. R., Lim, G. P. et al. 2008. Curcumin structure-function, bioavailability, and efficacy in models of neuroinflammation and Alzheimer's disease. *Journal of Pharmacology and Experimental Therapeutics* 326: 196–208.
29. Ma, Q. L., Yang, F., Rosario, E. R. et al. 2009. Beta-amyloid oligomers induce phosphorylation of tau and inactivation of insulin receptor substrate via c-Jun N-terminal kinase signaling: Suppression by omega-3 fatty acids and curcumin. *Journal of Neuroscience* 29: 9078–9089.
30. Hamaguchi, T., Ono, K., Yamada, M. 2006. Anti-amyloidogenic therapies: Strategies for prevention and treatment of Alzheimer's disease. *Cellular and Molecular Life Sciences* 63: 1538–1552.

31. Baum, L., Lam, C. W. K., Cheung, S. K. et al. 2008. Six-month randomized, placebo-controlled, double-blind, pilot clinical trial of curcumin in patients with Alzheimer's disease. *Journal of Clinical Psychopharmacology* 28: 110–113.
32. Hagl, S., Kocher, A., Schiborr, C. et al. 2015. Curcumin micelles improve mitochondrial function in neuronal PC12 cells and brains of NMRI mice – Impact on bioavailability. *Neurochemistry International* 89: 234–242.
33. Song, Z., Feng, R., Sun, M. et al. 2011. Curcumin-loaded PLGA-PEG-PLGA triblock copolymeric micelles: Preparation, pharmacokinetics and distribution *in vivo*. *Journal of Colloid and Interface Science* 354: 116–123.
34. Meng, F., Asghar, S., Gao, S. et al. 2015. A novel LDL-mimic nanocarrier for the targeted delivery of curcumin into the brain to treat Alzheimer's disease. *Colloids and Surfaces B: Biointerfaces* 134: 88–97.
35. Cheng, K. K., Yeung, C. F, Ho, S. W. et al. 2013. Highly stabilized curcumin nanoparticles tested in an *in vitro* blood–brain barrier model and in Alzheimer's disease Tg2576 mice. *Journal of the American Association of Pharmaceutical Scientists* 15: 324–336.
36. McClure, R., Yanagisawa, D., Stec, D. et al. 2015. Inhalable curcumin: Offering the potential for translation to imaging and treatment of Alzheimer's disease. *Journal of Alzheimer's Disease* 44: 283–295.
37. Hamrapurkar, P. D., Jadhav, K., Zine, S. 2011. Quantitative estimation of piperine in *Piper nigrum* and *Piper longum* using high performance thin layer chromatography. *Journal of Applied Pharmaceutical Science* 1: 117–120.
38. Subedee, L., Suresh, R. N., Jayanthi, M. K. et al. 2015. Preventive role of Indian black pepper in animal models of Alzheimer's disease. *Journal of Clinical and Diagnostic Research* 9: FF01–FF04.
39. Rathod, S. S., Rathod, V. K. 2014. Extraction of piperine from *Piper longum* using ultrasound. *Industrial Crops and Products* 58: 259–264.
40. Mengal, P., Mompon, B. 2006. Method and plant for solvent-free microwave extraction of natural products. *TrAC Trends in Analytical Chemistry* 47: 1–11.
41. Raman, G., Gaikar, V. G. 2002. Microwave-assisted extraction of piperine from *Piper nigrum*. *Industrial and Engineering Chemistry* Research 41: 2521–2528.
42. Khaw, K. Y., Parat, M. O., Shaw, P. N. et al. 2017. Solvent supercritical fluid technologies to extract bioactive compounds from natural sources: A review. *Molecules* 22: 1186–1208.
43. Schofield, K. 2013. *Hetero-aromatic nitrogen compounds: Pyrroles and pyridines*. Springer, New York. doi: https://doi.org/10.1007/978-1-4899-5892-1.
44. Dutta, S., Bhattacharjee, P. 2015. Enzyme-assisted supercritical carbon dioxide extraction of black pepper oleoresin for enhanced yield of piperine-rich extract. *Journal of Bioscience and Bioengineering* 120: 17–23.
45. Gigliarelli, G., Pagiotti, R., Persia, D. et al. 2017. Optimization of a Naviglio assisted extraction followed by determination of piperine content in *Piper longum* extracts. *Natural Products Research* 31: 214–217.
46. Etman, S. M., Elnaggar, Y. S., Abdelmonsif, D. A. et al. 2018. Oral brain-targeted microemulsion for enhanced piperine delivery in Alzheimer's disease therapy: *In vitro* appraisal, *in vivo* activity, and nanotoxicity. *Journal of the American Association of Pharmaceutical Scientists*, 19: 3698–3711.
47. Yusuf, M., Khan, M., Khan, R. A. et al. 2013. Preparation, characterization, *in vivo* and biochemical evaluation of brain targeted piperine solid lipid nanoparticles in an experimentally induced Alzheimer's disease model. *Journal of Drug Targeting* 21: 300–311.
48. Elnaggar, Y. S., Etman, S. M., Abdelmonsif, D. A. et al. 2015. Intranasal piperine-loaded chitosan nanoparticles as brain-targeted therapy in Alzheimer's disease: Optimization, biological efficacy, and potential toxicity. *Journal of Pharmaceutical Sciences* 104: 3544–3556.
49. Elnaggar, Y. S., Etman, S. M., Abdelmonsif, D. A. et al. 2015. Novel piperine-loaded Tween-integrated monoolein cubosomes as brain-targeted oral nanomedicine in Alzheimer's disease: Pharmaceutical, biological, and toxicological studies. *International Journal of Nanomedicine* 10: 5459–5473.
50. Priprem, A., Chonpathompikunlert, P., Sutthiparinyanont, S. et al. 2011. Antidepressant and cognitive activities of intranasal piperine-encapsulated liposomes. *Advances in Bioscience and Biotechnology* 2: 108–116.

51. Bhandari, A. K., Bisht, V. K., Negi, J. S. et al. 2013. 1,8-Cineole: A predominant component in the essential oil of large cardamom (*Amomum subulatum* Roxb.). *Journal of Medicinal Plants Research* 7: 1957–1960.
52. Morsy, N. F. 2015. A short extraction time of high quality hydrodistilled cardamom (*Elettaria cardamomum* L. Maton) essential oil using ultrasound as a pretreatment. *Industrial Crops and Products* 65: 287–292.
53. Ghosh, S., Bhattacharjee, P., Das, S. 2015. 1,8-cineol-rich cardamom seed (*Elettaria cardamomum*) extracts using green technologies and conventional extractions: Process analysis, phytochemical characterization, and food application. *Separation Science and Technology* 50: 1974–1985.
54. Paul, K., Bhattacharjee, P. 2018. Mass transfer and hydrodynamic study of supercritical carbon dioxide extraction of 1,8-cineole from small cardamom seeds. *Chemical Engineering Communications* 205: 1023–1033.
55. Paul, K., Bhattacharjee, P. 2018. Process optimization of supercritical carbon dioxide extraction of 1, 8-cineole from small cardamom seeds by response surface methodology: *In vitro* antioxidant, antidiabetic and hypocholesterolemic activities of extracts. *Journal of Essential Oil Bearing Plants* 21: 317–329.
56. Kamal, I. M., Das, A., Datta, S. et al. 2023. A Novel spice-antioxidant-based nano-vehicle as a putative green alternative of synthetic AChE inhibitor drugs. *Journal of Biomolecular Structure and Dynamics* doi: https://doi.org/10.1080/07391102.2023.2248274.
57. Auti, S. T., Kulkarni, Y. A. 2019. Neuroprotective effect of cardamom oil against aluminum induced neurotoxicity in rats. *Frontiers in Neurology* doi: https://doi.org/10.3389/fneur.2019.00399.
58. Ghahghaei, A., Bathaie, S. Z., Bahraminejad, E. 2012. Mechanisms of the effects of crocin on aggregation and deposition of A_{β}1–40 fibrils in Alzheimer's disease. *International Journal of Peptide Research and Therapeutics* 18: 347–351.
59. Ghahghaei, A., Bathaie, S. Z., Kheirkhah, H. et al. 2013. The protective effect of crocin on the amyloid fibril formation of A_{β}42 peptide *in vitro*. *Cellular and Molecular Biology Letters* 18: 328–339.
60. Papandreou, M. A., Kanakis, C. D., Polissiou, M. G. et al. 2006. Inhibitory activity on amyloid-β aggregation and antioxidant properties of *Crocus sativus* stigmas extract and its crocin constituents. *Journal of Agricultural and Food Chemistry* 54: 8762–8768.
61. Ghadrdoost, B., Vafaei, A. A., Rashidy-Pour, A. et al. 2011. Protective effects of saffron extract and its active constituent crocin against oxidative stress and spatial learning and memory deficits induced by chronic stress in rats. *European Journal of Pharmacology* 667: 222–229.
62. Batarseh, Y. S., Bharate, S. S., Kumar, V. et al. 2017. *Crocus sativus* extract tightens the blood-brain barrier, reduces amyloid β load and related toxicity in 5xfad mice. *ACS Chemical Neuroscience* 8: 1756–1766.
63. Akhondzadeh, S., Sabet, M. S., Harirchian, M. H. et al. 2010. A 22-week, multicenter, randomized, double-blind controlled trial of *Crocus sativus* in the treatment of mild-to-moderate Alzheimer's disease. *Psychopharmacology* 207: 637–643.
64. Akhondzadeh, S., Sabet, M. S., Harirchian, M. H. et al. 2010. Saffron in the treatment of patients with mild to moderate Alzheimer's disease: A 16-week, randomized and placebo-controlled trial. *Journal of Clinical Pharmacy and Therapeutics* 35: 581–588.
65. Farokhnia, M., Sabet, M. S., Iranpour, N. et al. 2014. Comparing the efficacy and safety of *Crocus sativus* L. with memantine in patients with moderate to severe Alzheimer's disease: A double-blind randomized clinical trial. *Human Psychopharmacology: Clinical and Experimental* 29: 351–359.
66. Tong, T., Jiang, Y., Guo, D. et al. 2018. Homogenate extraction of crocins from saffron optimized by response surface methodology. *Journal of Chemistry* doi: https://doi.org/10.1155/2018/9649062.
67. Karakani, A. M., Riazi, G., Ghaffari, S. M. et al. 2014. Inhibitory effect of crocin on aggregation of 1N/4R human tau protein *in vitro*. *Iranian Journal of Basic Medical Sciences* 18: 485–492.
68. Geromichalos, G. D., Lamari, F. N., Papandreou, M. A. et al. 2012. Saffron as a source of novel acetylcholinesterase inhibitors: Molecular docking and *in vitro* enzymatic studies. *Journal of Agricultural and Food Chemistry* 60: 6131–6138.
69. Rashedinia, M., Lari, P., Abnous, K. et al. 2015. Protective effect of crocin on acrolein-induced tau phosphorylation in the rat brain. *Acta Neurobiologiae Experimentalis* 75: 208–219.

70. Benkaci–Ali, F., Baaliouamer, A., Meklati, B. Y. et al. 2007. Chemical composition of seed essential oils from Algerian *Nigella sativa* extracted by microwave and hydrodistillation. *Flavour Fragrance Journal* 22: 148–153.
71. Amin, B., Hosseinzadeh, H. 2015. Black cumin (*Nigella sativa*) and its active constituent, thymoquinone: An overview on the analgesic and anti-inflammatory effects. *Planta Medica* doi: https://doi.org/10.1055/s-0035-1557838.
72. Azzubaidi, M. S., Saxena, A. K., Talib, N. K. et al. 2012. Protective effect of treatment with black cumin oil on spatial cognitive functions of rats that suffered global cerebrovascular hypoperfusion. *Acta Neurobiologiae Experimentalis* 72: 154–165.
73. Ismail, N., Ismail, M., Azmi, N. H. et al. 2017. Thymoquinone-rich fraction nano emulsion (TQRFNE) decreases $A_β40$ and $A_β42$ levels by modulating APP processing, up-regulating IDE and LRP1, and down-regulating BACE1 and RAGE in response to high fat/cholesterol diet-induced rats. *Biomedicine and Pharmacotherapy* 95: 780–788.
74. Piras, A., Rosa, B., Marongiua, S. et al. 2013. Chemical composition and *in vitro* bioactivity of the volatile and fixed oils of *Nigella sativa* L. extracted by supercritical carbon dioxide. *Industrial Crops and Products* 46: 317–323.
75. Ismail, N., Ismail, M., Mazlan, M. et al. 2013. Thymoquinone prevents β-amyloid neurotoxicity in primary cultured cerebellar granule neurons. *Cellular and Molecular Neurobiology* 33: 1159–1169.
76. Khan, A., Vaibhav, K., Javed, H. et al. 2012. Attenuation of Ab-induced neurotoxicity by thymoquinone via inhibition of mitochondrial dysfunction and oxidative stress. *Molecular and Cellular Biochemistry* 369: 55–65.
77. Cobourne-Duval, M. K., Taka, E., Mendonca, P. et al. 2016. The antioxidant effects of thymoquinone in activated BV-2 murine microglial cells. *Neurochemical Research* 41: 3227–3238.
78. Rishitha, N., Muthuraman, A., Saravanababu, C. 2020. Therapeutic evaluation of thymoquinone in the intracerebroventricular injection of L-cysteine induced vascular dementia in rats. *International Journal of Nutrition, Pharmacology, Neurological Diseases* 10: 120–127.
79. Pacifico, S., Gallicchio, M., Lorenz, P. et al. 2014. Neuroprotective potential of *Laurus nobilis* antioxidant polyphenol enriched leaf extracts. *Chemical Research in Toxicology* doi: https://doi.org/10.1021/tx5000415.
80. Ferreira, A., Proenca, C., Serralheiro, M. L. M. et al. 2006. The *in vitro* screening for acetylcholinesterase inhibition and antioxidant activity of medicinal plants from Portugal. *Journal of Ethnopharmacology* 108: 31–37.
81. Ayodeji, O. F., Goodness, I. O., Kayode, E. A. et al. 2021. Aqueous extracts of bay leaf (*Laurus nobilis*) and rosemary (*Rosmarinus officinalis*) inhibit iron-induced lipid peroxidation and key-enzymes implicated in Alzheimer's disease in rat brain- *in vitro*. *American Journal of Biochemistry and Biotechnology* doi: https://doi.org/10.3844/ajbbsp.9.22.
82. Brinza, I., Boiangiu, R. S., Hancianu, M. et al. 2021. Bay leaf (*Laurus nobilis* L.) incense improved scopolamine-induced amnesic rats by restoring cholinergic dysfunction and brain antioxidant status. *Antioxidants* doi: https://doi.org/10.3390/antiox10020259.
83. Msaada, K., Hosni, K., Taarit, M.B. et al. 2007. Changes on essential oil composition of coriander (*Coriandrum sativum* L.) fruits during three stages of maturity. *Food Chemistry* 102: 1131–1134.
84. Mohan, M., Yarlagadda, S., Chintala, S. 2015. Effect of ethanolic extract of *Coriandrum sativum* on tacrine induced or facial dyskinesia. *International Journal of Experimental Biology* 53: 292–296.
85. Mani, V., Parle, M., Ramasamy, K. et al. 2011. Reversal of memory deficits by *Coriandrum sativum* leaves in mice. *Journal of the Science of Food and Agriculture* 91: 186–192.
86. Caputo, L., Souza, L. F., Alloisio, S. et al. 2016. *Coriandrum sativum* and *Lavandula angustifolia* essential oils: Chemical composition and activity on central nervous system. *International Journal of Molecular Science* doi: https://doi.org/10.3390/ijms17121999.
87. Cioanca, O., Hritcu, L., Mihasan, M. et al. 2013. Cognitive-enhancing and antioxidant activities of inhaled coriander volatile oil in amyloid β (1–42) rat model of Alzheimer's disease. *Physiology and Behavior* 120: 193–202.
88. Wang, C., Cai, X., Hu, W. et al. 2019. Investigation of the neuroprotective effects of crocin via antioxidant activities in HT22 cells and in mice with Alzheimer's disease. *International Journal of Molecular Medicine* 43: 956–96.

89. Linardaki, Z. I., Orkoula, M. G. Kokkosis, A. G. et al. 2013. Investigation of the neuroprotective action of saffron (*Crocus sativus* L.) in aluminum-exposed adult mice through behavioral and neurobiochemical assessment. *Food and Chemical Toxicology* 52: 163–170.
90. Dashti-r, M. H., Zeinali, F., Anvari, M. et al. 2012. Saffron (*Crocus sativus* L.) extract prevents and improves d-galactose and nano2 induced memory impairment in mice. *EXCLI Journal* 11: 328–337.
91. Hosseinzadeh, H., Sadeghnia, H. R., Ghaeni, F. A. et al. 2012. Effects of saffron (*Crocus sativus* L.) and its active constituent, crocin, on recognition and spatial memory after chronic cerebral hypoperfusion in rats. *Phytotherapy Research* 26: 381–386.
92. Naghizadeh, B., Mansouri, M. T., Ghorbanzadeh, B. et al. 2013. Protective effects of oral crocin against intracerebroventricular streptozotocin-induced spatial memory deficit and oxidative stress in rats. *Phytomedicine* 20: 537–542.
93. Asadi, F., Jamshidi, A. H., Khodagholi, F. et al. 2015. Reversal effects of crocin on amyloid β-induced memory deficit: Modification of autophagy or apoptosis markers. *Pharmacology, Biochemistry and Behavior* 139: 47–58.
94. Khalili, M., Hamzeh, F. 2010. Effects of active constituents of *Crocus sativus* L., crocin on streptozocin-induced model of sporadic Alzheimer's disease in male rats. *Iranian Biomedical Journal* 14: 59–65.
95. Zhang, Y., Shoyama, Y., Sugiura, M. et al. 1994. Effects of *Crocus sativus* L. on the ethanol-induced impairment of passive avoidance performances in mice. *Biological and Pharmaceutical Bulletin* 17: 217–221.
96. Pitsikas, N., Sakellaridis, N. 2006. *Crocus sativus* L. extracts antagonize memory impairments in different behavioural tasks in the rat. *Behavioral Brain Research* 173: 112–115.
97. Beheshti, F., Hosseini, M., Vafaee, F. et al. 2016. Feeding of *Nigella sativa* during neonatal and juvenile growth improves learning and memory of rats. *Journal of Traditional and Complementary Medicine* 6: 146–152. doi: https://doi.org/10.1016/j.jtcme.2014.11.039.
98. Hosseini, M., Mohammadpour, T., Karami, R. et al. 2014. Effects of the hydro-alcoholic extract of *Nigella Sativa* on scopolamine-induced spatial memory impairment in rats and its possible mechanism. *Chinese Journal of Integrative Medicine* 21: 438–444.
99. Sahak, M. K., Mohamed, A. M., Hashim, N. H. et al. 2013. *Nigella sativa* oil enhances the spatial working memory performance of rats on a radial arm maze. *Evidence Based Complementary and Alternative Medicine*: 1180598. doi: https://doi.org/10.1155/2013/180598.
100. Imam, A., Ajao, M. S., Ajibola, M. I. et al. 2016. Black seed oil ameliorated scopolamine-induced memory dysfunction and corticohippocampal neural alterations in male Wistar rats. *Bulletin of Faculty of Pharmacy, Cairo University* 54: 49–57.
101. El-Marasy, S. A., El-Shenawy, S. M., El-Khatib, A. S. et al. 2012. Effect of *Nigella sativa* and wheat germ oils on scopolamine-induced memory impairment in rats. *Bulletin of Faculty of Pharmacy, Cairo University* 50: 81– 88.
102. Bargi, R., Asgharzadeh, F., Beheshti, F. et al. 2017. The effects of thymoquinone on hippocampal cytokine level, brain oxidative stress status and memory deficits induced by lipopolysaccharide in rats. *Cytokine* 96: 173–184.
103. Elibol, B., Terzioglu-Usak, S., Beker, M. et al. 2019. Thymoquinone (TQ) demonstrates its neuroprotective effect via an anti-inflammatory action on the Aβ (1–42)-infused rat model of Alzheimer's disease. *Psychiatry and Clinical Psychopharmacology* 29: 379–386.
104. Lotfi, M., Kazemi, S., Ebrahimpour, A. et al. 2022. Thymoquinone improved nonylphenol-induced memory deficit and neurotoxicity through its antioxidant and neuroprotective effects. *Molecular Neurobiology* 59: 3600–3616.
105. Poorgholam, P., Yaghmaei, P., Hajebrahimi, Z. 2018. Thymoquinone recovers learning function in a rat model of Alzheimer's disease. *Avicenna Journal of Phytomedicine* 8: 188–197.
106. Sabogal-Guáqueta, A. M., Osorio, E., Cardona Gómez, G. P. 2016. Linalool reverses neuropathological and behavioral impairments in old triple transgenic Alzheimer's mice. *Neuropharmacology* 102: 111–120.
107. Xu, P., Wang, K., Lu, C. et al. 2017. Protective effects of linalool against amyloid beta induced cognitive deficits and damages in mice. *Life Sciences* 174: 21–27.

108. Tsolaki, M., Karathanasi, E., Lazarou, I. et al. 2016. Efficacy and safety of *Crocus sativus* L. in patients with mild cognitive impairment: One year single-blind randomized, with parallel groups, clinical trial. *Journal of Alzheimer's Disease* 54: 129–133.
109. Bin Sayeed, M. S., Asaduzzaman, M. et al. 2013. The effect of *Nigella sativa* Linn. seed on memory, attention and cognition in healthy human volunteers. *Journal of Ethnopharmacology* 148: 780–786.
110. Messina, M., Lampe, J. W., Birt, D. F. et al. 2001. Reductionism and the narrowing nutrition perspective: Time for re-evaluation and emphasis on food synergy. *Journal of the Academy of Nutrition and Dietetics* 101: 1416–1419.
111. Liu, D., Shi, J., Ibarra, A. C. et al. 2008. The scavenging capacity and synergistic effects of lycopene, vitamin E, vitamin C, and β-carotene mixtures on the DPPH free radical. *LWT - Food Science and Technology* 41: 1344–1349.
112. Chakraborty, S., Bhattacharjee, P. 2018. Design of lemon-mustard nutraceutical beverages based on synergism among antioxidants and *in vitro* antioxidative, hypoglycaemic and hypocholesterolemic activities: Characterization and shelf life studies. *Journal of Food Measurement and Characterization* 12: 2110–2120.
113. Chakraborty, S., Paul, K., Mallick, P. et al. 2019. Consortia of bioactives in supercritical carbon dioxide extracts of mustard and small cardamom seeds lower serum cholesterol levels in rats: New leads for hypocholesterolaemic supplements from spices. *Journal of Nutritional Science* 8: e32: 1–15. doi: https://doi.org/10.1017/jns.2019.28.
114. Paul, K., Chakraborty, S., Mallick, P. et al. 2021. Supercritical carbon dioxide extracts of small cardamom and yellow mustard seeds have fasting hypoglycaemic effects: Diabetic rat, predictive iHOMA2 models and molecular docking study. *British Journal of Nutrition* 125: 377–388.
115. El-Naggar, T., Gomez-Serranillos, M. P., Palomino, O. M. et al. 2010. *Nigella sativa* L. seed extract modulates the neurotransmitter amino acids release in cultured neurons *in vitro*. *Journal of Biomedicine and Biotechnology* doi: https://doi.org/10.1155/2010/398312.
116. Mukhopadhyay, M. (Ed.). 2000. Fundamentals of supercritical fluids and phase equilibria. In *Natural Extracts using Supercritical Carbon Dioxide*. Chapter 2. CRC Press, Boca Raton. pp. 11–39.
117. Anatus, P. T., Warner, J. C. 2000. Principles of Green Chemistry. In *Green Chemistry: Theory and Practice*. Chapter 4. Oxford University Press, Oxford, UK. pp. 29–55.
118. Chakraborty, S and Bhattacharjee, P 2017. Supercritical carbon dioxide extraction of melatonin from Brassica campestris: *In vitro* antioxidant, hypocholesterolemic and hypoglycaemic activities of the extracts. *International Journal of Pharmaceutical Sciences and Research* 8: 2486–2495.
119. Chakraborty, S., Bhattacharjee, P., Chatterjee. N. et al. 2022. PEGylated nanoencapsulate of melatonin-rich supercritical CO_2 extract of yellow mustard seeds is an authentic lead for type 2 diabetes and cholesterol management. *Journal of Biologically Active Products from Nature* 12: 173–189. doi: https://doi.org/10.1080/22311866.2022.2029563.
120. Chatterjee, D., Bhattacharjee, P. 2013. Supercritical carbon dioxide extraction of eugenol from clove buds: Process optimization and packed bed characterization. *Food and Bioprocess Technology* 6: 2587–2599.
121. Dutta, S., Bhattacharjee, P. 2015. Enzyme-assisted supercritical carbon dioxide extraction of black pepper oleoresin for enhanced yield of piperine-rich extract. *Journal of Bioscience and Bioengineering* 120: 17–23.
122. Ghosh, S., Bhattacharjee, P., Das, S. 2015. 1,8-cineol-rich cardamom seed (*Elettaria cardamomum*) extracts using green technologies and conventional extractions: Process analysis, phytochemical characterization, and food application. *Separation Science and Technology* 50: 1974–1985.
123. Ghosh, S., Dutta, S., Ghosh, P. et al. 2016. Design of a polyherbal mix using SC-CO_2 extraction and encapsulation of the same using spray drying: Process characterization, phytochemical properties and shelf life study. *Journal of Food Process Engineering* 40: 1–15.
124. Dutta, S., Bhattacharjee, P. 2016. Modeling of supercritical carbon dioxide extraction of piperine from Malabar black pepper. Proceedings of International Conference in Advances in Bioprocess Engineering and Technology (ICABET 2016), *Materials Today: Proceedings* 3: 3238–3252.
125. Dutta, S., Bhattacharjee, P. 2017. Nanoliposomal encapsulates of piperine-rich black pepper extract obtained by enzyme-assisted supercritical carbon dioxide extraction. *Journal of Food Engineering* 201: 49–56.

126. Chakraborty, S., Paul, K., Mallick, P. et al. 2019. Consortia of bioactives in supercritical carbon dioxide extracts of mustard and small cardamom seeds lower serum cholesterol levels in rats: New leads for hypocholesterolaemic supplements from spices. *Journal of Nutritional Science* 8: e32: 1–15. doi: https://doi.org/10.1017/jns.2019.28.
127. Battle, C. E., Abdul-Rahim, A. H., Shenkin, S. D. et al. 2021. Cholinesterase inhibitors for vascular dementia and other vascular cognitive impairments: A network meta-analysis. *Cochrane Database Systematic Reviews.* doi: https://doi.org/10.1002/14651858.CD013306.
128. Schultz, C. 2003. Prodrugs of biologically active phosphate esters. *Bioorganic and Medicinal Chemistry* 11: 885–898.
129. Yiannopoulou, K. G., Papageorgiou, S. G. 2013. Current and future treatments for Alzheimer's disease. *Therapeutical Advances in Neurological Diseases* 6: 19–33.
130. Anonymous. 2022b, December, 22. Donepezil Uses, side effects & Warnings. Retrieved from www.drugs.com/mtm/donepezil.html.
131. Viswanathan, P., Muralidaran, Y., Ragavan, G. 2017. Challenges in oral drug delivery: A nano-based strategy to overcome. In *Nanostructures for Oral Medicine*. Chapter 7. Elsevier, Amsterdam. pp. 173–201.
132. Brglez, M. E., Knez, H. M., Škerget, M., Knez, T., and ren, U. 2016. Polyphenols: Extraction methods, antioxidative action, bioavailability and anticarcinogenic effects. *Molecules* 21: 901.
133. Martyasari, N. W. R., Ardiana, N., Ilhami, B. T. K., et al. 2021. The effect of extraction solvent polarity on cytotoxic properties of *Sargassum crassifolium* against B16–F10 melanoma cancer cell model. *IOP Conference Series: Earth and Environmental Science* doi: https://doi.org/10.1088/1755-1315/913/1/012105.
134. Paul, K., Bhattacharjee, P., Chatterjee, N., et al. 2019. Nanoliposomes of supercritical carbon dioxide extract of small cardamom seeds redresses type 2 diabetes and hypercholesterolemia. *Recent Patents on Biotechnology* 13: 284–303.
135. Colovic, M. B., Krstic, D. Z., Lazarevic-Pasti, T. D. et al. 2013. Acetylcholinesterase inhibitors: Pharmacology and toxicology. *Current Neuropharmacology* 11: 315–335.
136. Babitha, P. P., Sahila, M. M., Bandaru, S. et al. 2015. Molecular docking and pharmacological investigations of rivastigmine-fluoxetine and coumarin-tacrine hybrids against acetylcholine esterase. *Bioinformation* 11: 378–386.
137. Makarian, M., Gonzalez, M., Salvador, S. M. et al. 2022. Synthesis, kinetic evaluation and molecular docking studies of donepezil-based acetylcholinesterase inhibitors. *Journal of Molecular Structure* 5: 1247: 131425. doi: https://doi.org/10.1016/j.molstruc.2021.131425.
138. Elder, G. A., Gama Sosa, M. A., Gasperi, R. A. 2010. Transgenic mouse models of Alzheimer's disease. *Mount Sinai Journal of Medicine* 77: 69–81.
139. Lee, J. C., Park, J. H., Ahn, J. H. et al. 2018. Effects of chronic scopolamine treatment on cognitive impairment and neurofilament expression in the mouse hippocampus. *Molecular Medicine Reports* 17: 1625–1632.
140. Vorhees, C. V., Williams, M. T. 2006. Morris water maze: Procedures for assessing spatial and related forms of learning and memory. *Nature Protocols* 1: 848–858.
141. Palmer, D., Dumont, J. R., Dexter, T. D. et al. 2021. Touchscreen cognitive testing: Cross-species translation and co-clinical trials in neurodegenerative and neuropsychiatric disease. *Neurobiology of Learning and Memory* doi: https://doi.org/10.1016/j.nlm.2021.107443.
142. Anonymous. 2022c, December, 22. Bussey-Saksida Touch Systems – Lafayette Neuroscience. Retrieved from https://lafayetteneuroscience.com/listing/bussey-saksida-touch-systems.
143. Delotterie, D., Mathis, C., Cassel, J. C. et al. 2014. Optimization of touchscreen-based behavioral paradigms in mice: implications for building a battery of tasks taxing learning and memory functions. *PLoS ONE* doi: https://doi.org/10.1371/journal.pone.0100817.
144. Romberg, C., Mattson, M. P., Mughal, M. R. et al. 2011. Impaired attention in the 3xTgAD model of Alzheimer's disease assessed using a translational touchscreen method for mice: Rescue by donepezil (Aricept). *Journal of Neuroscience: The Official Journal of the Society for Neuroscience* 31: 3500–3507.
145. Kim, C. H., Heath, C. J., Kent, B. A. et al. 2015a. The role of the dorsal hippocampus in two versions of the touchscreen automated paired associates learning (PAL) task for mice. *Psychopharmacology* 232: 3899–3910.

146. Kim, C. H., Hvoslef-Eide, M., Nilsson, S. R. et al. 2015b. The continuous performance test (rCPT) for mice: A novel operant touchscreen test of attentional function. *Psychopharmacology* 232: 3947–3966.
147. Moser. 2014. Grid cells, Place cells and Memory, Nobel lecture, December 7, 2014 by May-Brit Moser, Norwegian University of Science and Technology (NTNU), Trondheim, Norway. Retrieved from www.nobelprize.org/prizes/medicine/2014/may-britt-moser/lecture. Accessed on December, 22, 2022.
148. Patel, A. B., de Graaf, R. A., Mason, G. F. et al. 2005. The contribution of GABA to glutamate/glutamine cycling and energy metabolism in the rat cortex *in vivo*. *Proceedings of the National Academy of Sciences* 102: 5588–5593.
149. Mishra, P. K., Kumar, A., Behar, K. L. et al. 2018. Sub anesthetic ketamine reverses neuronal and astroglial metabolic activity deficits in a social defeat model of depression. *Journal of Neurochemistry* 146: 722–734.
150. Aerts, L., Miccoli, B., Delahanty, A. et al. 2022. Do we still need animals? Surveying the role of animal-free models in Alzheimer's and Parkinson's disease research. *The EMBO Journal* doi: https://doi.org/10.15252/embj.2021110002.
151. Naqvi, F., Haider, S., Naqvi, F. et al. 2019. A comparative study showing greater effects of curcumin compared to donepezil on memory function in rats. *Pakistan Journal of Pharmaceutical Sciences* 32: 53–60.
152. Caraci, F., Sultana, J., Drago, F. et al. 2017. Clinically relevant drug interactions with anti-Alzheimer's drugs. *CNS and Neurological Disorders-Drug Targets (Formerly Current Drug Targets-CNS and Neurological Disorders)* 16: 501–513.
153. Savran, M., Halil, A. S. C. I., Sengeze, N. et al. 2019. The potential drug-drug interactions in Alzheimer patients' treatment. *Süleyman Demirel Üniversitesi Sağlık Bilimleri Dergisi*, 10: 381–388.
154. Akimoto, H., Negishi, A., Oshima, S., et al. 2020. Antidiabetic drugs for the risk of Slzheimer disease in patients with type 2 DM using FAERS. *American Journal of Alzheimer's Disease and Other Dementias®* 35: 1–11. doi: https://doi.org/10.1177/1533317519899546.
155. Ding, X. W., Li, R., Geetha, T. et al. 2020. Nerve growth factor in metabolic complications and Alzheimer's disease: Physiology and therapeutic potential. *Biochimica et Biophysica Acta (BBA)-Molecular Basis of Disease* doi: https://doi.org/10.1016/j.bbadis.2020.165858.
156. Pasqualetti, G., Tognini, S., Calsolaro, V. et al. 2015. Potential drug-drug interactions in Alzheimer patients with behavioral symptoms. *Clinical Interventions in Aging* 10: 1457.
157. Mehrpouya, M., Ataei, S., Nili-Ahmadabadi, A. 2017. Potential drug interactions with cholinesterase inhibitors in Alzheimer patients: A guideline for neurologists. *Journal of Applied Pharmaceutical Science* 7: 223–226.
158. Briguglio, M., Hrelia, S., Malaguti, M. et al. 2018. Food bioactive compounds and their interference in drug pharmacokinetic/pharmacodynamic profiles. *Pharmaceutics* 10: 277–299.

Appendix-I

List of Abbreviations

AChE	Acetylcholinesterase
AD	Alzheimer's disease
ADAS-cog	AD assessment scale-cognitive subscale
ADMET	Absorption, distribution, metabolism, excretion, and toxicity
Aβ	Amyloid beta
BBB	Blood-brain-barrier
CDR-SB	Clinical dementia rating scale: sums of boxes
CGN	Cerebellar granules
DPPH	2,2-diphenyl-1-picrylhydrazyl
EDX	Energy dispersive X-ray
EE	Enzyme-assisted solvent extraction

EILE	Enzyme-assisted ionic liquid extraction
EO	Essential oils
EPR	Electron paramagnetic resonance
FDA	Food and drug administration
FRAP	Ferric reducing antioxidant power assay
FST	Force swimming test
FT-IR	Fourier-transform infrared spectroscopy
GC-MS	Gas chromatography mass spectrometry
GRAS	Generally recognized as safe
HPLC	High performance liquid chromatography
I.P	Intraperitoneal
LC-MS	Liquid chromatography mass spectrometry
LD	Lethal Dose
MAE	Microwave-assisted solvent extraction
MCI	Mild cognitive impairment
MMSE	Mini-mental state examination
MWM	Morris water maze test
PA	Passive avoidance
PEG	Polyethylene glycol
PLGA	Poly lactic-co-glycolic acid
RAM	Radial arm maze
ROS	Reactive oxygen species
SC-CO_2	Supercritical carbon dioxide
SCF	Supercritical fluid
SE	Synergistic effect
SFE	Supercritical fluid extraction
SMP	Spatial memory performance
SOD	Superoxide dismutase
STZ	Streptozotocin
TPC	Total phenolic content
TQ	Thymoquinone
UAE	Ultrasonication-assisted solvent extraction
USC-CO_2	Ultrasound-assisted supercritical carbon dioxide
WME	Working memory error
XRD	X-Ray diffraction

5 Herbs with Nootropic Efficacy

Olakunle J. Onaolapo and Adejoke Y. Onaolapo

5.1 INTRODUCTION

Cognition encompasses all the different processes that are employed in learning, understanding, and retention of information [1.2]. It also refers to a capacity for processing information and application of knowledge. These processes include memory, attention, language, vigilance, executive functions, perception, and psychomotor functions [3, 4]. Cognitive ability has been broadly defined as an individual's ability to plan, reason, think abstractly, solve problems, understand complex ideas, and learn from experience [1]. Although noncognitive factors such as zeal, hard work, endurance, and grit have all been shown to contribute significantly to the achievement of success, the achievement of unusual success in life or academics has, however, been attributed to the individual's level of intelligence or cognitive ability [1, 5, 6]. Therefore, extensive research is dedicated to uncovering ways of improving or enhancing cognitive ability [6–9].

Cognitive enhancers or nootropics are chemical or natural compounds that can facilitate and/or improve attentional abilities, as well as the storage and retrieval of information in health; and attenuate cognitive deficits as is observed with normal ageing or disease [10–13]. Since 1972, when the term nootropics was coined following the discovery of the memory-enhancing potential of piracetam, hundreds of compounds have been examined for their nootropic ability [14]. Most drugs with nootropic ability enhance cognition through their ability to increase the brain's supply of oxygen as well as improve the levels and activities of neurotransmitters such as acetylcholine and or neurotransmitter enzymes in the brain [11, 14, 15]. The earliest known nootropic drug, piracetam, and others in its class, including oxiracetam and aniracetam, act through the glutamatergic/cholinergic systems, while drugs such as amphetamine, armodafinil and modafinil interact with the dopaminergic and serotonergic (5-hydroxy-tryptophan) systems [11].

The beneficial effects of synthetic nootropic agents like piracetam and related compounds in healthy individuals and persons with cognitive decline have been reported [11]. However, they have also been associated with undesirable adverse effects that limit their use. The growing demand for nootropic agents for the management of ageing and disease-related memory loss, as well as the need for improvement of cognitive abilities in the healthy [16, 17], is necessitating the continued search for newer sources of nootropics such as plants and plant-based compounds that may be associated with lesser side effects.

Plants and plant-derived compounds have been used for centuries in the maintenance of health as well as the prevention and management of disease [18]. Herbal medicines are becoming popular because they are generally believed to have limited or no adverse side effects [19]. A number of these plant-based compounds have also been touted by traditional medicine systems to be beneficial

DOI: 10.1201/9781003300502-5

in the improvement of cognition and in the management of ageing and disease-related memory impairment. In recent times, extensive research in this field has also provided scientific evidence to support the cognition-enhancing capacity of these plants [14, 18, 21–27].

In this chapter, we examine existing literature dealing with plants and plant-based compounds with preclinically or clinically reported nootropic properties. This is with a view to improving our understanding of the different compounds and the possible mechanisms through which they might enhance cognitive performance.

5.2 HERBAL MEDICINES IN HEALTH AND DISEASE

In the last few decades, the use of herbs and phytonutrients as nutritional supplements, nutraceuticals or medicines has increased exponentially. Globally, an increasing number of people in developed and developing economies now depend solely on these herbal remedies for the maintenance of health and the management of diseases [12, 28, 29]. Overall, there has been a global surge in the public interest and acceptance of natural compounds, with a number of these products being freely available in drug stores, food stores, and supermarkets.

There is anecdotal evidence demonstrating the ability of plants and plant-derived compounds to improve brain functions, including learning, memory, and attention. Approximately one hundred herbal medicinal plants are reported to have been used traditionally for the improvement of learning and memory function [30-32]. A global demand for increased workplace productivity as well as a rise in the prevalence of age-related memory impairment and dementias due to neurodegenerative disorders [33], have led to a surge in the demand for nootropics agents, particularly those from natural sources which are believed to cause little or no adverse effects [14, 34]. In the last few years, there has been growing preclinical and clinical evidence supporting the anecdotal reports of the nootropic abilities of several plants and plant bioactive compounds.

5.3 PLANT AND PLANT-BASED COMPOUNDS WITH COGNITION-ENHANCING ABILITIES

For centuries, the Ayurvedic, Chinese, and Native American medicine systems have documented the use of plants for the treatment of cognitive disorders, particularly in the elderly [35–37]. The results of several ethnopharmacological surveys have identified novel plants with the potential for ameliorating cognitive deficits or improving learning and memory function [35, 37–40]. Due to the fact that a number of the currently available conventional drugs were isolated from plants or derived from compounds isolated from plants, there has been increased research into developing or discovering novel plant and plant-based nootropic agents.

In the last two decades, a number of anticholinesterase alkaloids, such as galantamine, isolated from plants, have been investigated and approved for the treatment of cognitive deficits arising from Alzheimer's dementia [35, 37]. Galantamine is a bioactive compound derived from plants like *Lycoris radiata*, long used in traditional Chinese medicine for the treatment of memory loss in the elderly. Several plant species and their bioactive compounds have been reported to have pharmacological activities that are relevant to the improvement of memory and the management of cognitive deficits and are also showing promise in preclinical and clinical trials as potential nootropic agents [23, 40]. A number of these plants exert their nootropic abilities through the modulation of the cholinesterase system, activities at nicotinic receptors, and antioxidant potential [31, 32]. Several studies, scientific reports and texts have documented the potential or confirmed nootropic effects of a large number of plants, herbs, and plant bioactive compounds [31, 32]. The subsequent sections will review recent data on the nootropic effect of plants, herbs or bioactive compounds, as well as their respective mechanisms of action.

5.4 PLANTS AND HERBS WITH NOOTROPIC ABILITIES

Gingko biloba, Bacopa monieiri, Hypericum perforatum, Withania somnifera Centella asiatica Camellia sinensis, and *Panax ginseng* are plants with a centuries-long history of use in traditional medicine systems as memory enhancers. Extensive research (Table 5.1) of these plants and their standardized extracts have continued to provide evidence in support of not only their nootropic properties but also their adaptogenic and antioxidant potentials [14, 41]. Plants like *Camellia sinensis* have, for centuries, been used in the Chinese traditional medicine systems [18, 21]. A 2017 systematic review of the effects of *Camellia sinensis* on cognition and working memory reported that the nootropic effects of green tea could be attributed to the combined effects of caffeine and l-theanine, which are bioactive constituents of green tea [42]. There have also been reports that enriching green tea with (-)-gallocatechin gallate, another of its bioactive constituents, has the ability to improve age-related cognitive impairments through the restoration of long-term synaptic plasticity [43]. A formulation of green tea known as matcha tea, which is tea made from drying and powdering of tea leaves (*Camellia sinensis*), has also been shown to have nootropic effects. Studies have demonstrated that daily intake of matcha improved attention and executive functioning in healthy young, middle-aged, and older subjects. Its cognition-enhancing effects have been attributed to its catechin, theanine, and caffeine content [44, 45]. For centuries, the memory-enhancing effects of *Centella asiatica* have been reported. Results from a number of preclinical studies have demonstrated that *Centella asiatica* extracts have the capacity to enhance cognitive function in rodents [46, 47]. In β-amyloid overexpressing mice and aged mice, *Centella asiatica* improved memory function and executive functioning [46, 47]. Its memory-enhancing effects have been attributed to its ability to attenuate hippocampal mitochondrial dysfunction in these animals [46, 47]. *Bacopa monnieri*'s nootropic effects have also been reported, with its nootropic effects supported by clinical [48, 49] and preclinical studies [50]. By using a rodent model of Alzheimer's disease, Uabundit et al. [50] reported that an alcoholic extract of *Bacopa monnieri* reversed memory loss and mitigated Alzheimer's disease-induced loss of cholinergic neurons [50]. Similar effects have also been reported in a mouse model of vascular dementia [51]; in this study, however, the memory-enhancing and neuroprotective effects of *Bacopa monnieri* were attributed to bacopaside I, which acts via PKC and PI3K/Akt mechanisms [51]. Phan et al. [52] also observed an improvement in memory function in healthy adolescent mice. Several systematic reviews and metanalysis have also reported that there is ample scientific evidence in support of its use as a nootropic agent in healthy humans as well as subjects with cognitive deficits [48, 49, 53]. In aged humans, standardized *Bacopa monnieri* was associated with enhanced auditory verbal learning, delayed word recall, and executive functioning compared to those administered placebo [54]. For more than two decades, research has continued to demonstrate the nootropic effects of *Hypericum perforatum* [55–57]. The memory-enhancing effects of over-the-counter supplements containing *Hypericum perforatum* have also been examined [57]. The result of a single-dosage, randomized, double-blind, placebo-controlled study examining the effect of Remotiv (one of such supplements) on short-term memory performance in healthy adults revealed that acute treatment with small (250 mg) doses of the supplement was associated with improved short-term verbal memory [57]. *Centella asiatica* has several bioactive constituents, including asiaticoside, asiatic acid, madecassoside, and madecassic acid [58]. Its memory-enhancing effects have been attributed to its ability to inhibit acetylcholinesterase activity, reduce phospholipase A_2 activity and protect against the formation of ß-amyloid [58].

The development of novel enhancers of cognition from plants with antioxidant properties is also gaining attention (Table 5.1), and this has been attributed to the current understanding of the importance of oxidative stress in the development of cognitive impairment [59, 60]. Plants and herbs in this class with significant antioxidant capacity that have also been shown to have nootropic abilities include *Zingiber officinale*, *Curcuma longa, Murraya koenigii*, *Phyllanthus emblica, Ocimum tenuiflorum, Spirulina maxima, Cucumis melo,* and *Citrullus lanatus* [41, 61–64]. The nootropic effects of *Zingiber officinale* in preclinical and clinical studies have been reported [59, 65]. Saenghong et al. [59] reported that a standardized extract of *Zingiber officinale* improved working memory and cognitive function in healthy middle-aged women. In another study, *Zingiber officinale* was combined with *Cyperus rotundus* (another plant that had been reported to have a high polyphenol content and antioxidant capacity) in adult Wistar rats with age-related dementia and ethylcholine mustard aziridinium ion-induced memory deficits. Results revealed that the combination of extracts of the two plants was associated with improvement in memory, increased neuronal density, decreased acetylcholinesterase activity, decreased oxidative stress, and activation of hippocampal pERK1/2 [66]. *Spirulina maxima* is a microalga with high protein and polyphenols content and has been shown to exert memory-enhancing effects. Ethanolic extract of spirulina maxima was shown to ameliorate amyloid-β 1-42-induced cognitive deficits [67]. In other studies, *spirulina maxima* alone [68] or fermented with lactobacillus planetarium HY-08 [62] mitigated scopolamine-induced memory loss in mice. Its memory-enhancing effects have been attributed to its antioxidant activity [62, 68] and the inhibition of the increased phosphorylation of glycogen synthase kinase-3β caused by intracerebroventricular injection of $A\beta_{1-42}$ in mice [67] *Curcuma longa* is another herb with documented antioxidant properties. Also known as turmeric, *Curcuma longa* contains curcuminoids, which have antioxidant and neuroprotective properties. Studies have demonstrated its nootropic abilities [61, 69]. The results of the study evaluating the nootropic effect of its rhizome extract on trimethyltin-treated Sprague-Dawley rats revealed that at 200 mg/kg/body weight, *Curcuma longa* mitigated deficits in spatial memory performance and prevented the reduction in the number of pyramidal neurons in the hippocampal cornus ammonis [69]. In another study examining the effect of fermented *Curcuma longa* (using 5% Lactobacillus plantarum K154) on memory in mice, it was revealed that fermented *Curcuma longa* significantly attenuated scopolamine-induced impairment of memory in mice and prevented scopolamine-induced decrease in acetylcholinesterase activity in the hippocampus [61].

Standardized extracts of plants (Table 5.1) have also been examined for their memory-enhancing properties. Kim et al. [70] examined the effects of HemoHIM, a medicinal herbal formulation of three plants *Angelica gigas*, *Cnidium officinale*, and *Paeonia lactiflora* on scopolamine-induced memory loss in mice and reported attenuation of memory impairment through the inhibition of acetylcholinesterase activity, increase in acetylcholine levels; and upregulating of choline xcetyltransferase, muscarinic acetylcholine receptors and activation of the CREB and ERK signalling pathways [70]. Interest in the combined effect of these herbs is also increasing, as evidenced by research or clinical trials examining polyherbal formulations of their extracts [71–73]. Steiner et al. [71] examined the effects of Sailuotong, a standardized herbal formulation containing *Panax ginseng*, *Ginkgo biloba*, and *Crocus sativus* on a few healthy adults and reported a small but significant improvement in working memory scores in these healthy subjects [71]. Cave et al. [73] registered a randomized, double-blind, placebo-controlled clinical trial to study the effects of BioCeuticals Cognition Support Formula®, a standardized formulation of three herbs including *Bacopa monnieri, Ginkgo biloba, Panax ginseng,* and alpha-lipoic acid on cognition in older adults with subjective cognitive impairment. This was consequent upon verified memory-enhancing effects of the individual herbs [73]. Another standardized phytochemical formulation is BacoMind™ which is an enriched phytochemical composition derived from *Bacopa monnieri.* Acute administration of BacoMind™ was shown to enhance memory acquisition and retention in healthy rats and following scopolamine-induced amnesia in rats [74].

5.5 PLANT BIOACTIVE COMPOUNDS WITH NOOTROPIC ACTIVITIES

In the last few decades, bioactive compounds which include carotenoids, polyphenols, omega-3 fatty acids, vitamins, organic acids, phytosterols, nucleosides, and nucleotides, have continued to attract attention amongst researchers; largely due to our current understanding of their beneficial

TABLE 5.1
Plants and Herbs with nootropic abilities

Plant	Type of study	Effect	Mechanism of action	references
Camellia sinensis	Systematic review	Improved age-related cognitive impairments	Attributed the combined effects of caffeine and l-theanine, which are bioactive constituents of green tea. (-) -gallocatechin gallate another bioactive constituent restores long-term synaptic plasticity	[42, 43]
Powdered *Camellia sinensis* leaves (Matcha tea)	Healthy young, middle-aged and aged humans	Daily intake of matcha improved attention and executive functioning in healthy young, middle-aged and older subjects	Actions of catechin, theanine, and caffeine content	[44, 45]
Centella asiatica	Aged mice and β-amyloid overexpressing mice	Increased hippocampal synaptic density and improvement of memory and executive function	Attenuation of hippocampal mitochondrial dysfunction	[46, 47]
	Systematic review of six randomized clinical trials	Improvement of memory function	Inhibition of acetylcholinesterase activity, reduction of phospholipase A_2 (PLA_2) activity, protection against ß-amyloid formation, and protection from brain damage	[58]
Bacopa monnieri	Mouse model of Alzheimer's disease and vascular dementia	Reversal of memory loss and mitigation of Alzheimer's disease induced loss of cholinergic neuron	The high concentration of its active principle (bacopaside)I which acts via PKC and PI3K/Akt mechanisms	[50, 51]
Standardized *Bacopa monnieri*	Aged Humans	Enhanced auditory verbal learning test delayed word recall and stroop test		[54]
Hypericum perforatum	Single dosage, randomized, double-blind, placebo-controlled study	Acute low dose of *Hypericum perforatum* improved short-term verbal memory	Activities at amine neurotransmitters including the inhibition of the reuptake of serotonin, dopamine, norepinephrine	[56, 57]
Zingiber officinale (standardized extract)	Healthy middle-aged women	Improved working memory and cognitive function	Antioxidant properties	[59]
Zingiber officinale and Cyperus rotundus co-adminoistration	Wistar rats with age-related dementia and ethylcholine mustard aziridinium ion-induced memory deficits	Improvement in memory function, increased neuronal density, decreased acetylcholinesterase activity and decreased oxidative stress status together with activated pERK1/2 were observed in the hippocampus	Enhanced cholinergic function and decreased neurodegeneration and oxidative stress	[66]

(continued)

TABLE 5.1 (Continued)
Plants and Herbs with nootropic abilities

Plant	Type of study	Effect	Mechanism of action	references
Spirulina, maxima	Preclinical study	Ameliorate amyloid-β 1-42 induced cognitive deficits	and the inhibition of the increased	[67]
Spirulina, maxima fermented with lactobacillus		mitigated scopolamine-induced memory loss in mice	phosphorylation of glycogen synthase	[62]
		mitigated scopolamine-induced memory loss in mice	kinase-3β	[68]
planetarium HY-08			Antioxidant activity	
			Antioxidant activity	
Curcuma Longa rhizome extract	Trimethyltin -treated Sprague-Dawley rats	Mitigates deficits in the spatial memory performance and partially inhibits the reduction of the number of pyramidal neurons in the hippocampal cornus ammonis	Anti-inflammatory and antioxidant activity	[69]
Fermented Curcuma longa	Brain cell lines (rat glioma C6 and murine microglia BV2)	Prevention of oxidative stress-induced cell death, decrease in inflammatory markers and attenuation of scopolamine-induced memory loss	Inhibition of acetylcholinesterase activity	[61]
HemoHIM (a standardized formulation of three herbs)	Mice	Attenuation of scopolamine-induced memory loss	Inhibition of acetylcholinesterase activity, increase in acetylcholine (ACh) levels upregulating of Choline Acetyltransferase, Muscarinic ACh receptors and activation of the CREB and ERK signalling pathways	[70]
BioCeuticals Cognition Support Formula® a standardized formulation of three herbs	Proposed Randomized placebo-controlled clinical trial	One card learning and identification tests as co-primary outcome measures of visual short-term memory and attention; mean speed (ms), accuracy (%), and total number of hits, misses, and anticipations) compared with placebo	Combined effects of the three herbs	[73]
Sailuotong a standardized herbal formulation containing *Panax ginseng*, *Ginkgo biloba*, and *Crocus sativus*	Randomized placebo-controlled clinical trial	Resulted in small improvements in working memory	Combined effects of the three herbs	[71]

role in the prevention of several chronic diseases and more recently their ability to enhance memory as well as prevent and/or mitigate age-related cognitive decline. Bioactive compounds are present in small quantities in fruits, vegetables, and whole grains and have been known to have beneficial effects on health that exceed basic nutritional value [75]. They are molecules that can influence energy intake, reduce oxidative stress, and reduce inflammation [76, 77]. The results of numerous epidemiological studies have suggested that the consumption of diets rich in bioactive compounds, such as vitamins, polyphenols (flavonoids), and carotenoids, also positively impact brain health, the development of neurodegenerative diseases and age-related functional decline [75, 78]. A number of preclinical studies have also reported the memory-improving effects of a bioactive compound-rich diet in rodent brain function [79–81]. Bioactive compounds with nootropic abilities (Table 5.2) include bacosides, l-ascorbic acid, resveratrol, curcuminoids, berberine, lycopene, ginkgolides, epigallocatechin-3-gallate, ginsenosides, and loganin.

Bacosides are considered the putative bioactive constituents of *Bacopa monnieri* [82]. Studies have shown that purified bacosides or bacoside-rich extracts of *Bacopa monnieri* are efficient in enhancing acetylcholine synthesis, preventing the accumulation of lipofuscin accumulation and inhibiting lipid peroxidation [83]. Bacosides have been reported to exert their benefits through the suppression of lipid peroxidation, apoptosis, free radical scavenging and regulation of neurotransmitter levels and their activities [84]. Although the nootropic effects of *Bacopa monnieri* have been attributed to its bacoside content, with several over-the-counter supplements containing bacoside-enriched *Bacopa monnieri* extracts, the result of a recent study by Arora et al. [85] revealed that enrichment of *Bacopa monnieri* extracts with bacosides did not confer any significant nootropic benefit when compared to bacoside free extract of *Bacopa monnieri.*

Curcumin is an active constituent of *Curcuma longa*. Curcumin has potent antioxidant and anti-inflammatory properties and has also been linked to the ability of *Curcuma longa* to improve learning and memory function [86]. In recent times, several studies have examined the nootropic effects of curcumin itself in learning and memory paradigms [86–88]. Chang et al. [88] examined the memory-enhancing effects of curcumin in middle-aged and aged rhesus monkeys and observed a significant improvement in working memory in middle-aged compared to aged monkeys, corroborating the efficacy of curcumin in delaying working memory decline. The study also suggested that for effective prevention of age-related memory decline, interventions should commence in the early middle age period. Han et al. [86] examined the effects of curcumin on memory in transgenic mice and reported that dietary supplementation with curcumin was associated with shorter escape latencies and a significant increase in percent alternation when compared with control mice. Also, curcumin supplementation resulted in a decrease in the expression of high-mobility group box-1 protein, Toll-like receptors-4, advanced glycosylation end product-specific receptor and nuclear factor kappa B in the hippocampus of these transgenic mice. ElBini-Dhouib et al. [81] also examined the effects of curcumin in aluminium chloride-induced neurotoxicity and observed that curcumin significantly improved behavioural function, oxidative stress, and inflammatory markers in rats. Ortiz-Romero et al., [87] also examined the impact of the coadministration of curcumin and verapamil on memory domains in Williams-Beuren syndrome (a rare neurodevelopmental disorder characterized by a distinctive cognitive phenotype). The results showed that co-treatment prevented cognitive deficits through the recovery of altered gene expression in the cortex and the reduction of activated microglia [87]. Numerous mechanisms, including free radical scavenging, metal chelation, modulation of enzyme activities, and modulation of gene expression, have been suggested as possible mechanisms through which curcumin acts [86–88].

Green tea contains a large amount of catechins although when fully fermented, catechins undergo enzymatic oxidation to form theaflavins and thearubigins. Catechins like (–)-epigallocatechin-3-gallate (ECCG), (–)-epigallocatechin, (–)-epicatechin-3- gallate, and epicatechin make up more than two-thirds of all tea polyphenols. In brewed green tea, EGCG is the most abundant

TABLE 5.2
Plant Bioactive constituents with nootropic abilities

Plant	Study	Effect	References
Bacoside	200 mg/kg of bacoside was orally administered for 3 months to middle-aged and aged rats	Results suggested that bacosides may act as a potential therapeutic intervention in forestalling the deleterious effects of ageing	[83]
Curcumin	Amyloid β-induced cognitive deficiencies in a rat model of AD	Reduced oxidative stress, improvement of spatial memory and decreased the hallmarks of Aβ in AD	[81]
Curcumin and verapamil	Middle-aged and aged rhesus monkeys	Middle-aged subjects given curcumin exhibited better working memory performance and less neuronal excitability compared to control subjects.	[88]
	Transgenic mice	Shorter escape latencies, significant increase in percent alternation, a decrease in the expression of high-mobility group box-1 protein, Toll-like receptors-4, advanced glycosylation end product-specific receptor and nuclear factor kappa	[86]
	Williams-Beuren syndrome complete deletion (CD) mice	Co-treatment prevented cognitive deficits through the recovery of altered gene expression in the cortex and the reduction of activated microglia	[87]
Epigallocatechin-3-gallate	Aβ 25-35 induced Alzheimer's disease in mice	Decrease in the escape latency period, increase in the time at the target quadrant, decrease in the hyperphosphorylation of Tau, expression and activity of BACE1 expression and increase in acetylcholine levels	[91]
	Anaesthesia induced memory deficits in young mice	Significantly attenuate memory deficit, oxidative stress and cell apoptosis	[93]
	AD mouse model	Long-term oral administration of EGCG ameliorated spatial learning and memory impairments and rescued the reduction in synaptic proteins	[92]
	Williams-Beuren syndrome complete deletion (CD) mice	mitigate short-term memory loss in Williams-Beuren syndrome	[87]
Berberine	3×Tg AD mice	Significant improvement in spatial learning capacity and memory retention, attenuation of tau hyperphosphorylation via modulation of the activity of Akt/glycogen synthase kinase-3β and protein phosphatase 2A, simultaneously impacting cerebral blood flow	[94-96]
	Mice	Chronic administration at 5 mg/kg body weight improved long-term memory acquisition in mice, and increased concentration of parvalbumin-immunoreactive neurons and nerve fibres in the hippocampal cornus ammonis region	[97]
	Diabetic rats	clears amyloid β deposits and ameliorates spatial learning memory impairment through the activation of the cholinergic anti-inflammatory and insulin signalling pathways	[98]

Bilobalide	Rat model of vascular dementia and Alzheimer's dementia	Bilobalide protected against the development cognitive deficits and increased antioxidant status, decreased lipid peroxidation, decreased proinflammation and alleviated neuronal apoptosis in rats with vascular dementia. In rats with Aβ25-35 induced dementia, bilobalide also protected against Aβ25-35 induced learning and memory deficits and attenuated neuronal damage and apoptosis	[101, 102]
Ginkgolide B	Rat model of vascular dementia	Improved learning and memory function in rats with vascular dementia through the reduction of TLR4/NF-κB mediated neuroinflammation	[99]
Ginkgolide K	APP/PS1 mice	reversed memory impairment and downregulated the expression of mitochondrial calcium (Ca^{2+}) uniporter	[100]
Ginsenoside Rg5 or Rh3	Scopolamine-induced cognitive impairment in mice	reversed scopolamine e induced decrease in latency time, increased spontaneous alteration and shortened escape latencies. Decreased acetylcholinesterase activity, modulated the expression of hippocampal brain-derived neurotrophic factor and the activation of cAMP response element-binding protein	[103]
Ginsenoside Rg2	$A\beta_{25-35}$-induced memory deficits	Improved learning and memory through its ability to increase the ratio of Bcl-2/Bax ratio, attenuate the cleavage of caspase-3, and enhance the phosphorylation of Akt	Cui et al., 2021
Ginsenoside Rh1	Mice	Long-term administration enhanced learning and memory in the passive avoidance test and Morris water maze through the enhancement of hippocampal dentate gyrus cell survival and the upregulation of brain-derived neurotrophic factor expression	Hou et al., 2014

catechin, accounting for approximately 50% – 80% of the total catechin concentration [89, 90]. Recent understanding highlights the antioxidant and anti-inflammatory properties of EGCG with suggestions that it has the capacity to mitigate memory impairment and neurodegeneration in dementia [91]. Guo et al. [92] also examined the effectiveness of oral administration of ECCG in mitigating AD-related memory impairment. In the study using the senescence-accelerated mouse prone 8 AD model, it was observed that long-term oral administration of EGCG ameliorated spatial learning and memory impairments and rescued the reduction in synaptic proteins observed in the AD mouse model [92]. Also, in mice with anaesthesia-induced memory impairments, ECCG was observed to significantly attenuate memory deficit, oxidative stress and cell apoptosis [93]. ECCG has also been shown to mitigate short-term memory loss in Williams-Beuren syndrome [87]. Berberine is another polyphenolic compound with reported nootropic properties. Berberine is an isoquinoline alkaloid that has been isolated from *Rhizoma coptidis* (a Chinese herb with traditional medicinal value, particularly in the management of memory impairment) or *Berberis sibirica* [94–96]. Chen and colleagues observed that berberine was associated with significant improvement in spatial learning capacity and memory retention and the attenuation of tau hyperphosphorylation via modulation of the activity of Akt/glycogen synthase kinase-3β and protein phosphatase 2A [94, 95]. Szalak et al. [97] also observed that chronic administration of berberine at 5 mg/kg body weight improved long-term memory acquisition in mice and increased the concentration of parvalbumin-immunoreactive neurons and nerve fibres in the hippocampal cornus ammonis region. Berberine has also been shown to mitigate spatial learning and working memory deficits in diabetic rats through the modulation of cholinergic inflammatory pathways [98]. Ginkgolides are biologically active diterpenoids found in *Ginkgo biloba*. A number of these terpene lactones have been isolated, although ginkgolide B is the most abundant bioactive constituent of the *Ginkgo biloba* extract [99]. The nootropic effects of ginkgolide B and K have also been reported [99, 100]. Huang et al. [99] examined the effects of ginkgolide B in a rodent model of vascular dementia and observed that ginkgolide B significantly improved learning and memory function in rats with vascular dementia through the reduction of TLR4/NF-κB mediated neuroinflammation. Liu et al. [100] examined the effects of ginkgolide K on memory function in APP/PS1 mice and observed that ginkgolide K reversed memory impairment and downregulated the expression of mitochondrial calcium (Ca^{2+}) uniporter in the brain of these mice. Bilobalide is another bioactive constituent of *gingko biloba* leaves, which has documented nootropic effects [101, 102]. The results of two different studies that evaluated the effects of bilobalide on cognition in a mouse model of AD [101] and a mouse model of vascular dementia [102] observed that bilobalide protected against the development of cognitive deficits and improved antioxidant status. It also decreased lipid peroxidation, decreased proinflammation, and alleviated neuronal apoptosis in rats with vascular dementia [102]. In rats with Aβ25-35 induced dementia, bilobalide also protected against Aβ25-35 induced learning and memory deficits and attenuated neuronal damage and apoptosis [101]. Ginsenosides are a group of steroid glycosides and triterpene saponin, which are the main bioactive constituents of the *Panax ginseng* plant. The ginsenosides are largely believed to be responsible for the memory-enhancing effects of *Panax ginseng* [103]. A number of studies have reported the nootropic effects of different forms of ginsenosides [103]. Kim et al. [103] observed that ginsenosides Rg5 or Rh3 reversed the scopolamine-induced decrease in latency time in the passive avoidance test, increased spontaneous alteration in the Y-maze task, and shortened escape latencies in the Morris water maze. The effects on memory were attributed to its ability to decrease acetylcholinesterase activity, modulate the expression of hippocampal brain-derived neurotrophic factor and the activation of cAMP response element-binding protein [103]. Hou et al. [104] also reported that long-term administration of ginsenoside Rh1 enhanced learning and memory in the passive avoidance test and Morris water maze through the enhancement of hippocampal dentate gyrus cell survival and the upregulation of brain-derived neurotrophic factor expression [104]. The effects of ginsenoside Rg2 against $A\beta_{25-35}$-induced memory deficits were also examined by Cui et al. [105]; they observed that ginsenoside

Rg2 significantly improved learning and memory through its ability to increase the ratio of Bcl-2/Bax, attenuate the cleavage of caspase-3, and enhance the phosphorylation of Akt [105].

Although the nootropic abilities and neuroprotective effects of a number of these bioactive compounds have been documented, their low bioavailability and limited transport ability across the blood-brain barrier are among the drawbacks of their use [106, 107]. In the last few years, there has been increasing awareness regarding the use of polymeric nanoparticles and liposomes to improve the bioavailability of a number of these bioactive constituents [108]. They have also been beneficial in achieving targeted delivery. A number of the polyphenols have also been conjugated with receptors indirectly to enhance delivery across the blood-brain barrier [107, 108]. While there is still a dearth of scientific literature on the use of nanoparticle strategies for delivering bioactive compounds, bioactive constituents like curcumin and ginsenosides have been researched with promising results [109, 110]. Maiti et al. [110] examined the effects of solid lipid curcumin particles (SLCP) in one-year-old 5xFAD-and age-matched wild-type mice. The results showed that the acute daily intraperitoneal injection of SLCP was associated with improved permeability and better neuroprotective effect, suggesting that this could also enhance cognitive function. Hoppe et al [79] examined the effects of acute intraperitoneal injection of curcumin-loaded lipid-core nanocapsules (Cur-LNC) in rodents following Aβ(1-42)-induced cognitive impairment. Results showed that Cur-LNC reversed cognitive deficits at significantly lower doses compared to free curcumin [79]. In another study, curcumin-loaded nanostructured lipid carrier significantly increased brain concentration of curcumin and was associated with enhanced cognition and reduced oxidative stress in the hippocampal tissue [80].

5.5 CONCLUSION

Nootropic herbs and herbal products research and applications have come some way in recent years, especially with the identification and isolation of the bioactive molecules and the deciphering of their modes of action. It is also noteworthy to acknowledge the fact that it is possible that herbal nootropics may soon take centre stage in the management or possible prevention of human cognitive disorders. While challenges relating to delivery to target tissues are now being tackled, we still have a long way to go in this aspect. However, it brings a great sense of relief to know that a number of the answers we desperately seek have always been with us, and all we need to do is face the right direction for this.

TERMS AND ABBREVIATIONS

Aβ	amyloid β
AD	Alzheimer's disease
Akt	protein kinase B
APP/PSI	mice mouse/human amyloid precursor protein and a mutant human presenilin 1 mouse
Bcl-2	B cell lymphoma protein 2 family
Bax	Bcl-2-associated X protein
cAMP	cyclic adenosine monophosphate
ECCG	epigallocatechin-3-gallate
PKC	protein kinase C
PI3K	phosphatidylinositol 3-kinase

REFERENCES

1. Duckworth AL, Quirk A, Gallop R, et al.Cognitive and noncognitive predictors of success. Proc Natl Acad Sci U S A. 2019;116(47):23499–23504. doi: 10.1073/pnas.1910510116.
2. Lexico Online Dictionary Oxford University Press and Dictionary.com. Retrieved 6 May 2020.

3. Froestl W, Muhs A, Pfeifer A. Cognitive enhancers (nootropics). Part 1: drugs interacting with receptors. J Alzheimers Dis. 2012;32(4):793–887. doi: 10.3233/JAD-2012-121186.
4. Schifano F, Catalani V, Sharif S, Napoletano F, Corkery JM, Arillotta D, Fergus S, Vento A, Guirguis A. Benefits and harms of 'Smart Drugs' (Nootropics) in healthy individuals. *Drugs*. 2022 Apr;82(6):633–647. doi: 10.1007/s40265-022-01701-7.
5. Deary IJ. Intelligence. *Annu Rev Psychol*. 2012;63:453–482. doi: 10.1146/annurev-psych-120710-100353.
6. Titz C, Karbach J. Working memory and executive functions: effects of training on academic achievement. *Psychol Res*. 2014;78(6):852–868. doi: 10.1007/s00426-013-0537-1.
7. Karbach J, Strobach T, Schubert T. Adaptive working-memory training benefits reading, but not mathematics in middle childhood. *Child Neuropsychol*. 2015;21(3):285–301. doi: 10.1080/09297049.2014.899336.
8. Martin A, Booth JN, Laird Y, et al. Physical activity, diet and other behavioural interventions for improving cognition and school achievement in children and adolescents with obesity or overweight. *Cochrane Database Syst Rev*. 2018;1(1):CD009728. doi: 10.1002/14651858.CD009728.pub3.
9. Johann VE, Karbach J. Effects of game-based and standard executive control training on cognitive and academic abilities in elementary school children. *Dev Sci*. 2020;23(4):e12866. doi: 10.1111/desc.12866.
10. Froestl W, Muhs A, Pfeifer A. Cognitive enhancers (nootropics). Part 2: drugs interacting with enzymes. *Update. J Alzheimers Dis*. 2014;42(1):1–68. doi: 10.3233/JAD-140402.
11. Froestl W, Pfeifer A, Muhs A. Cognitive enhancers (Nootropics). Part 3: drugs interacting with targets other than receptors or enzymes. Disease-modifying drugs. Update 2014. *J Alzheimers Dis*. 2014;42(4):1079–1149. doi: 10.3233/JAD-141206.
12. Onaolapo AY, Obelawo AY, Onaolapo OJ. Brain ageing, cognition and diet: A review of the emerging roles of food-based nootropics in mitigating age-related memory decline. *Current Aging Science* 2019; 12 (1):2–14.
13. Sharif S, Guirguis A, Fergus S, Schifano F. The use and impact of cognitive enhancers among university students: A systematic review. *Brain Sci*. 2021 Mar 10;11(3):355. doi: 10.3390/brainsci11030355. PMID: 33802176; PMCID: PMC8000838.
14. Halder S, Anand U, Nandy S, et al. Herbal drugs and natural bioactive products as potential therapeutics: A review on pro-cognitives and brain boosters perspectives. *Saudi Pharm J*. 2021;29(8):879–907. doi: 10.1016/j.jsps.2021.07.003.
15. Crespo-Bujosa HB, Suárez Rodríguez RL Nootropics: Phytochemicals with neuroprotective and neurocognitive enhancing properties. *Eur J Clin Exp Med* 2019 (15):17 http://dx.doi.org/10.15584/ejcem.2019.3.9
16. Kulkarni R, Girish KJ, Kumar A. Nootropic herbs (Medhya Rasayana) in Ayurveda: An update. *Pharmacogn Rev*. 2012;6(12):147–153. doi: 10.4103/0973-7847.99949.
17. Agarwa P, Sharma B, Fatima A, et al. An update on Ayurvedic herb Convolvulus pluricaulis Choisy. *Asian Pac J Trop Biomed*. 2014;4(3):245–252. doi: 10.1016/S2221-1691(14)60240-9.
18. Onaolapo AY, Onaolapo OJ. Nutraceuticals and diet-based phytochemicals in type 2 diabetes mellitus: From whole food to components with defined roles and mechanisms. *Curr Diabetes Rev*. 2019a;16(1):12–25. doi: 10.2174/1573399814666181031103930.
19. Mollica A, Zengin G, Locatelli M, et al. An assessment of the nutraceutical potential of Juglans regia L. leaf powder in diabetic rats. *Food Chem Toxicol*. 2017;107(Pt B):554–564. doi: 10.1016/j.fct.2017.03.056.
20. Farooqui AA, Farooqui T, Madan A, Ong JH, Ong WY. Ayurvedic medicine for the treatment of dementia: Mechanistic aspects. *Evid Based Complement Alternat Med*. 2018;2018:2481076. doi: 10.1155/2018/2481076.
21. Onaolapo AY, Onaolapo OJ, Chapter 9 – Herbal Beverages and Brain Function in Health and Disease, Editor(s): Alexandru Mihai Grumezescu, Alina Maria Holban, *Functional and Medicinal Beverages*, Academic Press, 2019b, Pages 313–349, ISBN 9780128163979.
22. Onaolapo AY, Onaolapo OJ. African plants with antidiabetic potentials: beyond glycaemic control to central nervous system benefits. *Curr Diabetes Rev*. 2020;16(5):419–437. doi: 10.2174/1573399815666191106104941.

23. Lorca C, Mulet M, Arévalo-Caro C, et al. Plant-derived nootropics and human cognition: A systematic review. *Crit Rev Food Sci Nutr*. 2022:1–25. doi: 10.1080/10408398.2021.2021137.
24. Olofinnade AT, Alawode A, Onaolapo AY, et al. Lepidium meyenii supplemented diet modulates neurobehavioral and biochemical parameters in mice fed high-fat high-sugar diet. *Endocr Metab Immune Disord Drug Targets*. 2021a;21(7):1333–1343. doi: 10.2174/1871530320666200821155005.
25. Olofinnade AT, Onaolapo AY, Onaolapo OJ, et al. Corylus avellana L. modulates neurobehaviour and brain chemistry following high-fat diet. *Front Biosci (Landmark Ed)*. 2021b;26(3):537–551. doi: 10.2741/4906.
26. Olofinnade AT, Onaolapo AY, Onaolapo OJ, et al. Hazelnut modulates neurobehaviour and ameliorates ageing-induced oxidative stress, and caspase-3-mediated apoptosis in mice. *Curr Aging Sci*. 2021c;14(2):154–162.
27. Olofinnade AT, Onaolapo AY, Stefanucci A, et al. Cucumeropsis mannii reverses high-fat diet induced metabolic derangement and oxidative stress. *Front Biosci (Elite Ed)*. 2021d;13(1):54–76. doi: 10.2741/872.
28. Ekor M. The growing use of herbal medicines: issues relating to adverse reactions and challenges in monitoring safety. *Front Pharmacol*. 2014;4:177. doi: 10.3389/fphar.2013.00177.
29. Onaolapo OJ, Odeniyi AO, Onaolapo AY. Parkinson's disease: Is there a role for dietary and herbal supplements? *CNS Neurol Disord Drug Targets*. 2021;20(4):343–365. doi: 10.2174/1871527320666210218082954.
30. Adams M, Gmunder F, Hamburger M. Plants traditionally used in age related brain disorders—A survey of ethnobotanical literature. *J. Ethnopharmacol*. 2007;113:363–381. doi: 10.1016/j.jep.2007.07.016.
31. Kumar H, More SV, Han SD, et al. Promising therapeutics with natural bioactive compounds for improving learning and memory--a review of randomized trials. *Molecules*. 2012;17(9):10503–10539.
32. Zieneldien T, Kim J, Cao C. The multifaceted role of neuroprotective plants in Alzheimer's disease treatment. *Geriatrics (Basel)*. 2022;7(2):24. doi: 10.3390/geriatrics7020024.
33. Bello-Medina PC, González-Franco DA, Vargas-Rodríguez I, et al. Oxidative stress, the immune response, synaptic plasticity, and cognition in transgenic models of Alzheimer disease. *Neurologia*. 2019; 19, 30109–30104. doi: 10.1016/j.nrl.2019.06.002.
34. Anand U, Jacobo-Herrera N, Altemimi A, et al. A comprehensive review on medicinal plants as antimicrobial therapeutics: potential avenues of biocompatible drug discovery. *Metabolites*. 2019;9(11):258. doi: 10.3390/metabo9110258.
35. Howes MJ, Houghton PJ. Plants used in Chinese and Indian traditional medicine for improvement of memory and cognitive function. *Pharmacol Biochem Behav*. 2003;75(3):513–527. doi: 10.1016/s0091-3057(03)00128-x.
36. Koithan M, Farrell C. Indigenous native American healing traditions. *J Nurse Pract*. 2010;6(6):477–478. doi: 10.1016/j.nurpra.2010.03.016.
37. Natarajan S, Shunmugiah KP, Kasi PD. Plants traditionally used in age-related brain disorders (dementia): an ethnopharmacological survey. *Pharm Biol*. 2013;51(4):492–523. doi: 10.3109/13880209.2012.738423.
38. Shirbeigi L, Dalfardi B, Abolhassanzadeh Z, et al. Dementia etiologies and remedies in traditional Persian medicine; A review of medicinal plants and phytochemistry. *Curr Drug Metab*. 2018;19(5):414–423. doi: 10.2174/1389200218666170810170124.
39. Li JM, Zhao Y, Sun Y, et al. Potential effect of herbal antidepressants on cognitive deficit: Pharmacological activity and possible molecular mechanism. *J Ethnopharmacol*. 2020;257: 112830. doi: 10.1016/j.jep.2020.112830.
40. Malík M, Tlustoš P. Nootropic herbs, shrubs, and trees as potential cognitive enhancers. *Plants (Basel)*. 2023 Mar 18;12(6):1364. doi: 10.3390/plants12061364. PMID: 36987052; PMCID: PMC10056569.
41. Roe AL, Venkataraman A. The safety and efficacy of botanicals with nootropic effects. *Curr Neuropharmacol*. 2021;19(9):1442–1467. doi: 10.2174/157015
42. Mancini E, Beglinger C, Drewe J, et al. Green tea effects on cognition, mood and human brain function: A systematic review. *Phytomedicine*. 2017;34:26–37. doi: 10.1016/j.phymed.2017.07.008.
43. Ahn JW, Kim S, Ko S, Kim YH, Jeong JH, Chung S. Modified (-)-gallocatechin gallate-enriched green tea extract rescues age-related cognitive deficits by restoring hippocampal synaptic plasticity. *Biochem Biophys Rep*. 2022;29:101201. doi: 10.1016/j.bbrep.2022.101201.

44. Baba Y, Inagaki S, Nakagawa S, et al. Effects of daily matcha and caffeine intake on mild acute psychological stress-related cognitive function in middle-aged and older adults: A randomized placebo-controlled study. *Nutrients*. 2021;13(5):1700. doi: 10.3390/nu13051700.
45. Baba Y, Kaneko T, Takihara T. Matcha consumption maintains attentional function following a mild acute psychological stress without affecting a feeling of fatigue: A randomized placebo-controlled study in young adults. *Nutr Res*. 2021;88:44–52. doi: 10.1016/j.nutres.2020.12.024.
46. Gray NE, Zweig JA, Caruso M, et al. Centella asiatica increases hippocampal synaptic density and improves memory and executive function in aged mice. *Brain Behav*. 2018a;8(7):e01024. doi: 10.1002/brb3.1024.
47. Gray NE, Zweig JA, Caruso M, et al. Centella asiatica attenuates hippocampal mitochondrial dysfunction and improves memory and executive function in β-amyloid overexpressing mice. *Mol Cell Neurosci*. 2018b;93:1–9. doi: 10.1016/j.mcn.2018.09.002.
48. Pase MP, Kean J, Sarris J, et al. The cognitive-enhancing effects of Bacopa monnieri: a systematic review of randomized, controlled human clinical trials. *J Altern Complement Med*. 2012;18(7):647–652. doi: 10.1089/acm.2011.0367.
49. Brimson JM, Brimson S, Prasanth MI, et al. The effectiveness of Bacopa monnieri (Linn.) Wettst. as a nootropic, neuroprotective, or antidepressant supplement: analysis of the available clinical data. *Sci Rep*. 2021;11(1):596. doi: 10.1038/s41598-020-80045-2.
50. Uabundit N, Wattanathorn J, Mucimapura S, Ingkaninan K. Cognitive enhancement and neuroprotective effects of Bacopa monnieri in Alzheimer's disease model. *J Ethnopharmacol*. 2010;127(1):26–31. doi: 10.1016/j.jep.2009.09.056.
51. Le XT, Nguyet Pham HT, Van Nguyen T, et al. Protective effects of Bacopa monnieri on ischemia-induced cognitive deficits in mice: the possible contribution of bacopaside I and underlying mechanism. *J Ethnopharmacol*. 2015164:37–45. doi: 10.1016/j.jep.2015.01.041.
52. Pham HTN, Tran HN, Nguyen PT, et al. *Bacopa monnieri* (L.) Wettst. Extract improves memory performance via promotion of neurogenesis in the hippocampal dentate gyrus of adolescent Mice. *Int J Mol Sci*. 2020;21(9):3365. doi: 10.3390/ijms21093365.
53. Kongkeaw C, Dilokthornsakul P, Thanarangsarit P, et al. Meta-analysis of randomized controlled trials on cognitive effects of Bacopa monnieri extract. *J Ethnopharmacol*. 2014;151(1):528–535. doi: 10.1016/j.jep.2013.11.008.
54. Calabrese C, Gregory WL, Leo M, et al. Effects of a standardized Bacopa monnieri extract on cognitive performance, anxiety, and depression in the elderly: a randomized, double-blind, placebo-controlled trial. *J Altern Complement Med*. 2008;14(6):707–713. doi: 10.1089/acm.2008.0018.
55. Khalifa AE. Hypericum perforatum as a nootropic drug: enhancement of retrieval memory of a passive avoidance conditioning paradigm in mice. *J Ethnopharmacol*. 2001;76(1):49–57. doi: 10.1016/s0378-8741(01)00210-0.
56. Ben-Eliezer D, Yechiam E. Hypericum perforatum as a cognitive enhancer in rodents: A meta-analysis. *Sci Rep*. 2016;6:35700. doi: 10.1038/srep35700.
57. Yechiam E, Ben-Eliezer D, Ashby NJS, et al. The acute effect of Hypericum perforatum on short-term memory in healthy adults. *Psychopharmacology (Berl)*. 2019 Feb;236(2):613–623. doi: 10.1007/s00213-018-5088-0.
58. Puttarak P, Dilokthornsakul P, Saokaew S, et al. Effects of Centella asiatica (L.) Urb. on cognitive function and mood related outcomes: A systematic review and meta-analysis. *Sci Rep*. 2017;7(1):10646. doi: 10.1038/s41598-017-09823-9.
59. Saenghong N, Wattanathorn J, Muchimapura S, et al. Zingiber officinale improves cognitive function of the middle-aged healthy women. *Evid Based Complement Alternat Med*. 2012;2012:383062. doi: 10.1155/2012/383062.
60. Kandlur A, Satyamoorthy K, Gangadharan G. Oxidative stress in cognitive and epigenetic aging: A retrospective glance. *Front Mol Neurosci*. 2020;13(41). doi: 10.3389/fnmol.2020.00041.
61. Eun CS, Lim JS, Lee J, et al. The protective effect of fermented Curcuma longa L. on memory dysfunction in oxidative stress-induced C6 gliomal cells, proinflammatory-activated BV2 microglial cells, and scopolamine-induced amnesia model in mice. *BMC Complement Altern Med*. 2017;17(1):367. doi: 10.1186/s12906-017-1880-3.

62. Choi WY, Kang DH, Lee HY. Effect of fermented *Spirulina maxima* extract on cognitive-enhancing activities in mice with scopolamine-induced dementia. *Evid Based Complement Alternat Med.* 2018;2018:7218504. doi: 10.1155/2018/7218504.
63. Wahid S, Alqahtani A, Khan RA. Nootropic effects of *C. melo* and *C. lanatus* seed extracts. *Biomed Res Int.* 2020;2020:8823038. doi: 10.1155/2020/8823038.
64. Valotto Neto LJ, Reverete de Araujo M, Moretti Junior RC, Mendes Machado N, Joshi RK, Dos Santos Buglio D, Barbalho Lamas C, Direito R, Fornari Laurindo L, Tanaka M, Barbalho SM. Investigating the neuroprotective and cognitive-enhancing effects of *Bacopa monnieri*: A systematic review focused on inflammation, oxidative stress, mitochondrial dysfunction, and apoptosis. *Antioxidants (Basel)*. 2024 Mar 25;13(4):393. doi: 10.3390/antiox13040393.
65. Talebi M, İlgün S, Ebrahimi V, et al. Zingiber officinale ameliorates Alzheimer's disease and cognitive impairments: Lessons from preclinical studies. *Biomed Pharmacother.* 2021;133:111088. doi: 10.1016/j.biopha.2020.111088.
66. Sutalangka C, Wattanathorn J. Neuroprotective and cognitive-enhancing effects of the combined extract of Cyperus rotundus and Zingiber officinale. *BMC Complement Altern Med.* 2017;17(1):135. doi: 10.1186/s12906-017-1632-4.
67. Koh EJ, Kim KJ, Song JH, et al. Spirulina maxima extract ameliorates learning and memory impairments via inhibiting GSK-3β phosphorylation induced by intracerebroventricular injection of amyloid-β 1–42 in Mice. *Int J Mol Sci.* 2017;18(11):2401. doi: 10.3390/ijms18112401.
68. Ghanbari A, Vafaei AA, Naghibi nasab FS, et al. Spirulina microalgae improves memory deficit induced by scopolamine in male pup rats: Role of oxidative stress. *South African J Bot.* 2019;127: 220–225.
69. Yuliani S, Mustofa, Partadiredja G. Turmeric (Curcuma longa L.) extract may prevent the deterioration of spatial memory and the deficit of estimated total number of hippocampal pyramidal cells of trimethyltin-exposed rats. *Drug Chem Toxicol.* 2018;41(1):62–71. doi: 10.1080/01480545.2017.1293087.
70. Kim SK, Kwon DA, Kim YS, et al. Standardized extract (HemoHIM) protects against scopolamine-induced amnesia in a murine model. *Evid Based Complement Alternat Med.* 2021;2021:8884243. doi: 10.1155/2021/8884243.
71. Steiner GZ, Bensoussan A, Liu J, et al. Study protocol for a randomised, double-blind, placebo-controlled 12-week pilot phase II trial of Sailuotong (SLT) for cognitive function in older adults with mild cognitive impairment. *Trials.* 2018;19(1):522. doi: 10.1186/s13063-018-2912-0.
72. Steiner GZ, Yeung A, Liu JX, et al. The effect of Sailuotong (SLT) on neurocognitive and cardiovascular function in healthy adults: a randomised, double-blind, placebo controlled crossover pilot trial. *BMC Complement Altern Med.* 2016;16(15). doi: 10.1186/s12906-016-0989-0.
73. Cave AE, Chang DH, Münch GW, et al. Efficacy of Cognition Support Formula® on cognitive function in older adults with subjective cognitive impairment: a protocol for a 26-week, randomised, double-blind, placebo-controlled trial. *Trials.* 2019;20(1):345. doi: 10.1186/s13063-019-3431-3.
74. Kasture SB, Kasture VS, Joshua AI, et al. Nootropic activity of BacoMind™, an enriched phytochemical composition from Bacopa monnieri. *J Nat Remedies.* 2007;7(1):166–173.
75. Santos DI, Saraiva JMA, Vicente AA, et al., 2 – Methods for determining bioavailability and bioaccessibility of bioactive compounds and nutrients, Editor(s): Francisco J. Barba, Jorge Manuel Alexandre Saraiva, Giancarlo Cravotto, José M. Lorenzo, *Innovative Thermal and Non-Thermal Processing, Bioaccessibility and Bioavailability of Nutrients and Bioactive Compounds*, Woodhead Publishing, 2019, Pages 23–54.
76. Kamiloglu S, Tomas M, Ozdal T, et al. Chapter 2 – Bioactive component analysis, Editor(s): Charis M. Galanakis, *Innovative Food Analysis*, Academic Press, 2021, Pages 41–65, ISBN 9780128194935.
77. Adefegha SA, Oboh G, Oluokun OO, Chapter 11 – Food bioactives: the food image behind the curtain of health promotion and prevention against several degenerative diseases, Editor(s): Atta-ur-Rahman, *Studies in Natural Products Chemistry*, Elsevier, Volume 72, 2022, Pages 391–421, ISSN 1572–5995.
78. Schaffer S, Halliwell B. Do polyphenols enter the brain and does it matter? Some theoretical and practical considerations. *Genes Nutr.* 2012;7(2):99–109. doi: 10.1007/s12263-011-0255-5.

79. Hoppe JB, Coradini K, Frozza RL, et al. Free and nanoencapsulated curcumin suppress β-amyloid-induced cognitive impairments in rats: involvement of BDNF and Akt/GSK-3β signaling pathway. *Neurobiol Learn Mem*. 2013;106:134–144. doi: 10.1016/j.nlm.2013.08.001.
80. Sadegh Malvajerd S, Izadi Z, Azadi A, et al. Neuroprotective potential of curcumin-loaded nanostructured lipid carrier in an animal model of Alzheimer's disease: Behavioral and biochemical evidence. *J Alzheimers Dis*. 2019;69(3):671–686. doi: 10.3233/JAD-190083.
81. ELBini-Dhouib I, Doghri R, Ellefi A, et al. curcumin attenuated neurotoxicity in sporadic animal model of Alzheimer's disease. *Molecules*. 2021;26(10):3011. doi: 10.3390/molecules26103011.
82. Majumdar S, Basu A, Paul P, et al. Bacosides and Neuroprotection,Editor(s): Ramawat K., Mérillon JM., *Natural Products,* Springer, Berlin, Heidelberg. 2013 https://doi.org/10.1007/978-3-642-22144-6_157
83. Rastogi M, Ojha RP, Prabu PC, et al. Prevention of age-associated neurodegeneration and promotion of healthy brain ageing in female Wistar rats by long term use of bacosides. *Biogerontology*. 2012 Apr;13(2):183–195. doi: 10.1007/s10522-011-9367-y.
84. Sekhar VC, Viswanathan G, Baby S. Insights into the molecular aspects of neuroprotective Bacoside A and Bacopaside I. *Curr Neuropharmacol*. 2019;17(5):438–446. doi: 10.2174/1570159X16666180419123022.
85. Arora R, Kumar R, Agarwal A, et al. Bacosides enrichment does not improve the anti-amnesic effect of *Bacopa monnieri*: behavioural and biochemical evidences. *Nat Prod Res*. 2021 Dec;35(24):6054–6057. doi: 10.1080/14786419.2020.1817919.
86. Han Y, Chen R, Lin Q, et al. Curcumin improves memory deficits by inhibiting HMGB1-RAGE/TLR4-NF-κB signalling pathway in APPswe/PS1dE9 transgenic mice hippocampus. *J Cell Mol Med*. 2021;25(18):8947–8956. doi: 10.1111/jcmm.16855.
87. Chang W, Weaver CM, Medalla M, et al. Age-related alterations to working memory and to pyramidal neurons in the prefrontal cortex of rhesus monkeys begin in early middle-age and are partially ameliorated by dietary curcumin. *Neurobiol Aging*. 2022;109:113–124. doi: 10.1016/j.neurobiolaging.2021.09.012.
88. Ortiz-Romero P, González-Simón A, Egea G, et al. Co-treatment with verapamil and curcumin attenuates the behavioral alterations observed in Williams-Beuren syndrome mice by regulation of MAPK pathway and microglia overexpression. *Front Pharmacol*. 2021;12:670785. doi: 10.3389/fphar.2021.670785.
89. Kosińska A, Andlauer W. Chapter 12 – Antioxidant Capacity of Tea: Effect of Processing and Storage, Editor(s): Preedy V., *Processing and Impact on Antioxidants in Beverages*, Academic Press, 2014, Pages 109–120, ISBN 9780124047389.
90. Youn K, Ho C-T, Jun M Multifaceted neuroprotective effects of (-)-epigallocatechin-3-gallate (EGCG) in Alzheimer's disease: an overview of pre-clinical studies focused on β-amyloid peptide. *Food Sci Human Wellness*, 2022;11:483–493, https://doi.org/10.1016/j.fshw.2021.12.006
91. Nan S, Wang P, Zhang Y, et al Epigallocatechin-3-Gallate provides protection against Alzheimer's disease-induced learning and memory impairments in rats. *Drug Des Devel Ther*. 2021;15:2013–2024. doi: 10.2147/DDDT.S289473.
92. Guo Y, Zhao Y, Nan Y, et al. (-)-Epigallocatechin-3-gallate ameliorates memory impairment and rescues the abnormal synaptic protein levels in the frontal cortex and hippocampus in a mouse model of Alzheimer's disease. *Neuroreport*. 2017;28(10):590–597. doi: 10.1097/WNR.0000000000000803.
93. Ding L, Gao X, Hu J et al. (-)Epigallocatechin-3-gallate attenuates anesthesia-induced memory deficit in young mice via modulation of nitric oxide expression. *Mol Med Rep* 2018;18:4813–4820.
94. Huang M, Jiang X, Liang Y, et al. Berberine improves cognitive impairment by promoting autophagic clearance and inhibiting production of β-amyloid in APP/tau/PS1 mouse model of Alzheimer's disease. *Exp Gerontol*. 2017;91:25–33. doi: 10.1016/j.exger.2017.02.004.
95. Chen Y, Chen Y, Liang Y, et al. Berberine mitigates cognitive decline in an Alzheimer's Disease Mouse Model by targeting both tau hyperphosphorylation and autophagic clearance. *Biomed Pharmacother*. 2020;121:109670. doi: 10.1016/j.biopha.2019.109670.
96. Ye C, Liang Y, Chen Y, et al. Berberine improves cognitive impairment by simultaneously impacting cerebral blood flow and β-amyloid accumulation in an APP/tau/PS1 mouse model of Alzheimer's Disease. *Cells*. 2021;10(5):1161. doi: 10.3390/cells10051161.

97. Szalak R, Kukula-Koch W, Matysek M, et al. Effect of berberine isolated from barberry species by centrifugal partition chromatography on memory and the expression of parvalbumin in the mouse hippocampus proper. *Int J Mol Sci*. 2021;22(9):4487. doi: 10.3390/ijms22094487.
98. Wang K, Chen Q, Wu N, et al. Berberine ameliorates spatial learning memory impairment and modulates cholinergic anti-inflammatory pathway in diabetic rats. *Front Pharmacol*. 2019;10:1003. doi: 10.3389/fphar.2019.01003.
99. Huang L, Shi Y, Zhao L. Ginkgolide B alleviates learning and memory impairment in Rats with vascular dementia by reducing neuroinflammation *via* Regulating NF-κB Pathway. *Front Pharmacol*. 2021;12:676392. doi: 10.3389/fphar.2021.676392.
100. Liu H, Li Q, Zhang X, et al. Effect of ginkgolide K on calcium channel activity in Alzheimer's disease. *Exp Ther Med*. 2022;23(6):426. doi: 10.3892/etm.2022.11353.
101. Yin Y, Ren Y, Wu W, et al. Protective effects of bilobalide on Aβ(25–35) induced learning and memory impairments in male rats. *Pharmacol Biochem Behav*. 2013;106:77–84. doi: 10.1016/j.pbb.2013.03.005.
102. Li WZ, Wu WY, Huang H, et al. Protective effect of bilobalide on learning and memory impairment in rats with vascular dementia. *Mol Med Rep*. 2013;8(3):935–941. doi: 10.3892/mmr.2013.1573.
103. Kim EJ, Jung IH, Van Le TK, et al. Ginsenosides Rg5 and Rh3 protect scopolamine-induced memory deficits in mice. *J Ethnopharmacol*. 2013;146(1):294–299. doi: 10.1016/j.jep.2012.12.047.
104. Hou J, Xue J, Lee M, et al. Long-term administration of ginsenoside Rh1 enhances learning and memory by promoting cell survival in the mouse hippocampus. *Int J Mol Med* 2014; 33: 234–240.
105. Cui J, Shan R, Cao Y, et al. Protective effects of ginsenoside Rg2 against memory impairment and neuronal death induced by Aβ25–35 in rats. *J Ethnopharmacol*. 2021;266:113466. doi: 10.1016/j.jep.2020.113466.
106. Onaolapo OJ, Onaolapo AY. Chapter 5 – Caffeinated Beverages, Behavior, and Brain Structure, Editor(s): Grumezescu AM., Holban AM., *Caffeinated and Cocoa Based Beverages*, Woodhead Publishing, 2019c, Pages 163–207 ISBN 9780128158647, https://doi.org/10.1016/B978-0-12-815 864-7.00005-2
107. Mirhadi E, Rezaee M, Malaekeh-Nikouei B. Nano strategies for berberine delivery, a natural alkaloid of Berberis. *Biomed Pharmacother*. 2018;104:465–473. doi: 10.1016/j.biopha.2018.05.067.
108. Zhang Y, Lv C, Zhao G Ways to enhance the bioavailability of polyphenols in the brain: A journey through the blood-brain barrier. *Food Rev Int*. 2021. doi: 10.1080/87559129.2021.1888973.
109. Kim H, Lee JH, Kim JE, et al. Micro-/nano-sized delivery systems of ginsenosides for improved systemic bioavailability. *J Ginseng Res*. 2018 Jul;42(3):361–369. doi: 10.1016/j.jgr.2017.12.003.
110. Maiti P, Paladugu L, Dunbar GL. Solid lipid curcumin particles provide greater anti-amyloid, anti-inflammatory and neuroprotective effects than curcumin in the 5xFAD mouse model of Alzheimer's disease. *BMC Neurosci*. 2018;19(1):7. doi: 10.1186/s12868-018-0406-3.

6 Neuroprotective Potential of Some Traditional Indian Medicinal Plants in Alzheimer's and Parkinson's Disease

Bharti Chaudhry

6.1 INTRODUCTION

Neurodegenerative disorders afflict millions of people worldwide and impact not only individuals, but also families and societies. Neurodegenerative diseases constitute a group of chronic, progressive disorders, characterized by progressive loss of neurons in discrete areas of the central nervous system (CNS), resulting in deficits of specific brain functions performed by the affected CNS region.[1] The disorders of the nervous system, including Alzheimer's disease, Parkinson's disease, amyotrophic lateral sclerosis, multiple sclerosis and Huntington's disease, usually manifest in cognitive decline, dementia, movement dysfunction, and changes in personality and behaviour. Cognitive decline and senile dementia are emerging as serious global health concerns. As per the WHO Global Status Report on the public health response to dementia, September 2021 (www.who.int/publications-detail-redirect/9789240033245), an estimated 55.2 million people worldwide were living with dementia in the year 2019, and this number is expected to rise to 78 million in 2030 and 139 million in 2050, with a corresponding increase in the ageing population.

Alzheimer's disease (AD) is the most common form of dementia afflicting old people, which affects the brain cells, mainly in the hippocampus and cortex, leading to impairment in learning, thinking and memory. Alzheimer's disease is characterized by neuroinflammation, oxidative damage, accumulation of beta-amyloid plaques and neurofibrillary tangles (aggregates of hyperphosphorylated tau proteins) in the brain.[2,3] While excessive deposition of β-amyloid protein, occurs between the neurons (extracellular), tau protein hyperphosphorylation and accumulation occurs inside the neurons (intracellular). Alzheimer's disease disrupts the vital processes of neurons, including communication, metabolism and repair, leading to brain shrinkage (brain atrophy). The dysfunction and loss of cholinergic neurons in the basal forebrain is one of the most prominent components of the neuropathology of Alzheimer's disease.[4] The blood-brain disruption, mitochondrial impairment, hypometabolism, decline in acetylcholine levels, and a reduction in cerebral blood flow have been associated with AD.[5] The progression of AD is also associated with the depletion of neurotrophins that regulate the survival, differentiation and function of neurons.[6] Increasing evidence indicates a pathological synergy between β-amyloid and tau in Alzheimer's disease,[7,8] which may be the reason for the failure of drug therapies targeting either β-amyloid or tau.

Although a combination of genetics, lifestyle and environmental factors are implicated, the greatest risk factor for late-onset Alzheimer's is old age.[9] The percentage of people with Alzheimer's

DOI: 10.1201/9781003300502-6

dementia has been reported to significantly increase with age: 5.0% of people aged 65 to 74, 13.1% of people aged 75 to 84, and 33.2% of people aged 85 or older.[10] The pace at which the symptoms of dementia advance from mild to moderate to severe also differs from person to person. Cholinesterase inhibitors increase the availability of acetylcholine at synapses in the brain and are one of the drug therapies used in the treatment of Alzheimer's disease dementia. Despite several drugs approved by the U.S. Food and Drug Administration (FDA), including donepezil, rivastigmine, galantamine, memantine, and memantine combined with donepezil, there is no cure for Alzheimer's disease, as these drugs are unable to halt or reverse the progression of the disease. (www.nia.nih.gov/health/how-alzheimers-disease-treated)

Parkinson's disease (PD) is the second most common age-related neurodegenerative disorder, characterized by progressive loss of dopaminergic neurons and the presence of aggregates of α-synuclein in the brain.[11] It is a slow progressive disorder that affects movement, muscle control, and balance. The motor symptoms of the disease result from the death of cells in the pars compacta of substantia nigra, a region of the midbrain, leading to a deficit of the neurotransmitter dopamine. The aetiology of the disease is poorly understood, but the aggregates of misfolded α-synuclein proteins into Lewy bodies within the neurons constitute one of the important pathological hallmarks of PD.[12] Complex interactions between genetic, environmental and pathological factors, including neuroinflammation, oxidative stress and mitochondrial dysfunction are associated with the disease.[13] The clinical symptoms of the disease referred to as Parkinsonism include bradykinesia (slowness of movement), resting tremors, hypokinesia (decreased amplitude of movement), postural instability (imbalance and stooped posture), and cogwheel rigidity.[14] The non-motor symptoms of PD include sleep disturbances, cognitive decline with dementia, orthostatic hypotension and depression.[15] Global estimates in 2019 revealed over 8.5 million individuals with PD (www.who.int/news-room/fact-sheets/detail/parkinson-disease). The disease is clinically incurable, and disability and death due to PD are increasing globally.

Levodopa drug, a dopamine precursor, has been widely used for several years in the management of Parkinson's disease, often in combination with carbidopa. Levodopa works by crossing the blood-brain barrier (BBB), allows neurons to make new dopamine, and reduces the symptoms of slowness, stiffness and tremors. However, the drug does not stop or slow the progression of the disease, may cause side effects and may lose effectiveness over time. The adverse effects of the levodopa drug may include nausea, vomiting, dry mouth, dizziness and confusion, hallucinations, dyskinesia and psychosis in some patients. (https://medlineplus.gov/druginfo/meds/a601068.html).

The immune system plays a complex role in the progression of neurodegenerative disorders and exhibits both neurodestructive and neuroprotective responses.[6] The brain is prone to oxidative stress, lipid peroxidation, and chronic neuroinflammation triggered by the innate immune response of CNS plays a central role in the pathogenesis of Alzheimer's and Parkinson's disease. Neuroinflammation leads to a deficiency in neurotrophins, diminishing neuroplasticity.[16] Persistent inflammatory/immune response in the brain causes neurodegeneration as well as facilitates and exacerbates both amyloid and tau core pathologies of Alzheimer's disease.[17] Microglia and astrocytes, both types of glial cells, are the key innate immune cells in the CNS. While glia are essential for regulating homeostasis in the normal brain, their dysfunction contributes to neurodegeneration in many brain diseases, including Alzheimer's[18] and Parkinson's disease.[19] Increased microglial activation and astrocytes have been found in both postmortem AD and PD brains. Over-activated microglia (specialized macrophages in CNS) and increased expression of pro-inflammatory cytokines are common features of Alzheimer's[20] and Parkinson's disease.[21] The activated microglial cells are known to induce an excessive secretion of neurotoxic factors like nitric oxide (NO), reactive oxygen species (ROS), tumour necrosis factor-alpha (TNF-α) and interleukin-1beta (IL-1β), which lead to neuronal damage.

The functional and structural integrity of the BBB is pivotal to maintaining the homeostasis of the brain microenvironment.[22] The BBB is formed by microvascular endothelial cells lining the cerebral capillaries penetrating the brain and spinal cord and plays a critical role in controlling the influx and efflux of biological substances essential for the metabolic activity of the brain and neuronal function. However, the BBB constitutes an impediment to the treatment of neurological disorders, as it hinders the delivery of potential therapeutic drugs and diagnostic substances to the CNS.[23] Various neurodegenerative disease models, including rodents, *Drosophila melanogaster* (fruit fly), and *Caenorhabditis elegans* (nematode), as well as *in vitro* models of the blood-brain barrier, have been used to understand the pathogenesis of neurodegenerative disorders and evaluate the effects of drugs used for treatment.

6.2 THERAPEUTIC POTENTIAL OF INDIAN TRADITIONAL PLANTS

Conventional drugs available for neurodegenerative diseases provide symptomatic relief, do not prevent the progression of disease and may cause adverse effects. Ayurveda is the oldest, and the most practiced traditional system of medicine in India, well known for its preventive, restorative and holistic mode of treatment. Neuroprotective and nootropic medicinal plants used in Ayurveda are natural, safer therapeutics as alternative medicines for the alleviation of neurodegenerative, psychological and psychosomatic disorders. *Rasayana* therapy is a specialized branch of Ayurveda associated with the rejuvenation of both body and mind and promoting healthy longevity.[24] *Rasayana* herbs that are specific to the brain are called *Medhya-Rasayana*, which act as nervine tonics and have been used to enhance cognitive functions like memory, concentration, perception, and learning. Prevention of neuronal cell death and promotion of regeneration of neurons are key factors in mitigating the impact of neurodegenerative disorders. Recent research indicates that tau and amyloid beta proteins implicated in Alzheimer's disease are highly conserved proteins throughout evolution and may have crucial physiological functions. These functions, including neurogenesis, glucose metabolism, learning, memory, and DNA protection, may be lost during AD progression or may be unintentionally disrupted by tau- or amyloid beta-targeting therapies.[25] There is a paradigm shift from cure to prevention and natural treatment remedies. The dynamic balance of the network of pro-inflammatory cytokines and anti-inflammatory cytokines is essential to maintain the physiological function of inflammation in the body. Herbal medicine has been reported to be effective in the treatment of ageing-related diseases in many *in vitro*, *in vivo* and clinical trials.[26] With this backdrop, the article is intended to highlight the anti-inflammatory and neuroprotective potential of five traditional Indian plants in the prevention and inhibition of the progression of Alzheimer's and Parkinson's diseases. (See Figure 6.1)

6.2.1 *Bacopa monnnieri* (L.) Pennell (Family: *Plantaginaceae*)

Bacopa monnieri, commonly called Indian pennywort, water hyssop, or Brahmi, is a small, creeping, perennial herb that inhabits wetlands and muddy shores. It has long been used in the Ayurveda system of medicine as a brain rejuvenating (*Medhya-Rasayana*) drug for its potent nootropic, anxiolytic, anti-depressant, anti-convulsant, adaptogen and neuroprotective properties. It promotes healthy longevity, prevents ageing, sharpens intellect and attenuates mental deficits[24]. The herb has been extensively reviewed and has exhibited significant clinical potential in attenuation of neuroinflammation, anxiety, dementia, cognitive dysfunction, amnesia, and epilepsy.[27,28,29,30,31,32,33,34,35,36,37]

6.2.1.1 Neuroprotection in Alzheimer's Disease

Bacopa monnieri is known to protect the cells in the prefrontal cortex, hippocampus, and striatum against cytotoxicity and DNA damage implicated in Alzheimer's disease. It has been reported to reduce hippocampal β-amyloid deposition, protect the cholinergic neurons and reduce anticholinesterase activity comparable to donepezil, rivastigmine and galantamine.[34,37] Bacosides,

FIGURE 6.1 Representative picture of the five Indian traditional medicinal plants with potential use in Alzheimer's and Parkinson's disease (see text for more details).

the primary nootropic phytochemicals, protect the brain against oxidative damage and age-related cognitive decline via antioxidant neuroprotection. They are known to enhance nerve impulse transmission, repair damaged neurons by upregulating neuronal synthesis, and enhance nitric oxide-mediated cerebral vasodilation, resulting in the improvement of memory.[34,38] The therapeutic efficiency of *Bacopa monnieri* against cognitive impairment and oxidative damage induced by intracerebroventricular injection of streptozotocin in rat models has been demonstrated by amelioration of behavioural parameters, learning, memory and pathological abnormalities in the hippocampus through biochemical, immunohistochemical and histopathological assays.[39] The tea infusion and alkaloid extracts of *Bacopa monnieri* as well as bacoside A have been shown to significantly inhibit the release of pro-inflammatory cytokines, TNF-α and IL-6 from N9 microglial cell line in *in vitro* studies.[40] The herb was also found to inhibit the enzymes associated with inflammation in the brain (caspase-1 and 3 and matrix metalloproteinase-3) in the cell-free assay.

6.2.1.2 Neuroprotection in Parkinson's Disease

The neuroprotective and neurorescue effects of oral administration of *Bacopa monnieri* at 40 mg/kg body weight have been reported in 1-methyl-4-phenyl-1, 2, 3, 6 tetrahydropyridine (MPTP) induced Parkinson's disease mice models.[41] The results exhibit a significant recovery in behavioural parameters, dopamine level, glutathione level, lipid peroxides and nitrite level in PD models. Another study on the administration of *Bacopa monnieri* in Parkinson's disease animal models, resulted in suppression of levels of pro-inflammatory cytokines, decreased levels of α-synuclein, and reduction of ROS in the brain.[42] A recent review based on pre-clinical, evidence-based studies of *Bacopa monnieri* against Parkinson's disease has documented its neuroprotective potential through reduction of neuronal oxidative stress, degeneration of dopaminergic neurons, mitochondrial dysfunction as well as inflammation. It has also been shown to inhibit aggregation of alpha-synuclein and improve cognitive and learning behaviour.[43] Oral supplementation of *Bacopa monnieri* was found to significantly reverse the Paraquat-induced reduction of exploratory behaviour, gait abnormalities and motor impairment in mice models.[44] The herb also prevented the reduction in dopamine levels and reversed cholinergic activity in brain regions important for motor (striatum) pathology.

6.2.2 *Centella asiatica* (L.) Urban (Family: *Apiaceae*)

Centella asiatica, commonly called Asian pennywort, Gotu Kola, is a slender trailing herb with orbicular-reniform leaves, that exhibit anti-inflammatory, antioxidant, anti-convulsant, anti-depressant, anti-hypertensive, neuroprotective, cardioprotective, hepatoprotective, immunostimulant, antiviral, antibacterial, antifungal, anticancer and wound-healing properties. In Ayurveda, it is one of the main herbs for revitalization of the nerves and brain cells, widely used for cognitive dysfunction, memory enhancement, nervine disorders, blood purification, and promoting longevity. It protects the neurons from apoptosis and aids in the treatment of depression, anxiety, Alzheimer's and Parkinson's disease.[45]

The neuroprotective and cognitive-enhancing effects of *Centella asiatica* have been corroborated in several *in vivo* and *in vitro* model systems.[46,47,48,49,50] *Centella asiatica* is known to act by providing protection from oxidative stress, inhibition of acetylcholinesterase, enhancement of dendrite arborisation, increased synaptogenesis and protection against β-amyloid formation in the brain.[51, 52] *Centella* has been shown to improve cognitive function by improved learning and memory performance in aged mice and increased expression of mitochondrial and antioxidant response genes in the brain and liver of both young and old animals.[48] The neurotropic, neuroprotective properties are attributed to the active triterpene compounds, Asiatocoside, Madecassoside, and Asiatic acid. These phytochemicals have also been found to be highly BBB permeable, in BBB model of primary porcine brain endothelial cells.[53]

6.2.2.1 Neuroprotection in Alzheimer's Disease

A recent study demonstrated that three months of treatment with water extracts of *Centella asiatica* resulted in improved spatial and contextual memory as well as executive function in 5xFAD mouse model of amyloid beta accumulation.[54] The study also reported increased antioxidant gene expression and a decrease in Aβ plaque: amyloid-beta plaque burden. The neuroprotective effects of Asiaticoside in a β-amyloid (Aβ)-induced Alzheimer's disease rat models have been investigated by lateral intracerebroventricular injection of Aβ 1-42 oligomers.[55] Asiaticoside treatment resulted in improved learning and memory function, decreased Aβ deposition in the hippocampus and reversal of pathological changes. It was also found to reduce the elevated levels of pro-inflammatory cytokines, interleukin-6 and tumour necrosis factor-α, in the brains of Aβ-treated rats. In addition, it resulted in a significant decrease in the expression of caspases-3, whereas the expression of B-cell lymphoma-2 was significantly increased, in the Aβ-treated rats. The protective effects on Aβ-induced AD pathology, have been associated with the alleviation of mitochondrial injuries, anti-inflammatory activities, and influence on the expression levels of apoptosis-associated proteins.[55]

6.2.2.2 Neuroprotection in Parkinson's Disease

Mitochondrial dysfunction and ROS generation are known to cause dopaminergic neurodegeneration in Parkinson's disease. A standardized extract of *Centella asiatica* ECa233 (30 mg/kg) given orally in rotenone-induced Parkinsonism rats has demonstrated a significant increase in distances as well as a higher number and intensity of dopaminergic neurons in the substantia nigra and striatum in immunohistological staining studies.[56] ECa233 (30 mg/kg) also protected against mitochondrial complex I inhibition, decreased malondialdehyde levels, as well as increased antioxidant enzymes, superoxide dismutase and catalase expression in the brain tissue. Asiatic acid has been found to protect the dopaminergic neurons from neuroinflammation by suppressing Mitochondrial ROS production in LPS-induced BV2 microglia cells and 1-methyl-4-phenyl-pyridine (MPP+)-induced SH-SY5Y cells, in an *in vitro* model of PD.[57] Asiatic acid also inhibited activation of NLRP3 inflammasome in microglia cells and improved mitochondrial dysfunction in SH-SY5Y cells suggesting a promising clinical use of Asiatic acid for PD therapy.[57]

6.2.3 *Curcuma longa* L. (Family: *Zingiberaceae*)

Curcuma longa, commonly known as turmeric, is a perennial herb native to tropical South Asia. Curcumin, the most active constituent (hydrophobic polyphenol) in the rhizome, imparts the bright yellow colour as well as the potent antioxidant, anti-inflammatory, anti-carcinogenic, anticoagulant, antidiabetic, antiviral, antibacterial, antifungal, hypotensive and hypolipidaemic properties. It is widely acclaimed in Ayurveda for multiple health benefits against metabolic disorders, neurodegenerative diseases, cardiovascular diseases and inhibition of carcinogenesis.[58,59,60,24] Many studies in rodents and humans have shown that curcumin crosses the BBB.[61,62]

6.2.3.1 Neuroprotection in Alzheimer's Disease

Curcuma longa has long been known to slow the progression of Alzheimer's disease by removing amyloid-β plaques from the brain. Several *in vivo* and *in vitro* studies support the potential of curcumin in the modification of Alzheimer's disease pathology by inhibiting the formation and promoting the disaggregation of amyloid-β plaques, attenuation of hyperphosphorylated tau, binding copper, inhibition of acetylcholinesterase, as well as modification of microglial activity.[63] The beneficial effects of curcumin on cognition in Alzheimer's disease have been reviewed and documented in several pre-clinical and some clinical studies.[64] Curcumin has been shown to prevent cognitive impairment via an increase in the activity of antioxidant enzymes, activation of brain-derived neurotrophic factor-signalling, neutralization of cholinergic dysfunction[65], as well as downregulation of biomarkers of inflammation and increased neurogenesis.[64] Neurouroimaging studies support the potential of curcumin to reduce Aβ deposits in the brain[66] A recent systematic review to study the effects of curcumin in experimental models (*in vivo* and *in vitro*) of Alzheimer's disease revealed that curcumin reverses neurotoxic damage by preventing β-amyloid aggregation and tau protein hyperphosphorylation, as well as inhibiting oxidative stress, apoptosis and inflammation.[67] It has also been shown to decrease ROS levels and improve memory and cognitive deficits in animals with AD. Curcumin treatments of 50, 100 and 200 mg/ kg administered repeatedly (for seven consecutive days, once a day) to amyloid-beta-1-42 induced rodent models of Alzheimer's disease, resulted in improvement of cognitive functions in a dose-dependent manner, accompanied by increased BDNF (Brain-derived neurotrophic factor) and elevated levels of phosphorylated ERK (Extracellular signal-regulating kinase) in the hippocampus.[68] The neurotrophin BDNF plays an important role in neuronal structure and function in the CNS, involved in neuronal connectivity, synaptic plasticity and memory.[69] BDNF deficits have been reported in postmortem studies of AD patients as well as experimental AD animal models, and higher levels of BDNF are associated with a slower rate of cognitive decline in AD patients.

6.2.3.2 Neuroprotection in Parkinson's Disease

Dysregulation of salsonilol (an endogenous toxin) has been reported to contribute to the progressive development of PD. The neuroprotective potential and anti-apoptotic impact of the extract of *Curcuma longa* have been exhibited in salsonilol-induced toxic conditions in SH-SY5Y human neuroblastoma cells.[70] A systematic review of the literature on the efficacy of curcumin in toxin-based animal models of Parkinson's disease has revealed that the antioxidant property of curcumin protects substantia nigrans neurons and improves striatal dopamine levels. Curcumin was also found to be effective in reducing neuronal apoptosis, indicating its therapeutic potential as a neuroprotective drug for human PD patients.[71]

6.2.4 *Withania somnifera* (L.) Dunal. (Family: *Solanaceae*)

Withania somnifera, commonly called Indian Ginseng, Indian winter cherry, or Ashwagandha, is a potent rejuvenating drug of Ayurveda, the roots of which are used to relieve mental and physical fatigue and increase longevity. It is a powerful anti-inflammatory, antioxidant, and anti-arthritic, used as an analgesic for musculoskeletal disorders, a nervine tonic for Parkinson's disease, Alzheimer's disease, epilepsy, nervous exhaustion, anxiety, multiple sclerosis, senile dementia, and debility in old age.[45] It has been reported to mitigate various neurological disorders, including Alzheimer's and Parkinson's disease, Huntington's disease, tardive dyskinesia, ischaemic stroke, amyotrophic lateral sclerosis, bipolar disorder, schizophrenia, depression, and anxiety.[72,73,74] The anxiolytic potential of *Withania somnifera* has been shown in a randomized, double-blind, placebo-controlled trial of mildly anxious, 60 healthy adults via its moderating effects on the hypothalamus-pituitary-adrenal axis.[75] The predominant active phytochemicals are withanolides (steroidal lactones), including Withaferin A and Withanolide D. The protective effects of *Withania somnifera* extracts and/or its major constituents are reported to be mediated by restoration of mitochondrial and endothelial function, mitigation of apoptosis, inflammation and oxidative stress mechanisms.[72]

6.2.4.1 Neuroprotection in Alzheimer's Disease

The root extract of *Withania somnifera* has been shown to reverse β-amyloid and HIV induced neuro-pathogenesis in human neuronal cell lines.[76] A significant improvement in both immediate and general memory after treatment with Ashwagandha root extract has been seen in a clinical study of adults with mild cognitive impairment. The study also reported improvement in executive function, attention and information processing speed.[77] The neuroprotective effects of Withaferin A in AD are mediated by its ability to reduce β-amyloid plaque aggregation, tau protein accumulation, inhibition of acetylcholinesterase and butyrylcholinesterase activities, regulation of heat shock proteins, inhibition of oxidative and inflammatory components.[78] Withanolide A, Withanoside IV and Withanoside VI are known to induce axon regeneration and synaptic reconstruction in *in vitro* and *in vivo* model studies.[79] Withanone (WS-2) isolated from the roots of *Withania somnifera* and administered once daily to Wistar rats for a duration of 21 days was found to significantly improve cognitive impairment by inhibiting amyloid β-42. It also resulted in attenuation of elevated levels of pro-inflammatory cytokines, and enhancement of acetylcholine and glutathione activity.[80]

6.2.4.2 Neuroprotection in Parkinson's Disease

The neuroprotective role of *Withania somnifera* root extract in maneb-paraquat (herbicides) induced mouse model of Parkinsonism has been reported to counteract oxidative stress and improve behavioural, anatomical and biochemical deformities.[81] A standardized extract of *Withania somnifera* has been shown to markedly offset rotenone-induced locomotor deficits, oxidative impairment and neurotoxicity in *Drosophila melanogaster* model.[82] The protective effects of *Withania somnifera* root extract (KSM-66) ranging from 0.25 to 1 mg/ml on 6-hydroxydopamine (6-OHDA)-induced toxicity in the human neuroblastoma SH-SY5Y cell line have been shown to result in an increased viability of SH-SY5Y cells and an increased glutathione peroxidase and thioltransferase activity.[83] The extract also modulated oxidative response proteins, peroxiredoxin-I, VGF and vimentin proteins upon 6-OHDA pre/post treatments, controlled redox regulation via S-glutathionylation and increased mitochondrial Adenosine Triphosphate (ATP) levels.

6.2.5 *Nardostachys jatamansi* (D. Don) DC. (Family: *Caprifoliaceae*)

Nardostachys jatamansi, commonly called Spikenard or Jatamansi is a perennial herbaceous species endemic to the Himalayan ranges of India, China, Bhutan, Nepal, and Myanmar. The roots/ rhizomes of *Nardostachys jatamansi* are extensively used in Ayurveda and Unani systems of medicine as a sedative for the treatment of insomnia, convulsions, epilepsy and other central nervous system disorders. The plant has powerful antioxidant, antimicrobial, hepatoprotective, cardioprotective, vasodilator, hypotensive, hypolipidaemic, and bronchodilator properties.[84] It has several secondary metabolites, including terpenoids, lignans, aristolane- type and nardosinone-type sesquiterpenoids that exhibit anti- neuroinflammatory properties.[85] *Nardostachys jatamansi* has recently been assessed for The International Union for Conservation of Nature (IUCN) Red List of Threatened Species in 2020 and has been listed as critically endangered due to over-harvesting. (www.iucnredlist.org/species/50126627/88304158).

6.2.5.1 Neuroprotection in Alzheimer's Disease

The anti-Alzheimer's action of the ethanolic root extract of *Nardostachys jatamansi* has been studied against β amyloid toxicity *in vitro* using a mammalian cell culture system and *in vivo* using a *Drosophila* AD model.[86] The extract was found to inhibit Aβ-induced cell death in SH-SY5Y cells as well as suppressed cell death in the brain of Aβ42-expressing flies. It also rescued the neurological phenotypes of the model flies and reduced glial cell number, reactive oxygen species level, ERK phosphorylation, and nitric oxide level without affecting Aβ accumulation. The neuroprotective effects of Jatamansinol, a pyranocoumarin isolated from *Nardostachys jatamansi* have been studied against Aβ42 protein-induced neurotoxicity *Drosophila* AD model. Jatamansinol supplemented food was found to extend the lifespan, improve locomotor activity, enhance learning and memory, and reduce $A\beta_{42}$ protein expression in AD flies. Jatamansinol was also found to boost antioxidant enzyme activities, prevent $A\beta_{42}$ protein-induced oxidative stress, ameliorate eye degeneration, and inhibit cholinesterase activities in the AD *Drosophila* flies, suggesting its therapeutic potential against AD.[87]

6.2.5.2 Neuroprotection in Parkinson's Disease

The ethanolic extracts of *Nardostachys jatamansi* have been found to attenuate neuronal injury in 6-hydroxydopamine-induced (6-OHDA) rat models of Parkinson's.[88] The deficits in locomotor activity and coordination due to 6-OHDA injections were restored significantly by the plant extract in a dose-dependent manner. The dopamine levels and the activities of glutathione- dependent enzymes, catalase and superoxide dismutase in striatum reduced by lesioning were also restored as is evident in neurochemical, behavioural and immunohistochemical studies.[88] In a recent study, Nardosinone was identified to be the most effective anti-PD compound in *Nardostachys* based on the results of UHPLC-LTQ-Orbitrap-MS analysis of rat brains as well as behavioural tests in mouse PD models.[89] Network pharmacology analysis revealed dopamine receptor D2 (DRD2) as the key target of the nardosinone-PD interaction network, also verified by molecular docking and Western blotting techniques. The results demonstrated that nardosinone could restore the expression level of DRD2 that was decreased in the brain tissue of PD mice and that nardosinone may reduce the motor and cognitive symptoms in the animal PD model. Nardosinone was found to reverse Rotenone-Induced Motor and Cognitive Impairment in PD mice.[90] A combined transcriptome and proteome analysis also suggested that the anti-PD target of nardosinone is the slc38a2 gene and may involve the GABAergic synaptic pathway and cAMP-signalling pathway.

6.3 CONCLUSION

Neurodegenerative diseases are known to induce irreversible destruction of neuronal networks in the central nervous system, resulting in permanent functional impairment.[79] The pathological mechanisms of Alzheimer's and Parkinson's disease are still elusive, and no therapeutics are available for prevention, mitigation or reversal of the progression of the diseases. They progress slowly over a span of ten years, and the onset of neurodegeneration precedes clinical manifestations by several years. Lack of awareness and understanding of dementia results in stigmatization and barriers to diagnosis and care (WHO, 2021). Despite recent research on molecular mechanisms of neuroprotection, the current drugs provide only symptomatic treatment. Medicinal plants, with their long traditional usage, safety, efficacy and affordability, can prove to be promising therapeutic alternatives.

The role of chronic neuroinflammation in diseases of the central nervous system is well established. The identification of signal pathways that control the initiation, progression and termination of the inflammatory reaction and unravelling the complexities of communication between the immune, endocrine and nervous systems constitute important strategies for the treatment of neurodegenerative disorders.[1] Early prevention and management of neuroinflammation may inhibit the progression of neurodegenerative disorders. The fundamental doctrine of prevention of disease in Ayurveda and the use of *Rasayana* herbs is incomparable. The traditional Indian plants reviewed in this chapter, Indian pennywort, Asiatic pennywort, turmeric, Indian Ginseng, and Spikenard, have exhibited significant therapeutic potential against Parkinson's and Alzheimer's disease through various mechanisms including antioxidant neuroprotection, suppression of pro-inflammatory cytokines, acetylcholinesterase inhibition, β-amyloid reduction, decreased levels of α-synuclein, attenuation of hyperphosphorylated tau, enhancement of dendrite arborisation, increased synaptogenesis, mitochondrial stress regulation, increased cerebral blood flow, enhanced nerve impulse transmission and neurotransmitter modulation. They can protect the neurons against oxidative stress and inflammation, prevent apoptosis induced by neurotoxic agents, and promote neurogenesis through the upregulation of neurotrophic factors. Neurocognitive disorders are multifactorial pathological conditions, and the integration of conventional medicine with traditional pharmacological therapies can prevent chronic inflammation, restore healthy neuroendocrine functions, provide protection against geriatric cognitive decline and promote healthy longevity. However, the lack of large multicenter clinical trials, substitution and adulteration of herbs, limited bio-availability of drugs, optimum dosage for efficacy in humans and drug interactions with current medications remain major challenges in the acceptance of herbal drugs.

REFERENCES

1. Gao, H.M., Hong, J.S. 2008. Why neurodegenerative diseases are progressive: uncontrolled inflammation drives disease progression. *Trends in Immunology* 29(8):357–365.
2. Cheignon, C., Tomas, M., Bonnefont- Rousselot, D., Bureau, C., Collin, F. 2018. Oxidative stress and the amyloid beta peptide in Alzheimer's disease. *Redox Biology* 14, 450–464.
3. Teixeira, J.P., Castro, A.A., Soares, F.V., Cunha, E.F.F. Ramalho, T.C. 2019. Future therapeutic perspectives into the Alzheimer's disease targeting the oxidative stress hypothesis. *Molecules* 24, 4410.

4. Hampel, H., Mesulam, M.M., Cuello, A.C. 2018. The cholinergic system in the pathophysiology and treatment of Alzheimer's disease. *Brain* Jul 1;141(7):1917–1933.
5. Mendiola-Precoma, J., Berumen, L.C., Padilla, K., Garcia-Alcocer, G. 2016. Therapies for prevention and treatment of Alzheimer's disease. *BioMed Research International* 2589276. https://doi.org/10.1155/2016/2589276
6. Cooper, E.L., Ma, M.J. 2017. Alzheimer Disease: Clues from traditional and complementary medicine. *Journal of Traditional and Complementary Medicine* Jan 16;7(4):380–385.
7. Busche, M.A., Hyman, B.T. 2020. Synergy between amyloid-β and tau in Alzheimer's disease. *Nature Neuroscience* Oct;23(10):1183–1193.
8. Koller, E.J., Ibanez, K.R., Vo, Q. et al. 2022. Combinatorial model of amyloid β and tau reveals synergy between amyloid deposits and tangle formation. *Neuropathology and Applied Neurobiology* Feb;48(2):e12779.
9. National Institute on Aging. What causes Alzheimer's disease. Available at www.nia.nih.gov/health/what-causes alzheimers-disease. Accessed July16, 2022.
10. Alzheimer's disease facts and figures. *Alzheimers Dement.* 2022 Apr;18(4):700–789. doi: 10.1002/alz.12638.
11. Zeng, X.S., Geng, W.S., Jia, J.J., Chen, L., Zhang, P.P. 2018. Cellular and molecular basis of neurodegeneration in Parkinson disease. *Frontiers in Aging Neuroscience* Apr 17;10:109.
12. Poewe, W., Seppi, K., Marini, K., Mahlknecht, P. 2020. New hopes for disease modification in Parkinson's Disease. *Neuropharmacology* 2020 Jul;171:108085.
13. Kalia, L.V. and Lang, A.E. 2015. Parkinson's disease. *Lancet* 386:896–912.
14. Reich, S.G., Savitt, JM. 2019. Parkinson's disease. *Medical Clinics of North America* Mar;103(2):337–350. doi: 10.1016/j.mcna.2018.10.014.
15. Poewe, W., Seppi K., Tanner, C.M. et al. 2017. Parkinson disease. *Nature Reviews Disease Primers* Mar 23;3:17013.
16. Benedetto S., Müller, L., Wenger, E., Düzel, S., Pawelec, G. 2017. Contribution of neuroinflammation and immunity to brain aging and the mitigating effects of physical and cognitive interventions. *Neuroscience and Biobehavioral Reviews* Apr;75:114–128.
17. Kinney, J.W., Bemiller, S.M., Murtishaw, A.S., Leisgang, A.M., Salazar, A.M., Lamb, B.T. 2018. Inflammation as a central mechanism in Alzheimer's disease. *Alzheimers & Dementia* (NY) Sep 6;4:575–590.
18. Hansen D.V., Hanson J.E., Sheng M. 2018. Microglia in Alzheimer's disease. *Journal of Cell Biology* Feb 5;217(2):459–472.
19. Kam, T.I., Hinkle, J.T., Dawson, T.M., Dawson, V.L. 2020. Microglia and astrocyte dysfunction in Parkinson's disease. *Neurobiology of Disease* 2020 Oct;144:105028.
20. Wang, W.Y., Tan, M.S., Yu, J.T., Tan, L. 2015. Role of pro-inflammatory cytokines released from microglia in Alzheimer's disease. *Annals of Translational Medicine* 3(10):136.
21. Tanaka, J. 2020. Favorable and unfavorable roles of microglia and macrophages in the pathologic central nervous system. *Neuroimmunol Neuroinflammation* 7:73–91.
22. Kadry, H., Noorani, B., Cucullo, L. 2020. A blood-brain barrier overview on structure, function, impairment, and biomarkers of integrity. *Fluids and Barriers of the CNS* Nov 18;17(1):69.
23. Dominguez, A., Suarez-Merino, B., Goni-De-Cerio, B. 2014. Nanoparticles and blood-brain barrier: the key to central nervous system diseases. *Journal of Nanoscience and Nanotechnology* 14:766–779.
24. Chaudhry, B. 2021. Key rasayana plants of ayurveda. *Environment & We, An International Journal of Science and Technology* 16:53–73.
25. Kent, S.A., Spires-Jones, T.L., Durrant, C.S. 2020. The physiological roles of tau and Aβ: Implications for Alzheimer's disease pathology and therapeutics. *Acta Neuropathologica* Oct;140(4):417–447.
26. Liu Y, Weng W, Gao R, Liu Y. 2019. New insights for cellular and molecular mechanisms of aging and aging-related diseases: Herbal medicine as potential therapeutic approach. *Oxidative Medicine and Cellular Longevity* Dec 12;2019:4598167.
27. Russo, A., Borrelli, F. 2005. *Bacopa monnieri*, a reputed nootropic plant: an overview. *Phytomedicine* 12(4):305–317.
28. Calabrese, C., Gregory, W.L, Leo, M., Kraemer, D., Bone, K., Oken, B. 2008. Effects of a standardized *Bacopa monnieri* extract on cognitive performance, anxiety, and depression in the Elderly: A

randomized, double-blind, placebo-controlled trial. *Journal of Alternative and Complementary Medicine* 14,707–713.

29. Pase, M.P., Kean, J., Sarris, J., Neale, C., Scholey, A.B., Stough, C. 2012. The cognitive-enhancing effects of *Bacopa*: a systematic review of randomized, controlled human clinical trials. *Journal of Alternative and Complementary Medicine* 18,647–652.
30. Aguiar, S., Borowski, T. 2013. Neuropharmacological review of the nootropic herb *Bacopa monnieri*. *Rejuvenation Research* 16(4):313–326.
31. Kongkeaw, C., Dilokthornsakul, P., Thanarangsarit, P., Limpeanchob, N., Norman Scholfield C. 2014. Meta-analysis of randomized controlled trials on cognitive effects of *Bacopa monnieri* extract. *Journal of Ethnopharmacology* 151(1):528–535.
32. Stough, C., Singh, H., Zangara, A. 2015. Mechanisms, efficacy, and safety of *Bacopa monnieri* (Brahmi) for cognitive and brain enhancement. *Evidence-Based Complementary and Alternative Medicine: eCAM* 1–2. https://doi.org/10.1155/2015/717605
33. Kean, J.D., Downey, L.A., Stough, C. 2016. A systematic review of the Ayurvedic medicinal herb *Bacopa monnieri* in child and adolescent populations. *Complementary Therapies in Medicine* 29:56–62.
34. Chaudhari, K.S., Tiwari, N.R., Tiwari, R.R., Sharma, R.S. 2017. Neurocognitive effect of nootropic drug Brahmi (*Bacopa monnieri*) in Alzheimer's disease. *Annals of Neurosciences* 24(2):11–122.
35. Manap, A.S., Vijayabalan,. S., Madhavan, P. 2019. *Bacopa monnieri*, a neuroprotective lead in Alzheimer disease: a review on its properties, mechanisms of action, and preclinical and clinical studies. *Drug Target Insights* 13: 1177392819866412. https://doi.org/10.1177/1177392819866412.
36. Sukumaran, N.P., Amalraj, A., Gopi, S. 2019. Neuropharmacological and cognitive effects of *Bacopa monnieri* (L.) Wettst – A review on its mechanistic aspects. *Complementary Therapies in Medicine* Jun;44:68–82.
37. Shalini, V.T., Neelakanta, S.J., Sriranjini, J.S. 2021. Neuroprotection with *Bacopa monnieri* A review of experimental evidence. *Molecular Biology Reports.* Mar;48(3):2653–2668.
38. Simpson, T., Pase, M., Stough, C. 2015. *Bacopa monnieri* as an antioxidant therapy to reduce oxidative stress in the aging brain. *Evidence Based Complementary and Alternative Medicine: eCAM,* 615384. https://doi.org/10.1155/2015/615384
39. Khan, M.B., Ahmad, M., Ahmad, S. *et al.* 2015. Bacopa *monnieri* ameliorates cognitive impairment and neurodegeneration induced by intracerebroventricular-streptozotocin in rat: behavioral, biochemical, immunohistochemical and histopathological evidences. *Metabolic Brain Disease* Feb;30(1):115–127.
40. Nemetchek, M.D., Stierle, A.A., Stierle, D.B., Lurie, D.I. 2017. The Ayurvedic plant *Bacopa monnieri* inhibits inflammatory pathways in the brain. *Journal of Ethnopharmacology* 197:92–100.
41. Singh, B., Pandey, S., Rumman, M. *et al.* 2021. Neuroprotective and neurorescue mode of action of *Bacopa* monnieri (L.) Wettst in 1-Methyl-4-phenyl-1,2,3,6-tetrahydropyridine-induced Parkinson's disease: An in silico and in vivo study. *Frontiers in Pharmacology* 12:616413. https://doi.org/10.3389/fphar.2021.616413
42. Singh, B., Pandey, S., Rumman, M. *et al.* 2020. Neuroprotective effects of *Bacopa monnieri* in Parkinson's disease model. *Metabolic Brain Disease* 35:517–525.
43. Goyal, A., Gopika, S., Kumar, A., Garabadu, D. 2022. A comprehensive review on preclinical evidence based neuroprotective potential of *Bacopa monnieri* against Parkinson's disease. *Current Drug Targets* Mar 16. doi: 10.2174/1389450123666220316091734.
44. Krishna, G., Hosamani, R., Muralidhara. 2019. *Bacopa monnieri* supplements offset paraquat-induced behavioral phenotype and brain oxidative pathways in mice. *Central Nervous System Agents in Medicinal Chemistry* 19:57–66.
45. Chaudhry, B. 2019. *A handbook of common medicinal plants used in Ayurveda*. Kojo Press, Delhi, India.
46. Soumyanath A, Zhong Y.P, Henson E, Wadsworth T, Bishop J, Gold B.G, Quinn J.F. 2012. *Centella asiatica* extract improves behavioral deficits in a mouse model of Alzheimer's disease: Investigation of a possible mechanism of action. *International Journal of Alzheimers Disease* 2012:381974. doi: 10.1155/2012/381974.

47. Gray, N.E., Sampath, H., Zweig, J.A., Quinn, J.F., Soumyanath, A. 2015. *Centella asiatica* attenuates amyloid-ß-induced oxidative stress and mitochondrial dysfunction. *Journal of Alzheimers Disease* 45(3):933–946.
48. Gray, N.E., Harris, C.J., Quinn, J.F., Soumyanath, A. 2016. *Centella asiatica* modulates antioxidant and mitochondrial pathways and improves cognitive function in mice. *Journal of Ethnopharmacology* 180:78–86.
49. Gray, N.E., Zweig, J.A., Murchison, C. et al. 2017. *Centella asiatica* attenuates Aβ-induced neurodegenerative spine loss and dendritic simplification. *Neuroscience Letters* 2017 Apr 12;646:24–29.
50. Wong, J.H., Muthuraju, S., Reza, F. et al. 2019. Differential expression of entorhinal cortex and hippocampal subfields α-amino-3-hydroxy-5-methyl-4-isoxazolepropionic acid (AMPA) and N-methyl-D-aspartate (NMDA) receptors enhanced learning and memory of rats following administration of *Centella asiatica*. *Biomedicine and Pharmacotherapy* Feb;110:168–180.
51. Lokanathan, Y., Omar, N., Puzi, N.A.A., Saim,. A., Idrus, R.H. 2016. Recent updates in neuroprotective and neuroregenerative potential of *Centella asiatica*. *Malaysian Journal of Medical Sciences* 23(1):4–14.
52. Gray, N.E., Magana, A.A., Lak,. P. 2018. *Centella asiatica* – Phytochemistry and mechanisms of neuroprotection and cognitive enhancement. *Phytochemistry Reviews* 17(1):161–194.
53. Hanapi, N.A., Mohamad Arshad, A.S., Abdullah, J.M., Tengku Muhammad, T.S., Yusof, S.R. 2021. Blood-brain barrier permeability of asiaticoside, madecassoside and asiatic acid in porcine brain endothelial cell model. *Journal of Pharmaceutical Sciences* Feb;110(2):698–706.
54. Zweig, J.A., Brandes, M.S., Brumbach, B.H. et al. 2021. Prolonged treatment with *Centella asiatica* improves memory, reduces amyloid-β pathology, and activates NRF2-regulated antioxidant response pathway in 5xFAD Mice. *Journal of Alzheimers Disease* 81(4):1453–1468.
55. Zhang, Z., Li, X., Li, D. et al. 2017. Asiaticoside ameliorates β-amyloid-induced learning and memory deficits in rats by inhibiting mitochondrial apoptosis and reducing inflammatory factors. *Experimental and Therapeutic Medicine* Feb;13(2):413–420.
56. Teerapattarakan, N., Benya-Aphikul, H., Tansawat, R., Wanakhachornkrai, O., Tantisira, M.H., Rodsiri, R. 2018. Neuroprotective effect of a standardized extract of *Centella asiatica* ECa233 in rotenone-induced parkinsonism rats. *Phytomedicine* May 15;44:65–73.
57. Chen, D., Zhang, X.Y., Sun, J., Cong, Q.J. et al. 2019. Asiatic acid protects dopaminergic neurons from neuroinflammation by suppressing mitochondrial Ros production. *Biomolecules and Therapeutics* (Seoul) Apr 10;27(5):442–449.
58. Aggarwal, B.B., Harikumar, K.B. 2009. Potential therapeutic effects of Curcumin, the anti-inflammatory agent, against neurodegenerative, cardiovascular, pulmonary, metabolic, autoimmune and neoplastic diseases. *International Journal of Biochemistry and Cell Biology* 41(1):40–59.
59. Hewlings, S.J., Kalman, D.S. 2017. Curcumin: A review of its effects on human health. *Foods* 6(10):92.
60. Flores, G. 2017. *Curcuma longa* L. extract improves the cortical neural connectivity during the aging process. *Neural Regeneration Research* 12(6):875–880.
61. Dende, C., Meena, J., Nagarajan, P. et al. 2017. Nanocurcumin is superior to native curcumin in preventing degenerative changes in Experimental Cerebral Malaria. *Scientific Reports* 7(1):10062.
62. Reddy, P.H., Manczak, M., Yin, X., et al. 2018. Protective effects of indian spice curcumin against amyloid-β in Alzheimer's disease. *Journal of Alzheimers Disease* 61(3):843–866.
63. Tang, M., Taghibiglou, C. 2017. The mechanisms of action of curcumin in Alzheimer:s disease. *Journal of Alzheimer's Disease* 58(4):1003–1016.
64. Voulgaropoulou, S.D., van Amelsvoort, T.A.M.J., Prickaerts, J., Vingerhoets, C. 2019. The effect of curcumin on cognition in Alzheimer's disease and healthy aging: A systematic review of pre-clinical and clinical studies. *Brain Research* Dec 15; 1725:146476. https://doi.org/10.1016/j.brainres.2019.146476
65. Wu, X., Chen, H., Huang, C. et al. 2017. Curcumin attenuates surgery-induced cognitive dysfunction in aged mice. *Metabolic Brain Disease* 32:789–798.
66. Small, G.W., Siddarth, P., Li, Z.P. et al. 2018. Memory and brain amyloid and tau effects of a bioavailable form of curcumin in nondemented adults: a double-blind, placebo-controlled 18-month trial. *American Journal of Geriatric Psychiatry* 26(3):266–277.

67. Costa, I. M., Freire, M., de Paiva Cavalcanti, J., de Araújo, D. P. 2019. Supplementation with *Curcuma longa* reverses neurotoxic and behavioral damage in models of Alzheimer's disease: A Systematic Review. *Current Neuropharmacology* 17(5):406–421.
68. Zhang, L., Fang, Y., Xu,Y. 2015. Curcumin improves amyloid β-peptide (1–42) induced spatial memory deficits through BDNF-ERK signaling pathway. *PLoS One* Jun 26;10(6):e0131525. doi: 10.1371/journal.pone.0131525.
69. Lu, B., Nagappan, G., Lu, Y. 2014. BDNF and synaptic plasticity, cognitive function, and dysfunction. *Handbook of Experimental Pharmacology* 2014; 220:223–250. https://doi.org/10.1007/978-3-642-45106-5_9
70. Ma, X.W., Guo, R.Y. 2017. Dose-dependent effect of *Curcuma longa* for the treatment of Parkinson's disease. *Experimental and Therapeutic Medicine* 13(5):1799–1805.
71. Wang, X.S., Zhang, Z.R., Zhang, M.M., Sun, M.X., Wang, W.W., Xie, C.L. 2017. Neuroprotective properties of curcumin in toxin-base animal models of Parkinson's disease: a systematic experiment literatures review. *BMC Complementary and Alternative Medicine* Aug 17;17(1):412.
72. Dar, N.J., Ahmad, M. 2020. Neurodegenerative diseases and *Withania somnifera* (L.): An update. *Journal of Ethnopharmacology* 28, 256:112769.
73. Syed, A.A., Reza, M.I., Singh, P., Thombre, G.K., Gayen, J.R. 2021. *Withania somnifera* in neurological disorders: Ethnopharmacological evidence, mechanism of action and its progress in delivery systems. *Current Drug Metabolism* 22(7):561–571.
74. Paul, S., Chakraborty, S., Anand, U., Dey, S. et al. 2021. *Withania somnifera* (L.) Dunal (Ashwagandha): A comprehensive review on ethnopharmacology, pharmacotherapeutics, biomedicinal and toxicological aspects. *Biomedicine & Pharmacotherapy* Nov; 143:112175
75. Lopresti, A.L., Smith, S.J., Malvi, H., Kodgule, R. 2019. An investigation into the stress-relieving and pharmacological actions of an ashwagandha (*Withania somnifera*) extract: A randomized, double-blind, placebo-controlled study. *Medicine (Baltimore)* Sep;98(37):e17186.
76. Kurapati, K.R.V., Atluri, V.S.R., Samikkannu, T., Nair, M.P.N. 2013. Ashwagandha (*Withania somnifera*) reverses β-amyloid induced toxicity in human neuronal cells: Implications in HIV-associated neurocognitive disorders (HAND). *PLOS One* 8(10).
77. Choudhary, D., Bhattacharya, S. Bose, S. 2017. Efficacy and safety of Aswagandha (*Withania somnifera* (L.) Dunal) root extract in improving memory and cognitive functions. *Journal of Dietary Supplements* 14(6):599–612.
78. Das, R.; Rauf, A.; Akhter, S. et al. 2021. Role of Withaferin A and its derivatives in the management of Alzheimer's disease: Recent trends and future perspectives. *Molecules 26*, 3696. https://doi.org/10.3390/
79. Kuboyama, C. Tohda, K. Komatsu. 2014. Effects of Ashwagandha (roots of *Withania somnifera*) on neurodegenerative diseases. *Biological and Pharmaceutical Bulletin* 37: 892–897.
80. Pandey, A., Bani, S., Dutt, P. et al. 2018. Multifunctional neuroprotective effect of Withanone, a compound from *Withania somnifera* roots in alleviating cognitive dysfunction. *Cytokine* 2018 Feb; 102:211–221.
81. Prakash, J., Yadav, S.K., Chouhan, S., Singh, S.P. 2013. Neuroprotective role of *Withania somnifera* root extract in maneb-paraquat induced mouse model of parkinsonism. *Neurochemical Research* 38(5), 972–980.
82. Manjunath, M.J., Muralidhara. 2015. Standardized extract of *Withania somnifera* (Ashwagandha) markedly offsets rotenone-induced locomotor deficits, oxidative impairments and neurotoxicity in *Drosophila melanogaster*. *Journal of Food Science and Technology* Apr;52(4):1971–1981.
83. Wongtrakul, J., Thongtan, T., Kumrapich, B., Saisawang, C., Ketterman, A.J. 2021. Neuroprotective effects of *Withania somnifera* in the SH-SY5Y Parkinson cell model. *Heliyon* Oct 13;7(10):e08172. https://doi.org/10.1016/j.heliyon.2021.e08172
84. Sahu, R., Dhongade, H.J., Pandey, A. et al. 2016. Medicinal properties of *Nardostachys jatamansi* (A Review). *Oriental Journal of Chemistry* 32:859–866.
85. Kim, K.W., Yoon, C.S., Park, S.J., et al. 2021 Chemical analysis of the ingredients of 20% aqueous ethanol extract of *Nardostachys jatamansi* through phytochemical study and evaluation of Anti-neuroinflammatory component. *Evidence-Based Complementary and Alternative Medicine: eCAM* 5901653. https://doi.org/10.1155/2021/5901653

86. Liu, Q.F., Jeon, Y., Sung, Y.W. et al. 2018. *Nardostachys jatamansi* ethanol extract ameliorates Aβ42 cytotoxicity. *Biological and Pharmaceutical* Bulletin Apr 1;41(4):470–477.
87. Anupama, K.P., Shilpa, O., Antony, A., Raghu, S.V., Gurushankara, H.P. 2022. Jatamansinol from *Nardostachys jatamansi* (D.Don) DC. Protects $A\beta_{42}$-induced neurotoxicity in Alzheimer's disease *Drosophila* Model. *Neurotoxicology* May; 90:62–78.
88. Ahmad M., Yousuf S., Khan M.B., et al. 2006. Attenuation by *Nardostachys jatamansi* of 6-hydroxydopamine-induced parkinsonism in rats: behavioral, neurochemical, and immunohistochemical studies. *Pharmacology Biochemistry and Behaviour* Jan; 83(1):150–160.
89. Bian, L.H., Yao, Z.W., Zhao, C.B., Li, Q.Y., Shi, J.L., Guo, J.Y. 2021. Nardosinone alleviates Parkinson's disease symptoms in mice by regulating dopamine D2 receptor. *Evidence Based Complementary and Alternative Medicine* Aug 13;2021:6686965.
90. Bian, L.H., Yao, Z.W., Wang, Z.Y. et al. 2022 Nardosinone regulates the slc38a2 gene to alleviate Parkinson's symptoms in rats through the GABAergic synaptic and cAMPpathways, *Biomedicine & Pharmacotherapy* 153:113269.

7 Role of Herbal Drugs as Potential Therapeutics in Dementia

Soumya Mitra, Subhash C. Mandal, and Nilanjan Ghosh

7.1 INTRODUCTION

Effective survival of any individual has several aspects dependent on the physical and emotional quality of life. Any disruption in the balanced fabric of health can prove to be crucial. The harmony between the physical and emotional state of an individual is a major function of the brain. The brain is a complex organ that controls various functions such as thought processes, memory, emotion, touch, motor skills, temperature etc. Different parts of the brain are capable of controlling the different functions associated with life.

Memory, as commonly perceived by people, is the retention of information from the internal and external environment. According to Zlotnik et al., memory can be defined as the ability of an individual to record sensory stimuli, events, and pieces of information and can be recalled at any point in time as desired. Every individual has different levels of ability to recall supposed information, and thus, it becomes quite difficult to quantify such. Not only the retention of information but memory is also often associated with terms like cognition, intelligence, attention, concentration, and quality of life (QOL). According to the various systems of medicine, memory is often considered to be a combined result of the acquisition, retention, and recollection of information.

Acquisition is considered the ability of an individual to understand and analyse information, whereas retention is the ability of the individual to retain the acquired information in the brain. Lastly, recollection refers to the ability of the individual to recall any piece of desired information at will. These are the three major pillars of memory which can further hypothesize the concept of memory. Normally, **short-term memory (STM)** is created rapidly and is retained for a very short period of time (minutes to hours). **Long-term memory (LTM)**, on the other hand, is information that is retained for long periods of time (hours to days to years also). **Sensory memory** or **Iconic memory** is referred to as the ability of an individual to retain temporarily the huge amount of information that people encounter every day [1, 2].

7.2 HIPPOCAMPUS

Once an individual receives any information through sensory organs, neurons relay the information to the brain. The hippocampus, a structural extension of the temporal lobe of the cerebral cortex, receives the information and decides whether the information is worthy of retention. The hippocampus is located at the edge of the temporal lobe and is a curled S-shaped structure, which also makes it a part of the limbic system [3]. Being located at the posterior limbic system, it only accumulates a smaller volume of the cerebral cortex. The hippocampus is responsible for different primitive requirements like hunger, love, motivation, mood, etc., making it the 'Primitive Brain' [4]. The hippocampus has been thought to be an important instrument in the process of olfaction since it receives direct inputs from the olfactory bulb. Its role in learning and memory was investigated

DOI: 10.1201/9781003300502-7

only after the 1970s when Henry Gustav Molaison started developing anterograde and partial retrograde amnesia upon removal of the hippocampus during surgery [5]. The role of the hippocampus in the acquisition, storage, and consolidation of memory is carried out by virtue of different neurotransmitters, among which acetylcholine has been identified as the major proprietor.

The hippocampus has been one of the most investigated regions of the brain for its crucial involvement in various functions [6]. Neurodegeneration in the hippocampus continues to flourish in adult life, and a decrease in the rate of neurogenesis may ultimately result in hippocampal atrophy. The hippocampus is associated with different functions apart from learning and memory:

1. Spatial navigation [7].
2. Emotional behaviour [8].
3. Hypothalamic regulation of its functions [6].

The pivotal role of the hippocampus in learning and memory rests on a fine thread that balances its role in encoding the memories and their retrieval from the frontal lobe. There are two principal underlying mechanisms that functionalize the learning-memory loop, i.e., polysynaptic and direct pathways. The polysynaptic pathway plays a crucial role in the development of semantic memory, while spatial and episodic memory is consolidated by a direct intra-hippocampal pathway [9]. The hippocampus can also regulate motor functions and can affect hypothalamic functions as well. Hypothalamic adrenocorticoid hormone release may be affected by the hippocampus, which can be evident from elevated cortisol levels in individuals affected with hippocampal atrophy [10].

Hippocampal pyramidal cells of the CA1 region contain numerous diverse projections to different regions of the brain. One such important projection from the medial septal area is GABAergic and cholinergic by nature. This projection is thought to be of crucial importance for maintaining a physiological state, and its deformation or destruction may be of dire consequence, including severe memory impairment caused by disruption of the hippocampal θ (*theta*) rhythm. Another major contributing factor in the case of complete memory impairment is the destruction of the entorhinal region of the brain [11]. Bilateral hippocampal deformity leads to complete amnesia, whereas unilateral damage may not be as harmful.

7.3 ACETYLCHOLINE

The hippocampus is considered one of the major locations of cholinergic nicotinic and muscarinic receptors. The role of acetylcholine in learning and memory is well investigated and established by virtue of the 'cholinergic hypothesis' [12]. The role of both nicotinic and muscarinic cholinergic receptors is well established in encoding new memories, and selective antagonism of muscarinic receptors has been proven to inhibit the encoding process. However, the retrieval of old memories is not highly affected by the blocking of these receptors [13]. Cholinergic activators such as acetylcholine have been proven to be an important piece in the jigsaw that is memory. Pharmacological studies have established the role of acetylcholine in the learning and memory process. Higher amounts of acetylcholine have been often associated with encoding of new memories, whereas lower levels have often been considered as an important tool for memory consolidation [14].

Memory encoding in the brain by acetylcholine is carried out by four major molecular mechanisms. They are:

a. *Enhanced afferent input in relation to excitatory feedback*

Acetylcholine increases memory coding by enhancing the feedforward afferent inputs into the cerebral cortex, stimulating the cortical circuits in response to sensory stimuli. It also suppresses excitatory feedback mechanism responsible for retrieval of memory. Increased nicotinic excitatory afferent input and presynaptic muscarinic excitation feedback may be the underlying mechanisms for this [15, 16].

b. Enhancement of the θ (theta) rhythm oscillation

Acetylcholine increases encoding of memory by enhancing θ rhythm oscillations responsible for memory encoding within the hippocampus. Studies have established that memory encoding may be enhanced by modulation of the network functionality between encoding and retrieval phase of memory within each θ rhythm cycle [17].

c. Increased persistent spiking

Cortical neurons in entorhinal region have been shown to exhibit increased persistent spiking upon the influence of acetylcholine which is considered to be an important mechanism in acquiring new information [18].

d. Enhanced long-term potentiation by cholinergic system

Acetylcholine can also enhance memory encoding by increasing long-term potentiation (LTP) of information in different regions such as hippocampus, entorhinal cortex etc. Studies have revealed the role of nicotinic receptors in enhancement of LTP [19].

7.4 DEMENTIA

Dementia, more commonly known as loss of memory, may be considered as a syndrome rather than a disease itself. It may be defined as the decrease in an individual's level of cognition to such an extent that it hampers the daily functioning of the domestic, occupational, and social space of the individual. The diminished cognizance in an individual may be caused by different other disorders that ultimately result in dementia. It is thus conceptualized as the inability of an individual to carry out daily functioning as a result of neurodegeneration [20]. The major underlying conditions progressing to dementia may be natural or as an effect of external factors. Ageing, genetic profile, and different systemic vascular diseases are considered to be the major proprietors in causing dementia. Dementia, as simply described, may be classified as reversible or irreversible disorders that result from a number of factors. Dementia has been one of the leading disorders in the elderly population for the last few decades, and it counts to around 35.6 million in 2010, which will increase to 66 million in 2030. With an annual spike of 7.7 million new cases, the number of individuals suffering from dementia is set to reach almost 115 million subjects by 2050, and it is estimated to bear 10% of the total disease burden of the worldwide population [21].

Dementia has become synonymous with Alzheimer's disease, which can be better explained in terms that one of the major syndromes associated with Alzheimer's disease is dementia. The underlying factors that result in dementia are

a. Alzheimer's disease (AD).
b. Vascular dementia.
c. Mixed dementia.
d. Dementia with Lewy bodies.
e. Dementia in Parkinson's disease.

Memory impairment is often caused by external factors like stress, ageing and emotions, while conditions like Alzheimer's disease, Parkinson's disease or Schizophrenia may also contribute as underlying mechanisms in developing dementia. In individuals with Alzheimer's disease, the hippocampal deformity or degeneration may account for memory impairment when the brain becomes incapable of storing and processing new information in the form of memory. AD is commonly seen in the elderly population over 65 years of age and may be caused by the destruction of the neurons. Another major well-established hypothesis behind developing dementia is the dysfunction of the cholinergic system in the brain, which results in memory impairment [22]. Drugs that have already been approved for the treatment of dementia often provide temporary relief of the symptoms and can cause adverse effects. Thus, there is always a need for drugs that can treat the underlying conditions and provide permanent solutions without major adverse effects. This chapter aims to provide an overview of such drugs from natural sources, which have been used since ancient times as nootropics or memory enhancers.

7.5 NOOTROPICS

Nootropics or smart drugs are referred to as a class of drugs that enhance the functionality of the brain. These cognitive enhancers can be used to treat cognitive deficits that appear in disorders leading up to dementia. These are non-toxic agents that exhibit neuroprotective action against different underlying conditions leading to dementia. Froestl et al., in 2014, proposed a classification of the nootropics based on their mechanisms of action, citing 19 categories. Of the 19 categories, few have been found to be established as primary contenders as memory-enhancing agents. The categories are depicted in Figure 7.1 [23–25].

During the search for nootropic drugs, many synthetic molecules have been investigated, but none have been able to deliver the desired effects. In some cases, though the molecules seemed pharmacologically promising, they exhibited various adverse effects. Nature, on the other hand, has always been resourceful towards providing humans with pharmacologically relevant and important molecules. Due to their presence in plants and animals, primarily the only challenge humans had to overcome was its identification and collection. Ancient civilizations like the Egyptian, Greek, Indus Valley, and Chinese explored nature in search of pharmacologically potential herbs. The ancient systems of medicine have always looked into nature for remedies, and humans have never come back empty-handed. In ancient civilizations, memory boosters (nootropics) have been well

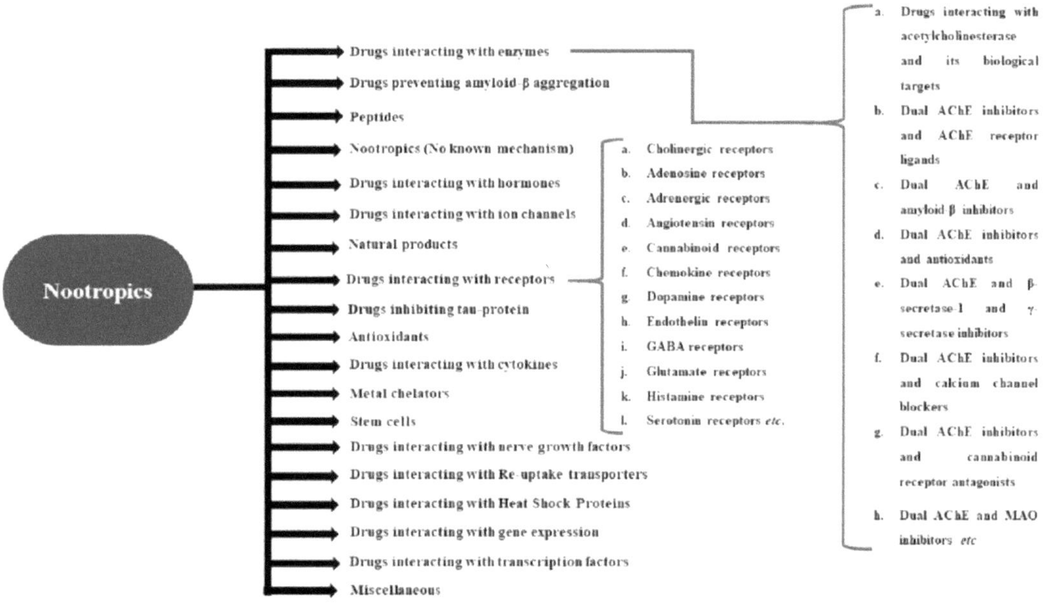

FIGURE 7.1 Classification of nootropics.

investigated, and different herbs have been identified. In the modern era also, studies have been conducted to identify nootropics or memory enhancers from natural sources. Plants like Brahmi, Jatamanshi, and Shankhapushpi have been extensively investigated for their memory-enhancing properties. Natural nootropics often exhibit a multi-faceted approach to combatting the progress of neurodegenerative disorders. This chapter aims to provide an overview of the nootropics from natural sources along with their probable underlying mechanisms.

7.5.1 Natural Nootropics

Different secondary metabolites from natural sources have been investigated extensively for their possible functionality in combatting dementia. They have been frequently mentioned and often used in ethnomedicine for centuries. However, only recently have they been investigated to determine the probable underlying mechanisms, as shown in Figure 7.2. Studies have been conducted to unravel the multimodal suitable approach in the management of dementia in subjects.

i. *Prevention of pathogenic protein aggregation*

   Alzheimer's disease is often characterized by the formation of neurofibrillary tangles and extracellular plaques, which result from protein accumulation. Mutation and misfolded proteins are often responsible for deformed β-sheet by faulty polymerization and protein pathways [26]. Protein accumulation causes neurotoxicity, and phytochemicals have been shown to act by inhibiting the synthesis of amyloidogenic monomers and preventing fibrillar aggregation. Synthesis of amyloidogenic monomers can be prevented by inhibiting the cleavage of amyloid precursors by β-secretase and γ-secretase [27]. Several phytochemicals have also shown the tendency to convert amyloid precursor proteins into non-toxic aggregates by expressing α-secretase [28].

   Other studies have confirmed the influence of different phytochemicals in the formation of soluble Aβ oligomers, which in turn compromise the integrity of the β-sheet motif and

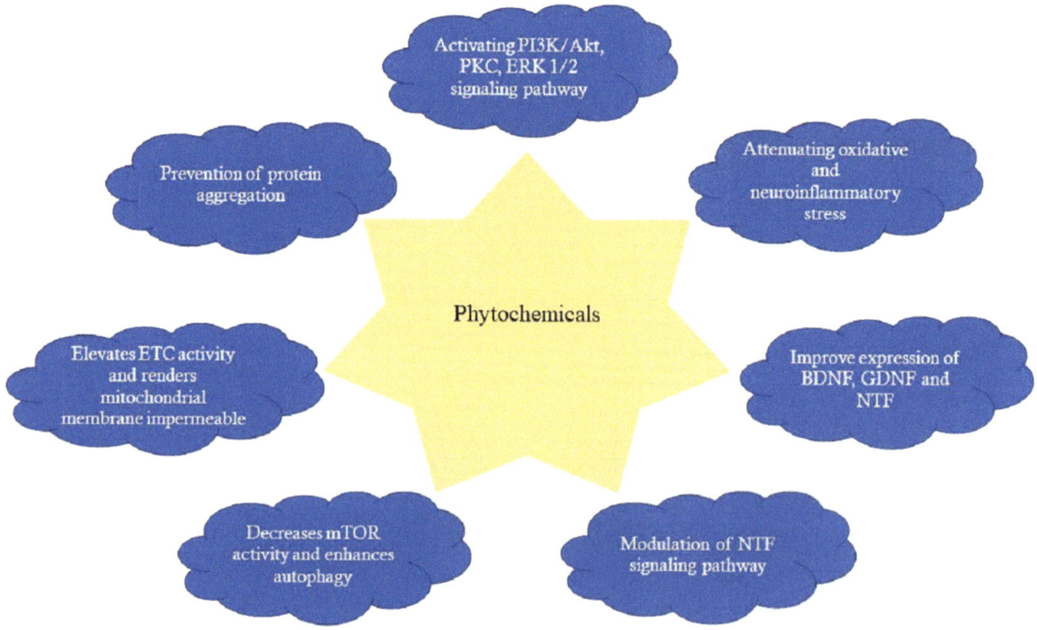

FIGURE 7.2 Probable mechanisms of action of different phytochemicals.

subsequently inhibit the fibril formation [29]. Tau (τ) proteins are another soluble CNS protein, found abundantly in the nervous system, which hyper phosphorylates to form insoluble oligomers. These Tau oligomers aggregate to form pair helical filaments (PHFs) which cause synaptic impairment in AD. Several phytochemicals have been able to inhibit the aggregation and dissolve the PHFs in cells resulting in inhibition of the aggregate formation [30, 31].

Modulation of the cellular signalling pathway: Phytochemicals from different natural sources have been instrumental in regulating various pro-survival mitogen-activated protein kinase (MAPK) cellular signalling pathways. These pathways play an important role in retaining cellular functionality, memory formation, and synaptic plasticity. Secondary metabolites are capable of enhancing the expression of PI3K/Akt and PKC signalling pathways by attaching to the ATP binding site of the receptors and enzymes. Several studies have proven that phytochemicals such as flavonoids are capable of increasing catalase activity, ROS levels and calcium influx, as well as diminishing caspase-3 activity. Flavonoids have also been investigated for their ability to activate Akt/JNK pathways at very small doses [32]. Studies have revealed the roles of different phytochemicals in improving neuronal connectivity in the hippocampus by triggering PKCγ [33]. Polyphenolic compounds have also been seen to exhibit PKC activation, thereby regulating neuronal differentiation, learning, and memory. They can also regulate Keap-1-Nrf-antioxidant response element mediated expression of the antioxidant enzymes [34].

Neuroprotection mediated by neurotrophins: Neurotrophins are one of the major factors for the effective functioning and development of the neurons. The severity of neurodegenerative disorders increases with ageing, and it can be attributed to the improper functioning of the brain-derived neurotrophic factors (BDNF) and glial-cell-derived neurotrophic factors (GDNF) [35]. In neurodegenerative disorders, BDNF and GDNF are considered to be therapeutic targets that become inadvertently inaccessible due to the blood-brain barrier. However, several flavonoid-like phytochemicals are capable of passing across the BBB and exerting their neurotrophic activity by interacting with neurotrophic factor receptors [36].Monoamine oxidases (MAO), a family of enzymes found generally on the outer membranes of the mitochondria, are responsible for the removal of the different types of neurotransmitters from the brain. It is responsible for maintaining homeostasis of the neurotransmitters and also influences signalling pathways to regulate neuronal survival and death [37]. MAO-B inhibitors have been able to activate pathways that lead to neuroprotection. Several studies have suggested the capability of flavonoids in inhibiting MAO as well as elevating the expression of BDNF, while other phytochemicals have been proven to inhibit MAO-B activity and increase GDNF expression [38].

ii. *Modulation of Pro-survival neurotrophins*

Phytochemicals from different sources elevate the BDNF, GDNF, and NTF levels in subjects with different neurodegenerative disorders like Alzheimer's disease, Parkinson's disease etc. Flavonoids and polyphenols are the frontrunners in the aspect of elevating the expression of different neurotrophins, which enhance the longevity of the neurones in the brain [39]. Additionally, polyphenolic secondary metabolites are also capable of increasing Trk expression-mediated neuroprotection and subsequent neurogenesis [40].

iii. *Effects on mitochondria*

Several phytochemicals have exhibited their effects on biogenesis, functioning and apoptosis of mitochondria which concomitantly regulate the neurodegeneration in ageing individuals. Mitochondrial biogenesis is regulated by peroxisome proliferator-activated receptor-gamma coactivator 1 alpha (PGC-1α) and nuclear respiratory factors (NRFs). PGC-1α also stimulates the expression of NRFs by modulating the release of deacetylase Sirutin-1 (Sirt-1), while

PGC-1α expression is mediated through the AMPK pathway. Several polyphenols are capable of elevating AMPK and Sirt-1 levels in 0 to promote biogenesis of the mitochondria, which in turn maintain the oxidative balance in the brain [41, 42].

Mitochondrial functioning has been established as one of the major factors in ameliorating neurodegenerative disorders in patients with Alzheimer's disease, Parkinson's disease etc. Aggregation of the β-amyloid plaques and tau (τ) proteins in mitochondria have been identified as major diagnostic characteristics in these disorders. Intramitochondrial aggregation of the proteins results from a downstream phenomenon, where inhibition of complex-IV and dysregulation of complex-I occur in the electron transport chain (ETC). Phytochemicals such as polyphenols are capable of regulating the ETC to prevent anomalies leading to an imbalance between oxidants and antioxidants [43, 44]. Mitochondrial apoptosis has been identified as one of the major underlying conditions in bringing about neurodegeneration in individuals. The apoptosis cell signalling is considered to be one of the most important approaches in the treatment. Phenylpropanoids and polyphenols have shown the ability to provide protective actions by reducing mitochondrial cellular permeability and thus inhibiting the apoptosis pathway [45].

A list has been assorted to showcase the herbs that have been well-established as nootropics from natural sources to combat dementia.

7.5.2 *Allium sativum* (Family: *Amaryllidaceae*)

Allium sativum, commonly known as garlic, is one of the most widely used food components around the globe. The bulbous flowering plant is often used as a flavouring agent in different food products due to its volatile oil content. It has been widely used for its anticarcinogenic, antioxidant, antimicrobial, antiseptic, and antihypertensive activities. Garlic contains Allicin (S-(2-propenyl)-2-1-sulfinothioate) as the principal chemical component, which is derived from Alliin upon the influence of the enzyme Alliinase [46]. Garlic has been considered a prime candidate in the treatment of dementia in Alzheimer's disease due to its ability to prevent amyloid-β protein plaque formation. Subsequently, it can also hinder the protein deposition and apoptosis-mediated disease progression in Alzheimer's disease. Studies have revealed its potential to act as a natural neuroprotectant, and its regular intake might help in hindering neurodegenerative disease progression [47].

Further studies have confirmed the role of *Allium sativum* in impeding the cognitive decline in amyloid-β induced Alzheimer's disease subjects which is also simultaneously helped by ameliorating mitochondrial dysfunction. Garlic may reduce mitochondrial dysfunction by upregulation of the antioxidant enzymes which prove to be beneficial in the therapy of dementia [48]. Other studies have shown its potential as a memory enhancer by inhibiting the AchE activity, which is one of the major mechanisms involved in the treatment of dementia. On the other hand, garlic can also reduce oxidative stress-induced injury in the CNS, which can be considered one of the major reasons for the aetiology of dementia [49].

7.5.3 *Asparagus racemosus* (Family: *Asparagaceae*)

Asparagus racemosus, commonly known as Shatavari, is one of the most abundantly found herbs in the Northern Himalayas. The plant has been used for centuries in different traditional systems of medicine and still finds its importance in modern medicine. Shatavari contains saponins such as Shatavarin I, Shatavarin IV, Shatavarin V, and Racemosides A-C as major phytochemical components from the roots of the plant. It has been widely used as an adaptogen, antimicrobial, antiulcer, galactagogue, and neuroprotective [50].

Asparagus has been investigated for its neuroprotective activities in animals with hippocampal and striatal lesions in the brain. Kainic acid induced lesions result in elevated lipid peroxidation, increased protein carbonyl content and decreased glutathione content and glutathione peroxidase activity. The study proved the effects of Asparagus root extract by enhancing glutathione peroxidase activity and glutathione levels [51]. Other studies have confirmed the role of ethanolic extract of the roots in neuroprotective action and reducing oxidative stress that leads to neurodegeneration [52]. Asparagus roots are also capable of exerting neuroprotective action by inhibiting AchE and MAO-B levels in the brain. Additionally, it has been investigated for its neuroprotective action through activation of BDNF release in the brain. It is also believed to prevent α-synuclein aggregation [53]. Sarsapogenin, a steroidal secondary metabolite from the roots of the herb, has been reported to possess anti-amyloidogenic activity, which prevents the formation of the Aβ proteins by inhibiting the cleaving of its precursor [54]. Nootropic and anti-amnesic property of the roots of the plant was investigated in the methanol extract with the animals introduced in the Morris water maze, which have shown significant results with decreased cognitive impairment [55].

7.5.4 *Benincasa hispida* (Family: *Cucurbitaceae*)

Benincasa hispida, commonly known as wax gourd or Chalkumra (Bengali), is a well-known fruit used in the diet of tropical Indians. The fruit has been used for a long time due to its different pharmacological activities in the Ayurveda and traditional Chinese systems of medicine. The fruits show antipyretic, anti-inflammatory, antibacterial and neuroprotective activities, among several others. The wide spectrum of pharmacological functions is contributed by the presence of different phytochemical constituents, which include alkaloids, polyphenols, flavonoids, steroids, polysaccharides, terpenes, and terpenoids [56].

Studies have been carried out to investigate the neuroprotective activities of the fruit. Apart from showing considerable anticonvulsant activity of the methanolic extract, it has been proven to exhibit activity in Alzheimer's disease. The aqueous extract of the pulp of the fruit is proven to lower lipid peroxidation levels and overcome cognitive deficits associated with Alzheimer's disease. The study has also revealed the role of the fruits in regulating acetyl choline, catalase, and superoxide dismutase levels [57]. Another different study has revealed the role of the fruit in exhibiting neuroprotective activity by inhibiting amyloid-β aggregation, an important pathological marker in Alzheimer's disease. The fruits have shown their capability in preventing plaque formation by downregulating Keap-1/Nrf2 mediated oxidative stress and inflammation, which is considered to be one of the pathological conditions leading to neurodegeneration in Alzheimer's disease. The same study has revealed that ethanolic extract of the fruits of the plant elevates AchE levels in the hippocampus where antioxidant gene expression also seems to increase as in test animals affected by Alzheimer's disease [58].

7.5.5 *Clitoria ternatea* (Family: *Fabaceae*)

Clitoria ternatea, commonly known as Aparajita in Bengali, is one of the most abundantly found ornamental herbs in India. The plant has been used in the Ayueveda and other traditional system of medicine for centuries. The herb is used for a plethora of wide spectrum of activities, that from anti-inflammatory, antidiabetic, local anaesthetic, platelet aggregation to antimicrobial potential [59]. However, perhaps its most important pharmacological activity lies in its potential to enhance brain functionality.

Clitoria ternatea, also known as Shankhapushpi, has been widely used for its potential role in enhancing learning and memory, exerting neuroprotective action. Different pharmacological activities of the herbs are contributed by a wide variety of phytochemicals, which include kaempferol, myricetin, delphinidin, taraxerol, and cinnamic acid derivatives [60]. Studies conducted to investigate the different CNS activities have been extensively investigated to report the role of acetylcholine elevating role in the hippocampus [61]. Different studies have also confirmed its potential to ameliorate the neurodegenerative role of AchE in the brain, which is considered to be one of the major underlying mechanisms of Alzheimer's disease. Root extract of the herb significantly diminishes the AchE level in the CA1 region of the hippocampus to prevent vascular dementia [62]. Another possible mechanism by which the herb may exhibit its neuroprotective action may include its ability to augment cellular antioxidant properties. Polyphenolics such as flavonoids are well established in promoting antioxidant defence mechanisms in the CNS, subsequently providing neuroprotective action [63].

7.5.6 *Cyperus rotundus* (Family: *Cyperaceae*)

The use of herbs from the *Cyperus* genus from the family Cyperaceae has been going on for ages. Commonly known as nut grass or Nagarmotha, this plant has been used widely in different traditional systems of medicine for a very long time. Native to wetlands of tropical and subtropical regions, the plant is abundantly found in the Indian subcontinent. It has found uses in the remedy of stomach and bowel disorders, used as a diuretic, for menstrual irregularities, and for digestion, among many others [64]. The major chemical components of the plant include quercetin and chlorogenic acid, while thiazole and decanoic acid derivatives are other constituents of the plant [65].

Cyperus rotundus, a nervine tonic from Ayurveda, has been exceptional in exhibiting neuroprotective activity in response to neurological cognitive deficits. The oligomeric flavonoids from the plant have been able to diminish the neurological deficits while ameliorating the panicogenic tendencies. It reduces malondialdehyde (MDA) and elevates superoxide dismutase along with glutathione content in the brain. Total oligomeric flavonoids have been able to provide protective action in neurodegeneration by increasing glutamate content and elevating glutamine synthase and Na^+ -K^+ATPase to nullify oxidative stress and excitotoxicity during neurodegenerative disorders [66]. Further studies have confirmed its neuroprotective action in hypoxic rat models to ameliorate cognitive deficits [67]. Additionally, the combined extract of *Cyperus rotundus* and *Zingiber officinalis* have been established as major supplements in the treatment of dementia. The extract has been proven to reduce memory impairment by different mechanisms. The possible mechanism includes reducing oxidative stress, which subsequently increases pERK1/2 in the hippocampus and also decreasing AchE activity, which acts as one major mechanism in attenuating neurodegenerative deficits in dementia [68].

7.5.7 *Embelia ribes* (Family: *Primulaceae*)

Embelia ribes, commonly known as Vidanga or false black pepper, is one of the widely investigated herbs in Ayurveda as a memory booster. It has been well-versed in the treatment of different human ailments like rhinitis, haemorrhage, and epilepsy, among others. The plant is a rich source of secondary metabolites, which contribute to its different pharmacological activities. Embelin, a natural alkyl substituted benzoquinone, accounts for its different activities like anti-inflammatory, hepatoprotective, analgesic, antioxidant, and wound healing, among others [69].

Due to its use in Ayurveda as a memory-boosting drug, a series of studies were conducted to confirm its efficacy in dementia. Embelin, the major chemical constituent of the plant, significantly decreases the activity of AchE to enhance the functionality of the cholinergic nervous system to enhance memory. It has been found to modulate the expression of BDNF, CREB1, and catalase m-RNA levels, which decreases as a marker in Alzheimer's disease [70]. Other studies have also supported the hypothesis of Embelin being able to provide neuroprotective action in neurodegenerative diseases like Alzheimer's by exerting antioxidant properties. Embelin has been reported to show anxiolytic effects by exerting agonistic effects GABA/benzodiazepine receptor complex [71].

7.5.8 *Huperzia serrata* (Family: *Lycopodiaceae*)

The toothed clubmoss or firm moss, as commonly known, is a well-known plant in eastern Asian countries like China, Japan, Vietnam etc. *Huperzia serrata* is widely used for its nootropic effects and has been in use for several centuries. Though, initially, it was unknown which particular secondary metabolite was responsible for the nootropic activity, later Huperzine A was found to be responsible for the activities. Huperzine A, a lycopodium alkaloid, has been accounted for the crucial nootropic activity of the plant and has been approved as a dietary supplement [72].

Huperzine A has shown moderate inhibition of AchE, which is considered to be one of the major therapeutic targets for the treatment of Alzheimer's disease. The same study also confirmed the inactive nature of the compound against butyrylcholinesterase (BuchE) in the same dose. Additional studies have reported its efficacy in inhibiting the formation of amyloid-β plaque formation along with mild antioxidant properties [73]. Huperzine A is also an NMDA receptor antagonist, which also favours its neuroprotective activity. Huperzine A, a quinolizidine alkaloid, has been effective in regulating glutamate levels in the brain, which plays an important role in the onset of Alzheimer's disease-mediated dementia [74]. A different study also reported the role of secondary metabolites from an epiphytic fungus; *Colletotrichum spp.*, to exert neuroprotective action in Alzheimer's disease. The fungus from *Huperzia serrata* contains different chemical components that show potent AchE inhibitory effects in studies [75].

7.5.9 *Marsilea minuta* (Family: *Marsileaceae*)

Marsilea minuta, commonly known as dwarf water clover or Sushnisak in Bengali, is an aquatic or semi-aquatic plant that can be found in the waterlands of the Indian subcontinent. The plant is a very well-known dietary component among the native ethnicities. The major chemical component of the herb is marsiline, while other components from the plant include polyphenols, flavonoids, terpenoids, anthraquinones, and coumarins. *Marsilea minuta* has been used in Ayurveda for a very long time for its antipyretic, analgesic, antidiabetic, antimicrobial, hepatoprotective, antitussive, anti-amnesic, antifertility, and antioxidant properties [76].

The standardized extract of the leaves has been shown to exhibit antidepressant activity by downregulating the 5-HT2A receptor in the frontal cortex. The plant is often investigated for its potential as a sedative and antidepressant [77]. *Marsilea minuta* has been investigated for its fabled anti-amnesic activity extensively, and reports suggest that its standardized ethanolic extract is able to exert memory-enhancing properties. The nootropic potential of the plant may be a result of its ability to interfere with the cholinergic nervous system. It is reported to inhibit Ach metabolism, mediated by AchE, by interfering with the enzyme Acetylcholinesterase. The role of nicotinic and muscarinic receptors of the cholinergic nervous system is well established, and targeting the AchE enzyme has been a therapeutic approach for nootropics [78].

7.5.10 *Piper betel* (Family: *Piperaceae*)

Since ancient times, *Piper betel* leaves have been used for different religious, recreational and medicinal aspects in the Indian subcontinent. The popular mouth freshener has been consumed for its pungent and aromatic properties. Additionally, it has been effective in showing carminative, antimicrobial, antioxidant, free radical scavenging and neuroprotective actions. The major phytochemical components of the leaves, the most widely used part of the plant, include a plethora of diverse components that are responsible for exhibiting the different pharmacological activities. They consist of different terpenoids, volatile oils, and tannins like cadinene, eugenol, chavicol, methyl chavicol, pinene, β-sitosterol, ursolic acid, chavibetol among many others [79, 80].

Traditionally, it has been consumed by chewing the leaves, which is said to be effective for its carminative and flavouring properties. The juice of the leaves has been used in Ayurveda to treat mild infections in the eyes and night blindness. However, these uses are still to be validated, and their activity in combatting Alzheimer's disease-induced dementia is well established. The terpenoids and phenylpropanoids from the leaves have been instrumental in improving scopolamine-induced amnesia in a dose-dependent manner [81]. Other studies have confirmed the role of chavicol and its derivatives in ameliorating cognitive deficits in Alzheimer's disease. This may be carried out by down-regulation of the pro-inflammatory cytokines like IL-6, TNF-α, lipid peroxides, IL-1β, γ and β- secretase, which are known to cause neuroinflammation and ultimate neurodegeneration. Hydroxychavicol, a major component of the leaves of Piper betel, has been able to successfully reverse the effects of AchE to establish cholinergic transmission in the brain, which is considered to be an important therapeutic approach towards combatting dementia in Alzheimer's disease [82].

7.5.11 *Polygala tenuifolia* (Family: *Polygalaceae*)

The traditional Chinese system of medicine has been one of the richest sources of crude drugs from natural sources, along with Ayurveda. *Polygala tenuifolia*, a common Chinese herb known as Yuanzhi, is a widely used herb in the Chinese system of medicine. It has been used in light of its possible effects on the brain. It has been widely investigated for its nootropic properties, and presently, studies are being conducted to establish its cognition-boosting properties. The aerial parts, as well as the roots of the plant, contain diverse phytochemical constituents, which are probably responsible for its neuroprotective activities. Studies have revealed the presence of saponins such as tenuifolin, onjisaponins; xanthones such as onjixanthone, polygalaxanthones; oligosaccharides like tenuifolisides and other components in the herbs which have been instrumental in its different pharmacological activities [83].

Polygalasaponin XXXII was investigated for its cognition-boosting effects, which may be mediated by the activation of the ERK/MAPK pathway and expression of BDNF. The activity may also be dependent on CREB activation, which acts as one of the major components in the acquisition and retention of memory. Polygalasaponin XII may also contribute to its cognition-enhancing property by expressing BDNF, whose level declines in ageing patients with Alzheimer's disease [84]. Senegenin, another saponin from the plant, is also able to promote neuroprotection by activating PI3K/Akt signalling cascade, which prolongs neuronal survival. Another possible mechanism for Senegenin is by preventing amyloid-β aggregation, tau (τ) phosphorylation [85]. Additional studies have also revealed the role of the aerial parts of the plant in ameliorating neurodegeneration by elevating BDNF levels along with exhibiting antioxidant effects through increasing superoxide dismutase and glutathione levels while reducing malonaldehyde levels. The activity is also attributed to the anti-inflammatory potential of the herb, which includes down-regulation of the pro-inflammatory cytokines like IL-6 and IL-1β [86]. 3,6′-Disinapoyl sucrose (DISS) has been proven to exhibit excellent hippocampal neurogenesis, activating the amyloid precursor protein/ neuronal stem cell-mediated proliferation [87].

7.5.12 *Salvia officinalis* (Family: *Lamiaceae*)

The popular seasoning and flavouring herb *Salvia officinalis*, also known as sage, has been used for a very long time. Sage is one of the most widely used condiments in Asian and Latin American delicacies due to its pleasant aroma, which comes from the essential oils in the aerial parts of the shrub. The plant has been extensively used in folk medicine for its potential in rheumatism, gout, inflammation, dizziness, diarrhoea, and paralysis treatment. However, recent studies have mostly been conducted to establish its nootropic properties [88].

The plant is a rich source of diverse phytochemicals, which include phenolic acids like caffeic acid, rosmarinic acid, salvianolic acid, and lithospermic acid; flavonoids like apigenin, luteolin, kaempferol, and quercetin; terpenoids like thujone, caryophyllene, carnosic acid and ursolic acid among others [89]. The aromatic components of the plants have exhibited cognition-enhancing properties by modulating the cholinergic nervous system. Volatile oils like thujone have been effective in inhibiting AchE activity, which plays a crucial role in the progression of dementia. Nicotinic and muscarinic transmission has been a well-documented therapeutic approach in dementia [90]. A different study has been conducted to observe the synergistic effect of the different species of sages (*Salvia officinalis* and *Salvia lavandulaefolia*). The study found the role of these hydroalcoholic extracts in increasing the calmodulin-dependent protein kinase II expression, which plays an important role in memory and other neuronal functions like neuronal metabolism, synaptic plasticity and neuronal signal transduction [91].

7.5.13 *Vitis vinifera* (Family: *Vitaceae*)

Vitis vinifera or the common grapevine, also known as Angur in Bengali, is a flowering plant that grows in the Mediterranean region and several European as well as Asian countries. The plant is a well-known source of flavonoids and has been used for centuries due to its nutritional component. The common grape is rich in terms of phenolics, flavonoids, and stilbenes. Catechin, epicatechin, oleanolic acid, betulinic acid, resveratrol, viniferin, and gallic acid are some of its major phytochemical constituents, among others. Various parts of the plant have been examined to report antioxidant, hepatoprotective, anti-inflammatory, and neuroprotective properties [92].

Grapevine has been extensively investigated for its potential in the treatment of Alzheimer's disease. Initial studies have reported the adaptogenic and nootropic properties of the ethanol extract of fruits and seeds. Findings suggested the inhibition of the free radicals has played a key role in providing the neuroprotective property of the plant [93]. A different study has been conducted to isolate novel stilbenes from the aerial parts of the plant and investigate their potential in Alzheimer's disease [94]. The study revealed that Vitisinol C, an oligostibene from the shoot of the plant, has potent inhibitory activity towards the aggregation of amyloid-β protein fibrillation. The flavonoids from the herb are also capable of enhancing learning and memory. The underlying mechanism for their bioactivity may include expression of p-CREB, BDNF and synaptotagmin-1(SYT1) in the hippocampal region. BDNF and SYT1 promote neurodegeneration and increase synaptic plasticity, which is considered to be a great therapeutic approach in combatting neurodegenerative disorders like Alzheimer's disease [95].

Structures of a few crucial phytochemicals have been shown in Figure 7.3a and Figure 7.3b, which potentially show nootropic effects.

An additional list has been incorporated to enumerate the more common crude drugs that have been used in neurodegenerative disorders that result in dementia in Table 7.1.

7.6 CONCLUSION

Neurodegenerative disorders have continued to be one of the major contributors to cognitive impairment in the global population. Dementia, as already described, is a result of different neurological disorders and hence can be considered to play a major role in the increasing global disease burden. The complex pathophysiological conditions that lead to dementia are often related to diseases, and its prevention seems almost impossible. However, extensive research has provided some succour in this aspect, and few drugs from natural sources have been identified as probable candidates. Drugs of synthetic origin fail to cater for the general requirement of functioning through multiple modes and are often responsible for adverse effects that prove to be harmful in the long run. Naturally occurring nootropics are considered to be more suitable candidates to act through multiple modes

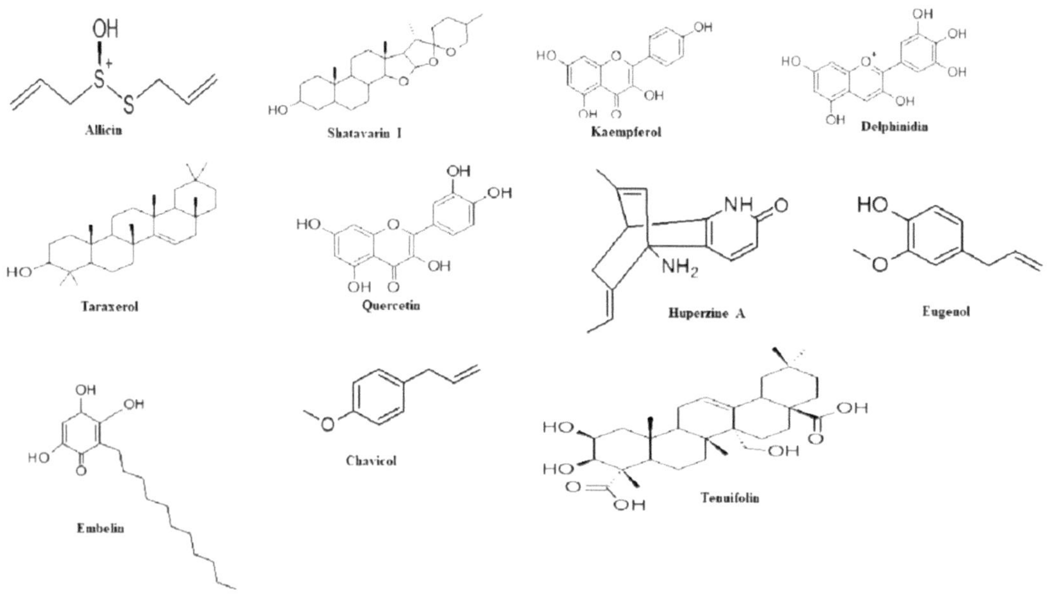

FIGURE 7.3a Structures of some secondary metabolites from the plant sources exerting nootropic activity.

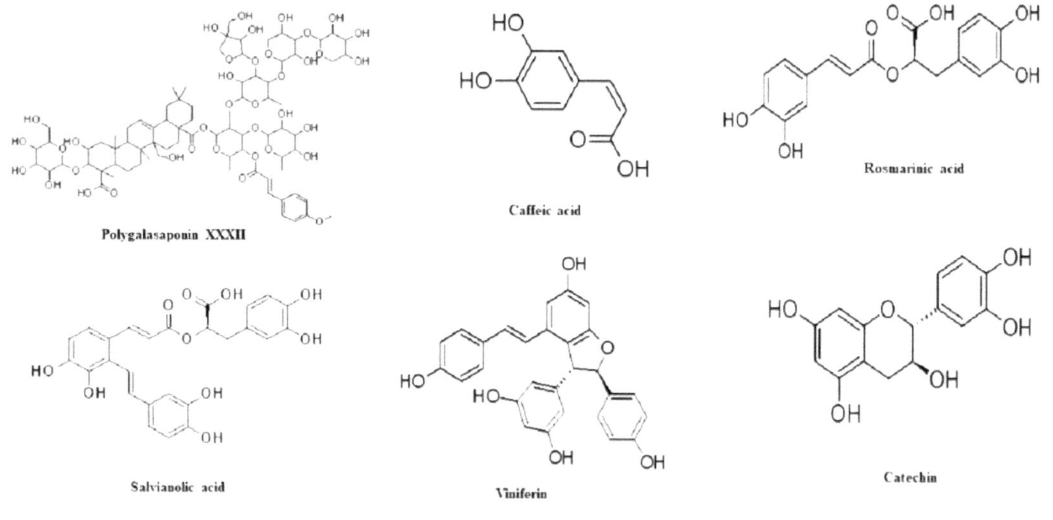

FIGURE 7.3b Structures of some secondary metabolites from the plant sources exerting nootropic activity.

and impede disease progression. Traditional systems of medicines have contributed immensely towards identification and provided the general idea of natural nootropics.

Nootropic drugs from natural sources have been in use for centuries, owing to the contribution of ancient civilizations. In recent years, these herbs have been extensively investigated to identify the phytopharmaceuticals responsible for exerting neuroprotective action and its underlying mechanisms of action. It can be concluded from the different studies that plants and their phytoconstituents often function through more than one mechanism, which has been proven to be useful in mitigation, management and treatment of neurodegenerative diseases. Disorders associated with ageing, such as Alzheimer's disease, have been regressed by prevention of amyloid-β deposition and

TABLE 7.1
List of some well-known herbs used as memory boosters

Plant	Family	Chemical constituents	Mechanism of action
Acorus calamus [96]	*Acoraceae*	β-asarone, phenylpropanoids, lectins, saponins, flavonoids, alkaloids, sesquiterpenes, α-asarone etc.	β-asarone is an AChE inhibitor, thus enhancing ACh activity. Asarones in general exhibits anti-inflammatory activity which downregulates inflammatory cytokines.
Centella asiatica [97, 98]	*Apiaceae*	Asiatic acid, Asiaticosside, Madecassic acid, Madecassoside.	Improves blood circulation in the brain. Ameliorates Scopolamine induced memory impairment through inhibition of AChE. Prevents amyloid-β protein aggregation on hippocampus. Asiaticoside also exerts neuroprotective action by inhibiting p38 MAPK pathway.
Ilex paraguariensis [99]	*Aquifoliaceae*	Polyphenol (Chlorogenic acid), tannins, xanthines (Caffeine, Theophylline, Theobromine), Flavonoids (Kaempferol and Quercetin), Saponins and Vitamins (A, C, B1, B12, E).	Provides neuroprotective action by preventing amyloid-β aggregation. Inhibits pro-inflammatory cytokines IL-6, IL-1β to ameliorate inflammation.
Celastrus paniculatus [100]	*Celastraceae*	Triterpenoid Pristimerin, Sesquiterpenoid polyalcohol esters (Celapnin, Polyalcohol A-D, Malkanguniol, Malkanginnol), Polyaclcohol (Paniculatusdiol), Alkaloids celastrine, Paniculatin, Calapagine, Celapanine, Celapaginine, Sterols (β-amyrin, β-sitosterol), saturated fatty acids (Palmitic, Stearic and Lignosteric acid), unsaturated fatty acids (Oleic, Linoleic, Linolenic acid), Phenolic triterpenoids (Paniculatadiol, Celastriol).	Whole plant extract has been used as anti-epileptic. Seed oil is used as brain tonic as neuroprotective agent, while improving learning and memory. Decreases noradrenaline, serotonin and dopamine levels in the brain. Also reduces AChE activity.
Convolvulus pluricaulis [101]	*Convolvulaceae*	β-sitosterol, Ascorbic acid, Silane, Decanoic acid, Carbohydrates, Vitamin E, Valeric acid, Cinnamic acid, Squalene, Linoleic acid, Pthalic acid, Gums and several other alkaloids.	Prominent memory enhancing drug, a psycostimulant and tranquilliser. Brain tonic from the whole plant is believed to be memory enhancer and also increases intellect. Consumption of the drug prevents memory loss. Aq. extract of the plant has been proven to downregulate τ- protein, enhances locomotor functions and increases longevity of the human microtubule associated protein τ (hMAPτ).

(*continued*)

TABLE 7.1 (Continued)
List of some well-known herbs used as memory boosters

Plant	Family	Chemical constituents	Mechanism of action
Cornus officinalis [102, 103]	*Cornaceae*	Anthocyanin, Flavonoid (Quercetin), Phenolic acids (trans-cinnamic and benzoic acid), Triterpenoids and Carbohydrates. Cornel iridoid glycosides (CIG) are the main constituents of the plant. Polyphenolic such as telimagrandin I and tellimagrandin II are also present.	CIG has the ability to promote neurogenesis and also promote neurological functionality in ischaemic rats. CIG also improves memory loss in Fimbria-fornix transaction. Reduces loss of neurone in hippocampus. Anti-amnesic and neuroprotective action of the plant is due to its antioxidant and radical scavenging properties. The polyphenols also exhibit neuroprotective activity in AD by downregulating different enzymes responsible for development of the disease.
Ginkgo biloba [104]	*Ginkgoaceae*	Terpenoids (Bilobalide, Ginkgolide A, B, C, J, M), Flavonoids like (Kaempferol, Quercetin).	EGb761, an aqueous extract of the plant is widely used for its effects on neurological systems. It can interact with the cholinergic system and promotes regeneration of neuronal stem cells. It also enhances blood flow in the brain and subsequently promotes supply of nutrients in the brain. Ginkgo also eliminates free radicals from the body, thereby improving memory and other cognitive functions. Bilobalide has shown neuroprotective action in cerebral ischaemia. EGb761 exhibits antioxidant activity and downregulate hyperphosphorylation of τ proteins in addition to the protection against Aβ-induced neurotoxicity; making it a prime candidate in the treatment of AD.
Hypericum perforatum [105, 106]	*Hyperiaceae*	Flavonoids (Rutin, Amentoflavone, Hyperoside, Quercetin, biapigenin, isoquercetin, kaempferol), phenolic compounds, acylated phloroglucinol (hyperforin), naphthodianthrones.	Rutin prevents cytotoxicity and accumulation of β-amyloids, mitigates mitochondrial damage, ROS, oxidative stress and reduces NO and pro-inflammatory cytokines production. Ameliorates spatial and learning memory. Helps in generation of nervous system, specifically the myellin sheath surrounding the nerves.
Crocus sativus [107, 108]	*Iridaceae*	α-crocin, crocetin, picrocrocin, safranal, water-soluble carotenoids, vitamins (riboflavin and thiamine), carbohydrates, low concentration of monoterpene aldehydes and its glucosides, lipids, polypeptides, proteins, flavonoids (quercetin, kaempferol, isorhamnetin), anthocyanins (petunidin, malvidin and delphinidin), starch, amino acids, gums and minerals.	Neuroprotective (Anti-Alzheimer, anti-Parkinson), learning and memory enhancer, anxiolytic and hypnotic, in cerebral ischaemia, memory deficits and brain damage, against Morphine dependency mitigation. Significantly reverses Scopolamine induced dementia. Aqueous extract significantly up-regulates brain-derived neurotrophic factor (BDNF) (both transcriptomic and protein levels) and cyclic-AMP response element binding protein (CREB) (only protein levels) in the hippocampus.

TABLE 7.1 (Continued)
List of some well-known herbs used as memory boosters

Plant	Family	Chemical constituents	Mechanism of action
Gastrodia elata [109]	*Orchidaceae*	Gastrodin (4-hydroxybenzyl alcohol-4-O-β-D-glucopyranoside).	Suppresses intracellular and extracellular Aβ levels in a dose dependent manner. This is associated with the mitigation of β and γ-secretase activities.
Phyllanthus emblica [110, 111]	*Phyllanthaceae*	Vitamin C, tannins, oils, polyphenols (gallic acid, pyrogallol, ellagic acid, punigluconin, chebulinic acid, apigenin, luteolin, quercetin, phyllembin), amino acids and diiferent sugars.	Amla shows positive effect on scopolamine and diazepam induced memory loss. It increases memory and learning. Lowers cholesterol and decreases cholinesterase activity. It is a potent antioxidant which prevents oxidative damage to the brain.
Bacopa monnieri [112]	*Plantaginaceae*	Alkaloids (Brahmine, Herpestine), steroidal saponins (hersaponin, bacopaside I-XII, bacosaponins, bacosides A, B and monnierin).	Learning and memory was enhanced. Amnesia induced by electroschock treatment or scopolamine was ameliorated. Protein kinase C activity in the hippocampus was enhanced which acts as nootropic. It has inhibitory action against AChE.
Camellia sinensis [113]	*Theaceae*	Polyphenols (chlorogenic acid, gallic acid, caffeic acid), alkaloids (theophylline, caffeine, theobromine), flavonoids (quercetin, kaempferol and myricetin), catechins and epigallocatechin 3 gallate (ECGC).	Neuprotective, reduces production of β-amyloid monomers: major contributor in formation of amyloid plaque in the brain; resulting in AD. Reduces learning and memory deficits. Due to its antioxidant property, it can prevent neurodegeneration caused by oxidative stress which is induced by free radicals.
Curcuma longa [114]	*Zingiberaceae*	Curcuminoids (Curcumin).	Curcumin inhibits aggregation of Aβ or promotes disaggregation of the same at low concentration. It can also decrease Aβ toxicity by inhibiting its synthesis by interrupting with Presenilin I (PS1) expression and glycogen synthase kinase 3β (GSK3β). In cell lines, it has shown neuroprotective activity by attenuating quinoprotein development and MAPK expression. Increases cognitive function by restoring low IGF1 levels. Shows antiepileptic action by antioxidant activity.

hyperphosphorylation of tau (τ) proteins. A plethora of polyphenols and flavonoids have been identified in preventing oxidative damage to the brain by exerting antioxidant properties. Furthermore, phytochemicals from different sources have been able to modulate the cholinergic nervous system and potentiate the role of acetylcholine in learning and memory. This chapter has been dedicated to enumerating different herbs that act in multiple modes in alleviating the neurological conditions that lead to cognitive impairment and, subsequently, dementia.

REFERENCES

1. S. Halder, U. Anand, S. Nandy, P. Oleksak, S. Qusti, E.M. Alshammari, G. El-Saber Batiha, E.P. Koshy, A. Dey, Herbal drugs and natural bioactive products as potential therapeutics: A review on pro-cognitives and brain boosters perspectives, *Saudi Pharm J*, 29 (2021) 879–907. https://doi.org/10.1016/j.jsps.2021.07.003.
2. S. Rho, M. Kang, B. Choi, D. Sim, J. Lee, E. Lee, C. Cho, J.W. Oh, S. Park, S. Ko, M. Shin, M. Hong, H. Bae, Effects of Yukmijihwang-tang derivatives (YMJd), a memory enhancing herbal extract, on the gene-expression profile in the rat hippocampus, *Biol Pharm Bull*, 28 (2005) 87–93. https://doi.org/10.1248/bpb.28.87.
3. K.S. Anand, V. Dhikav, Hippocampus in health and disease: An overview, *Ann Indian Acad Neurol*, 15 (2012) 239–46. https://doi.org/10.4103/0972-2327.104323.
4. P.E. Gilbert, A.M. Brushfield, The role of the CA3 hippocampal subregion in spatial memory: a process oriented behavioral assessment, *Prog Neuropsychopharmacol Biol Psychiatry*, 33 (2009) 774–81. https://doi.org/10.1016/j.pnpbp.2009.03.037.
5. H.J. Markowitsch, M. Pritzel, The neuropathology of amnesia, *Prog Neurobiol*, 25 (1985) 189–287. https://doi.org/10.1016/0301-0082(85)90016-4.
6. M. Koehl, D.N. Abrous, A new chapter in the field of memory: Adult hippocampal neurogenesis, *Eur J Neurosci*, 33 (2011) 1101–14. https://doi.org/10.1111/j.1460-9568.2011.07609.x.
7. F. Stella, E. Cerasti, B. Si, K. Jezek, A. Treves, Self-organization of multiple spatial and context memories in the hippocampus, *Neurosci Biobehav Rev*, 36 (2012) 1609–25. https://doi.org/10.1016/j.neubiorev.2011.12.002.
8. H. Toyoda, X.Y. Li, L.J. Wu, M.G. Zhao, G. Descalzi, T. Chen, K. Koga, M. Zhuo, Interplay of amygdala and cingulate plasticity in emotional fear, *Neural Plast*, 2011 (2011) 813749. https://doi.org/10.1155/2011/813749.
9. A. Knapman, S.F. Kaltwasser, D. Martins-de-Souza, F. Holsboer, R. Landgraf, C.W. Turck, M. Czisch, C. Touma, Increased stress reactivity is associated with reduced hippocampal activity and neuronal integrity along with changes in energy metabolism, *Eur J Neurosci*, 35 (2012) 412–22. https://doi.org/10.1111/j.1460-9568.2011.07968.x.
10. I. Morgado-Bernal, Learning and memory consolidation: linking molecular and behavioral data, *Neuroscience*, 176 (2011) 12–9. https://doi.org/10.1016/j.neuroscience.2010.12.056.
11. A.M. Stranahan, M.P. Mattson, Selective vulnerability of neurons in layer II of the entorhinal cortex during aging and Alzheimer's disease, *Neural Plast*, 2010 (2010) 108190. https://doi.org/10.1155/2010/108190.
12. A.V. Terry, Jr. J.J. Buccafusco, The cholinergic hypothesis of age and Alzheimer's disease-related cognitive deficits: recent challenges and their implications for novel drug development, *J Pharmacol Exp Ther*, 306 (2003) 821–7. https://doi.org/10.1124/jpet.102.041616.
13. A. Atri, S. Sherman, K.A. Norman, B.A. Kirchhoff, M.M. Nicolas, M.D. Greicius, S.C. Cramer, H.C. Breiter, M.E. Hasselmo, C.E. Stern, Blockade of central cholinergic receptors impairs new learning and increases proactive interference in a word paired-associate memory task, *Behav Neurosci*, 118 (2004) 223–36. https://doi.org/10.1037/0735-7044.118.1.223.
14. M.E. Hasselmo J. McGaughy, High acetylcholine levels set circuit dynamics for attention and encoding and low acetylcholine levels set dynamics for consolidation, *Prog Brain Res*, 145 (2004) 207–31. https://doi.org/10.1016/s0079-6123(03)45015-2.
15. L.M. Giocomo M.E. Hasselmo, Nicotinic modulation of glutamatergic synaptic transmission in region CA3 of the hippocampus, *Eur J Neurosci*, 22 (2005) 1349–56. https://doi.org/10.1111/j.1460-9568.2005.04316.x.
16. K.E. Vogt W.G. Regehr, Cholinergic modulation of excitatory synaptic transmission in the CA3 area of the hippocampus, *J Neurosci*, 21 (2001) 75–83. https://doi.org/10.1523/jneurosci.21-01-00075.2001.
17. C.J. Siok, J.A. Rogers, B. Kocsis, M. Hajós, Activation of alpha7 acetylcholine receptors augments stimulation-induced hippocampal theta oscillation, *Eur J Neurosci*, 23 (2006) 570–4. https://doi.org/10.1111/j.1460-9568.2005.04560.x.
18. B.J. Young, T. Otto, G.D. Fox, H. Eichenbaum, Memory representation within the parahippocampal region, *J Neurosci*, 17 (1997) 5183–95. https://doi.org/10.1523/jneurosci.17-13-05183.1997.

19. S.V. Ovsepian, R. Anwyl, M.J. Rowan, Endogenous acetylcholine lowers the threshold for long-term potentiation induction in the CA1 area through muscarinic receptor activation: in vivo study, *Eur J Neurosci*, 20 (2004) 1267–75. https://doi.org/10.1111/j.1460-9568.2004.03582.x.
20. S.A. Gale, D. Acar, K.R. Daffner, *Dementia, Am J Med*, 131 (2018) 1161–9. https://doi.org/10.1016/j.amjmed.2018.01.022.
21. M. Wortmann, Dementia: a global health priority – highlights from an ADI and World Health Organization report, *Alzheimers Res Ther*, 4 (2012) 40. https://doi.org/10.1186/alzrt143.
22. J.A. Dumas, P.A. Newhouse, The cholinergic hypothesis of cognitive aging revisited again: cholinergic functional compensation, *Pharmacol Biochem Behav*, 99 (2011) 254–61. https://doi.org/10.1016/j.pbb.2011.02.022.
23. W. Froestl, A. Muhs, A.J.J.o.A.s.D. Pfeifer, Cognitive enhancers (nootropics). Part 1: Drugs interacting with receptors. *Update 2014*, 41 (2014) 961–1019.
24. W. Froestl, A. Muhs, A.J.J.o.A.s.D. Pfeifer, Cognitive enhancers (nootropics). Part 2: Drugs interacting with enzymes. *Update 2014*, 42 (2014) 1–68.
25. W. Froestl, A. Pfeifer, A.J.J.o.A.s.D. Muhs, Cognitive enhancers (Nootropics). Part 3: drugs interacting with targets other than receptors or enzymes. *Disease-Modifying Drugs. Update 2014*, 42 (2014) 1079–1149.
26. K. Brueggen, E. Kasper, S. Ochmann, H. Pfaff, S. Webel, W. Schneider, S. Teipel, Cognitive rehabilitation in Alzheimer's disease: A controlled intervention trial, *J Alzheimers Dis*, 57 (2017) 1315–24. https://doi.org/10.3233/jad-160771.
27. M.T. Kabir, M.A. Sufian, M.S. Uddin, M.M. Begum, S. Akhter, A. Islam, B. Mathew, M.S. Islam, M.S. Amran, G. Md Ashraf, NMDA receptor antagonists: Repositioning of memantine as a multitargeting agent for Alzheimer's therapy, *Curr Pharm Des*, 25 (2019) 3506–3518. https://doi.org/10.2174/1381612825666191011102444.
28. M. Kostomoiri, A. Fragkouli, M. Sagnou, L.A. Skaltsounis, M. Pelecanou, E.C. Tsilibary, A.K. Tzinia, Oleuropein, an anti-oxidant polyphenol constituent of olive promotes α-secretase cleavage of the amyloid precursor protein (AβPP), *Cell Mol Neurobiol*, 33 (2013) 147–54. https://doi.org/10.1007/s10571-012-9880-9.
29. A. Thapa, S.D. Jett, E.Y. Chi, Curcumin attenuates amyloid-β aggregate toxicity and modulates amyloid-β aggregation pathway, *ACS Chem Neurosci*, 7 (2016) 56–68. https://doi.org/10.1021/acschemneuro.5b00214.
30. H.J. Wobst, A. Sharma, M.I. Diamond, E.E. Wanker, J. Bieschke, The green tea polyphenol (-)-epigallocatechin gallate prevents the aggregation of tau protein into toxic oligomers at substoichiometric ratios, *FEBS Lett*, 589 (2015) 77–83. https://doi.org/10.1016/j.febslet.2014.11.026.
31. M. Pickhardt, Z. Gazova, M. von Bergen, I. Khlistunova, Y. Wang, A. Hascher, E.M. Mandelkow, J. Biernat, E. Mandelkow, Anthraquinones inhibit tau aggregation and dissolve Alzheimer's paired helical filaments in vitro and in cells, *J Biol Chem*, 280 (2005) 3628–35. https://doi.org/10.1074/jbc.M410984200.
32. S.L. Hwang, G.C. Yen, Modulation of Akt, JNK, and p38 activation is involved in citrus flavonoid-mediated cytoprotection of PC12 cells challenged by hydrogen peroxide, *J Agric Food Chem*, 57 (2009) 2576–82. https://doi.org/10.1021/jf8033607.
33. C. Menard, S. Bastianetto, R. Quirion, Neuroprotective effects of resveratrol and epigallocatechin gallate polyphenols are mediated by the activation of protein kinase C gamma, *Front Cell Neurosci*, 7 (2013) 281. https://doi.org/10.3389/fncel.2013.00281.
34. S. Qin, F. Deng, W. Wu, L. Jiang, T. Yamashiro, S. Yano, D.X. Hou, Baicalein modulates Nrf2/Keap1 system in both Keap1-dependent and Keap1-independent mechanisms, *Arch Biochem Biophys*, 559 (2014) 53–61. https://doi.org/10.1016/j.abb.2014.03.011.
35. V.K. Gupta, S. Singh, A. Agrawal, N.J. Siddiqi, B. Sharma, Phytochemicals mediated remediation of neurotoxicity induced by heavy metals, *Biochem Res Int*, 2015 (2015) 534769. https://doi.org/10.1155/2015/534769.
36. K.A. Youdim, B. Shukitt-Hale, J.A. Joseph, Flavonoids and the brain: interactions at the blood-brain barrier and their physiological effects on the central nervous system, *Free Radic Biol Med*, 37 (2004) 1683–93. https://doi.org/10.1016/j.freeradbiomed.2004.08.002.

37. M. Naoi, W. Maruyama, M. Shamoto-Nagai, Type A and B monoamine oxidases distinctly modulate signal transduction pathway and gene expression to regulate brain function and survival of neurons, *J Neural Transm (Vienna),* 125 (2018) 1635–50. https://doi.org/10.1007/s00702-017-1832-6.
38. S. Carradori, M.C. Gidaro, A. Petzer, G. Costa, P. Guglielmi, P. Chimenti, S. Alcaro, J.P. Petzer, Inhibition of human monoamine oxidase: Biological and molecular modeling studies on selected natural flavonoids, *J Agric Food Chem*, 64 (2016) 9004–11. https://doi.org/10.1021/acs.jafc.6b03529.
39. S. Neshatdoust, C. Saunders, S.M. Castle, D. Vauzour, C. Williams, L. Butler, J.A. Lovegrove, J.P. Spencer, High-flavonoid intake induces cognitive improvements linked to changes in serum brain-derived neurotrophic factor: Two randomised, controlled trials, *Nutr Healthy Aging*, 4 (2016) 81–93. https://doi.org/10.3233/nha-1615.
40. S. De Nicoló, L. Tarani, M. Ceccanti, M. Maldini, F. Natella, A. Vania, G.N. Chaldakov, M. Fiore, Effects of olive polyphenols administration on nerve growth factor and brain-derived neurotrophic factor in the mouse brain, *Nutrition*, 29 (2013) 681–7. https://doi.org/10.1016/j.nut.2012.11.007.
41. D.C. Nieman, A.S. Williams, R.A. Shanely, F. Jin, S.R. McAnulty, N.T. Triplett, M.D. Austin, D.A. Henson, Quercetin's influence on exercise performance and muscle mitochondrial biogenesis, *Med Sci Sports Exerc*, 42 (2010) 338–45. https://doi.org/10.1249/MSS.0b013e3181b18fa3.
42. C. Sandoval-Acuña, J. Ferreira, H. Speisky, Polyphenols and mitochondria: an update on their increasingly emerging ROS-scavenging independent actions, *Arch Biochem Biophys*, 559 (2014) 75–90. https://doi.org/10.1016/j.abb.2014.05.017.
43. A.H. Schapira, Mitochondria in the aetiology and pathogenesis of Parkinson's disease, *Lancet Neurol*, 7 (2008) 97–109. https://doi.org/10.1016/s1474-4422(07)70327-7.
44. D. Valenti, D. De Rasmo, A. Signorile, L. Rossi, L. de Bari, I. Scala, B. Granese, S. Papa, R.A. Vacca, Epigallocatechin-3-gallate prevents oxidative phosphorylation deficit and promotes mitochondrial biogenesis in human cells from subjects with Down's syndrome, *Biochim Biophys Acta,* 1832 (2013) 542–52. https://doi.org/10.1016/j.bbadis.2012.12.011.
45. A. Camilleri, C. Zarb, M. Caruana, U. Ostermeier, S. Ghio, T. Högen, F. Schmidt, A. Giese, N. Vassallo, Mitochondrial membrane permeabilisation by amyloid aggregates and protection by polyphenols, *Biochim Biophys Acta*, 1828 (2013) 2532–43. https://doi.org/10.1016/j.bbamem.2013.06.026.
46. G. El-Saber Batiha, A. Magdy Beshbishy, G.W. L, Y.H.A. Elewa, A.A.-S. A, M.E. Abd El-Hack, A.E. Taha, M.A.-E. Y, H. Prasad Devkota, Chemical constituents and pharmacological activities of garlic (Allium sativum L.), *A Review, Nutrients*, 12 (2020) https://doi.org/10.3390/nu12030872.
47. S. Tasnim, P.S. Haque, M.S. Bari, M.M. Hossain, S.M. Islam, M. Shahriar, M.A. Bhuiyan, M.S. Bin Sayeed, Allium Sativum L. improves visual memory and attention in healthy human volunteers, *Evid Based Complement Alternat Med*, 2015 (2015) 103416. https://doi.org/10.1155/2015/103416.
48. N. Saini, M. Kadian, A. Khera, A. Aggarwal, A. Kumar, Therapeutic potential of Allium Sativum against the Aβ((1–40))-induced oxidative stress and mitochondrial dysfunction in the Wistar rats, *Am J Neurodegener Dis*, 10 (2021) 13–27.
49. D. Mukherjee, S. Banerjee, Learning and memory promoting effects of crude garlic extract, *Indian J Exp Biol*, 51 (2013) 1094–100.
50. R. Singh, Asparagus racemosus: a review on its phytochemical and therapeutic potential, *Nat Prod Res*, 30 (2016) 1896–908. https://doi.org/10.1080/14786419.2015.1092148.
51. M.S. Parihar T. Hemnani, Experimental excitotoxicity provokes oxidative damage in mice brain and attenuation by extract of Asparagus racemosus, *J Neural Transm (Vienna),* 111 (2004) 1–12. https://doi.org/10.1007/s00702-003-0069-8.
52. M.S. Uddin, M. Asaduzzaman, A. Mamun, M.A. Iqbal, F. Wahid, R.K.J.J.A.D.P. Rony, *Neuroprotective activity of Asparagus racemosus Linn. against ethanol-induced cognitive impairment and oxidative stress in rats brain: Auspicious for controlling the risk of Alzheimer's disease*, 6 (2016) 1–2.
53. S. Majumdar, S. Gupta, S.K. Prajapati, S. Krishnamurthy, Neuro-nutraceutical potential of Asparagus racemosus: A review, *Neurochem Int*, 145 (2021) 105013. https://doi.org/10.1016/j.neuint.2021.105013.
54. P. Kashyap, K. Muthusamy, M. Niranjan, S. Trikha, S. Kumar, Sarsasapogenin: A steroidal saponin from Asparagus racemosus as multi target directed ligand in Alzheimer's disease, *Steroids*, 153 (2020) 108529. https://doi.org/10.1016/j.steroids.2019.108529.

55. R. Ojha, A.N. Sahu, A.V. Muruganandam, G.K. Singh, S. Krishnamurthy, Asparagus recemosus enhances memory and protects against amnesia in rodent models, *Brain Cogn*, 74 (2010) 1–9. https://doi.org/10.1016/j.bandc.2010.05.009.
56. M.T. Islam, C. Quispe, D.M. El-Kersh, M.C. Shill, K. Bhardwaj, P. Bhardwaj, J. Sharifi-Rad, M. Martorell, R. Hossain, A. Al-Harrasi, A. Al-Rawahi, M. Butnariu, L.S. Rotariu, H.A.R. Suleria, Y. Taheri, A.O. Docea, D. Calina, W.C. Cho, A literature-based update on Benincasa hispida (Thunb.) Cogn.: Traditional uses, nutraceutical, and phytopharmacological profiles, *Oxid Med Cell Longev*, 2021 (2021) 6349041. https://doi.org/10.1155/2021/6349041.
57. C. Roy D.J.W.J.o.P.R. Mazumdar, *A Nootropic Effect of Benincasa Hispida on Ach and ChAT Activity in Colchicine induced Experimental Rat Model of Alzheimer's Disease: Possible Involvement of Antioxidants*, 3 (2014) 1282–96.
58. D. Rapaka, V.R. Bitra, R. Ummidi, A. Akula, Benincasa hispida alleviates amyloid pathology by inhibition of Keap1/Nrf2-axis: Emphasis on oxidative and inflammatory stress involved in Alzheimer's disease model, *Neuropeptides*, 88 (2021) 102151. https://doi.org/10.1016/j.npep.2021.102151.
59. P.K. Mukherjee, V. Kumar, N.S. Kumar, M. Heinrich, The Ayurvedic medicine Clitoria ternatea--from traditional use to scientific assessment, *J Ethnopharmacol*, 120 (2008) 291–301. https://doi.org/10.1016/j.jep.2008.09.009.
60. G. Chakraborthy, V. Kumar, S. Gupta, A. Kumar, N. Gautam, L.J.J.o.A.P.S. Kumari, Research, *Phytochemical and Pharmacological Aspects of Clitoria Ternatea-A review*, (2018) 3–9.
61. K.S. Rai, K.D. Murthy, K.S. Karanth, K. Nalini, M.S. Rao, K.K. Srinivasan, Clitoria ternatea root extract enhances acetylcholine content in rat hippocampus, *Fitoterapia*, 73 (2002) 685–9. https://doi.org/10.1016/s0367-326x(02)00249-6.
62. T. Damodaran, P.S. Cheah, V. Murugaiyah, Z. Hassan, The nootropic and anticholinesterase activities of Clitoria ternatea Linn. root extract: Potential treatment for cognitive decline, *Neurochem Int*, 139 (2020) 104785. https://doi.org/10.1016/j.neuint.2020.104785.
63. M.S. Uddin, A. Al Mamun, M.T. Kabir, M. Jakaria, B. Mathew, G.E. Barreto, G.M. Ashraf, Nootropic and anti-Alzheimer's actions of medicinal plants: molecular insight into therapeutic potential to alleviate Alzheimer's neuropathology, *Mol Neurobiol*, 56 (2019) 4925–44. https://doi.org/10.1007/s12035-018-1420-2.
64. A.M. Pirzada, H.H. Ali, M. Naeem, M. Latif, A.H. Bukhari, A. Tanveer, Cyperus rotundus L.: Traditional uses, phytochemistry, and pharmacological activities, *J Ethnopharmacol*, 174 (2015) 540–60. https://doi.org/10.1016/j.jep.2015.08.012.
65. A.G. Sunil, K.S. Kesavanarayanan, P. Kalaivani, S. Sathiya, V. Ranju, R.J. Priya, B. Pramila, F.D. Paul, J. Venkhatesh, C.S. Babu, Total oligomeric flavonoids of Cyperus rotundus ameliorates neurological deficits, excitotoxicity and behavioral alterations induced by cerebral ischemic-reperfusion injury in rats, *Brain Res Bull*, 84 (2011) 394–405. https://doi.org/10.1016/j.brainresbull.2011.01.008.
66. D. Jebasingh, D. Devavaram Jackson, S. Venkataraman, E. Adeghate, B. Starling Emerald, The protective effects of Cyperus rotundus on behavior and cognitive function in a rat model of hypoxia injury, *Pharm Biol*, 52 (2014) 1558–69. https://doi.org/10.3109/13880209.2014.908395.
67. C. Sutalangka, J. Wattanathorn, Neuroprotective and cognitive-enhancing effects of the combined extract of Cyperus rotundus and Zingiber officinale, *BMC Complement Altern Med*, 17 (2017) 135. https://doi.org/10.1186/s12906-017-1632-4.
68. B.S. Thippeswamy, P. Nagakannan, B.D. Shivasharan, S. Mahendran, V.P. Veerapur, S. Badami, Protective effect of embelin from Embelia ribes Burm. against transient global ischemia-induced brain damage in rats, *Neurotox Res*, 20 (2011) 379–86. https://doi.org/10.1007/s12640-011-9258-7.
69. S. Bhuvanendran, Y. Kumari, I. Othman, M.F. Shaikh, Amelioration of cognitive deficit by embelin in a scopolamine-induced Alzheimer's disease-like condition in a rat model, *Front Pharmacol*, 9 (2018) 665. https://doi.org/10.3389/fphar.2018.00665.
70. P. Saini, L. Lakshmayya, V.S. Bisht, Anti-Alzheimer activity of isolated karanjin from Pongamia pinnata (L.) pierre and embelin from Embelia ribes Burm.f, *Ayu*, 38 (2017) 76–81. https://doi.org/10.4103/ayu.AYU_174_16.
71. M. Afzal, G. Gupta, I. Kazmi, M. Rahman, G. Upadhyay, K. Ahmad, F. Imam, M. Pravez, F.J.B. Anwar, A. Pathology, Evaluation of anxiolytic activity of embelin isolated from *Embelia Ribes*, 2 (2012) 45–47.

72. X. Ma, D.R. Gang, In vitro production of huperzine A, a promising drug candidate for Alzheimer's disease, *Phytochemistry*, 69 (2008) 2022–8. https://doi.org/10.1016/j.phytochem.2008.04.017.
73. D. Kim Thu, D.T. Vui, N.T. Ngoc Huyen, D.K. Duyen, B. Thanh Tung, The use of Huperzia species for the treatment of Alzheimer's disease, *J Basic Clin Physiol Pharmacol*, 31 (2019) https://doi.org/10.1515/jbcpp-2019-0159.
74. M.R. Howes, R. Fang, P.J. Houghton, Effect of Chinese herbal medicine on Alzheimer's disease, *Int Rev Neurobiol*, 135 (2017) 29–56. https://doi.org/10.1016/bs.irn.2017.02.003.
75. Z. Li, N. Ma, P.J. Zhao, Acetylcholinesterase inhibitory active metabolites from the endophytic fungus Colletotrichum sp. *YMF432, Nat Prod Res*, 33 (2019) 1794–7. https://doi.org/10.1080/14786419.2018.1434648.
76. R.J. Sajini, S. Prema, K.J.I.J.o.P.S. Chitra, Research, Phytoconstituents, *Pharmacological Activities Of Marsilea Minuta L.(Marsileaceae)-An Overview*, 10 (2019) 1582–7.
77. S.K. Bhattamisra, V.K. Khanna, A.K. Agrawal, P.N. Singh, S.K. Singh, Antidepressant activity of standardised extract of Marsilea minuta Linn, *J Ethnopharmacol*, 117 (2008) 51–7. https://doi.org/10.1016/j.jep.2008.01.012.
78. S.K. Bhattamisra, P.N. Singh, S.K. Singh, Effect of standardized extract of Marsilea minuta on learning and memory performance in rat amnesic models, *Pharm Biol*, 50 (2012) 766–72. https://doi.org/10.3109/13880209.2011.632421.
79. F. Fazal, P.P. Mane, M.P. Rai, K.R. Thilakchand, H.P. Bhat, P.S. Kamble, P.L. Palatty, M.S. Baliga, The phytochemistry, traditional uses and pharmacology of Piper Betel. linn (Betel Leaf): A pan-asiatic medicinal plant, *Chin J Integr Med*, (2014) https://doi.org/10.1007/s11655-013-1334-1.
80. A. Atiya, B.N. Sinha, U.R. Lal, Bioactive phenylpropanoid analogues from Piper betle L. var. birkoli leaves, *Nat Prod Res*, 31 (2017) 2604–11. https://doi.org/10.1080/14786419.2017.1285297.
81. V.P. Saka, S. Babu, V.J.J.o.C. Himaja, P. Sciences, *Effect of Aqueous Piper Betle Leaf Extract against Scopolamine Induced Amnesia on Albino Rats*, 10 (2017) 116–20.
82. A. Pandey, S. Bani, Hydroxychavicol inhibits immune responses to mitigate cognitive dysfunction in rats, *J Neuroimmunol*, 226 (2010) 48–58. https://doi.org/10.1016/j.jneuroim.2010.05.031.
83. X. Zhao, Y. Cui, P. Wu, P. Zhao, Q. Zhou, Z. Zhang, Y. Wang, X. Zhang, Polygalae Radix: A review of its traditional uses, phytochemistry, pharmacology, toxicology, and pharmacokinetics, *Fitoterapia*, 147 (2020) 104759. https://doi.org/10.1016/j.fitote.2020.104759.
84. W. Xue, J.F. Hu, Y.H. Yuan, J.D. Sun, B.Y. Li, D.M. Zhang, C.J. Li, N.H. Chen, Polygalasaponin XXXII from Polygala tenuifolia root improves hippocampal-dependent learning and memory, *Acta Pharmacol Sin*, 30 (2009) 1211–9. https://doi.org/10.1038/aps.2009.112.
85. X. Deng, S. Zhao, X. Liu, L. Han, R. Wang, H. Hao, Y. Jiao, S. Han, C. Bai, Polygala tenuifolia: a source for anti-Alzheimer's disease drugs, *Pharm Biol*, 58 (2020) 410–16. https://doi.org/10.1080/13880209.2020.1758732.
86. D. Zhang, X. Wang, R. Li, L. Wang, Z. Zhou, Q. Fu, Y. Cao, M. Fang, Extract of the aerial part of polygala tenuifolia attenuates d-Galactose/NaNO2-induced learning and memory impairment in Mice, *Planta Med*, 86 (2020) 1389–99. https://doi.org/10.1055/a-1212-3212.
87. X.F. Wang, H.H. Xiao, Y.T. Wu, L. Kong, J.C. Chen, J.X. Yang, X.L. Hu, Active constituent of Polygala tenuifolia attenuates cognitive deficits by rescuing hippocampal neurogenesis in APP/PS1 transgenic mice, *BMC Complement Med Ther*, 21 (2021) 267. https://doi.org/10.1186/s12906-021-03437-5.
88. A. Ghorbani M. Esmaeilizadeh, Pharmacological properties of Salvia officinalis and its components, *J Tradit Complement Med*, 7 (2017) 433–40. https://doi.org/10.1016/j.jtcme.2016.12.014.
89. A.L. Lopresti, Salvia (Sage): A review of its potential cognitive-enhancing and protective effects, *Drugs R D*, 17 (2017) 53–64. https://doi.org/10.1007/s40268-016-0157-5.
90. S.U. Savelev, E.J. Okello, E.K. Perry, Butyryl- and acetyl-cholinesterase inhibitory activities in essential oils of Salvia species and their constituents, *Phytother Res*, 18 (2004) 315–24. https://doi.org/10.1002/ptr.1451.
91. A.L. Dinel, C. Lucas, D. Guillemet, S. Layé, V. Pallet, C. Joffre, Chronic supplementation with a mix of Salvia officinalis and Salvia lavandulaefolia improves morris water maze learning in normal adult C57Bl/6J Mice, *Nutrients*, 12 (2020) https://doi.org/10.3390/nu12061777.

92. M. Nassiri-Asl H. Hosseinzadeh, Review of the pharmacological effects of Vitis vinifera (Grape) and its bioactive constituents: An update, *Phytother Res*, 30 (2016) 1392–403. https://doi.org/10.1002/ptr.5644.
93. S. Sreemantula, S. Nammi, R. Kolanukonda, S. Koppula, K.M. Boini, Adaptogenic and nootropic activities of aqueous extract of Vitis vinifera (grape seed): an experimental study in rat model, *BMC Complement Altern Med*, 5 (2005) 1. https://doi.org/10.1186/1472-6882-5-1.
94. N. Chaher, K. Arraki, E. Dillinseger, H. Temsamani, S. Bernillon, E. Pedrot, J.C. Delaunay, J.M. Mérillon, J.P. Monti, J.C. Izard, D. Atmani, T. Richard, Bioactive stilbenes from Vitis vinifera grapevine shoots extracts, *J Sci Food Agric,* 94 (2014) 951–4. https://doi.org/10.1002/jsfa.6341.
95. L. Ma, H. Xiao, J. Wen, Z. Liu, Y. He, F. Yuan, Possible mechanism of Vitis vinifera L. flavones on neurotransmitters, synaptic transmission and related learning and memory in Alzheimer model rats, *Lipids Health Dis*, 17 (2018) 152. https://doi.org/10.1186/s12944-018-0708-6.
96. P.K. Mukherjee, V. Kumar, M. Mal, P.J. Houghton, In vitro acetylcholinesterase inhibitory activity of the essential oil from Acorus calamus and its main constituents, *Planta Med*, 73 (2007) 283–5. https://doi.org/10.1055/s-2007-967114.
97. S. Chen, Z.J. Yin, C. Jiang, Z.Q. Ma, Q. Fu, R. Qu, S.P. Ma, Asiaticoside attenuates memory impairment induced by transient cerebral ischemia-reperfusion in mice through anti-inflammatory mechanism, *Pharmacol Biochem Behav*, 122 (2014) 7–15. https://doi.org/10.1016/j.pbb.2014.03.004.
98. A. Soumyanath, Y.P. Zhong, E. Henson, T. Wadsworth, J. Bishop, B.G. Gold, J.F. Quinn, Centella asiatica extract improves behavioral deficits in a mouse model of Alzheimer's Disease: Investigation of a possible mechanism of action, *Int J Alzheimers Dis*, 2012 (2012) 381974. https://doi.org/10.1155/2012/381974.
99. R.Y. Gan, D. Zhang, M. Wang, H. Corke, Health benefits of bioactive compounds from the genus Ilex, a source of traditional caffeinated beverages, *Nutrients*, 10 (2018) https://doi.org/10.3390/nu10111682.
100. M. Bhanumathy, M.S. Harish, H.N. Shivaprasad, G. Sushma, Nootropic activity of Celastrus paniculatus seed, *Pharm Biol*, 48 (2010) 324–7. https://doi.org/10.3109/13880200903127391.
101. P.A. Kizhakke, S. Olakkaran, A. Antony, K.S. Tilagul, P.G. Hunasanahally, Convolvulus pluricaulis (Shankhapushpi) ameliorates human microtubule-associated protein tau (hMAPτ) induced neurotoxicity in Alzheimer's disease Drosophila model, *J Chem Neuroanat*, 95 (2019) 115–122. https://doi.org/10.1016/j.jchemneu.2017.10.002.
102. J. Huang, Y. Zhang, L. Dong, Q. Gao, L. Yin, H. Quan, R. Chen, X. Fu, D. Lin, Ethnopharmacology, phytochemistry, and pharmacology of Cornus officinalis Sieb. *et Zucc, J Ethnopharmacol,* 213 (2018) 280–301. https://doi.org/10.1016/j.jep.2017.11.010.
103. K.Y. Lee, S.H. Sung, S.H. Kim, Y.P. Jang, T.H. Oh, Y.C. Kim, Cognitive-enhancing activity of loganin isolated from Cornus officinalis in scopolamine-induced amnesic mice, *Arch Pharm Res*, 32 (2009) 677–83. https://doi.org/10.1007/s12272-009-1505-6.
104. Q. Yuan, C.W. Wang, J. Shi, Z.X. Lin, Effects of Ginkgo biloba on dementia: An overview of systematic reviews, *J Ethnopharmacol*, 195 (2017) 1–9. https://doi.org/10.1016/j.jep.2016.12.005.
105. N. Zirak, M. Shafiee, G. Soltani, M. Mirzaei, A. Sahebkar, Hypericum perforatum in the treatment of psychiatric and neurodegenerative disorders: Current evidence and potential mechanisms of action, *J Cell Physiol*, 234 (2019) 8496–508. https://doi.org/10.1002/jcp.27781.
106. A.E. Khalifa, Hypericum perforatum as a nootropic drug: enhancement of retrieval memory of a passive avoidance conditioning paradigm in mice, *J Ethnopharmacol*, 76 (2001) 49–57. https://doi.org/10.1016/s0378-8741(01)00210-0.
107. K.I. Avgerinos, C. Vrysis, N. Chaitidis, K. Kolotsiou, P.G. Myserlis, D. Kapogiannis, Effects of saffron (Crocus sativus L.) on cognitive function. A systematic review of RCTs, *Neurol Sci*, 41 (2020) 2747–54. https://doi.org/10.1007/s10072-020-04427-0.
108. G. D'Onofrio, S.M. Nabavi, D. Sancarlo, A. Greco, S. Pieretti, Crocus Sativus L. (Saffron) in Alzheimer's disease treatment: Bioactive effects on cognitive impairment, *Curr Neuropharmacol*, 19 (2021) 1606–16. https://doi.org/10.2174/1570159x19666210113144703.
109. Y. Ding, X. Bao, L. Lao, Y. Ling, Q. Wang, S. Xu, p-Hydroxybenzyl alcohol prevents memory deficits by increasing neurotrophic factors and decreasing inflammatory factors in a mice model of Alzheimer's Disease, *J Alzheimers Dis*, 67 (2019) 1007–19. https://doi.org/10.3233/jad-180910.

110. H.O. Malve, S.B. Raut, P.A. Marathe, N.N. Rege, Effect of combination of Phyllanthus emblica, Tinospora cordifolia, and Ocimum sanctum on spatial learning and memory in rats, *J Ayurveda Integr Med*, 5 (2014) 209–15. https://doi.org/10.4103/0975-9476.146564.
111. M. Vasudevan M. Parle, Memory enhancing activity of Anwala churna (Emblica officinalis Gaertn.): An ayurvedic preparation, *Physiol Behav*, 91 (2007) 46–54. https://doi.org/10.1016/j.physbeh.2007.01.016.
112. S. Banerjee, U. Anand, S. Ghosh, D. Ray, P. Ray, S. Nandy, G.D. Deshmukh, V. Tripathi, A. Dey, Bacosides from Bacopa monnieri extract: An overview of the effects on neurological disorders, *Phytother Res*, 35 (2021) 5668–79. https://doi.org/10.1002/ptr.7203.
113. J. Williams, D. Sergi, A.J. McKune, E.N. Georgousopoulou, D.D. Mellor, N. Naumovski, The beneficial health effects of green tea amino acid l-theanine in animal models: Promises and prospects for human trials, *Phytother Res*, 33 (2019) 571–83. https://doi.org/10.1002/ptr.6277.
114. A. Bhat, A.M. Mahalakshmi, B. Ray, S. Tuladhar, T.A. Hediyal, E. Manthiannem, J. Padamati, R. Chandra, S.B. Chidambaram, M.K. Sakharkar, Benefits of curcumin in brain disorders, *Biofactors*, 45 (2019) 666–89. https://doi.org/10.1002/biof.1533.

8 Marine Plant Species

Nutritional and Medicinal Value for Brain Health

Shaileyee Das

8.1 INTRODUCTION

Extrapyramidal and pyramidal movement disorders, as well as cognitive and behavioural impairments, are the most common clinical manifestations of neurodegenerative illnesses (NDs). Only a few patients have pure abnormalities, with the majority having a mix of clinical symptoms [1]. Even though particular protein accumulations and anatomic frailty are commonly associated with neurodegenerative diseases, neurodegenerative diseases are part of numerous essential processes characterized by advanced neuronal malfunction, like proteotoxic anxiety and its associated aberrations in the autophagosomal/lysosomal and ubiquitin proteasomal processes, cell death, neuroinflammation, and oxidative stress. Furthermore, protein irregularities that identify neurodegenerative disorders can exist prior to the emergence of medical symptoms [2; 3; 4; 5; 6; 7; 8; 9], and an individual might have more than a neurodegenerative disease process [10; 11].

Tauopathies, amyloidoses, transactivation, α-synucleinopathies response TDP-43 proteinopathies are the most frequent neurodegenerative illnesses. The key histopathologic topographies required in determining a detailed neuropathologic analysis are abnormal protein conformations in various illnesses, as well as their cellular and neuroanatomical distribution. Tau in astrocytic plaques, thorn-shaped astrocytes, and tufted astrocytes are examples of protein accretions inside astrocytes. Tau in α-synuclein and coiled bodies in glial cytoplasmic inclusions are two proteins found in oligodendroglia. The protein has an aberrant structure with amyloid-like characteristics in many cases. The majority form filaments and β-pleated sheets are abundant in their secondary structures [12; 13].

Alzheimer's disease (AD) [14; 15], Parkinson's disease [16], dementia with Lewy bodies (DLB) [17], amyotrophic lateral sclerosis (ALS) [18], frontotemporal lobar degeneration with TDP-43 inclusions (FTLD-TDP) [19], and chronic traumatic encephalopathy (CTE) [20] have all had their own staging schemes developed. The increase in average life expectancy is one of the most significant factors causative to the spread of these diseases. Ageing is a common element in a wide range of disorders, yet there have been few therapeutic breakthroughs in the search for a cure. NDs are the world's sixth biggest cause of death, and their frequency rises in lockstep with age [21].

8.2 AMYLOIDOSES

Amyloids are fibrous proteins that are insoluble in water with unique structural features, such as a -sheet-rich secondary structure that allows dyes, such as Congo red, to attach to them. Almost all prevalent neurodegenerative illnesses contain protein abnormalities with amyloid features; extracellular protein collections, which are the crucial characteristic of non-neural amyloidoses, are present in only a minority of these diseases [22]. In neurodegenerative illnesses, amyloid-like filamentous

DOI: 10.1201/9781003300502-8

aggregates were commonly observed in the cytoplasm of neurons and glia. Amyloid plaques and amyloid angiopathy can be found in the brain parenchyma and blood vessel walls, respectively [23]. The kind or species of amyloid, the phase of the disease, and the localization of the deposition in the brain all influence the morphologic variety β -amyloid or Aβ, which is the most common form of amyloidosis [24]. In terms of peptide fragment size and posttranslational modifications, Aβ secluded from the human brain is biochemically varied [25, 26, 27, 28, 29, 30].

8.3 PRION DISEASES

In prion disorders, the transition of the cellular function's prion protein, PrPC, into the infectious form, PrPSc, is the most important molecular mechanism. The word 'prion' was invented in 1982 by Stanley B. Prusiner to define a communicable mediator that triggered scrapie in goats; sheep seemed to be totally constituted of protein [31]. Prion disorders are categorised as infectious diseases due to their transmissibility [32].

These diseases have a variety of clinical and pathologic characteristics, but they all show an accumulation of pathogenic PrP, which is associated with neuronal and synaptic loss, microvacuolation, and gliosis. Since these are communicable to certain other humans or experimental animal models, these sometimes are called transmissible spongiform encephalopathies.

The genetics and transmission route of prion disorders are used to classify them [33]. CJD is typically fast-advancing dementia with minimal if any, brain shrinkage. Spongiform alteration must not be disordered as superficial laminar spongiosis, that can occur in a variety of neurodegenerative diseases, including FTLD [34], or confined spongiform pathology, which is common in limbic regions and the temporal lobe in DLB [34; 35, 36].

8.4 TAUOPATHIES

Disorders characterised by uncontrolled buildup of tau protein in neurons and glia. They are further classified as primary or secondary [37; 38; 39; 40]. AD is the most obvious example, although there are others, like familial British dementia [41] and GSS [42]. Tau is a microtubule-related phosphoprotein found in abundance in axons and is tangled in microtubule polymerization and stability [43; 44; 45]. Tau undergoes a variety of posttranslational changes in addition to phosphorylation, including nitration, ubiquitination, glycation, and acetylation, all of which are associated with aberrant tau accumulation in degenerative tauopathies [46; 47; 48; 49]. Alterations in the tau protein gene, MAPT, were discovered in FTDP-17T [50], demonstrating that aberrant tau protein can induce neurodegeneration.

Another mRNA splicing of exons 2, 3, and 10 generates six main isoforms of tau protein [51]. Three isoforms with four preserved 32-amino-acid repeats (4R tau) are produced by another splicing of exon 10, while three isoforms with 3R tau in the region important for microtubules are formed by another splicing of exon 10. The amount and tau isoforms vary by cell type and illness, and they even segregate in some investigational cell models [52]. In glial cells, for example, 4R tau makes up the majority of tau. In neurodegenerative tauopathies, the proportional levels of 3R and 4R tau that gather vary [44; 51; 45]. Progressive supranuclear palsy, argyrophilic grain disease, cortico-basal degeneration, and various types of FTDP-17T are all 4R tauopathies. Some types of FTDP-17T and Pick's Disease are 3R tauopathies, whereas CTE, AD, tangle predominate dementia, some types of FTDP-17T, and Guam Parkinson dementia complex, are 3R/4R tauopathies [53].

8.5 ALZHEIMER'S DISEASE

Usually, there is also amyloid angiopathy. Noncompact, scattered amyloid deposits, dubbed 'pre-amyloid' deposits [54] for their weak or inattentive staining with standard amyloid dyes, are assumed to represent the earliest amyloid plaques. Dense cores can form in amyloid deposits, which are frequent in the major motor and visual cortices, depending on the brain region. Dystrophic neurites, as well as astroglial processes and microglia, make up the cellular components of amyloid plaques. Microglia are particularly common when dense deposits are present [55]. Tau-immunoreactive dystrophic neurites ('neuritic plaques') are the most clinically relevant kind of plaque in Alzheimer's disease [56, 57, 58]. Using a staging method proposed [15], the most current modification of neuropathological criteria [59] incorporates a calculation of topographic spreading before density of amyloid accumulations.

AD is categorised as a secondary tauopathy [60]. Mutations in MAPT, on the other hand, cause tauopathies but not Alzheimer's disease. Varying atrophy of multimodal connection cortices in the frontal, parietal, and temporal lobes, as well as occipital lobe atrophy, are macroscopic abnormalities in AD. The extent of cortical atrophy is linked to the location and compactness of tau pathology, as well as the clinical syndrome, but not to amyloid plaques [61; 62].

The inclusion-bearing neuron dies when NFTs mature, leaving an extracellular NFT, and tau masses can be found in neuronal perikarya as NFTs, dystrophic neuronal processes, neuritic plaques, and ubiquitous neuritic processes through afflicted grey matter. 'Neuropil threads' has been coined to describe the latter [63, 64, 65]. In ageing and AD, NFTs progress in an anatomically hierarchical manner, with proposing six stages [66, 67, 68]. Although most Alzheimer's disease cases follow the staging model, up to 20% of instances do not [69, 70, 71]. Limbic-predominant AD is a slowly progressive condition characterised by a severe amnestic state in the elderly, whereas hippocampal-sparing AD is more common in younger people and can manifest as isolated cortical syndromes [71]. Although the distribution of neurofibrillary disease differs between these forms of AD, no changes in the location of amyloid plaques have been found yet [72].

8.6 CHRONIC TRAUMATIC ENCEPHALOPATHY (CTE)

CTE has been known for many years as the result of chronic repetitive brain damage [73, 74]. The clinical condition is varied, although it is frequently linked to behavioural and cognitive characteristics [75]. Although it is classed as a primary tauopathy, it differs from other tauopathies in that it has an etiologic environmental component. Because of its link to repetitive head traumas in contact sports like American football, this condition has gained a lot of attention recently [76]. In some circumstances, the cavum septum pellucidum is fenestrated. Tau that builds up in CTE NFTs is comparable to tau that builds up in AD, especially 3R/4R tau [77]. NFTs have a different distribution than AD, with early neurofibrillary disease in the superficial cortical layers [78] and later medial temporal lobe involvement [79].

CTE has just been given new neuropathologic criteria [80]. CTE has been classified into four phases; although the staging system needs to be confirmed, preliminary research suggests that CTE can be discovered in nonprofessional athletes with other neurodegenerative illnesses [79, 81].

Neuronal tau pathology in a specific distribution is the hallmark lesion in CTE. ARTAG (ageing related tau astrogliopathy) is most found as thorn-shaped astrocytes in the mediobasal forebrain's subpial and perivascular spaces [82, 83]. However, astrocytes that are unclear or grainy can be seen in grey matter locations like the amygdala.

8.7 PROGRESSIVE SUPRANUCLEAR PALSY (PSP)

PSP is a tauopathy that most commonly manifests as nonconforming parkinsonism with vertical gaze palsy and early unexplained falls, although additional clinical manifestations are becoming more prevalent [84; 85]. PSP is characterised pathologically by aberrant tau inside neurons and glia. PSP differs from other 4R tauopathies biochemically because it is abundant in 4R tau isoforms [44]. The motor cortex and neostriatum have a lot of tufted astrocytes, whereas other affected areas have a lot less. Neuronal loss is linked to neuronal tau pathology rather than astroglial tau pathology. Tau builds up in the afflicted brain regions' neuronal and glial processes ('threads').

8.8 CORTICO-BASAL DEGENERATION (CBD)

Cortico-basal Degeneration is a tauopathy that causes deterioration in the association cortices and the basal ganglia. It was first identified as a separate entity in 1968 [87, 88] and invented with the term 'CBD' in 1989. Atypical parkinsonism affects many patients. Cortical neurons with chromatolytic-like alterations, often known as 'ballooned neurons', were the focus of the early neuropathologic reports. Only later was tau pathology recognised as a significant component [89; 90]. In addition to varying tau staining, inflated neurons have neurofilament and a-B-crystallin immunoreactivity [91]. They are seen in cortices with localised atrophy, amygdala, cingulate gyrus, and claustrum, where they are linked to other age-related conditions like argyrophilic grin sickness (AGD). Astrocytic plaques, which are characteristic tau astrocytic lesions, are common in afflicted cortices and the neostriatum [90, 92].

Patients with symmetrical cortical impairments are becoming documented, even though clinical symptoms linked with CBD are frequently asymmetric [91, 93, 94]. Clinically, equal CBD patients are hard to distinguish from PSP patients, although they have distinct biochemical and pathologic results [95]. Argyrophilic Grain Disease, often known as Pick's Disease, is another kind of ND.

8.9 SYNUCLEINOPATHIES

The presynaptic protein (α-synuclein) aggregates in neurons and glia in this family of neurodegenerative disorders (mostly oligodendroglia). It is a 140-amino-acid protein that was first identified as an NACP (non-amyloid component of senile plaques) in Alzheimer's disease [96]. It is possible that it is involved in synaptic vesicle trafficking [97, 98].

Multiple system atrophy (MSA) is characterised by abnormal buildup of a-synuclein in addition to Lewy body diseases [100]. The aggregates in MSA are usually found in oligodendrocytes, whereas they are mostly found in neurons in Lewy body disorders. In addition to perikaryal Lewy bodies, a-synuclein accumulates in neurites ('Lewy neurites')[101]. For unknown reasons, men are more likely to develop Lewy body diseases. Many individuals experience autonomic dysfunction, sleep difficulties, and mental symptoms.

Lewy bodies are classified into two histologic types based on their major anatomic location: brainstem-type and spinal-cord-type. Lewy bodies include around 300 distinct proteins, according to proteomic investigations [102], with α-synuclein being the defining hallmark. Neurofilament [103] and the ubiquitin-binding protein p62/sequestosome are two further components that have been identified [104]. Antibodies to tau reveal immunoreactivity in a subgroup of Lewy bodies [105]. The causes of aberrant α-synuclein conformation are unknown. However, numerous posttranslational changes, like truncation, phosphorylation, and oxidative damage, are suspected [106, 107].

[108] suggested a pathological arrangement of Lewy body illnesses to describe patients with psychiatric or cognitive abnormalities rather than mainly Parkinsonism. He classified Lewy body illness into three categories: brainstem predominate, diffuse cortical, and limbic (transitional). Dementia is the most common symptom of diffuse cortical dementia. The Kosaka arrangement has been included in DLB diagnostic criteria [109], and it is also useful for classifying Lewy-related pathology in the context of Alzheimer's disease [59]. When observed in the setting of Parkinson's disease (PD), Lewy bodies and Lewy neurites, commonly referred to as 'Lewy-related pathology', exhibit a characteristic distribution, prompting Braak to suggest a PD staging plan [110]. The PD stage system needs to change, according to Brain Net Europe [111].

Despite the fact that these staging or classification procedures are effective, up to 50% of older adults with Lewy illness are misclassified. There is no single neurodegenerative mechanism in Lewy disorders, as evidenced by the various stages and categorization systems. Initial a-synuclein aggregates in Parkinson's disease can be discovered in the brainstem or the peripheral autonomic nervous system, and they are often found in the olfactory and limbic systems in AD and DLB [112; 113]. [114] presented a change created on cluster analysis of many instances of Alzheimer's disease with Lewy bodies. The cerebral cortex is involved in most cases of DLB and PD with dementia

[115]. It is important to remember that numerous Lewy body illnesses, particularly DLB and PD with dementia, feature AD-like pathophysiology [116]. The development of the underlying coexisting AD is thought to be responsible for the jaggedly partial PD with dementia [117]. Furthermore, the DLB diagnostic criteria take both AD-type, Lewy-related and diseases into account when predicting the possibility of the pathology being linked to the DLB clinical condition [109]. Multiple System Atrophy is another kind of ND.

8.10 TDP-43 (PROTEINOPATHIES)

TDP-43 is regulating gene RNA metabolism, splicing, and stress granules in addition to transcriptional repression [118, 119]. TDP-43 is also observed in the mainstream of elderly instances of hippocampal sclerosis, that would be characterised by discriminating neuronal death in the CA1 sector of the hippocampus, which is linked to amnesia, despite the fact that it is not pathognomonic [119, 120, 123, 124].

8.11 APPROVED THERAPIES

Despite the impact on society, Alois Alzheimer [1] reported the first instance of AD in 1907, and current medications merely dismiss symptoms, with really no disease-modifying therapies or treatments (DMTs) now available [4,6].

Again, for treatment of Alzheimer's disease, five medicines have been licenced (Figure 8.1). Tacrine, donepezil, galantamine, rivastigmine, are four acetylcholinesterase inhibitors (AChEI). Memantine, an N-methyl-D-aspartate (NMDA) receptor antagonist, is the sixth drug [11,12]. Antipsychotic and antidepressant medications are also used to control behavioural issues.

Early in the disease process, cholinergic systems are impacted, subsequent in a reduction in acetylcholine in neurons. This also causes enzymes intricate in the manufacture and breakdown of acetylcholine [13]. The goal of AChEIs is to boost acetylcholine levels by blocking the enzyme acetylcholinesterase (AChE), which is responsible for its breakdown [13,14]. The three types of AChEIs used in clinical practice (donepezil, rivastigmine, and galantamine) have been shown to be similarly effective in treating mild to moderate AD in studies [15]. However, this class of medications is linked to a number of side effects, including an increased risk of bradycardia, and syncope. In comparison to galantamine, and rivastigmine, donepezil has fewer gastro-intestinal side effects [13].

Together, amyloid plaques and NFT have the potential to overstimulate glutamate, the central nervous system's major excitatory a.a., resulting in stimulation of the NMDA receptor, which can lead to excitotoxicity and neuronal death. Through modulation of glutamate activity, memantine, an NMDA receptor antagonist, lowers calcium input into the cell [16]. The most common side effects are headache, dizziness, and disorientation, while agitation can occur in a small number of people [13]. When compared to monotherapy, studies of combination therapy showed improvements in cognitive purposes, behavioural signs, and purposeful independence in people with mild to severe Alzheimer's disease [17,18].

Even though the development of innovative therapeutics for AD is critical, the outcomes so far have not met expectations; it should be underlined that the causes of this scenario are not due to a lack of understanding of the biology and pathophysiology of CNS disorders [19].

As previously stated, Aβ peptide is generated by the act of β- and γ-secretase, and when Aβ peptide accumulates in the CNS [12,20,21]. Some compounds can prevent the Aβ peptide from aggregating or binding to Aβ monomers, keeping them in a non-fibrillar state.

MAO A is involved in the deamination of neurotransmitters such as serotonin, dopamine, noradrenaline. MAO B is primarily responsible for the breakdown of benzylamine and phenyl ethylamine. Equally isoforms are linked to AD pathogenesis and could be targeted. MAO A is linked to toxicity and neuronal death, while MAO B is linked to the Aβ peptide [12, 13, 22].

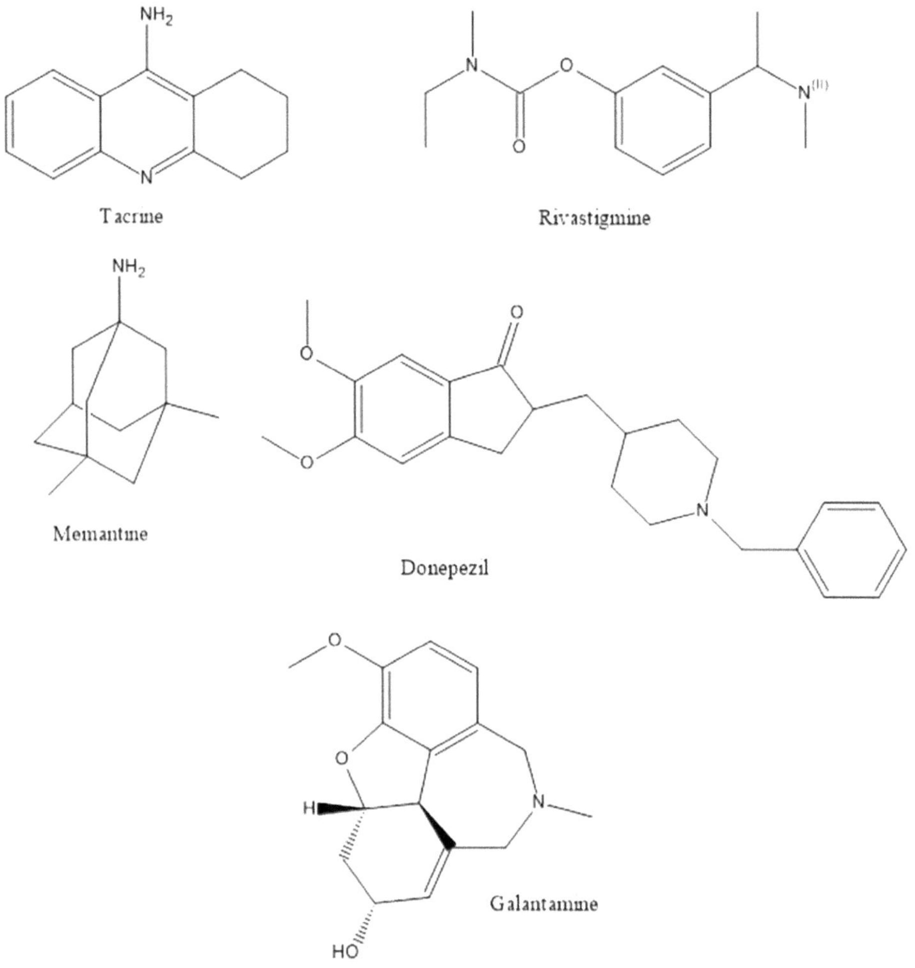

FIGURE 8.1 Five medicines have been licenced for the treatment of Alzheimer's disease: Tacrine, Donepezil, Galantamine, and Rivastigmine, are four acetylcholinesterase inhibitors (AChEI). Memantine, an NMDA receptor antagonist.

According to [23], In 2019, 132 agents were enrolled in 156 anti-AD therapy trials, with 28 in phase III, 74 in phase II, and 30 in phase I. 19 Agents targeted cognitive enhancement, 14 targeted neuropsychiatric and behavioural indications, and 96, the common, aimed to convert a DMT with amyloid as their primary target.

In 42 phase III clinical studies, 28 drugs are currently being evaluated. Eight drugs (AVP-786, nabilone, escitalopram, AXS-05, methylphenidate, brexpiprazole, zolpidem, and mirtazapine) target neuropsychiatric symptoms such apathy, agitation, and sleep disturbances, while three molecules are being explored as symptomatic cognitive enhancers (guanfacine, octohydroaminoacridine succinate, ginkgo biloba). Three have neuroprotective qualities (E-EPA, ACB101, BHV4157), and three have anti-inflammatory capabilities (masitinib, COR388, and Losartan + amlodipine + atorvastatin) in relation to DMT. Nine of the drugs are anti-amyloid (aducanumab, solanezumab, cad106, gantenerumab, crenezumab, E2609, etc.), one is anti-tau (TRx0237), and one is both anti-amyloid and anti-tau [23].

Drug development in AD has centred on DMT therapies in recent years, with the goal of changing the disease process by affecting distinctive brain alterations. Several entrants failed to go through clinical trials, and no DMTs were found to make a substantial alteration in phase III clinical trials or acquired marketing authorisation [24]. The amyloid hypothesis is the most researched; however, due to a lack of success in developing specific medicines, it has been abandoned in favour of other approaches, like combination therapy [23]. A one-target high-affinity medication method is unlikely to be beneficial in AD. Since AD is a neurodegenerative disease involving several pathways, it is currently assumed that finding a medicine that can control multiple target pathways is the best approach [25]. The methodologies for identifying a promising treatment for Alzheimer's disease will be explored in this chapter. Marine natural products (MNPs) have outperformed other types of natural products (NPs) in clinical trials. Two further emerging techniques in AD medication research, polypharmacology and drug repurposing, will be highlighted together with the most promising MNPs in the pipeline for AD treatment. There will also be examples of drugs from this approach that have shown to have potential activity and are currently being researched.

8.12 MARINE NATURAL PRODUCTS

The seas, which span 70% of the earth's surface and are a source of primary minerals and bioactive compounds with a wide range of applications, are the cradle of life and home to a broad range of marine species [35,36]. More than 10,000 natural compounds with biotechnological potential have been identified from marine creatures, primarily from bottom-of-the-food-chain organisms (fungi, micro- and macroalgae, bacteria, sponges, gorgonians, bryozoans, sea cucumbers, nudibranchs, tunicates, and sea hares, among other marine organisms) [37]. As secondary metabolites involved in defensive/offensive mechanisms, these bioactive chemicals are the product of these species' required adaptations to survive in this 'primordial soup' [38]. Microorganisms such as bacteria, protozoa, fungi, and microalgae currently account for 38% of marine chemicals with pharmacological action [31]. About 20% comes from marine sponges and corals, with a focus on the Porifera phylum, which includes fish, molluscs, macroalgae, echinoderms, and coelenterates, with fewer papers on this topic [39].

Secondary metabolites from both terrestrial and marine organisms are valuable sources for novel medication development. Approximately half of all medications put into the market in the last 20 years were derived straight from tiny molecules of natural origin [28]. Current research has discovered that, midst NP's many biological actions, neuroprotective activity is a particularly beneficial trait that can help to avoid and mitigate brain illnesses [10]. The popular medications derived from natural creatures come from land-based bases [28]. Marine species, on the other hand, can be regarded as the most recent source of bioactive chemicals from an industrial standpoint. Because of the connection of a vast genetic variety of species with an enormous molecular and ecological diversity, oceans offer an exclusive and abundant basis of valuable active chemicals, notably for the pharmaceutical company. Marine creatures must constantly adjust to changing physicochemical qualities like pressure, salt content, temperature, radiation exposure, light penetration, and oxygen concentration in order to thrive. As a result, they can produce a wide range of metabolites, from simple molecules to more complicated compounds such as peptides and enzymes, which are commonly utilised to immobilise and trap predators [29].

The difficulties with marine chemicals stem from their complicated chemical structures, which are difficult to manufacture, and the fact that they are only available in small quantities from natural sources [30]. Since new sample collection techniques, spectrometric techniques, and separation procedures have been developed, the exploration of the marine environment has increased [31].

Marine chemicals are increasingly being studied, foremost for the development of novel chemical structures with intriguing beneficial properties, such as neuroprotective, immune-modulatory, anti-inflammatory, analgesic, and anticancer properties [28,32].

Alkaloids, phenolic chemicals, terpenoids, carotenoids, sterols, and polysaccharides derived from bacteria, fungi, algae, and weeds have a variety of biological activity and health advantages. Tables 8.1, 8.2, and 8.3 show the structure of various chemical compounds isolated from marine

TABLE 8.1
The structure of various chemical compounds isolated from marine sources of bacteria

Bacteria				
Marine source	**Scientific name**	**Source of collection**	**Chemical compound isolated**	**Nature of chemical**
marine bacteria	*Rapidithrix thailandica*	isolated from a decomposed wood, which was collected from the Andaman Sea in Thailand	Marinoquinoline A	Alkaloids (quinoline derivative)
strains TISTR 1742	*R. thailandica*	isolated from a submerged sea grass blade, which was collected from the Andaman Sea	3-(2′-aminophenyl)-pyrrole	Alkaloids (pyrrole derivatives)
SH5.13.2		isolated from a submerged sea grass blade, from a debris, which was collected from the Gulf of Thailand	2,2-dimethylpyrrolo-1,2-dihydroquinoline	
marine bacteria	*Saccharomonospora sp.*	isolated from the saline culture of a marine bacteria	saccharochlorines A	Phenolics (chlorinated phenolic metabolites)
			saccharochlorines B	
marine cyanobacterium	*Symploca sp.*		Tasiamide B	Peptides (containing eight amino acid residues)

Chemical structure	Mechanism of action	Results of animal studies	References
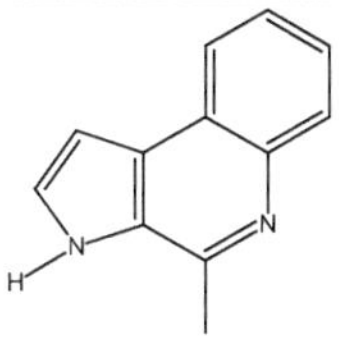	show an effective AChE inhibitory activity, IC_{50} = 4.9 μM (IC_{50} of galantamine = 0.6 μM)		127
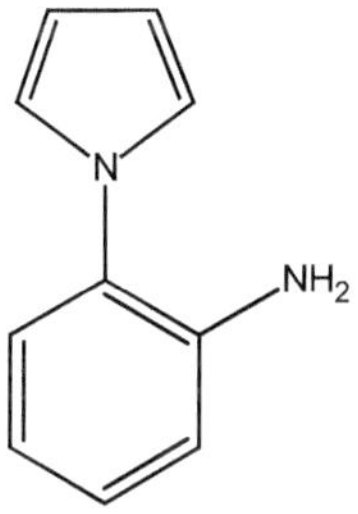 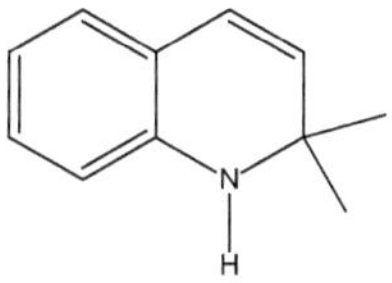	enzyme inhibition showed = < 30 % at 0.1 g/L)		127
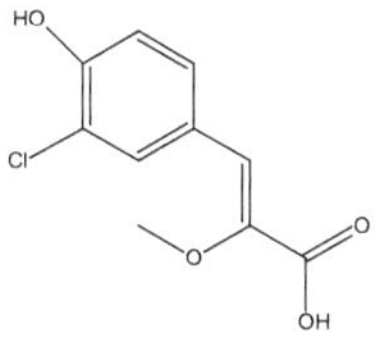 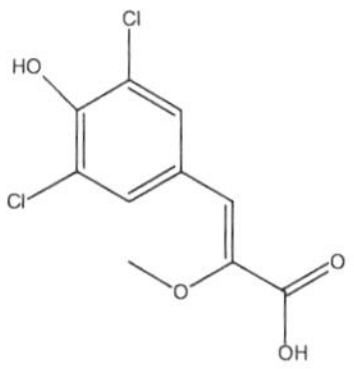	Showed weak inhibition of BACE-1 activity = 42.4% and 32.0% inhibition, respectively concentration was 50 μM		128
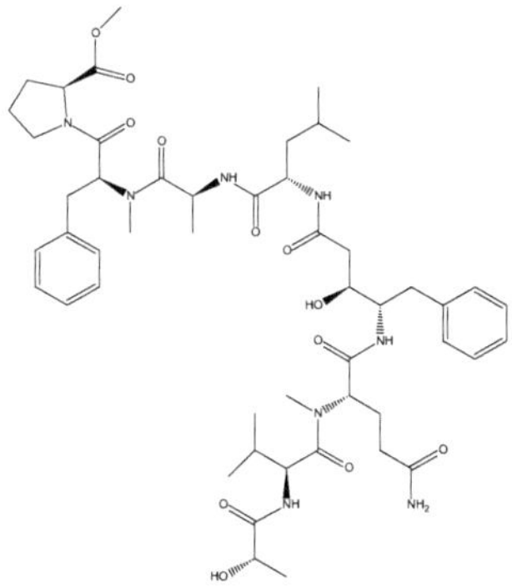	showed inhibition of BACE-1 with IC_{50} = 353 nM		129, 130

(*continued*)

TABLE 8.1 (Continued)
The structure of various chemical compounds isolated from marine sources of bacteria

Bacteria				
Marine source	**Scientific name**	**Source of collection**	**Chemical compound isolated**	**Nature of chemical**
marine cyanobacterium	*Lyngbya sp*		Tasiamide F	Peptides (containing eight amino acid residues)
	Streptomyces griseus	isolated from a deep-sea sediment which was collected from the Canary Basin	benzoxacystol	1,4-benzoxazine
Gram-positive bacteria	*Amycolatopsis, Salinispora*		Rifamycin	
	Streptomyces		Anhydroexfolia-mycin	
			Undecylprodi-gioisin	
Phylum Actinobacteria	*Streptomyces caniferus*		Caniferolide A	Polyol macrolides

Chemical structure	Mechanism of action	Results of animal studies	References
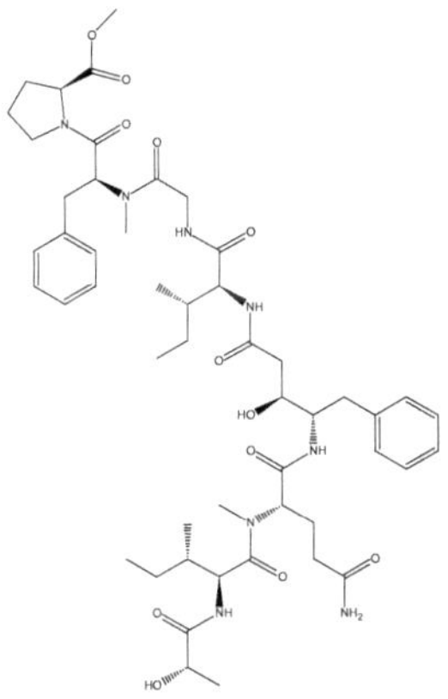	inhibited BACE-1 activity (IC_{50} = 690 nM)		131
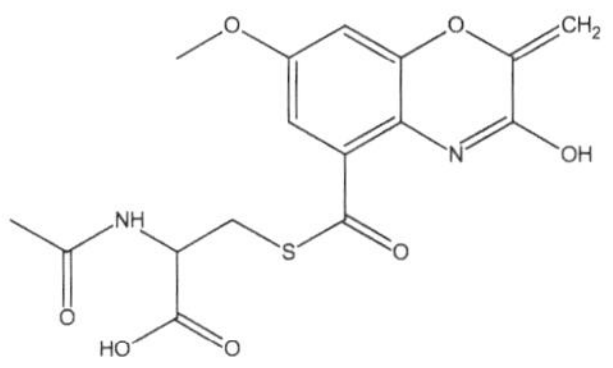		*in vitro* inhibitory activity showed against GSK-3β with IC_{50} = 1.35 μM	132
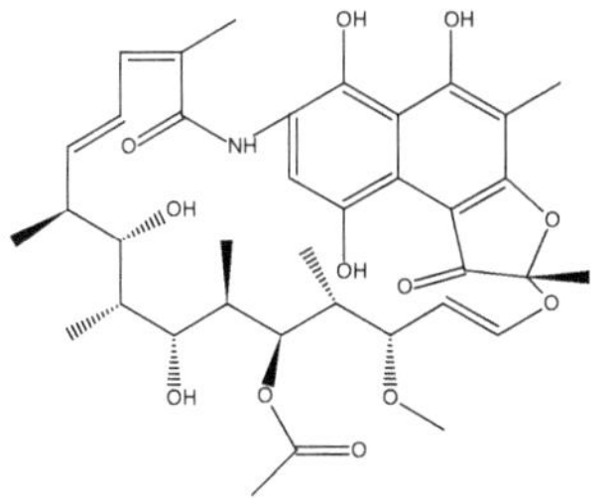	predictable anti-infectious activity, also displayed substantial neuroprotective activity in several experiments		133, 134
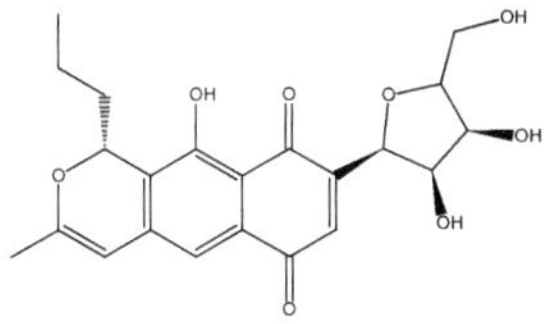 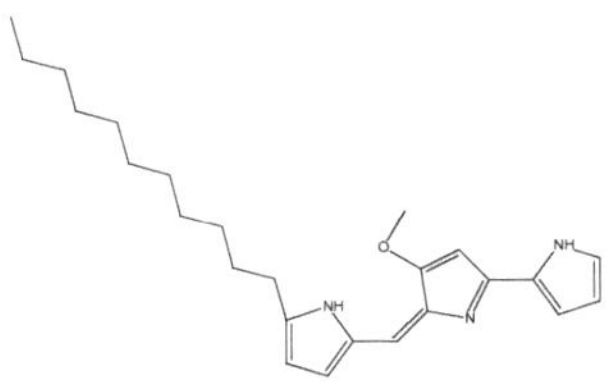	In an oxidative damage model, their effects were demonstrated in primary cortical neurons, suggesting their capacity to decrease the levels of reactive oxygen species (ROS) and to improve the antioxidant defences via elevation of the glutathione levels and catalase activity.		135
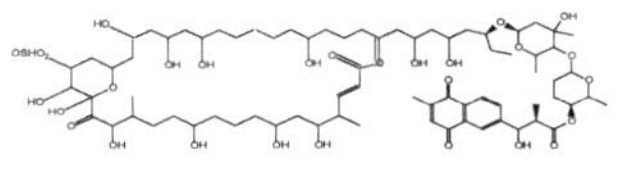	Anti-inflammatory and antioxidant action. Blockade of the BACE-1 enzyme.		136

TABLE 8.2
The structure of various chemical compounds isolated from marine sources of fungus

		Fungus		
Sl. No.	**Marine source**	**Scientific name**	**Source of collection**	**Chemical compound isolated**
1.	TK-43, an endophytic fungus	*Acrostalagmus luteoalbus*	obtained from a marine green alga *Codium fragile*, collected in Turkey	Acrozines A
				Acrozines B
				Acrozines C
2.	racemic mixture from a marine-derived fungus	*Penicillium commune* 366606	isolated from seawater, collected in China	Penicillamine
3.	mangrove endophytic fungus	*Penicillium chemesinum ZH$-E2*	Isolated from the stem of the mangrove plant *Kandelia candel*, and isolated from the ethyl acetate extract of cultures, which was collected from the South China Sea.	O-desmethylter-phenyllin
				3-hydroxy-6′-O-desmethylterphenyllin
				3″-deoxy-6′-O-desmethylcandidusin B
				3,3″-dihydroxy-6′-O-desmethylterphenyllin 6′-O-desmethylcandidusin B

Nature of chemical	Chemical structure	Mechanism of action	References
Alkaloids (N-methoxyindolediketopiperazine derivatives)	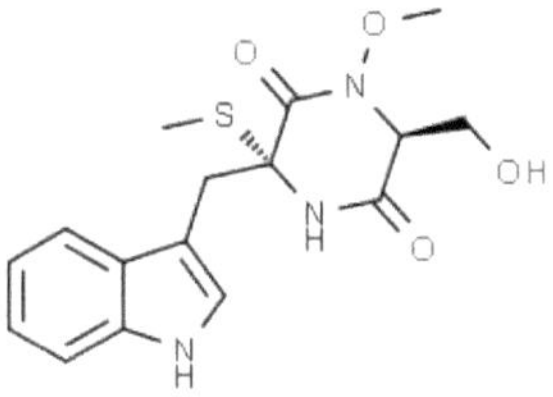	anti-AChE activity (IC50 = 9.5 μM)	137
	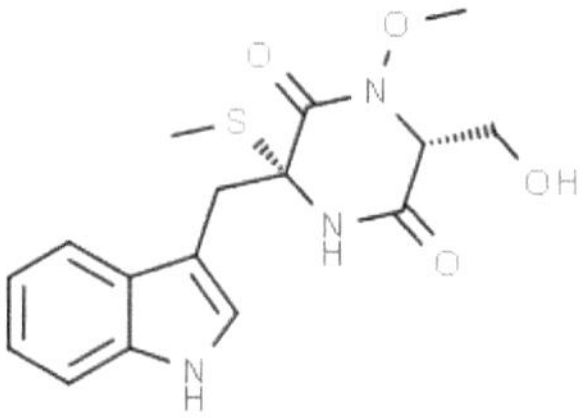	anti-AChE activity (IC50 = 60.7 μM)	
		anti-AChE activity (IC50 = 130.5μM)	
Alkaloids	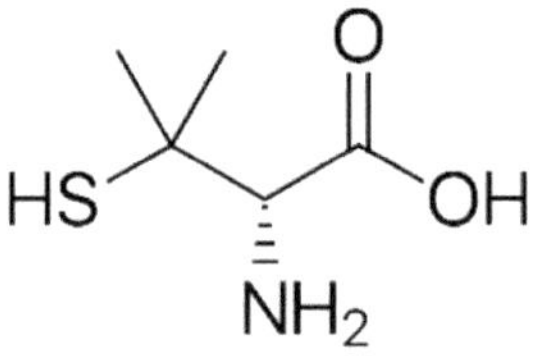	potent anti-AChE activity (32.4% inhibition)	138
Phenolics (p-terphenyls)	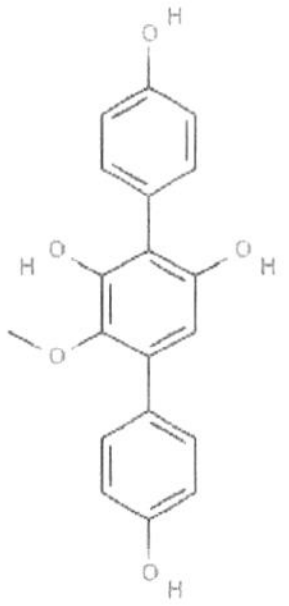 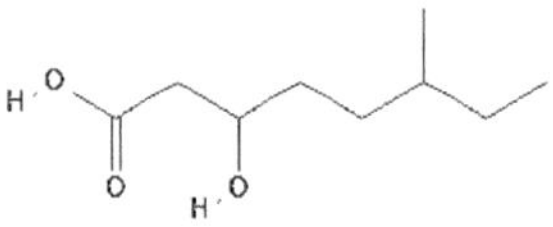 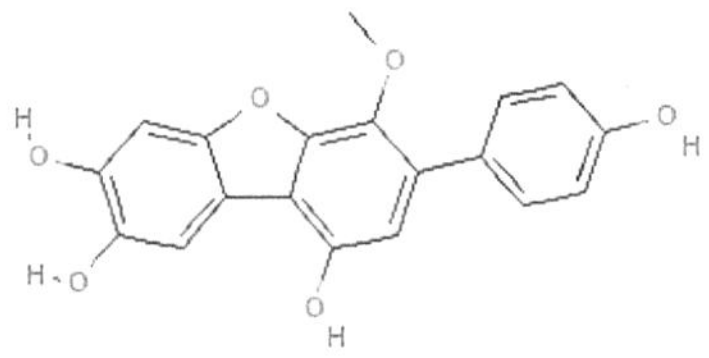	inhibitory activity against AChE with IC_{50} = 7.8 and 5.2 μM,	139

(*continued*)

TABLE 8.2 (Continued)
The structure of various chemical compounds isolated from marine sources of fungus

Fungus				
Sl. No.	**Marine source**	**Scientific name**	**Source of collection**	**Chemical compound isolated**
4.	Seaweed allied fungus	*Aspergillus unguis* DLEP2008001	collected from the South China Sea, and from the plasma-induced mutant of the fungus	Aspergillusidone A
5.	marine sponge-associate fungus	*Talaromyces sp.* strain LF458	isolated from tissues of a marine sponge *Axinella verrucose,* extracts of a culture broth and mycelia, which was collected at Punta di Fetovaia, Isle of Elba (Mediterranean Sea) Italy	Talaromycesone A
				Talaroxanthenone
6.	deep sea-derived fungus	*Ochroconis sp.* FS449	methanol extract of a fermentation broth, isolated from a sediment collected from the Indian Ocean	Ochuscins B Ochuscins D Ochuscins G
7.	Isolated from a culture of a mangrove-allied fungus	*Colletotrichum sp.* SCSIO KcB3- 2	obtained from a mangrove plant *K. candel*, which was collected in Guangdong Province, China	Bisabolanoic acid A
8.	Isolated from a culture of an endophytic fungus	*Aspergillus flavus*	isolated from the inner tissue of a marine red alga *Corallina officinalis*, which was collected from the coast of Yantai, China	3β,4α-dihydroxy-26-methoxyergosta-7,24(28)-dien-6-one
				Episterol,
				(22E,24R)-ergosta-7,22-dien-3β,5α,6α-triol

Nature of chemical	Chemical structure	Mechanism of action	References
Phenolics	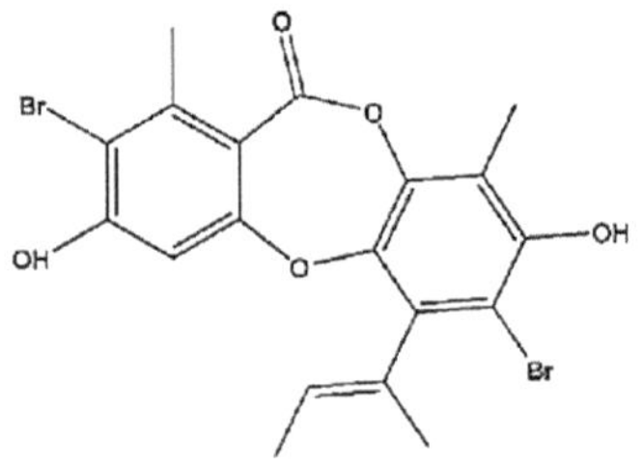	AChE inhibition with $IC_{50} = 56.8\ \mu M$	140
Phenolics (anti-AChE metabolites viz. oxaphenalenone dimer)	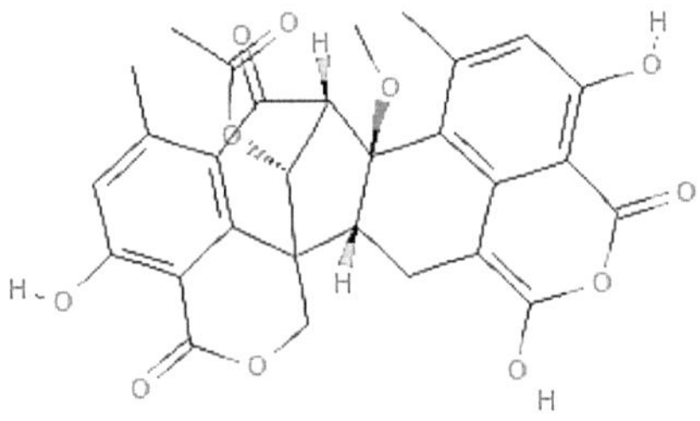	$IC_{50} = 7.49\ \mu M$	141
Phenolics (isopentenyl xanthenone)		AChE with $IC_{50} = 1.61\ \mu M$	
Phenolics (highly oxygenated usnic acid derivatives)		displayed moderate anti-AChE activity, with $IC_{50} = 50$ to $75\ \mu M$	142
Terpenoids (polychiral bisabolane sesquiterpene)		moderate inhibition of AChE activity with $IC_{50} = 2.2\ \mu M$	143
Terpenoids (Sterols)	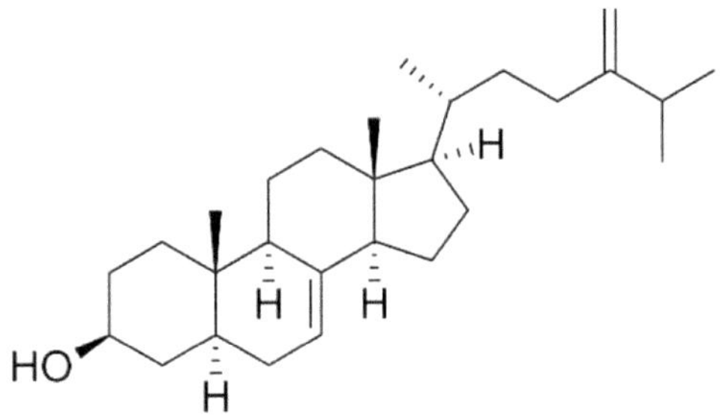	AChE inhibitory activity	144

(*continued*)

TABLE 8.2 (Continued)
The structure of various chemical compounds isolated from marine sources of fungus

Fungus				
Sl. No.	**Marine source**	**Scientific name**	**Source of collection**	**Chemical compound isolated**
				(22E,24R)-ergosta-5,22- dien-3β-ol
				(22E,24R)-ergosta-4,6,8(14),22-tetraen-3-one
9.	Marine originated fungus	*Aspergillus sp.* SCS-KFD66	isolated from a bivalve mollusc *Sanguinolaria chinensis*, which was collected from Haikou Bay, China	Asperpenes D Asperpenes E
10.	Marine originated fungus	*A. terreus* BCC51799	isolated from a decayed wood sample from the Gulf of Thailand	Territrem B
11.	Marine originated fungus	*A. terreus* SCSGAF0162	isolated from a gorgonian coral *Echinogorgia aurantiaca*, which was collected from the South China Sea	Territrems C

Nature of chemical	Chemical structure	Mechanism of action	References
	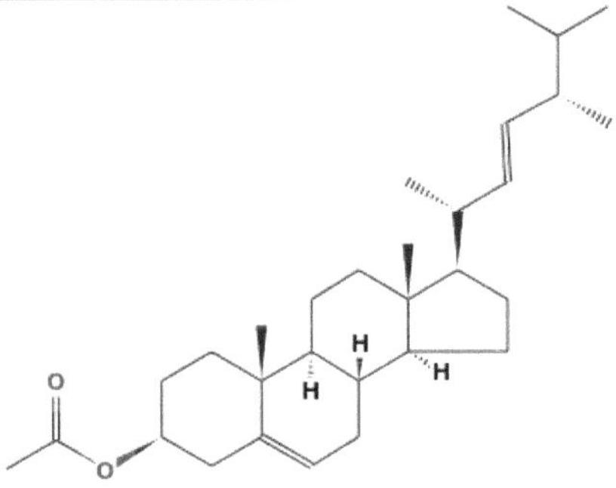 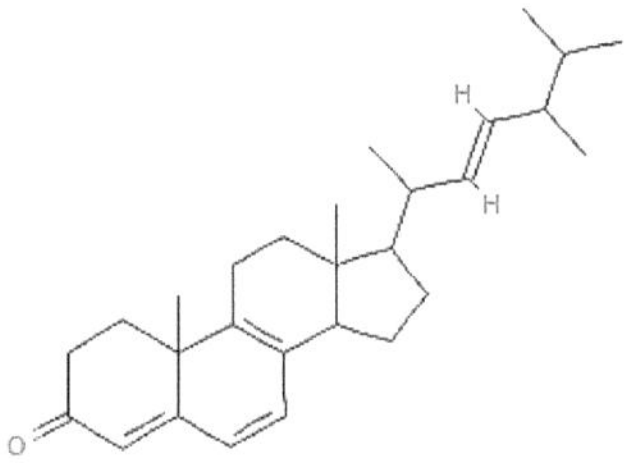		
Terpenoids (Meroterpenoids)		weak inhibitory activity against AChE	145
Terpenoids (Meroterpenoids)	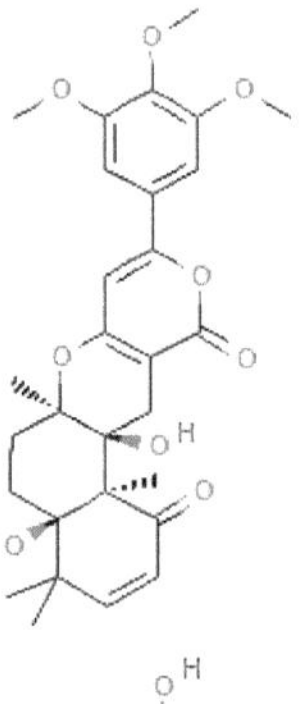	anti-AChE activity with IC_{50} = 0.071 μM (IC_{50} of galantamine	146
Terpenoids (Meroterpenoids)	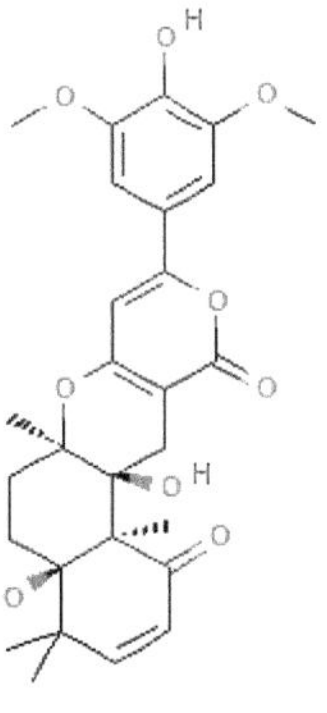	anti-AChE activity	147

(*continued*)

TABLE 8.2 (Continued)
The structure of various chemical compounds isolated from marine sources of fungus

Fungus				
Sl. No.	**Marine source**	**Scientific name**	**Source of collection**	**Chemical compound isolated**
				Territrems D Territrems E Arisugacin Terreulactone C
12.	endophytic fungus	*Penicillium sp.* SK5GW1L	isolated form leaves of a mangrove tree *K. candel*	five α-pyrone meroterpenoids 3-epiarigsugacin E
				Arisugacin B Arisugacin D
13.	endophytic fungus	*A. flavus*		(8E,12Z)-10,11-dihydroxyoctadeca-8,12-dienoic acid

Nature of chemical	Chemical structure	Mechanism of action	References
Terpenoids (Meroterpenoids)	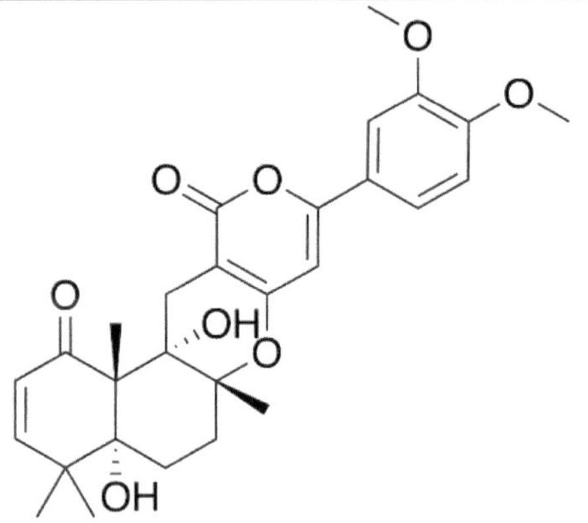 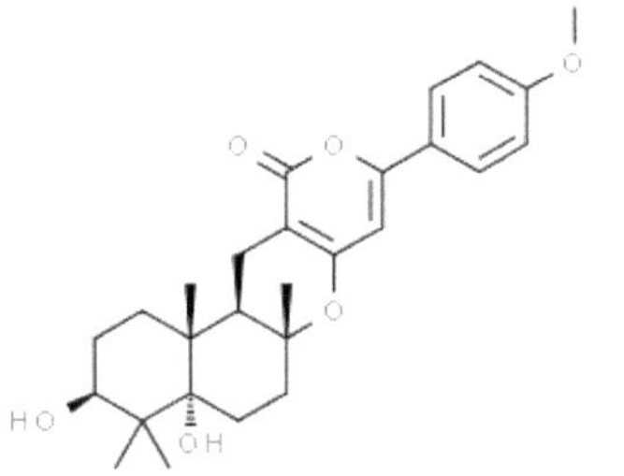 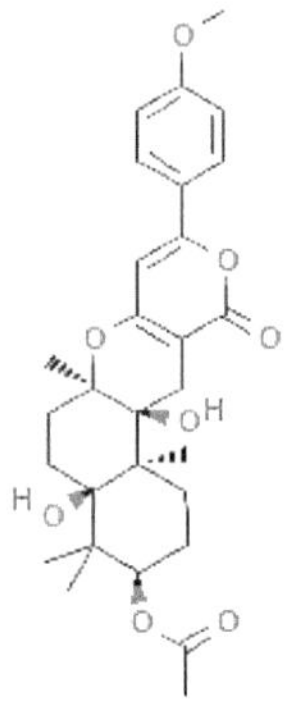	potent inhibitory activity	148
Marine Fatty Acids	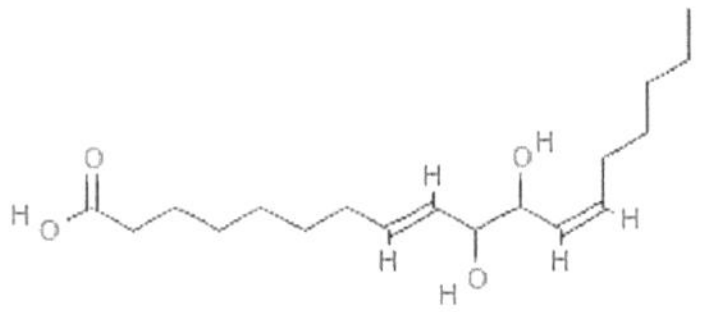	weak AChE inhibitory activity	149

(*continued*)

TABLE 8.2 (Continued)
The structure of various chemical compounds isolated from marine sources of macroalgae

		Fungus		
Sl. No.	**Marine source**	**Scientific name**	**Source of collection**	**Chemical compound isolated**
14.	marine originated fungus	*Biscogniauxia mediterranea* strain LF657	isolated from a sediment collected in Eastern Mediterranean Sea	cyclo-(L-Phe-L-Leu-L-Val-L-Leu-L-Leu)
15.	Marine originated fungus	*Biscogniauxia mediterranea*	isolated from a sediment collected in the Eastern Mediterranean Sea (2800 m depth)	Biscogniauxone
16.	Marine originated fungus	*Penicillium sp.* KF620	isolated from the water sample of the North Sea	Eutypoids B Eutypoids C Eutypoids D Eutypoids E
17.	marine-derived fungus	*Aspergillus sp. strains* LF666	isolated from a deep-sea sediment sample which was collected from the Mediterranean Sea	Pannorin
18.	marine-derived fungus	*Xylaria sp.*	obtained by mangrove fungus of the South China Sea coast.	Xyloketal B

Nature of chemical	Chemical structure	Mechanism of action	References
cyclopentapeptide		exhibited anti-AChE activity with a minimal inhibition concentration (MIC) = 5.87 μM. Huperzine A	150
Alkaloids (isopyrrolonaphthoquinone)		inhibit GSK-3β, with an IC_{50} = 8.04 μM.	150
Phenolics (Aromatic butenolides)	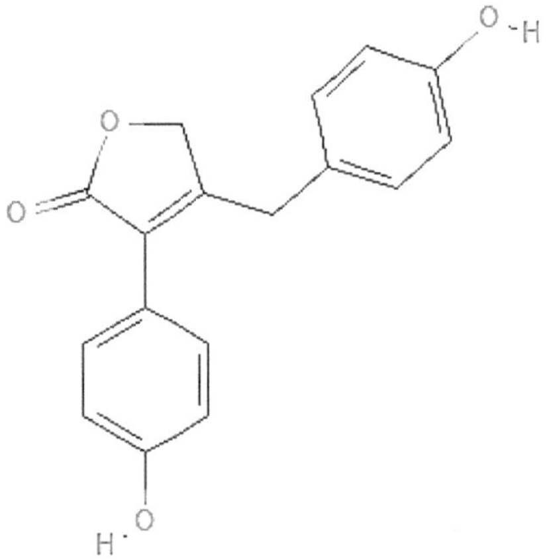	moderate inhibitory activity against GSK-3β with IC_{50} = 0.78, 0.67, 4.08, and 2.26 μM, respectively	151
oxygenated benzo- coumarins	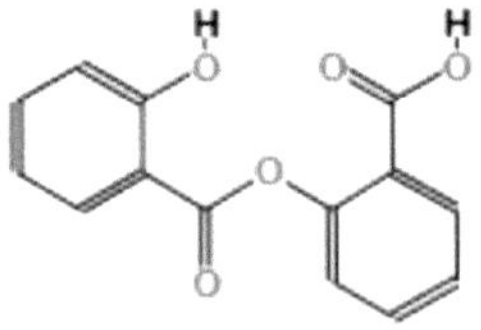		151
	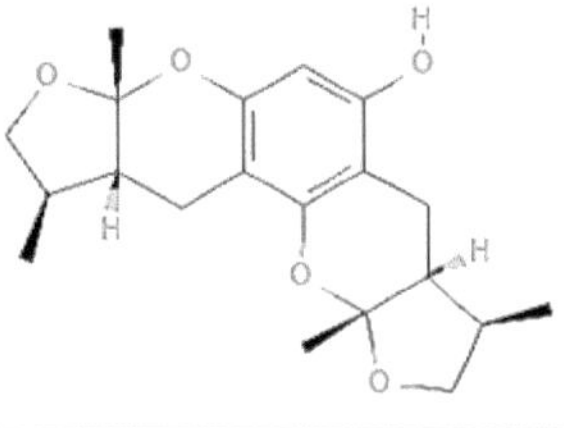	showed that the modulation of elements in the inflammatory pathway could regulate neurons state	152

TABLE 8.3
The structure of various chemical compounds isolated from marine sources of macroalgae

Macroalgae					
Sl. No.	**Marine source**	**Scientific name**	**Source of collection**	**Chemical compound isolated**	**Nature of chemical**
1.	brown alga	*Ecklonia maxima*	The aqueous methanol extract and organic solvent fractions of this algae, collected on the West coast of South Africa	phloroglucinol	Phenolics
				dibenzo[1,4] dioxine- 2,4,7,9-tetraol	
				eckol	
2.	red alga	*Grateloupia elliptica*	collected offshore of Jeju Island, Korea	6,6′-bieckol	Phenolics
3.	red alga	*Symphyocladia latiuscula*	collected in Korean water, by using a Multiple Enzyme Targets approach	2,3,6-tribromo-4,5-dihydroxybenzyl alcohol	Phenolics

Chemical structure	Mechanism of action	Results of animal studies	Outcomes of clinical Studies	References
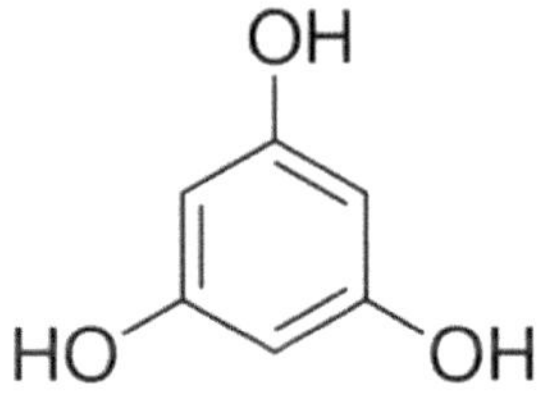 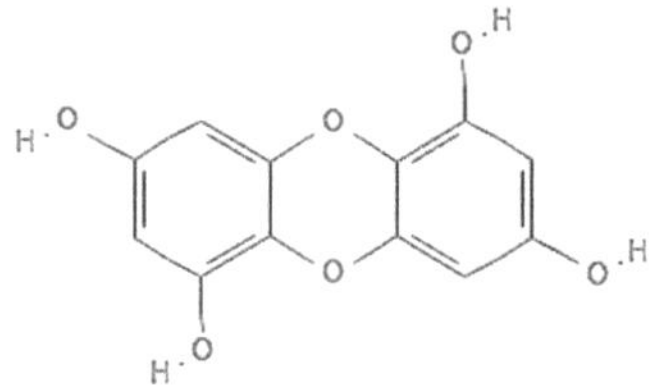 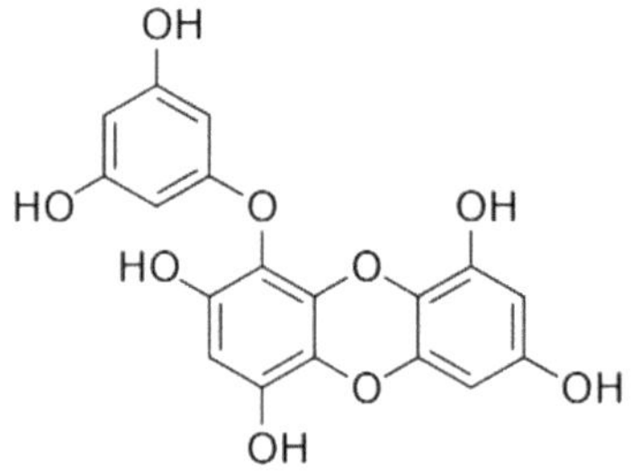	displayed anti-AChE inhibitory activity with IC_{50} = 84.48 and 76.70 μM, respectively			153
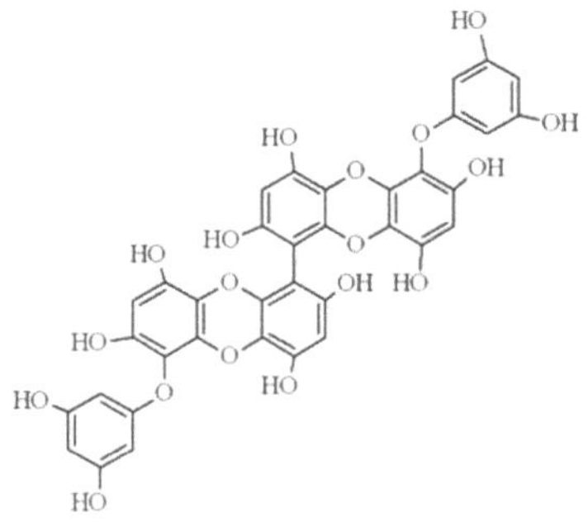	moderate inhibitory activity toward AChE and BuChE compared to donepezil; however, its anti-BuChE activity (IC_{50} = 27.4 μM) was more potent than its anti-AChE activity (IC_{50} = 44.5 μM)			154
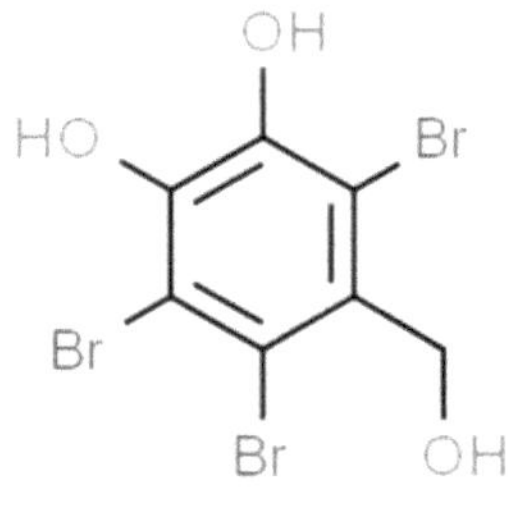	inhibited both AChE and BuChE activities, (IC_{50} =7.31 for AChE and 8.95 μM for BuChE)			155

(continued)

TABLE 8.3 (Continued)
The structure of various chemical compounds isolated from marine sources of macroalgae

Macroalgae					
Sl. No.	**Marine source**	**Scientific name**	**Source of collection**	**Chemical compound isolated**	**Nature of chemical**
				2,3,6- tribromo-4,5-dihydroxybenzyl methyl ether	
				bis-(2,3,6-tribromo-4,5-dihydroxybenzyl) ether	
4.	isolated from different populations of a red alga	*Laurencia dendroidea*	collected from the coast of Brazil	(-) elatol	Terpenoids (five halogenated sesquiterpenes)
				(+)-elatol	
				(+)-obtusol	
				(-)dendroidol	
				(-) cartilagineol	

Chemical structure	Mechanism of action	Results of animal studies	Outcomes of clinical Studies	References
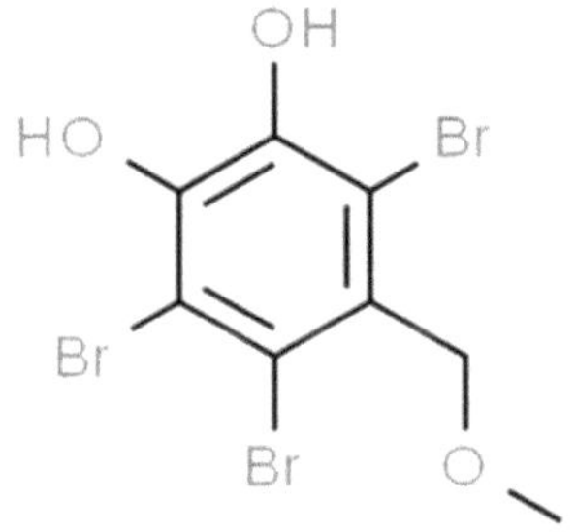	inhibited both AChE and BuChE activities, (IC_{50} = 9.61 μM for AChE and 14.41 μM for BuChE)			
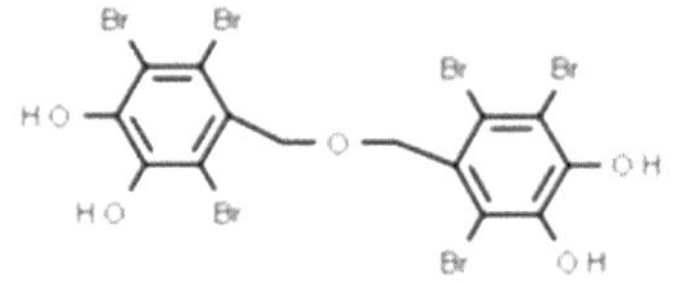	inhibited both AChE and BuChE activities, (IC_{50} = 2.66 μM for AChE and 4.03 μM for BuChE),			
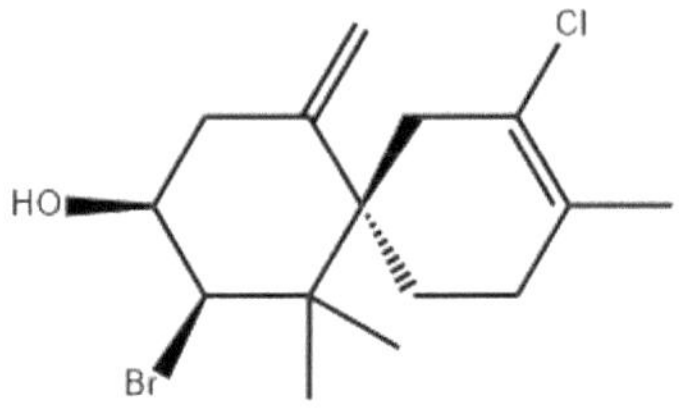	AChE inhibitory activity			156
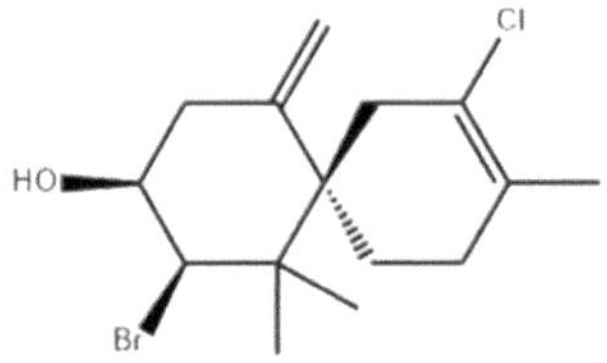	AChE inhibitory activity			
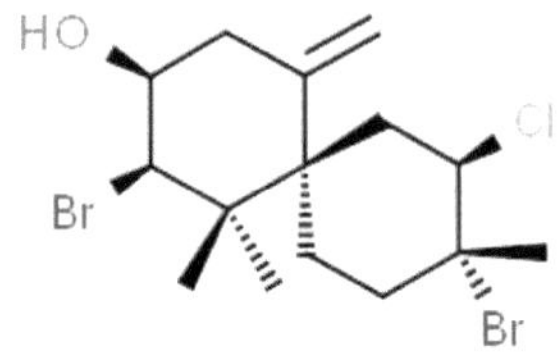	AChE inhibitory activity			
	AChE inhibitory activity			
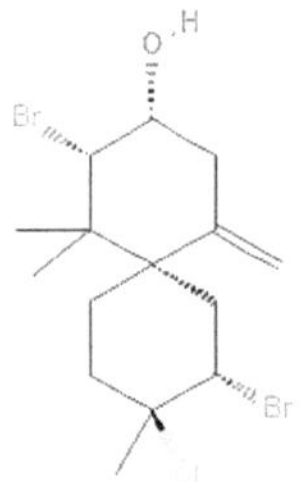	AChE inhibitory activity			

(continued)

TABLE 8.3 (Continued)
The structure of various chemical compounds isolated from marine sources of macroalgae

Sl. No.	Marine source	Scientific name	Source of collection	Chemical compound isolated	Nature of chemical
			Macroalgae		
5.	red alga	*Gelidiella acerosa*		phytol	Terpenoids (Diterpenes)
6.	isolated from the extract of a sea weed	*Sargassum horridum, E. stolonifera*		fucosterol	Terpenoids (Sterols)
7.	edible brown seaweeds	*S. horneri, Undaria pinnatifida and E. bicyclis*		Fucoxanthin	Terpenoids (Tetraterpenes, natural carotenoid)
8.	brown alga	*Ecklonia maxima*	collected from the West coast of South Africa	sargaquinoic acid	Terpenoids (Meroterpenoids, plastoquinones)
				sargachromenol	

Chemical structure	Mechanism of action	Results of animal studies	Outcomes of clinical Studies	References
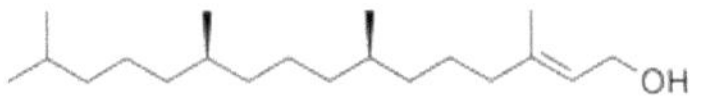	inhibited AChE and BuChE activities with IC_{50} = 95.27 and 2.07 μg/mL, respectively			157
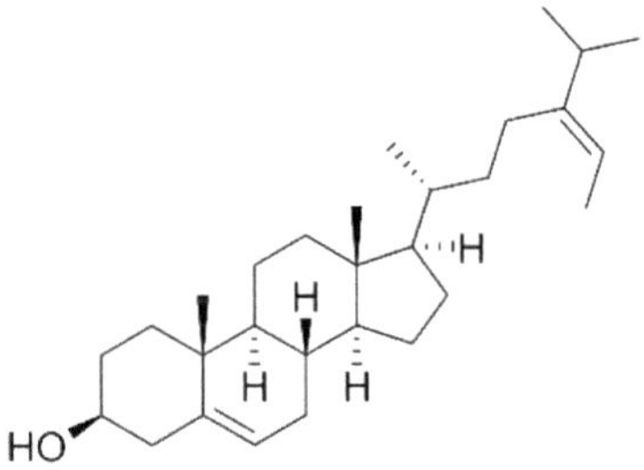	AChE inhibitory activity			158, 159
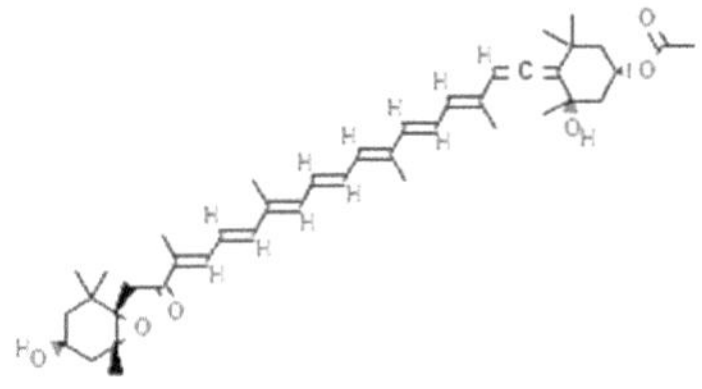		at 100 and 200 mg/kg, and donepezil significantly reduced the scopolamineinduced increase of AChE activity in the hippocampus of ICR mice. In an *in vitro* AChE activity assay, 88 was shown to directly inhibit AChE activity with an IC_{50} = 81.2 μM		159, 160, 161
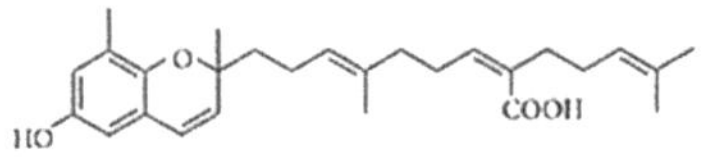	moderate anti-AChE activity, potent BuChE inhibitory activity			162, 163
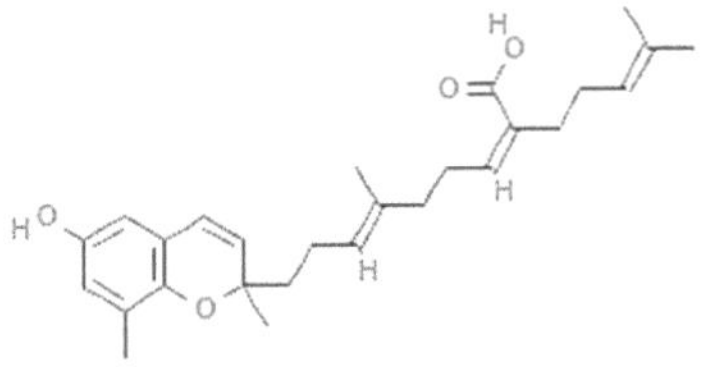	moderate anti-AChE activity			

(*continued*)

TABLE 8.3 (Continued)
The structure of various chemical compounds isolated from marine sources of macroalgae

Sl. No.	Marine source	Scientific name	Source of collection	Chemical compound isolated	Nature of chemical
Macroalgae					
				Sulfated polysaccharides	polysaccharides
9.		*S. serratifolium*	collected along the coast of Busan, South Korea	sargahydroquinic acid	Terpenoids (Meroterpenoids)
10.	edible brown seaweed	*Hizikia fusiforme*		oleic acid palmitic acid myristic acid arachidonic acid 11,14,17- eicosatrienoic acid	Fatty acids
11.	edible brown macroalgae	*Undaria pinnatifida*	purchased from the market in Korea	A glycoprotein MW of approximately 10 kDa, consisting of carbohydrate (42.53%) and protein components (57.47%)	glycoprotein
				FX	Pigments
				Fucoidan	Polysaccharides

Chemical structure	Mechanism of action	Results of animal studies	Outcomes of clinical Studies	References
		In vitro: stimulation of anti-oxidant activities (increase in anti-oxidant enzymes and glutathione content) in hippocampal cell line with Zn-induce damage		164
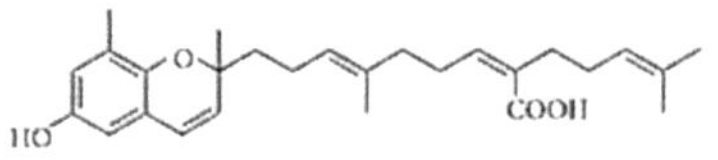	moderate anti-AChE activity			162
	anti-AChE activity with IC_{50} = 1 mg/mL			163
	inhibited AChE and BuChE activities with IC_{50} = 63.56 and 99.03 μM, respectively			165
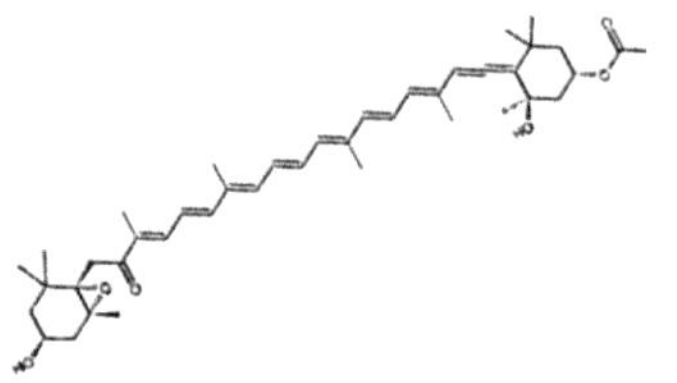		Reduction of oxidative stress in rat hippocampal neurons; Increase in neuron survival in traumatic brain injury models		166, 167
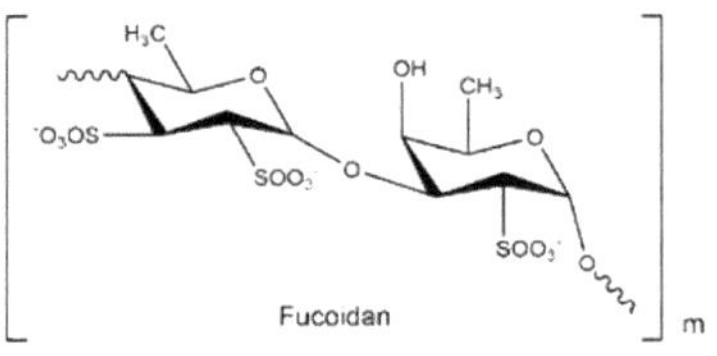		*In vitro*: reduction of A β and hydrogen peroxide-mediated cytotoxicity in PC12 cells		168

(*continued*)

TABLE 8.3 (Continued)
The structure of various chemical compounds isolated from marine sources of macroalgae

Macroalgae					
Sl. No.	**Marine source**	**Scientific name**	**Source of collection**	**Chemical compound isolated**	**Nature of chemical**
12.		*Eisenia bicyclis*	collected from the coast of South Korea	Fucofuroeckol-b	Phenolics (eckol derivative)
13.	green alga	*Caulerpa racemose*	collected from the Zhanjiang coastline in the East China Sea	racemosins A	Alkaloids (bis-indole)
				racemosins B	
				caulerpin	

Chemical structure	Mechanism of action	Results of animal studies	Outcomes of clinical Studies	References
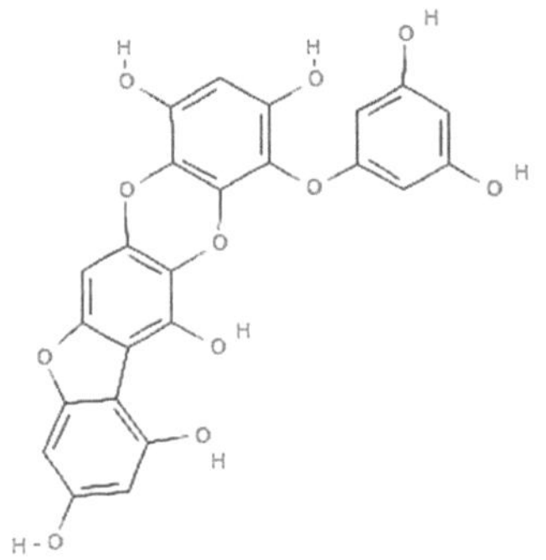	a strong β-secretase inhibition with IC_{50} = 16.1 μM. Kinetic study revealed a noncompet- itive inhibition pattern			169
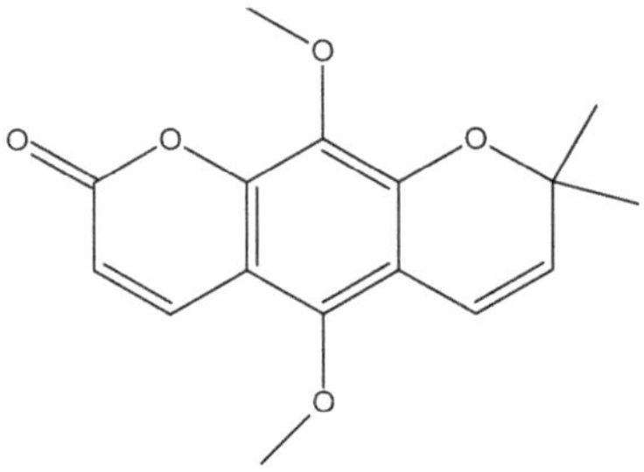	exhibited neuroprotection with 14.6% increase in cell viability at a concentration of 10 μM			170
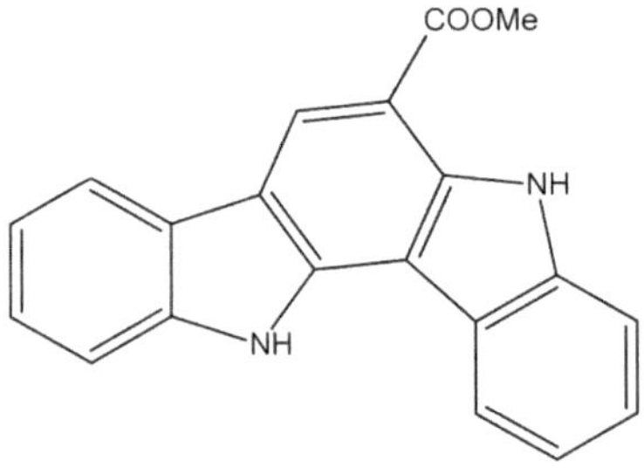	weak neuroprotective activity with 5.5% increase in cell viability at a concentration of 10 μM,			
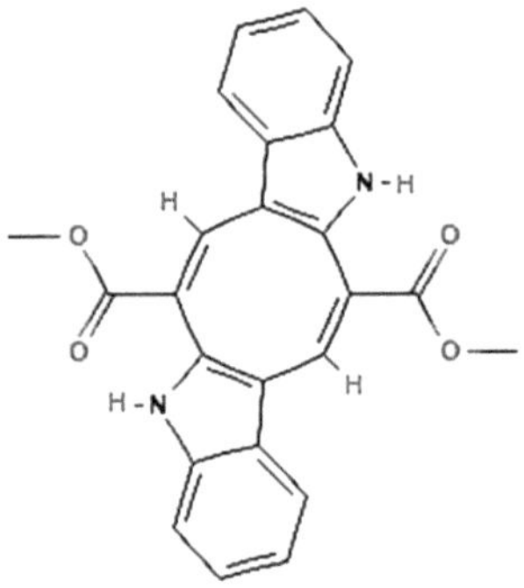	moderate neuroprotective activity with 8.1% increase in cell viability at a concentration of 10 μM,			

(*continued*)

TABLE 8.3 (Continued)
The structure of various chemical compounds isolated from marine sources of macroalgae

Macroalgae					
Sl. No.	**Marine source**	**Scientific name**	**Source of collection**	**Chemical compound isolated**	**Nature of chemical**
14.	green alga	*E. stolonifera*		dioxinodehydroeckol	Phenolics
				dieckol	
				phlorofucofuroeckol-A	
15.	green alga	*Caulerpa racemosa*	collected from the Zhanjiang coastline in the East China Sea	α-tocospirone	Terpenoids (Meroterpenoids, α-tocopherol derivative)
16.	South African brown macroalga	*Ecklonia maxima*		eckmaxo	Phenolics (phlorotannin)
17.	microalga	*Scenedesmus sp.*		1,2-di-O-palmitoyl-3-O-(6J-deoxy-6J-sulfo-D-glycopyranosyl)-glycerol	sulfolipids

Chemical structure	Mechanism of action	Results of animal studies	Outcomes of clinical Studies	References
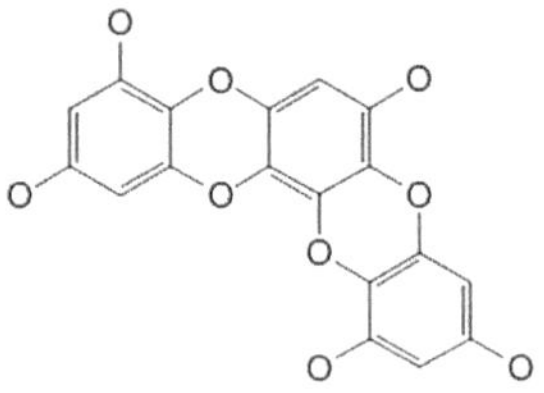 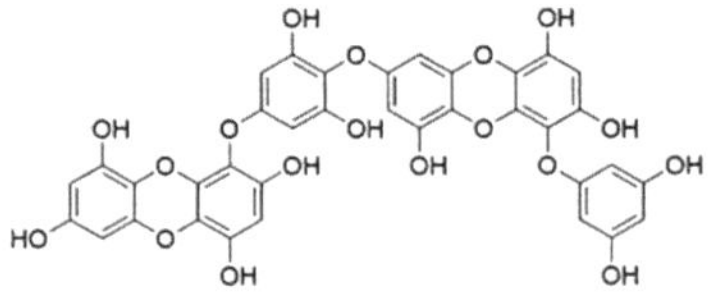	dose-dependent inhibitory effects on Aβ self-aggregation with IC_{50} values	inhibit Aβ self-aggregation by an *in vitro* ThT assay		171
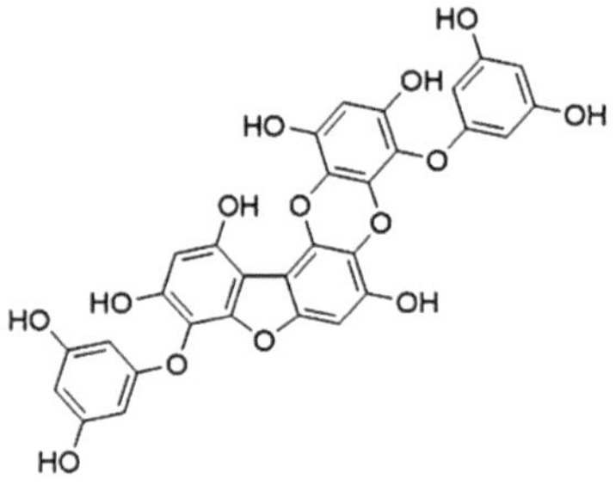	displayed a significant effect against Aβ induced cell damage in SH-SY5Y cells with 13.6% increase in cell viability at 10 μM			172
	showed an ability to produce anti-Aβ plaque neuroprotective effects in SH-SY5Y cells by preventing Aβ peptide-induced neuronal apoptosis, and an increase in ROS			174
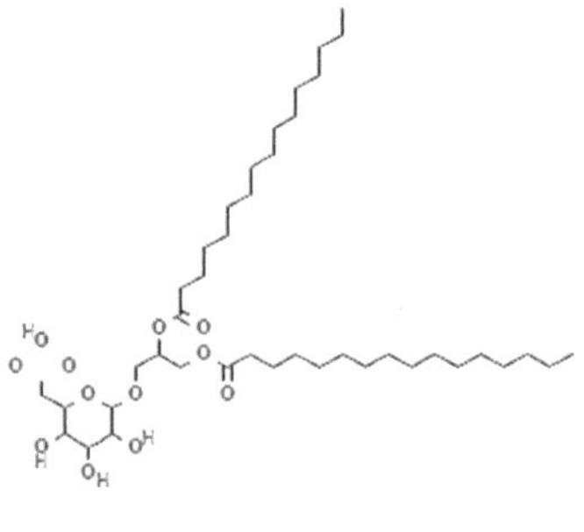	displayed QCs inhibition ranging from 76% to 81%			175

(*continued*)

TABLE 8.3 (Continued)
The structure of various chemical compounds isolated from marine sources of macroalgae

Macroalgae					
Sl. No.	**Marine source**	**Scientific name**	**Source of collection**	**Chemical compound isolated**	**Nature of chemical**
				1-O-palmitoyl-2-O-linolenyl-3-O-(6J-deoxy-6J-sulfo-D-glycopyranosyl)-glycerol	
				1-O-linolenyl-2-O-palmitoyl-3-O-(6J-deoxy-6J-sulfo-D-glycopyranosyl)-glycerol	
18.	seaweed	*Fucus vesiculosus*		fucophlorethol	phlorotannins
				tetrafucol	
				trifucodiphlorethol	
19.	Green algae	*Caulerpa Cupressoides*		Lectins	Glycoproteins

Chemical structure	Mechanism of action	Results of animal studies	Outcomes of clinical Studies	References
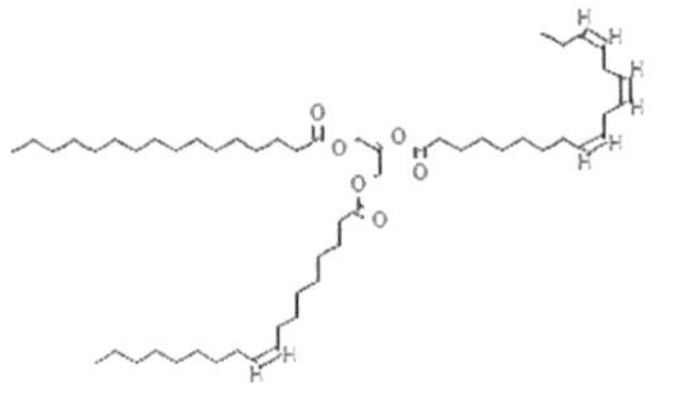				
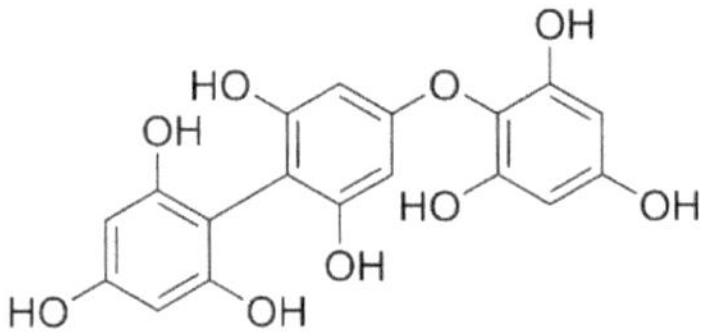	showed a promising activity with EC_{50} = 0.169 and 0.278 mg/m			176
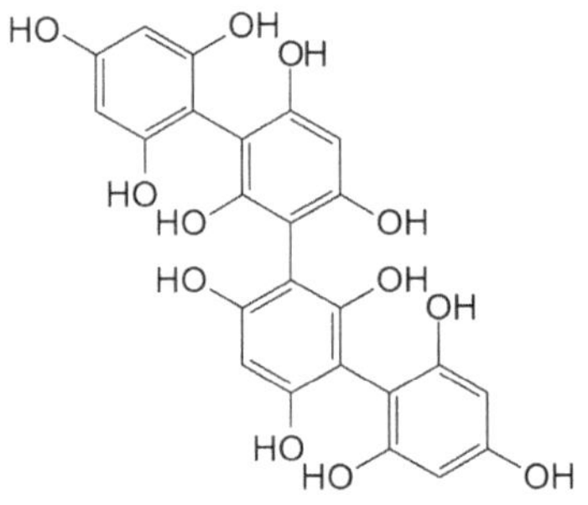				
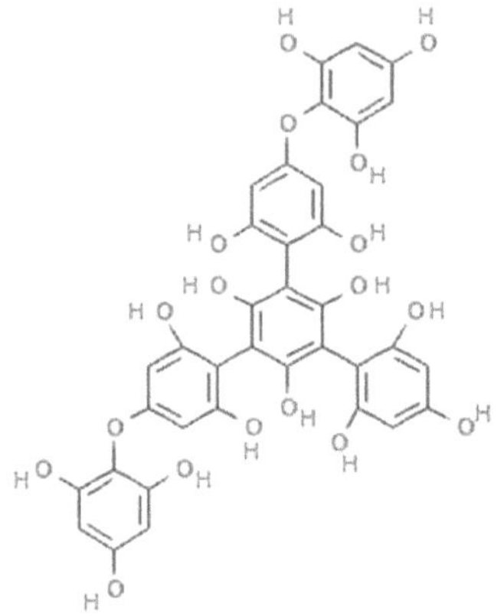				
		In vivo: anti-nociceptive and anti-inflammatory activity in Swiss mice		177

(*continued*)

TABLE 8.3 (Continued)
The structure of various chemical compounds isolated from marine sources of macroalgae

Macroalgae					
Sl. No.	**Marine source**	**Scientific name**	**Source of collection**	**Chemical compound isolated**	**Nature of chemical**
20.	Microalgae	*Haematococcus pluvialis*		AXT	Pigments
21.	Red algae	*Hypnea musciformis*		κ-carrageenan	Polysaccharides
22.	brown alga	*Saccharina japonica*		Fucoidan	Polysaccharides
23.	seaweed	*Codium Tomentosum*	collected from the Atlantic Ocean by high-pressure liquid chromatography analysis	different organic acid, such as oxalic acid, aconitic acid, ketoglutaric acid, pyruvic acid, malic acid, malonic and fumaric acids, and a great variety of volatile compounds, such as phenolic compounds	
24.	brown algae	*Ascophyllum nodosum*		tetrafucol A	phlorotannin

Chemical structure	Mechanism of action	Results of animal studies	Outcomes of clinical Studies	References
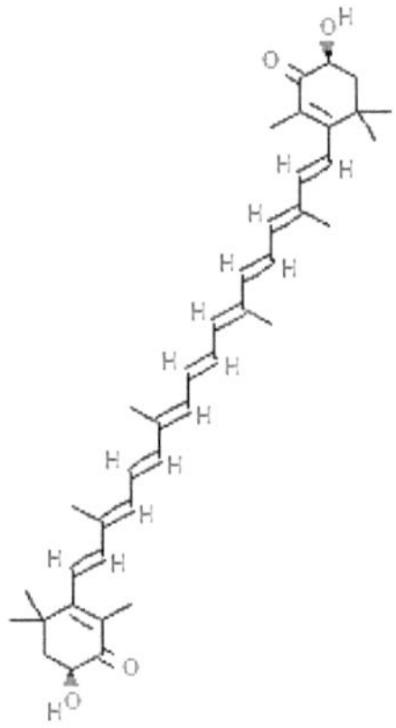		*In vitro*: protection from 6-OHDA-induced apoptosisandinhibition of mitochondrial impairment in SH-SY5Y	*In vivo*: antidepressant effects, anti-oxidant activity (mediated by an increasing in GSH and superoxide dismutase)	178, 179, 180
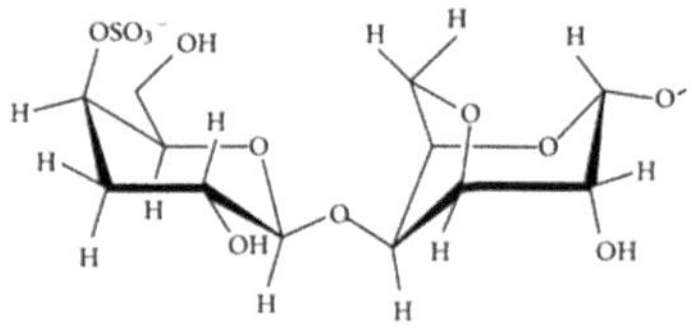		*In vitro*: anti-oxidant and cytoprotective effects against 6-OHDA-induced neurotoxicity in SH-5YSY models		181
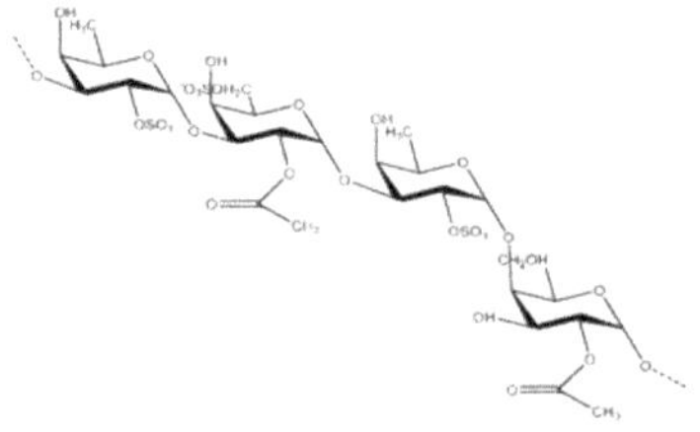	fucoidan treatment significantly improved the motor impairment in a MPTP mice model of PD	showed a protective effect in 1-methyl-4-phenyl-1,2,3,6-tetrahydropyridine (MPTP) C57BL/6 mic		182
	its ability to scavenge the reactive oxygen and nitrogen species			183
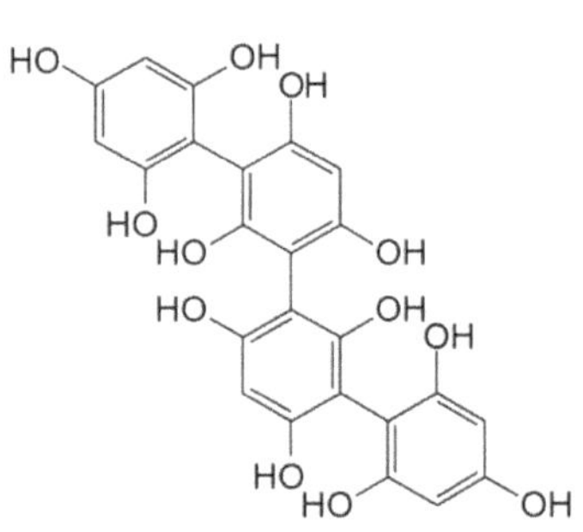				184

(*continued*)

TABLE 8.3 (Continued)
The structure of various chemical compounds isolated from marine sources of macroalgae

Macroalgae					
Sl. No.	**Marine source**	**Scientific name**	**Source of collection**	**Chemical compound isolated**	**Nature of chemical**
25.	coldwater seaweeds	*Porphyra yezoensis*		Porphyrin	
26.	Brown seaweeds	*Laminaria sp.*		Laminarin	polysaccharides
27.	green algae	*Ulva lactuca, Enteromorpha prolifera*			Sulfated oligosaccharides
28.	red marine algae	*Dictyota Dichotoma* and any algae		Homotaurine	natural amino acid

Chemical structure	Mechanism of action	Results of animal studies	Outcomes of clinical Studies	References
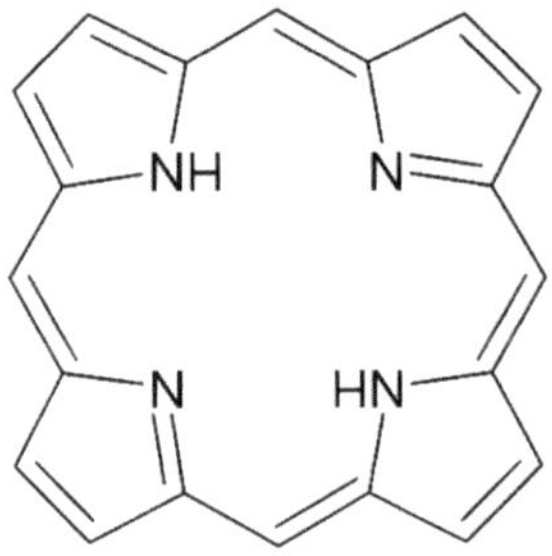	suppressed NO generation in LPS-stimulated RAW264.7 cells by downregulation of iNOS expression			185
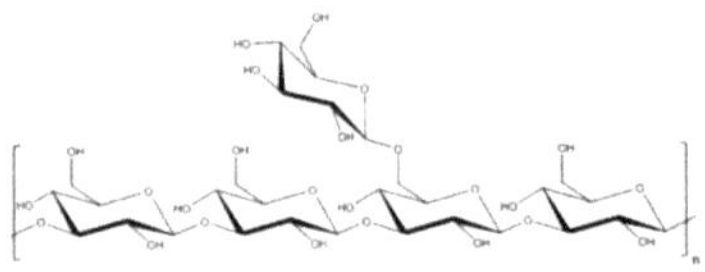	neuroprotective properties in cerebral ischaemic events through gliosis reduction and controlling pro-inflammatory microglia			186, 187
		decreased pro-inflammatory agents, reduced p53 and fork-head box protein O1 (FoxO1) genes, and caused overexpression of Sirt1 gene in SAMP8 mice		188
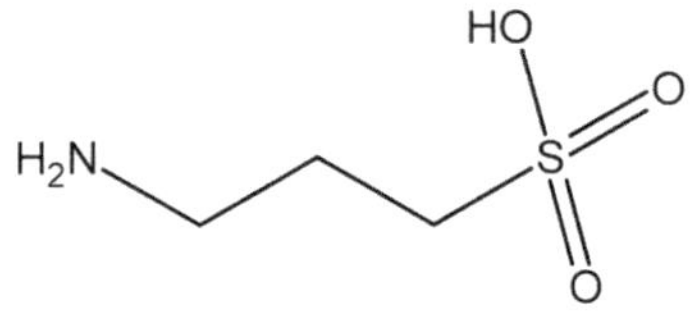	preclinical studies show that tramiprosate decreases the formation of A oligomers and the deposition of amyloid fibrils as plaques in an AD mouse model		In a phase II clinical trial, it was confirmed that tramiprosate safely decreases concentrations of A 42 in the cerebrospinal fluid (CSF) of individuals with mild to moderate AD. Furthermore, a pooled analysis of the two phase III trials among 2025 mild to moderate AD patients	189

(*continued*)

TABLE 8.3 (Continued)
The structure of various chemical compounds isolated from marine sources of macroalgae

Macroalgae					
Sl. No.	**Marine source**	**Scientific name**	**Source of collection**	**Chemical compound isolated**	**Nature of chemical**
29.	Brown Algae	*Dictyopteris undulata*		zonarol	Sesquiterpene
30.	Brown Algae	*Marginariella boryana*		Sulfated fucans	
31.	Brown Algae	*Ishige okamurae*		Diphlorethohydroxycarmalol (DPHC)	
32.	Brown Algae	*Zonaria spiralis*		Spiralisone A	
33.				Chromone 6	

Chemical structure	Mechanism of action	Results of animal studies	Outcomes of clinical Studies	References
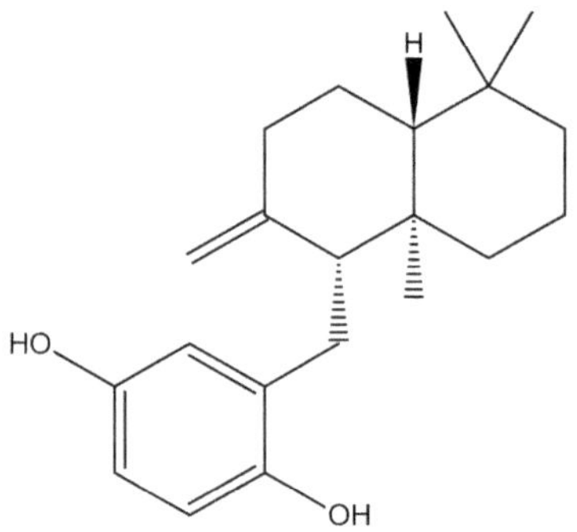		Activates the Nrf2/ ARE pathway, induces phase-2 enzymes, and protects neuronal cells from oxidative stres		190
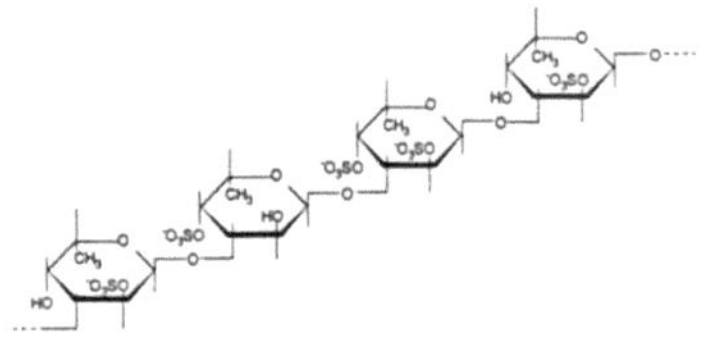		Prevents the accumulation of A		191
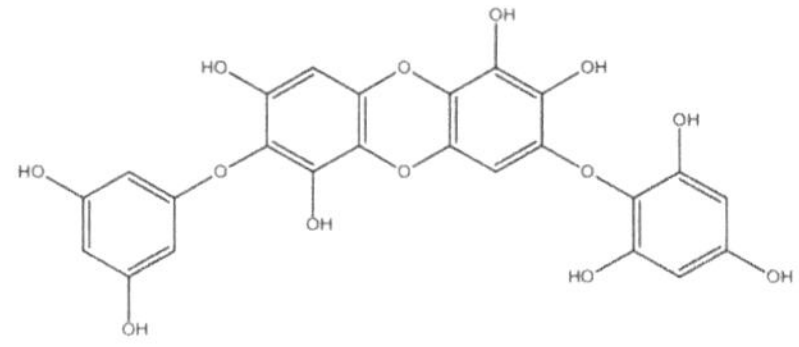		Neuroprotection against hydrogen peroxide (H_2O_2)-induced oxidative stress in murine hippocampal neuronal cells at IC_{50} = 50 μM		192
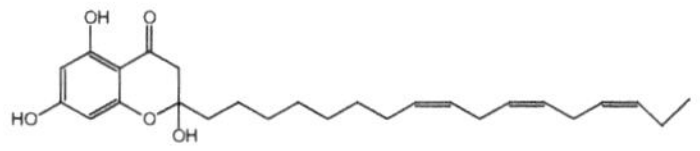		Kinases inhibitory to CDK5/p25, CK1, and GSK3 at IC_{50} = 10.0, <10 μM, and <10 μM, respectively		193
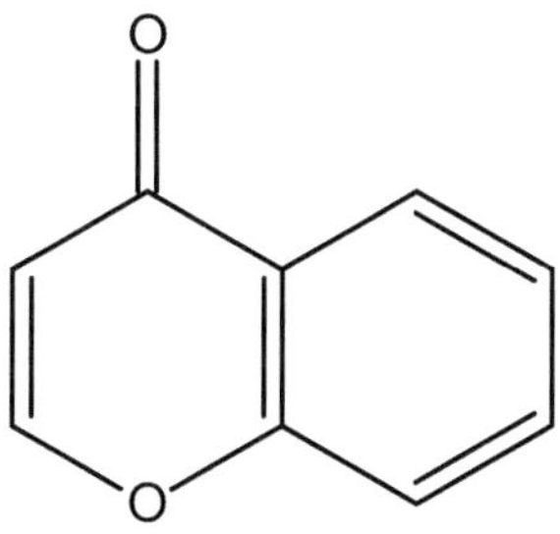				

(*continued*)

TABLE 8.3 (Continued)
The structure of various chemical compounds isolated from marine sources of macroalgae

Sl. No.	Marine source	Scientific name	Source of collection	Chemical compound isolated	Nature of chemical
			Macroalgae		
34.	Red Algae	*Eucheuma denticulatum*		Iota-carrageenan	
35.	Red Algae	*Chondracanthus acicularis*		Lambda-carrageenan	

sources of bacteria, fungus, algae, and weeds, as well as the aforementioned candidate marine natural products capable of treating and neuroprotectant for Parkinson's disease, Alzheimer's disease, and other brain disorders.

8.13 CONCLUSION

Marine chemicals and their byproducts have taken more medical and pharmaceutical research space in the last ten years. One of the causes is the huge quantity of secondary metabolites available, as well as their variation, due to the diverse unfavourable conditions of the oceans and the nearly unlimited number of creatures that inhabit them. The antioxidant and anti-inflammatory characteristics, which are particularly useful in the treatment of neurodegenerative illnesses, are among the dissimilar medicinal actions on which the research is focused. After years of research, the carotenoids AXT and FX are now being cast as food complements and prospective pharmaceutical dealings for Parkinson's disease and AD. Pharmaceutical research employs the active chemicals acquired as a source for the construction of medications that are more detailed than the original compound, in addition to the direct usage of compounds created by various aquatic creatures. Sodium oligomannate, a seaweed-derived oligosaccharide that was the first to be licenced as dealing for AD in China, is a significant specimen of molecular design from marine-derived chemicals. On this premise, MNPs appear to be a viable 'library' of natural substances for developing new treatments for neurodegenerative illnesses as adjuvants to novel therapy, enhancing the efficiency of predictable medications and applying synergistic effects. Neurodegenerative illnesses have a significant unmet demand for safe, effective, and new treatments. Because of technological advancements in sample gathering as well as purification and characterisation of the products, it is now possible to generate viable biologically active chemicals from marine sources.

Chemical structure	Mechanism of action	Results of animal studies	Outcomes of clinical Studies	References
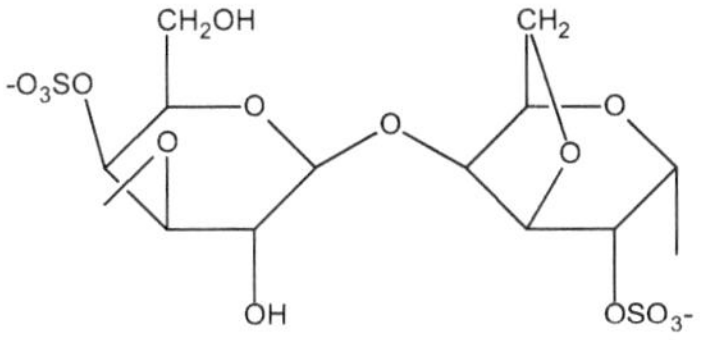		Inhibits superoxide radicals, hydroxyl radicals, and lipid peroxidation at IC_{50} = 0.332, 0.281, and 0.830 mg.mL^{-1}, respectively		194
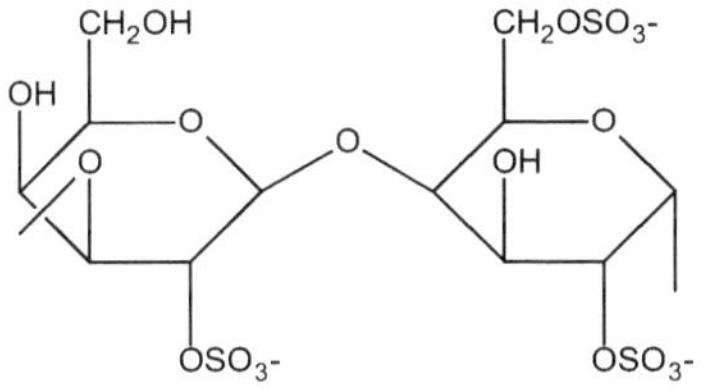		Inhibits superoxide radicals, hydroxyl radicals, and lipid peroxidation at IC_{50} = 0.046, 0.357, and 2.267 mg.mL^{-1}, respectively		194

REFERENCES

1. Dugger, BN, Dickson, DW., Cold Spring Harb. *Perspect. Biol*, 9: a028035, (2017).
2. Gibb WR, Lees AJ., *J Neurol Neurosurg Psychiatry*, 51: 745–752, (1988).
3. Sparks DL, Danner FW, Davis DG, Hackney C, Landers T, Coyne CM. *J Neuropathol Exp Neurol*, 53: 37–42, (1994).
4. Schmitt FA, Davis DG, Wekstein DR, Smith CD, Ashford JW, Markesbery WR., *Neurology*, 55: 370–376, (2000).
5. Adler CH, Connor DJ, Hentz JG, Sabbagh MN, Caviness JN, Shill HA, Noble B, Beach TG. *Mov Disord*, 25: 642–646, (2010).
6. Evidente VG, Adler CH, Sabbagh MN, Connor DJ, Hentz JG, Caviness JN, Sue LI, Beach TG. *Parkinsonism Relat Disord*, 17: 365–371, (2011).
7. Frigerio R, Fujishiro H, Ahn TB, Josephs KA, Maraganore DM, DelleDonne A, Parisi JE, Klos KJ, Boeve BF, Dickson DW, et al. *Neurobiol Aging*, 32: 857–863, (2011).
8. Milenkovic I, Kovacs GG. *Clin Neuropathol*, 32: 69–72, (2013).
9. Dugger BN, Hentz JG, Adler CH, Sabbagh MN, Shill HA, Jacobson S, Caviness JN, Belden C, Driver-Dunckley E, Davis KJ. *J Neuropathol Exp Neurol*,73: 244–252, (2014c).
10. Uchikado H, DelleDonne A, Ahmed Z, Dickson DW. *J Neuropathol Exp Neurol*, 65: 387–395, (2006a).
11. Dugger BN, Adler CH, Shill HA, Caviness J, Jacobson S, Driver-Dunckley E, Beach TG., *Parkinsonism Relat Disord*, 20: 525–529, (2014a).
12. Alafuzoff I, Ince PG, Arzberger T, Al-Sarraj S, Bell J, Bodi I, Bogdanovic N, Bugiani O, Ferrer I, Gelpi E. *Acta Neuropathol*, 117: 635–652, (2009a).
13. Alafuzoff I, Thal DR, Arzberger T, Bogdanovic N, Al-Sarraj S, Bodi I, Boluda S, Bugiani O, Duyckaerts C, Gelpi E, et al. "*Acta Neuropathol*, 117: 309–320, (2009b).
14. Braak H, Braak E., *Acta Neuropathol*, 82: 239–259, (1991).
15. Thal DR, Rub U, Orantes M, Braak H. *Neurology*, 58: 1791–1800, (2002).

16. Braak H, Del Tredici K, Rub U, de Vos RA, Jansen Steur EN, Braak E. *Neurobiol Aging*, 24: 197–211, (2003).
17. Kosaka K, Yoshimura M, Ikeda K, Budka H. *Clin Neuropathol*, 3: 185–192, (1984).
18. Brettschneider J, Del Tredici K, Toledo JB, Robinson JL, Irwin DJ, Grossman M, Suh E, Van Deerlin VM, Wood EM, Baek Y, et al. *Ann Neurol*, 74: 20–38, (2013).
19. Brettschneider J, Del Tredici K, Irwin DJ, Grossman M, Robinson JL, Toledo JB, Lee EB, Fang L, Van Deerlin VM, Ludolph AC, et al. *Acta Neuropathol*, 127: 423–439, (2014).
20. McKee AC, Stern RA, Nowinski CJ, Stein TD, Alvarez VE, Daneshvar DH, Lee HS, Wojtowicz SM, Hall G, Baugh CM, et al. *Brain*, 136: 43–64, (2013).
21. Alzheimer's Association Report. 2020, 16, 391–460, (2020).
22. Westermark P, Benson MD, Buxbaum JN, Cohen AS, Frangione B, Ikeda S, Masters CL, Merlini G, Saraiva MJ, Sipe JD. *Amyloid*, 12: 1–4, (2005).
23. Dickson DW. *J Neuropathol Exp Neurol*, 56: 321–339, (1997).
24. Kang J, Lemaire HG, Unterbeck A, Salbaum JM, Masters CL, Grzeschik KH, Multhaup G, Beyreuther K, Muller-Hill B. *Nature*, 325: 733–736, (1987).
25. Kuo YM, Kokjohn TA, Beach TG, Sue LI, Brune D, Lopez JC, Kalback WM, Abramowski D, Sturchler-Pierrat C, Staufenbiel M, et al. *J Biol Chem*, 276: 12991–12998, (2001).
26. Josephs KA, Tsuboi Y, Cookson N, Watt H, Dickson DW. *Arch Neurol*, 61: 1579–1584, (2004).
27. Goldfarb LG, Brown P, McCombie WR, Goldgaber D, Swergold GD, Wills PR, Cervenakova L, Baron H, Gibbs CJ Jr, Gajdusek DC. *Proc Natl Acad Sci*, 88: 10926–10930, (1991).
28. Hsiao KK, Cass C, Schellenberg GD, Bird T, Devine-Gage E, Wisniewski H, Prusiner SB. *Neurology*, 41: 681–684, (1991).
29. Vidal R, Frangione B, Rostagno A, Mead S, Revesz T, Plant G, Ghiso J. *Nature*, 399: 776–781, (1999).
30. Vidal R, Revesz T, Rostagno A, Kim E, Holton JL, Bek T, Bojsen-Moller M, Braendgaard H, Plant G, Ghiso J, et al. *Proc Natl Acad Sci*, 97: 4920–4925, (2000).
31. Prusiner SB. *Science*, 216: 136–144, (1982).
32. Ellison D. *Mosby, Edinburgh*, (2013).
33. Prusiner SB. *Proc Natl Acad Sci*, 95: 13363–13383, (1998).
34. Kertesz A, Hudson L, Mackenzie IR, Munoz DG. *Neurology*, 44: 2065–2072, (1994).
35. Sherzai A, Edland SD, Masliah E, Hansen L, Pizzo DP, Corey-Bloom J. *Alzheimer Dis Assoc Disord*, 27: 157–161, (2013).
36. Budka H, Aguzzi A, Brown P, Brucher JM, Bugiani O, Gullotta F, Haltia M, Hauw JJ, Ironside JW, Jellinger K, et al., Brain *Pathol*, 5: 459–466, (1995).
37. Lee VM, Goedert M, Trojanowski JQ. *Annu Rev Neurosci*, 24: 1121–1159, (2001).
38. Dickson DW, Kouri N, Murray ME, Josephs KA. *J Mol Neurosci*, 45: 384–389, (2011).
39. Spillantini MG, Goedert M. *Lancet Neurol*, 12: 609–622, (2013).
40. Kovacs GG, "Invited review: Neuropathology of tauopathies: Principles and practice", Neuropathol *Appl Neurobiol*, 41: 3–23, (2015).
41. Holton JL, Ghiso J, Lashley T, Rostagno A, Guerin CJ, Gibb G, Houlden H, Ayling H, Martinian L, Anderton BH, et al. *Am J Pathol*, 158: 515–526, (2001).
42. Ghetti B, Dlouhy SR, Giaccone G, Bugiani O, Frangione B, Farlow MR, Tagliavini F. *Brain Pathol*, 5: 61–75, (1995).
43. Binder LI, Frankfurter A, Rebhun LI. *J Cell Biol*, 101: 1371–1378, (1985).
44. Buee L, Bussiere T, Buee-Scherrer V, Delacourte A, Hof PR. *Brain Res Brain Res Rev*, 33: 95–130, (2000).
45. Mandelkow EM, Mandelkow E. *Cold Spring Harb Perspect Med*, 2: a006247, (2012).
46. Morishima M, Ihara Y.*Dementia*, 5: 282–288, (1994).
47. Alonso AC, Li B, Grundke-Iqbal I, Iqbal K. *Curr Alzheimer Res*, 5: 375–384, (2008).
48. Martin L, Latypova X, Terro F. *Neurochem Int*, 58: 458–471, (2011).
49. Cook C, Stankowski JN, Carlomagno Y, Stetler C, Petrucelli L. *Alzheimers Res Ther*, 6: 29, (2014).
50. Hutton M, Lendon CL, Rizzu P, Baker M, Froelich S, Houlden H, Pickering-Brown S, Chakraverty S, Isaacs A, Grover A, et al. *Nature*, 393: 702–705, (1998).
51. Andreadis A. *Biochim Biophys Acta*, 1739: 91–103, (2005).

52. Woerman AL, Aoyagi A, Patel S, Kazmi SA, Lobach I, Grinberg LT, McKee AC, Seeley WW, Olson SH, Prusiner SB. *Proc Natl Acad*, Sci 10.1073/pnas.1616344113, (2016).
53. Mackenzie IR, Neumann M, Bigio EH, Cairns NJ, Alafuzoff I, Kril J, Kovacs GG, Ghetti B, Halliday G, Holm IE, et al. *Acta Neuropathol*, 119: 1–4, (2010).
54. Tagliavini F, Giaccone G, Frangione B, Bugiani O. "Preamyloid deposits in the cerebral cortex of patients with Alzheimer's disease and nondemented individuals", *Neurosci Lett*, 93: 191–196, (1988).
55. Fukumoto H, Asami-Odaka A, Suzuki N, Iwatsubo T., "Association of Aβ 40-positive senile plaques with microglial cells in the brains of patients with Alzheimer's disease and in non-demented aged individuals", *Neurodegeneration*, 5: 13–17, (1996).
56. Dickson DW, "The pathogenesis of senile plaques", *J Neuropathol Exp Neurol*, 56: 321–339, (1997).
57. Hyman BT, Trojanowski JQ. *J Neuropathol Exp Neurol*, 56: 1095–1097, (1997).
58. Mirra SS, Heyman A, McKeel D, Sumi SM, Crain BJ, Brownlee LM, Vogel FS, Hughes JP, van Belle G, Berg L. *Neurology*, 41: 479–486, (1991).
59. Montine TJ, Phelps CH, Beach TG, Bigio EH, Cairns NJ, Dickson DW, Duyckaerts C, Frosch MP, Masliah E, Mirra SS, et al. *Acta Neuropathol*, 123: 1–11, (2012).
60. Hardy J, Duff K, Hardy KG, Perez-Tur J, Hutton M. *Nat Neurosci*, 1: 355–358, (1998).
61. Josephs KA, Whitwell JL, Ahmed Z, Shiung MM, Weigand SD, Knopman DS, Boeve BF, Parisi JE, Petersen RC, Dickson DW, et al. *Ann Neurol*, 63: 204–212, (2008b).
62. Whitwell JL, Josephs KA, Murray ME, Kantarci K, Przybelski SA, Weigand SD, Vemuri P, Senjem ML, Parisi JE, Knopman DS, et al. *Neurology*, 71: 743–749, (2008).
63. Masters CL, Simms G, Weinman NA, Multhaup G, McDonald BL, Beyreuther K. 1985. *Proc Natl Acad Sci* 82: 4245–4249, (1985).
64. Grundke-Iqbal I, Iqbal K, Quinlan M, Tung YC, Zaidi MS, Wisniewski HM. *J Biol Chem*, 261: 6084–6089, (1986).
65. Braak H, Braak E, Grundke-Iqbal I, Iqbal K. *Neurosci Lett*, 65: 351–355, (1986).
66. Goedert M, Klug A, Crowther RA. *J Alzheimers Dis*, 9: 195–207, (2006).
67. Espinoza M, de Silva R, Dickson DW, Davies P. *J Alzheimers Dis*, 14: 1–16, (2008).
68. Braak H, Braak E. *Acta Neuropathol*, 82: 239–259, (1991).
69. Braak H, Thal DR, Ghebremedhin E, Del Tredici K. *J Neuropathol Exp Neurol*, 70: 960–969, (2011).
70. Crary JF, Trojanowski JQ, Schneider JA, Abisambra JF, Abner EL, Alafuzoff I, Arnold SE, Attems J, Beach TG, Bigio EH, et al. *Acta Neuropathol*, 128: 755–766, (2014).
71. Murray ME, Graff-Radford NR, Ross OA, Petersen RC, Duara R, Dickson DW. *Lancet Neurol*, 10: 785–796, (2011).
72. Janocko NJ, Brodersen KA, Soto-Ortolaza AI, Ross OA, Liesinger AM, Duara R, Graff-Radford NR, Dickson DW, Murray ME. *Acta Neuropa*thol, 124: 681–692, (2012).
73. Montenigro PH, Corp DT, Stein TD, Cantu RC, Stern RA. *Annu Rev Clin Psychol*, 11: 309–330, (2015).
74. Roberts GW. *Lancet*, 2: 1456–1458, (1988).
75. Stern RA, Daneshvar DH, Baugh CM, Seichepine DR, Montenigro PH, Riley DO, Fritts NG, Stamm JM, Robbins CA, McHale L, et al. 81: 1122–1129, (2013).
76. McKee AC, Cantu RC, Nowinski CJ, Hedley-Whyte ET, Gavett BE, Budson AE, Santini VE, Lee HS, Kubilus CA, Stern RA. *J Neuropathol Exp Neurol*, 68: 709–735, (2009).
77. Schmidt ML, Zhukareva V, Newell KL, Lee VM, Trojanowski JQ. *Acta Neuropathol*, 101: 518–524, (2001).
78. Hof PR, Bouras C, Buee L, Delacourte A, Perl DP, Morrison JH, *Acta Neuropathol*, 85: 23–30, (1992).
79. McKee AC, Stern RA, Nowinski CJ, Stein TD, Alvarez VE, Daneshvar DH, Lee HS, Wojtowicz SM, Hall G, Baugh CM, et al., *Brain*, 136: 43–64, (2013).
80. McKee AC, Cairns NJ, Dickson DW, Folkerth RD, Dirk Keene C, Litvan I, Perl DP, Stein TD, Vonsattel JP, Stewart W, et al. *Acta Neuropathol*, 131: 75–86, (2015).
81. Bieniek KF, Ross OA, Cormier KA, Walton RL, Soto-Ortolaza A, Johnston AE, DeSaro P, Boylan KB, Graff-Radford NR, Wszolek ZK, et al. *Acta Neuropathol*, 130: 877–889, (2015).

82. Kovacs GG, Ferrer I, Grinberg LT, Alafuzoff I, Attems J, Budka H, Cairns NJ, Crary JF, Duyckaerts C, Ghetti B, et al. *Acta Neuropathol*, 131: 87–102, (2016).
83. Schultz C, Ghebremedhin E, Del Tredici K, Rub U, Braak H. *Neurobiol Aging*, 25: 397–405, (2004).
84. Williams DR, Lees AJ. *Lancet Neurol*, 8: 270–279, (2009).
85. Dickson DW, Ahmed Z, Algom AA, Tsuboi Y, Josephs KA. *Curr Opin Neurol*, 23: 394–400, (2010).
86. Arai T, Ikeda K, Akiyama H, Nonaka T, Hasegawa M, Ishiguro K, Iritani S, Tsuchiya K, Iseki E, Yagishita S, et al. *Ann Neurol*, 55: 72–79, (2004).
87. Rebeiz JJ, Kolodny EH, Richardson EP Jr. *Arch Neurol*, 18: 20–33, (1968).
88. Gibb WR, Luthert PJ, Marsden CD. *Brain*, 112: 1171–1192, (1989).
89. Ikeda K, Akiyama H, Haga C, Kondo H, Arima K, Oda T. *Neurosci Lett*, 174: 157–159, (1994).
90. Feany MB, Dickson DW. *Am J Pathol*, 146: 1388–1396, (1995).
91. Dickson DW, Bergeron C, Chin SS, Duyckaerts C, Horoupian D, Ikeda K, Jellinger K, Lantos PL, Lippa CF, Mirra SS, et al. *J Neuropathol Exp Neurol*, 61: 935–946, (2002).
92. Spillantini MG, Goedert M. *Lancet Neurol*, 12: 609–622, (2013).
93. Armstrong MJ, Litvan I, Lang AE, Bak TH, Bhatia KP, Borroni B, Boxer AL, Dickson DW, Grossman M, Hallett M, et al. *Neurology*, 80: 496–503, (2013).
94. Hassan A, Whitwell JL, Boeve BF, Jack CR Jr, Parisi JE, Dickson DW, Josephs KA. *Parkinsonism Relat Disord* 16: 208–214, (2010).
95. Kouri N, Whitwell JL, Josephs KA, Rademakers R, Dickson DW. *Nat Rev Neurol*, 7: 263–272, (2011).
96. Ueda K, Fukushima H, Masliah E, Xia Y, Iwai A, Yoshimoto M, Otero DA, Kondo J, Ihara Y, Saitoh T. *Proc Natl Acad Sci*, 90: 11282–11286, (1993).
97. Sudhof TC, Rizo J. *Cold Spring Harb Perspect Biol*, 3: a005637, (2001).
98. Polymeropoulos MH, Lavedan C, Leroy E, Ide SE, Dehejia A, Dutra A, Pike B, Root H, Rubenstein J, Boyer R, et al., *Science*, 276: 2045–2047, (1997).
99. Spillantini MG, Schmidt ML, Lee VM, Trojanowski JQ, Jakes R, Goedert M. *Nature*, 388: 839–840, (1997).
100. Lantos PL. *J Neuropathol Exp Neurol*, 57: 1099–1111, (1998).
101. Irizarry MC, Growdon W, Gomez-Isla T, Newell K, George JM, Clayton DF, Hyman BT. J *Neuropathol Exp Neurol*, 57: 334–337, (1998).
102. Leverenz JB, Umar I, Wang Q, Montine TJ, McMillan PJ, Tsuang DW, Jin J, Pan C, Shin J, Zhu D, et al. *Brain Pathol*, 17: 139–145, (2007).
103. Galvin JE, Lee VM, Baba M, Mann DM, Dickson DW, Yamaguchi H, Schmidt ML, Iwatsubo T, Trojanowski JQ. *Ann Neurol*, 42: 595–603, (1993).
104. Kuusisto E, Parkkinen L, Alafuzoff I. *J Neuropathol Exp Neurol*, 62: 1241–1253, (2003).
105. Ishizawa T, Mattila P, Davies P, Wang D, Dickson DW. *J Neuropathol Exp Neurol*, 62: 389–397, (2003).
106. Dugger BN, Dickson DW. *Acta Neuropathol*, 120: 633–639, (2010).
107. Dickson DW. *Curr Opin Neurol,* 14: 423–432, (2001).
108. Kosaka K, Yoshimura M, Ikeda K, Budka H. *Clin Neuropathol*, 3: 185–192, (1994).
109. McKeith IG, Dickson DW, Lowe J, Emre M, O'Brien JT, Feldman H, Cummings J, Duda JE, Lippa C, Perry EK, et al. *Neurology*, 65: 1863–1872, (2005).
110. Braak H, Del Tredici K, Rub U, de Vos RA, Jansen Steur EN, Braak E., *Neurobiol Aging*, 24: 197–211, (2003).
111. Alafuzoff I, Ince PG, Arzberger T, Al-Sarraj S, Bell J, Bodi I, Bogdanovic N, Bugiani O, Ferrer I, Gelpi E, et al. *Acta Neuropathol*, 117: 635–652, (2009a).
112. Uchikado H, Lin WL, DeLucia MW, Dickson DW. *J Neuropathol Exp Neurol*, 65: 685–697, (2006b).
113. Beach TG, Adler CH, Lue L, Sue LI, Bachalakuri J, Henry-Watson J, Sasse J, Boyer S, Shirohi S, Brooks R, et al. *Acta Neuropathol*, 117: 613–634, (2009).
114. Toledo JB, Gopal P, Raible K, Irwin DJ, Brettschneider J, Sedor S, Waits K, Boluda S, Grossman M, Van Deerlin VM, et al. *Acta Neuropathol*, 131: 393–409, (2016).
115. Halliday G, Hely M, Reid W, Morris J, *Acta Neuropathol*,115: 409–415, (2008).
116. Dugger BN, Adler CH, Shill HA, Caviness J, Jacobson S, Driver-Dunckley E, Beach TG. *Parkinsonism Relat Disord*, 20: 525–529, (2014a).

117. Dugger BN, Serrano GE, Sue LI, Walker DG, Adler CH, Shill HA, Sabbagh MN, Caviness JN, Hidalgo J, Saxon-Labelle M, et al, *J Parkinsons Dis*, 2: 57–65, (2012).
118. Ratti A, Buratti E. *J Neurochem*, 138: 95–111, (2016).
119. Neumann M, Sampathu DM, Kwong LK, Truax AC, Micsenyi MC, Chou TT, Bruce J, Schuck T, Grossman M, Clark CM, et al. *Science*, 314: 130–133, (2006).
120. Amador-Ortiz C, Lin WL, Ahmed Z, Personett D, Davies P, Duara R, Graff-Radford NR, Hutton ML, Dickson DW. *Ann Neurol*, 61: 435–445, (2007).
121. Hu WT, Josephs KA, Knopman DS, Boeve BF, Dickson DW, Petersen RC, Parisi JE. *Acta Neuropathol*, 116: 215–220, (2008).
122. Josephs KA, Murray ME, Whitwell JL, Parisi JE, Petrucelli L, Jack CR, Petersen RC, Dickson DW. *Acta Neuropathol*, 127: 441–450, (2014).
123. Fujishiro H, Uchikado H, Arai T, Hasegawa M, Akiyama H, Yokota O, Tsuchiya K, Togo T, Iseki E, Hirayasu Y. *Acta Neuropathol*, 117: 151–158, (2009).
124. Arnold SJ, Dugger BN, Beach TG. 2013. *Acta Neuropathol*, 126: 51–57, (2013).
125. Lomen-Hoerth C, Anderson T, Miller B. *Neurology*, 59: 1077–1079, (2002).
126. Ince PG, Highley JR, Kirby J, Wharton SB, Takahashi H, Strong MJ, Shaw PJ. 2011. *Acta Neuropathol*, 122: 657–671, (2011).
127. Sangnoi Y, Sakulkeo O, Yuenyongsawad S, Kanjana-opas A, Ingkaninan K, Plubrukarn A, Suwanborirux K. *Mar. Drugs*, 6, 578–586, (2008).
128. Le TC, Katila N, Park S, Lee J, Yang I, Choi H, Choi D-Y, Nam S-J. *Bioorg. Med. Chem. Lett.*, 2020, 30, 127145, (2020).
129. Williams PG, Yoshida WY, Moore RE, Paul VJ. *J. Nat. Prod.*, 66, 1006–1009, (2003).
130. Liu Y, Zhang W, Li L, Salvador LA, Chen T, Chen W, Felsenstein KM, Ladd TB, Price AR, Golde TE. *J. Med. Chem.*, 55, 10749–10765, (2012).
131. Al-Awadhi FH, Ratnayake R, Paul VJ, Luesch H, Tasiamide F. *Bioorg. Med. Chem.*, 24, 3276–3282, (2016).
132. Nachtigall J, Schneider K, Bruntner C, Bull AT, Goodfellow M, Zinecker H, Imhoff JF, Nicholson G, Irran E, Süssmuth RD, et al. *J. Antibiot.*, 64, 453–457, (2011).
133. Kim TK, Hewavitharana AK, Shaw PN, Fuerst JA. *Appl. Environ. Microbiol.*, 72, 2118–2125, (2006).
134. Kilic Ü, Kilic E, Lingor P, Yulug B, Bähr M. Acta Neuropathol. Finder V.H, Glockshuber R. *Neurodegener. Dis.*, 4, 13–27, (2007).
135. Alvarino R, Alonso E, Lacret R, Oves-Costales D, Genilloud O, Reyes F, Alfonso A, Botana LM. *Mol. Pharm.*, 16, 1456–1466, (2019).
136. Cao J, Li X-M, Meng L-H, Konuklugil B, Li X, Li H-L, Wang B-G. *Bioorg. Chem.*, 90, 103030, (2019).
137. He J-B, Ji Y-N, Hu D-B, Zhang S, Yan H, Liu X-C, Luo H-R, Zhu H-J. *Tetrahedron Lett.*, 55, 2684–2686, (2014).
138. Huang H, Feng X, Xiao ZE, Liu L, Li H, Ma L, Lu Y, Ju J, She Z, Lin Y. *J. Nat. Prod.*, 74, 997–1002, (2011).
139. Yang W-C, Bao H-Y, Liu Y-Y, Nie Y-Y, Yang J-M, Hong P-Z, Zhang Y. Molecules, 23, 2245, (2018).
140. Wu B, Ohlendorf B, Oesker V, Wiese J, Malien S, Schmaljohann R, Imhoff JF. *Mar. Biotechnol.*, 17, 110–119, (2015).
141. Chen S, Liu H, Ye W, Li S, Li D, Liu Z, Zhang. *Tetrahedron*, 76, 131066, (2020).
142. Li K-L, Dai Y, She J-L, Zeng Y-B, Dai H-F, Ou S-L, Zhou X-F, Liu Y-H. J. *Asian Nat. Prod. Res.*, 1–8, (2021).
143. Qiao MF, Ji NY, Miao FP, Yin X. *Magn. Reson. Chem.*, 49, 366–369, (2011).
144. An C-L, Kong F-D, Li Y, Ma Q-Y, Xie Q-Y, Yuan J-Z, Zhou L-M, Dai H-F, Yu Z-F, Zhao Y-X. *J. Asian Nat. Prod. Res.*, 23, 1–6, (2020).
145. Bunbamrung N, Intaraudom C, Dramae A, Komwijit S, Laorob T, Khamsaeng S, Pittayakhajonwut P. *Tetrahedron*, 76, 131496, (2020).
146. Nong X-H, Wang Y-F, Zhang X-Y, Zhou M-P, Xu X-Y, Qi S-H. *Mar. Drugs*, 12, 6113–6124, (2014).
147. Ding B, Wang Z, Huang X, Liu Y, Chen W, She Z. *Nat. Prod. Res.*, 30, 2805–2812, (2016).
148. Qiao MF, Ji NY, Miao FP, Yin XL. *Magn. Reson. Chem.*, 49, 366–369, (2011).

149. Wu B, Wiese J, Schmaljohann R, Imhoff JF. *Mar. Drugs*, 14, 204, (2016).
150. Schulz D, Ohlendorf B, Zinecker H, Schmaljohann R, Imhoff JF. *J. Nat. Prod.*, 74, 99–101, (2011).
151. Lin Y, Wu X, Feng S, Jiang G, Luo J, Zhou S, Vrijmoed LLP, Jones EBG, Krohn K, Steingröver K. et al., *J. Org. Chem.*, 66, 6252–6256, (2001).
152. Kannan RR, Aderogba MA, Ndhlala, AR, Stirk WA, Van Staden J. *Food Res. Int.*, 54, 1250–1254, (2013).
153. Lee BH, Choi BW, Lee SY. *Ocean Polar Res.*, 39, 45–49, (2017).
154. Paudel P, Seong SH, Zhou Y, Park HJ, Jung HA, Choi JS. *ACS Omega*, 4, 12259–12270, (2019).
155. Gonçalves KG, da Silva LL, Soares AR, Romeiro NC. *Algal Res.*, 52, 102130, (2020).
156. Syad AN, Devi KP. *Neurol. Res.*, 37, 14–22, (2015).
157. Castro-Silva E, Bello M, Hernández-Rodríguez M, Correa-Basurto J, Murillo-Alvarez J, Rosales-Hernandez M, Muñoz- Ochoa, MJ. *Biomol. Struct. Dyn.*, 37, 3259–3268, (2018).
158. Jung HA, Ali MY, Choi RJ, Jeong, HO, Chung, HY, Choi JS. *Food Chem. Toxicol.*, 89, 104–111, (2016).
159. Miyashita K, Function of marine carotenoids. Yoshikawa, T., Ed.; *Forum Nutr. Basel: Karger*, Switzerland, 61, 136–146, (2009).
160. Lin J, Huang L, Yu J, Xiang S, Wang J, Zhang J, Yan X, Cui W, He S, Wang Q. *Mar. Drugs*, 14, 67, (2016).
161. Choi BW, Ryu G, Park SH, Kim ES, Shin J, Roh SS, Shin HC, Lee BH. *Phytother. Res.*, 21, 423–426, (2007).
162. Seong SH, Ali MY, Kim H-R, Jung HA, Choi JS, *Med. Chem.*, 25, 3964–3970, (2017).
163. Tello E, Castellanos L, Arevalo-Ferro C, Duque C, *J. Nat. Prod.*, 72, 1595–1602, (2009).
164. Rafiquzzaman S, Kim EY, Lee JM, Mohibbullah M, Alam MB, Moon IS, Kim J-M, Kong I-S, *Food Res. Int.*, 77, 118–124, (2015).
165. Mohibbullah M, Haque MN, Khan MNA, Park I-S, Moon IS, Hong, Y-K, *J. Appl. Phycol.*, 30, 3243–3252, (2018).
166. Zhang L, Wang H, Fan Y, Gao Y, Li X, Hu Z, Ding K, Wang Y, Wang X. *Sci. Rep.*, 7, 46763, (2017).
167. Alghazwi M, Smid S, Karpiniec S, Zhang W. *Int. J. Biol. Macromol.*, 122, 255–264, (2019).
168. Lee JK, Byun H-G. "A novel BACE inhibitor isolated from Eisenia bicyclis exhibits neuroprotective activity against β-amyloid toxicity.", *Fish. Aquat. Sci.*, 21, 38, (2018).
169. Liu D-Q, Mao S-C, Zhang H-Y, Yu X-Q, Feng M-T, Wang B, Feng L-H, Guo Y-W. *Fitoterapia*, 91, 15–20, (2013).
170. Seong SH, Paudel P, Jung HA, Choi JS. *Mar. Drugs*, 17, 600, (2019).
171. Yang P, Liu D-Q, Liang T-J, Li J, Zhang H-Y, Liu A-H, Guo Y-W, Mao S-C. *Bioorg. Med. Chem.*, 23, 38–45, (2015).
172. Wang J, Zheng J, Huang C, Zhao J, Lin J, Zhou X, Naman C.B.;Wang, N, Gerwick WH, Wang Q, et al. *Neurosci.*, 9, 1349–1356, (2018).
173. Hielscher-Michael S, Griehl C, Buchholz M, Demuth H-U, Arnold N. , Wessjohann LA. *Mar. Drugs*, 14, 203, (2016).
174. Liu H, Gu L. *J. Agric. Food Chem.*, 60, 1326–1334, (2012).
175. Vanderlei ESO, Patoilo KKNR, Lima NA, Lima APS, Rodrigues JAG, Silva LMCM, Lima MEP, Lima V, Benevides NMB. *Int. Immunopharmacol.*, 10, 1113–1118, (2010).
176. Liu X, Shibata T, Hisaka S, Osawa T, *Brain Res.*, 1254, 18–27, (2009).
177. Fakhri S, Yosifova Aneva I, Farzaei MH, Sobarzo-Sánchez E. *Molecules*, 24, 2640, (2019).
178. Ikeda Y, Tsuji S, Satoh A, Ishikura M, Shirasawa T, Shimizu T. *J. Neurochem.*, 107, 1730–1740, (2008).
179. Souza RB, Frota AF, Silva J, Alves C, Neugebauer AZ, Pinteus S, Rodrigues JAG, Cordeiro EMS, de Almeida RR, Pedrosa R, et al. *Int. J. Biol. Macromol.*, 112, 1248–1256, (2018).
180. Luo D, Zhang Q, Wang H, Cui Y, Sun Z, Yang J, Zheng Y, Jia J, Yu F, Wang X. et al., *Eur. J. Pharmacol.*, 617, 33–40, (2009).
181. Celikler S, Vatan O, Yildiz G, Bilaloglu R. *Food. Chem. Toxicol.*, 47, 796–801, (2009).
182. Hannan MA, Dash R, Haque MN, Mohibbullah M, Sohag AAM, Rahman MA, Uddin MJ, Alam M, Moon IS. *Mar. Drugs*, 18, 347, (2020).
183. Jiang Z, Hama Y, Yamaguchi K, Oda T. *J. Biochem*, 151, 65–74, (2011).
184. Shang Q, Jiang H, Cai C, Hao J, Li G, Yu G. *Carbohydr. Polym.*, 179, 173–185, (2018).

185. Zargarzadeh M, Amaral AJ, Custódio CA, Mano JF. *Carbohydr. Polym.*, 232, 115774, (2020).
186. Liu X-Y, Liu D, Lin G-P, Wu Y-J, Gao L-Y, Ai C, Huang Y-F, Wang M-F, El-Seedi HR, Chen X-H, et al., *Int. J. Biol. Macromol*, 139, 342–351, (2019).
187. Tsolaki M. *Hell. J. Nucl. Med.*, 22, 82–94, (2019).
188. Shimizu H, Koyama T, Yamada S, Lipton SA, Satoh T, Zonarol. *Biochem. Biophys. Res. Commun.*, 457, 718–722, (2015).
189. Wozniak M, Bell T, Dénes Á, Falshaw R, Itzhaki R. *Int. J. Biol. Macromol*, 74, 530–540, (2015).
190. Yoon NY, Lee SH, Yong L, Kim SK. *J. Funct. Foods*, 1, 331–335, (2009).
191. Bauer S, Jin W, Zhang F, Linhardt R.J. *Mar. Drugs*, 19, 89, (2021).
192. Rocha De Souza MC, Marques CT, Guerra Dore CM, Ferreira Da Silva, FR, Oliveira Rocha HA, Leite EL. *J. Appl. Phycol.,* 19, 153–160, (2007).

9 Walnut (*Juglans regia* L.) as Brain Food

Atanu Bhattacharjee, Akula Ramakrishna, and Kunal Bhattacharya

9.1 INTRODUCTION

Growing populations and longer average lifespans are causing a significant ageing population in many nations. The prevalence of neurodegenerative illnesses has increased as a result of this demographic transition. It has been estimated that over 35 million folks across the globe are suffering from dementia, and approximately 7 million Americans have been given a diagnosis of dementia or another neurologic condition (1). The quality of life is significantly affected by neurodegenerative disorders, which also places a heavy economic liability on both victims and their relatives. Due to the synchronization of brain cells, the progression of the disease converted from mild to moderate to extreme within 5–9 years (2).

Chronic inflammation and oxidative stress exposure, a loss of protective signalling, and a buildup of harmful proteins are all factors in neurodegenerative disorders. These elements set off a chain reaction of changed chemical events that damage or kill brain cells as we age. Memory, cognitive, and motor function deficiencies result from altered interneuronal connections caused by individual cell damage. Neurodegenerative illnesses like Alzheimer's disease (AD), Parkinson's disease, dementia etc., are eventually caused by these cellular changes (2,3). Free radical-induced oxidative stress triggers inflammatory mediators affecting major organs, which may lead to neurodegenerative disorders. Therefore, the primary goal is to provide systemic defence against free radicals and inflammatory mediators like cytokinin, prostaglandin, etc., in order to provide a shield to the brain to maintain its healthy status and function smoothly without hiccups.

Neurodegeneration may be halted or prevented by dietary changes. A large community-based, case-control cohort study involving 194 AD patients and 1790 non-AD patients found that the Mediterranean diet, which contains high amounts of fruits, vegetables, cereals, and fish and low amounts of alcohol, red meat, and dairy products, significantly lowers the risk of developing AD (4). In a cohort trial with 1367 patients > 65 years of age, a related inquiry demonstrated a lower risk of dementia with higher flavonoid-rich food consumption (5). The Mediterranean diet has been shown to reduce the risk of dementia and death in AD patients, according to a few other research. These studies also highlight the critical importance of fruit and nut bioactive components for maintaining cognitive function (6,7).

Linoleic acid (LA; 18:2n6), -linolenic acid (ALA; 18:3n3), and other polyphenolics, phytosterols, and micronutrients are abundant in English walnuts (*Juglans regia* L.) (8,9). Experiments have demonstrated that feeding old animals walnut supplements can enhance their memory, cognition, and motor performance. Despite the fact that the majority of this research has focused on walnuts' high PUFA content, walnuts' considerable polyphenol content also contributes significantly to lowering inflammation and oxidative stress in the ageing brain (10,11). This review discusses recent studies on the effects of dietary walnuts on brain health, cognition, and motor performance.

DOI: 10.1201/9781003300502-9

9.2 PHYTOCONSTITUENTS OF WALNUT

In terms of weight, walnuts have around 65% fat. Walnuts, like other nuts, get the majority of their energy from fat. They are, therefore, a calorie-dense, high-energy food. Walnuts are high in fat and calories, but research shows that when they are substituted for other items in the diet, they do not raise the risk of obesity. Triacylglycerols make up the majority of the oil's components; free fatty acids, diacylglycerols, monoacylglycerols, sterols, sterol esters, and phosphatides are only trace amounts of each [12]. Walnut oil contains significant amounts of oleic (18:1), linoleic (18:2), and linolenic (18:3) fatty acids. Linoleic acid, an omega-6 fatty acid, is the most prevalent. They also contain a sizable amount of alpha-linolenic acid (ALA), a beneficial omega-3 lipid. About 8–14 % of the total fat content is made up of this. Actually, the only nuts with considerable ALA content are walnuts (13,14,15). ALA is thought to be particularly advantageous for heart health (16). Additionally, it helps to modify the composition of blood fats and lessen inflammation. The long-chain omega-3 fatty acids EPA and DHA, which have been associated with a number of health advantages, are precursors of ALA (17, 18).

9.2.1 Amino Acids

Amino acids present in walnuts are a great source of protein required for a healthy diet. According to reports, walnuts have a protein concentration that ranges from 13.6 to 18.1 gm per 100 g of DM. Walnuts have significant quantities of arginine and a comparatively low amount of lysine. Since arginine may be turned into nitric oxide, a powerful vasodilator that can prevent platelet adhesion and aggregation, the high quantities of arginine in walnuts have already been recognized as a benefit. A low ratio of lysine/arginine may be considered a prophylaxis factor against atherosclerosis in-vivo (20, 21).

9.2.2 Dietary Fibre and Phytosterols

Walnuts may offer cardioprotective activity due to high fibre content ranging from 3.1 to 5.2 g/ 100 g dry matter in different varieties. A subsequent amount of phytosteroids were detected in walnuts. It was reported that moderate intake of walnut powder reduces serum total cholesterol in vivo via decrease in their absorption (22).

9.2.3 Minerals and Vitamins

Walnuts are high in a variety of vitamins and minerals (Table 9.1), including:

9.2.4 Vitamin E

Walnuts are high in a kind of vitamin E known as **α**-tocopherol. Vitamin E has strong antioxidant properties and helps to prevent lipid peroxidation (23).

9.2.5 Vitamin B6

This vitamin may help to boost the immune system and maintain neurological function. Anaemia can be caused by a lack of vitamin B6 (23).

9.2.6 Vitamin B9

Folic acid, often known as folate, serves a variety of biological activities. A lack of folic acid during pregnancy may result in birth abnormalities (23).

TABLE 9.1
Vitamins and minerals essential for healthy brain cells

B Complex	Necessary for the production of neurotransmitters, neuronal cell membranes, membrane fluidity and nerve impulse transmission between brain cells	References
Vitamin E	Natural antioxidant and protect neurons via reducing oxidative stress and free radical scavenging activity	23
Magnesium and Calcium	Regulation of nerve impulse transmission between brain cells; Ca^{2+} influx to a nerve terminal induces neurotransmitter release	24
Zinc	Prosthetic element for enzyme activation; structural component of neuropeptides in brain cell	25
Iron	Carrier of oxygen to the brain cells essential for their survival and vitality; helps in neurotransmitter synthesis	25
Phytonutrients (such as carotenoids and flavonoids)	Neuro-protective function through its role as an antioxidant	27,28

9.2.7 Minerals

Copper: Copper is an essential trace element in the human body. Copper serves as a prosthetic group that is permanently bound to many critical enzymes involved in the creation of ATP, iron metabolism, neuropeptide activation, and neurotransmitters formation. It also plays a crucial role to maintain catecholamine balance, myelination, and signal transduction in between two neurons in the CNS. It has been reported that abnormal copper-protein interactions may generate free radicals via the Haber-Weiss reaction resulting in mitochondrial damage, DNA breakage, and neuronal injury which may be associated with neurological disorders like depression, Alzheimer's etc. (25).

Phosphorus: Phosphorus, a mineral found mostly in bones, accounts for around 1% of our total weight. It serves several roles in the organism (24).

Manganese: The largest concentrations of this trace element are found in nuts, whole grains, fruits, and vegetables. Other plant compounds in walnuts contain a rich variety of bioactive plant components. They are extremely high in antioxidants, which are concentrated in the thin, brown skin (24).

Ellagic acid derivatives: Ellagic acid is a polyphenolic compound. This antioxidant is abundant in walnuts, as are other related chemicals such as ellagitannins. Ellagic acid may lower the risk of heart disease and aid in the suppression of cancer development (26).

Catechin: Catechin is a flavonoid antioxidant with potential health benefits. It may also benefit heart health (6).

Melatonin: It is a tryptophan base and neurohormone aiding in the regulation of the body clock. It is also a potent antioxidant, which may lower the risk of heart disease (27,28).

9.3 THERAPEUTIC BENEFITS OF BRAIN FOOD

9.3.1 Cognitive Benefit

In a clinical trial with 64 human volunteers, it was observed that the administration of walnuts (10 gm/day; 8 weeks) improves verbal reasoning abilities and motor function by 11.2 %. This effect may be associated with antioxidant potential of the fruit (28, 29).

9.3.2 Ageing and Metabolic Effects on Brain Health

It has been postulated that the human brain consumes a higher amount of oxygen comparatively as it requires enhanced metabolic reactions. These physiological responses lead to generate disproportionate amounts of reactive oxygen species (ROS) and reactive nitrogen species (RNS). The generated ROS and RNS are scavenged by antioxidant enzymes like oxydo-reductase, SOD, catalase etc. extrinsically, whereas intrinsic amino acids act as molecular quenchers at the cellular level and thereby maintain the haemostatics (30,31,32). Another theory depicts oxidative stress triggers pro-inflammatory mediators and causes lipid peroxidation (LPO), which results in neuroinflammation followed by synchronization of healthy brain cells. All these resulted in impairment in various cognitive and behavioural abilities (33, 34).

Learning is the acquisition of information and skills, and subsequent retention of that information is known as memory. Thus, cognition is a process of storing a sequence of information in a systematic manner by which one becomes aware of one's surroundings, objects and thoughts (35). During the process of learning and memory formation, the brain undergoes various physical and chemical changes known as synaptic plasticity. It involves various signal transduction pathways and induction of gene expression, which results in the formation of new synapses between nerve cells (36). The main characteristic of memory formation in animals and human beings is its progression from a short-lived liable form to a long-lasting stable form (37). This process undergoes a continuous remodelling with time and new experiences. Principally, the memory process consists of three different stages, viz. registration, consolidation and retrieval. Registration is the process of sensory perception and the ability to act on the information perceived (behavioural responses), which involves a change in brain activity (electrical and electrochemical) and is referred to as short-term memory. Consolidation is the process of conversion of registered short-term information to long-term memory, which involves physicochemical changes in neuronal networks (38,39). The stored information is made accessible by the process known as retrieval (Figure 9.1).

The consolidation hypothesis postulates that memory storage proceeds through temporally distinct phases, each disrupted by a subset of agents like electro-convulsive shock, etc. The retrieval hypothesis postulates that amnesic agents like scopolamine, etc., disrupt memory recall rather than storage, as the effect of these agents diminishes over time, resulting in the appearance of normal

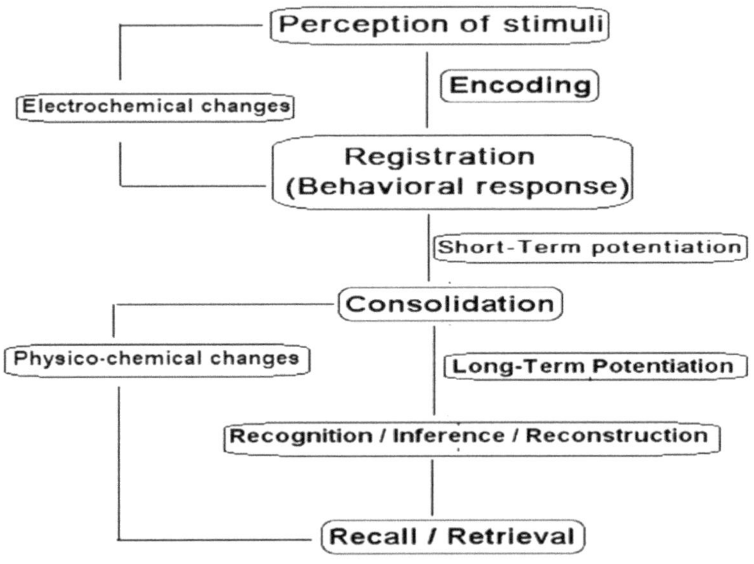

FIGURE 9.1 Multiple stages of cognition process.

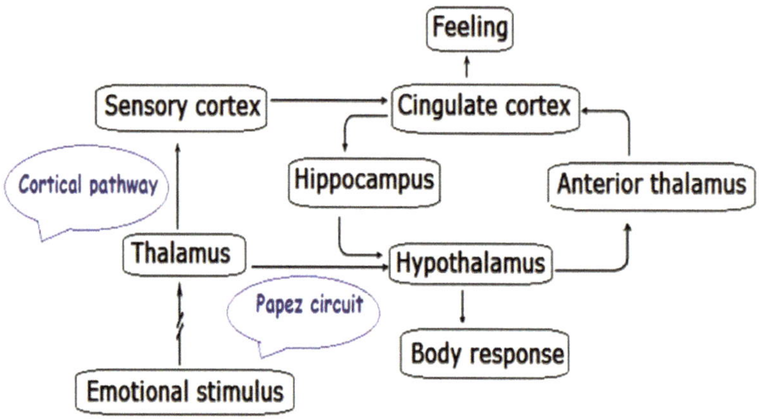

FIGURE 9.2 Major pathways of learning and memory process.

memory retention (40). The consolidation of information is mediated by limbic structures, with the hippocampal formation playing a vital role in the memory process (41). There are two major pathways involved in the mechanism of learning and memory process viz. Papez pathway and cortical pathway. The Papez circuit is a neural circuit connecting the hypothalamus to the limbic lobe and controls emotional aspects of behaviour. The Papez circuit goes through the following neural pathway:

hippocampal formation → fornix → mammillary bodies → mammillothalamic tract → anterior thalamic nucleus → cingulum → entorhinal cortex → hippocampal formation.

The second pathway leading from the cortical association areas via cingulate gyrus to the hippocampus via thalamic nucleus and enables information to be stored, presumably allowing it to reverberate for some time for body response (Figure 9.2) (42).

Memory function is vulnerable to a variety of pathological conditions, including neurodegenerative diseases like AD. Dementia, an age-related mental problem, is a characteristic symptom of AD (43).

It has been reported that the composition of PUFA decreases in neuronal membranes during ageing, which causes neuronal function to directly hamper neuronal networking symmetry. Due to a decline in PUFA level, neuronal functions are affected, which ultimately leads to altered neurosignal transmission within the brain. Walnut is a rich source of PUFA and therefore may increase PUFAs content in brain neuronal cells and thereby maintain healthy status of brain neuronal cells (44, 45).

9.4 BRAIN CELL PROTECTIVE ACTIVITY OF WALNUT PHYTOCHEMICALS: *IN-VITRO* STUDIES

Walnuts are a rich source of nutrients and bioactive phytochemicals. It contains a high amount of ω-3-fatty acids and other PUFAs (Table 9.2 and Figure 9.3). Collectively they boost brain health and function by providing nourishment and free radical scavenging property (46,47).

In human liver cells, **α**- Lenoleic acid present in walnuts is converted to DHA through enzyme-mediated sequential desaturation process (Figure 9.1). Both EPA and DHA are vital for brain health because they not only reduce oxidative stress and modify immunological function but also preserve neuronal membrane integrity, gene expression, and neurogenesis (49).

The polyphenolic compounds, vitamin E, carotenoids, and minerals, along with PUFAs, provide defence against ROS/RNS-induced toxicity and protection through inhibition of LPO via increased plasma antioxidant capacity. Polyphenols promote neuronal calcium homeostasis in the striatum and hippocampus, regions of the brain crucial for primary and secondary memory functions (50).

TABLE 9.2
Fatty acids of walnuts

Fatty acid	Amount/100g
Lenileic acid (LA)	38 g
α-Lenileic acid (ALA)	9.0 g
Saturated fatty acid (palmitic acid)	4.4 g
Monounsaturated fatty acid (oleic acid)	8.7 g
Polyunsaturated fatty acid (oleic acid)	20.4 g

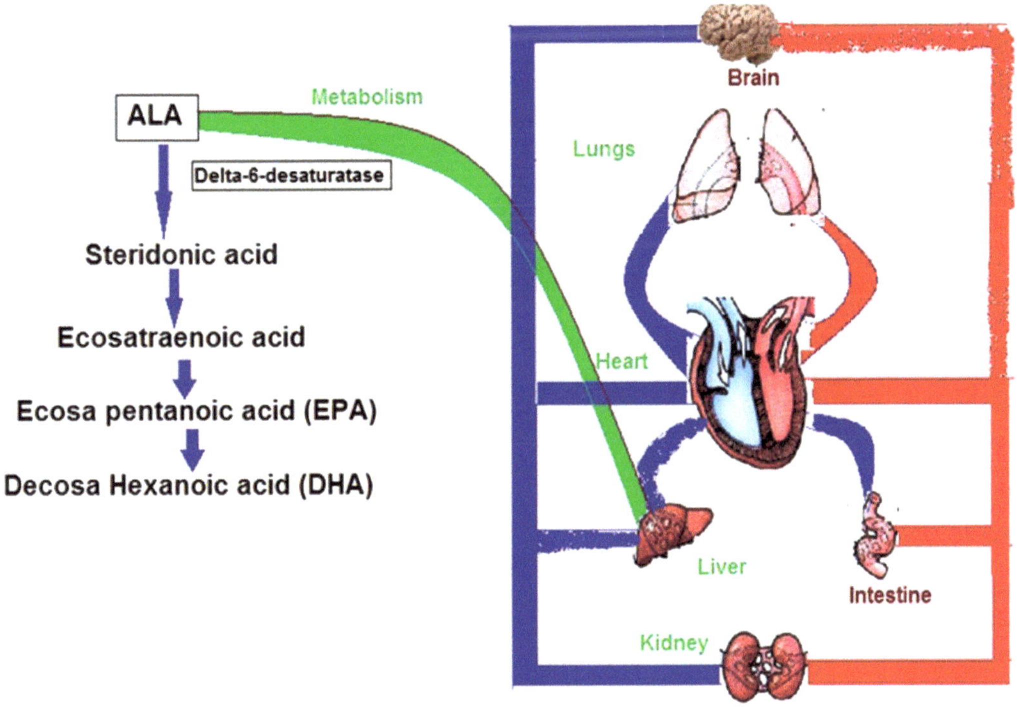

FIGURE 9.3 Metabolism of α-Lenoleic acid to Decosa hexaenoic acid in liver cells.

Walnut is enriched with tryptophan base melatonin. Endogenous melatonin is a primary hormone to regulate circadian rhythms (51, 52).

It also possesses antioxidant and neuro-protective roles in AD. Melatonin plays a protective role in cholinergic neuron degeneration and in anti-inflammation. It also provides defence against amyloid β (Aβ) induced neurotoxicity, which are considered as two vital pathological hallmarks of AD. Due to melatonin-rich content, intake of walnuts increased blood melatonin concentrations, thereby showing prophylactic activity against neural damage in the brain and subsequently preventing the early stages of AD (53,54,55).

Walnut extract protects the brain from a range of stresses. Scientific studies revealed the anti-inflammatory and antioxidant potential of walnut oil via reducing lipopolysaccharide-induced microglial cell activation in BV-2 mice. Walnut oil significantly inhibits the synthesis of inflammatory mediators like NO production and TNF-**α**, a pro-inflammatory mediator. Walnut extract protected against LPS-induced loss of calcium recovery (56). Another similar type of study also predicted that PUFAs of walnuts attenuate neuro-inflammation via reducing inflammatory stress through phospholipase D2-mediated internalization of Toll-like receptor 4 (TLR-4) (57).

A scientific study depicted the anti-amyloid activity of walnut extract via antagonizing Aβ-induced oxidative stress and cytotoxicity in PC12 cells of rat adrenal medulla (58).

9.5 BRAIN CELL PROTECTIVE ACTIVITY OF WALNUT PHYTOCHEMICALS: *IN-VIVO* AND CLINICAL STUDIES

Recent in-vivo and human trials suggested that long-term dietary supplementation with walnuts may provide a prophylactic effect against mild cognitive impairment and AD.

9.5.1 Animal Studies

Aβ is produced by the proteolysis of amyloid precursor protein (APP). APP-transgenic mice (AD-tg) animals (dose 28–42 g/day for 14 months p.o.) showed improvement in memory and learning behaviour, motor coordination, and the results were comparable with placebo drugs i.e. a control diet without walnuts (59). In another study, adult Wistar rats (aged 19 months old), a diet with 6% walnuts showed improvement in cognitive and motor activities (60).

9.5.2 Clinical Trials in Humans

In two different PREDIMED clinical trials viz. Prevención con Dieta Mediterránea, healthy adults were subjected to a Mediterranean diet (dose: 30 g mixed nuts/day (15 g walnuts, 7.5 g hazelnuts, and 7.5 g almonds)) showed elevated nootropic activities compared to the control group on a low-fat diet (61).

During clinical trials, two different sets of human volunteers viz. i) 522 adults (mean age: 74.6 years) and 447 adults (mean age: 66.9 years) were considered. In the first study, the effect of the Mediterranean diet on global cognitive performance was assessed by the Mini-Mental State Examination (MMSE) and Clock Drawing Test (CDT) after 6.5 years of dietary intervention without baseline evaluation. The second study examined rates of memory and behavioural changes with time by comparing scores from six different neuropsychological tests after 4.1 years of Mediterranean diet (62).

In another clinical trial (human volunteer no.: 15,467 women; mean age: 74 years), the long-term intake of walnuts demonstrated better cognitive performance. The National Health and Nutrition Examination Survey (NHANES) of an adult population (two different age groups (20–59 years; 60 years and older)) in the United States found that walnut eating improved cognitive scores. Another study of 64 young adults examined the effects of short-term (8-week) dietary supplementation with walnuts on cognitive performance (63). Memory, mood states, verbal reasoning, and non-verbal reasoning were assessed at baseline and at the end of the 8-week nutritional intervention period. There was a significant increase in inferential verbal reasoning in the subjects on the walnut-enriched diet, but there were no significant differences in mood, memory, or non-verbal reasoning. This may be because dietary supplementation of walnuts occurred for only eight weeks in this study (64). The potential neuroprotective mechanism of walnuts as brain food is shown in Figure 9.4.

9.6 CONCLUSION

Inflammation and oxidative stress both play key roles in the ageing process, dementia, and many neurodegenerative disorders. Walnuts contain different antioxidant and anti-inflammatory components that may synergistically exhibit anti-inflammatory and antioxidant potential. Walnuts reduce oxidative stress by scavenging ROS/RNS and increase antioxidant defence, thereby reducing LPO (65). Even in the absence of particular neurodegenerative illnesses, this process of brain

FIGURE 9.4 Potential neuroprotective mechanism of walnuts as brain food.

ageing happens. Although most chronic neurodegenerative illnesses cannot presently be treated, early intervention can safeguard cognitive function in old age and may prevent or postpone the onset of devastating neurodegenerative disorders (66, 67). Dietary treatments are safe and, more significantly, help to sustain the healthy neuronal environment of the brain. The polyphenolic compounds viz. flavonoids, catechin, ellgic acid etc., along with tocopherols and melatonin, can reduce oxidative stress and inflammation; furthermore, PUFAs help maintain neuronal membrane integrity and attenuate protein aggregation involved in AD (68–70). Dietary intervention with walnuts showed significant improvement in memory and behavioural symptoms associated with dementia and AD and also provided prophylaxis to delay the progression of different neurodegenerative diseases (62). In humans, the inclusion of walnuts in the diet improved cardiovascular health, which is itself a risk factor for neurodegenerative diseases and age-related cognitive decline (71, 72). Taken together, this evidence suggests that the integration of walnuts into a healthy diet could be an effective means of prolonging health spans, slowing the processes of brain ageing, and reducing the risk of chronic neurodegenerative disease.

REFERENCES

1. Livingston, G.; Sommerlad, A.; Orgeta, V.; Costafreda, S.G.; Huntley, J.; Ames, D.; Ballard, C.; Banerjee, S.; Burns, A.; Cohen-Mansfield, J.; et al. *Lancet* 2017, 390, 2673–2734.
2. Huang, H.C.; Jiang, Z.F. *J. Alzheimers Dis*. 2009, 16, 15–27.
3. Birben, E.; Sahiner, U.M.; Sackesen, C.; Erzurum, S.; Kalayci, O. *World Allergy Organ. J.* 2012, 5, 9–19.
4. Chauhan, V.; Chauhan, A. *Pathophysiology* 2006, 13, 195–208.
5. Bonda, D.J.; Wang, X.; Perry, G.; Nunomura, A.; Tabaton, M.; Zhu, X.; Smith, M.A. *Neuropharmacology* 2010, 59, 290–294.

6. Wang, X.; Wang, W.; Li, L.; Perry, G.; Lee, H.G.; Zhu, X. *Biochim. Biophys. Acta* 2014, 1842, 1240–1247.
7. Galasko, D.; Montine, T.J. *Biomark. Med.* 2010, 4, 27–36.
8. Schwab, C.; McGeer, P.L. *J. Alzheimers Dis.* 2008, 13, 359–369.
9. Agostinho, P.; Cunha, R.A.; Oliveira, C. *Curr. Pharm. Des.* 2010, 16, 2766–2778.
10. Wang, W.Y.; Tan, M.S.; Yu, J.T.; Tan, L. *Ann. Transl. Med.* 2015, 3, 136.
11. Schrag, M.; Mueller, C.; Zabel, M.; Crofton, A.; Kirsch, W.M.; Ghribi, O.; Squitti, R.; Perry, G. *Neurobiol. Dis.* 2013, 59, 100–110.
12. Pratico, D.; Clark, C.M.; Liun, F.; Rokach, J.; Lee, V.Y.; Trojanowski, J.Q. *Arch. Neurol.* 2002, 59, 972–976.
13. Torres, L.L.; Quaglio, N.B.; de Souza, G.T.; Garcia, R.T.; Dati, L.M.; Moreira, W.L.; Loureiro, A.P.; de Souza-Talarico, J.N.; Smid, J.; Porto, C.S.; et al. *J. Alzheimers Dis.* 2011, 26, 59–68.
14. Cai, Z.; Zhao, B.; Ratka, A. *Neuromol. Med.* 2011, 13, 223–250.
15. Walker, D.G.; Link, J.; Lue, L.F.; Dalsing-Hernandez, J.E.; Boyes, B.E. *J. Leukoc. Biol.* 2006, 79, 596–610.
16. Combs, C.K.; Karlo, J.C.; Kao, S.C.; Landreth, G.E. *J. Neurosci.* 2001, 21, 1179–1188.
17. Pasinetti, G.M.; Aisen, P.S. *Neuroscience* 1998, 87, 319–324.
18. Montine, T.J.; Sidell, K.R.; Crews, B.C.; Markesbery, W.R.; Marnett, L.J.; Roberts, L.J., 2nd; Morrow, J.D. *Neurology* 1999, 53, 1495–1498.
19. Galasko, D.; Montine, T.J. *Biomark. Med.* 2010, 4, 27–36.
20. Schwab, C.; McGeer, P.L. J. *Alzheimers Dis.* 2008, 13, 359–369.
21. Agostinho, P.; Cunha, R.A.; Oliveira, C. *Curr. Pharm. Des.* 2010, 16, 2766–2778.
22. Wang, W.Y.; Tan, M.S.; Yu, J.T.; Tan, L. *Ann. Transl. Med.* 2015, 3, 136.
23. Schrag, M.; Mueller, C.; Zabel, M.; Crofton, A.; Kirsch, W.M.; Ghribi, O.; Squitti, R.; Perry, G. *Neurobiol. Dis.* 2013, 59, 100–110.
24. Pratico, D.; Clark, C.M.; Liun, F.; Rokach, J.; Lee, V.Y.; Trojanowski, J.Q. *Arch. Neurol.* 2002, 59, 972–976.
25. Torres, L.L.; Quaglio, N.B.; de Souza, G.T.; Garcia, R.T.; Dati, L.M.; Moreira, W.L.; Loureiro, A.P.; de Souza-Talarico, J.N.; Smid, J.; Porto, C.S.; et al. *J. Alzheimers Dis.* 2011, 26, 59–68.
26. Cai, Z.; Zhao, B.; Ratka, A. *Neuromol. Med.* 2011, 13, 223–250.
27. Walker, D.G.; Link, J.; Lue, L.F.; Dalsing-Hernandez, J.E.; Boyes, B.E. *J. Leukoc. Biol.* 2006, 79, 596–610.
28. Combs, C.K.; Karlo, J.C.; Kao, S.C.; Landreth, G.E. *J. Neurosci.* 2001, 21, 1179–1188.
29. Pasinetti, G.M.; Aisen, P.S. *Neuroscience* 1998, 87, 319–324.
30. Montine, T.J.; Sidell, K.R.; Crews, B.C.; Markesbery, W.R.; Marnett, L.J.; Roberts, L.J., 2nd; Morrow, J.D. *Neurology* 1999, 53, 1495–1498.
31. Anderson, K.J.; Teuber, S.S.; Gobeille, A.; Cremin, P.; Waterhouse, A.L.; Steinberg, F.M. *J. Nutr.* 2001, 131, 2837–2842.
32. Fukuda, T.; Ito, H.; Yoshida, T. *Phytochemistry* 2003, 63, 795–801.
33. Halvorsen, B.L.; Carlsen, M.H.; Phillips, K.M.; Bohn, S.K.; Holte, K.; Jacobs, D.R., Jr.; Blomhoff, R. *Am. J. Clin. Nutr.* 2006, 84, 95–135.
34. Reiter, R.J.; Manchester, L.C.; Tan, D.X. *Nutrition* 2005, 21, 920–924.
35. Chauhan, A.; Chauhan, V. In *Brain Aging and Therapeutic Interventions*; Thakur, M.K., Rattan, S.I.S., Eds.; Springer: Dordrecht, The Netherland, 2012; pp. 239–252.
36. Ren, J.; Chung, S.H. *J. Agric. Food Chem.* 2007, 55, 5073–5080.
37. Erdinest, N.; Shmueli, O.; Grossman, Y.; Ovadia, H.; Solomon, A. *Investig. Ophthalmol. Vis. Sci.* 2012, 53, 4396–4406.
38. Reifen, R.; Karlinsky, A.; Stark, A.H.; Berkovich, Z.; Nyska, A. *J. Nutr. Biochem.* 2015, 26, 1632–1640.
39. Mishra, N.; Dubey, A.; Mishra, R.; Barik, N. *Food Chem. Toxicol.* 2010, 48, 3316–3320.
40. Muthaiyah, B.; Essa, M.M.; Lee, M.; Chauhan, V.; Kaur, K.; Chauhan, A. *J. Alzheimers Dis.* 2014, 42, 1397–1405.
41. Willis, L.M.; Shukitt-Hale, B.; Cheng, V.; Joseph, J.A. *Br. J. Nutr.* 2009, 101, 1140–1144.

42. Martinez-Lapiscina, E.H.; Clavero, P.; Toledo, E.; Estruch, R.; Salas-Salvado, J.; San Julian, B.; Sanchez-Tainta, A.; Ros, E.; Valls-Pedret, C.; Martinez-Gonzalez, M.A. *J. Neurol. Neurosurg. Psychiatry* 2013, 84, 1318–1325.
43. Valls-Pedret, C.; Sala-Vila, A.; Serra-Mir, M.; Corella, D.; de la Torre, R.; Martinez-Gonzalez, M.A.; Martinez-Lapiscina, E.H.; Fito, M.; Perez-Heras, A.; Salas-Salvado, J.; et al. Mediterranean diet and age-related cognitive decline: A randomized clinical trial. *JAMA Intern. Med.* 2015, 175, 1094–1103.
44. O'Brien, J.; Okereke, O.; Devore, E.; Rosner, B.; Breteler, M.; Grodstein, F. Long-term intake of nuts in relation to cognitive function in older women. *J. Nutr. Health Aging* 2014, 18, 496–502.
45. Arab, L.; Ang, A. A cross sectional study of the association between walnut consumption and cognitive function among adult US populations represented in NHANES. *J. Nutr. Health Aging* 2015, 19, 284–290.
46. Pribis, P.; Bailey, R.N.; Russell, A.A.; Kilsby, M.A.; Hernandez, M.; Craig, W.J.; Grajales, T.; Shavlik, D.J.; Sabate, J. Effects of walnut consumption on cognitive performance in young adults. *Br. J. Nutr*. 2012, 107, 1393–1401.
47. Chauhan, N.; Wang, K.C.; Wegiel, J.; Malik, M.N. Walnut extract inhibits the fibrillization of amyloid beta-protein, and also defibrillizes its preformed fibrils. *Curr. Alzheimer Res*. 2004, 1, 183–188.
48. Muthaiyah, B.; Essa, M.M.; Chauhan, V.; Chauhan, A. Protective effects of walnut extract against amyloid beta peptide-induced cell death and oxidative stress in PC12 cells. *Neurochem. Res*. 2011, 36, 2096–2103.
49. Pandareesh, M.D.; Chauhan, V.; Chauhan, A. Walnuts supplementation in the diet reduces oxidative damage and improves antioxidant status in transgenic mouse model of Alzheimer's disease. *J. Alzheimer Dis*. 2018, 64, 1295–1305.
50. Torabian, S.; Haddad, E.; Rajaram, S.; Banta, J.; Sabate, J. Acute effect of nut consumption on plasma total polyphenols, antioxidant capacity and lipid peroxidation. *J. Hum. Nutr. Diet* 2009, 22, 64–71.
51. Choi, J.G.; Park, G.; Kim, H.G.; Oh, D.S.; Kim, H.; Oh, M.S. In vitro and in vivo neuroprotective effects of walnut (Juglandis Semen) in models of Parkinson's disease. *Int. J. Mol. Sci.* 2016, 17, 108.
Dotson, V.M.; Beydoun, M.A.; Zonderman, A.B. Recurrent depressive symptoms and the incidence of dementia and mild cognitive impairment. *Neurology* 2010, 75, 27–34.
52. Singh-Manoux, A.; Dugravot, A.; Fournier, A.; Abell, J.; Ebmeier, K.; Kivimaki, M.; Sabia, S. *JAMA Psychiatry* 2017, 74, 712–718.
53. Arab, L.; Guo, R.; Elashoff, D *Nutrients* 2019, 11, 275.
54. Sanchez-Villegas, A.; Martinez-Gonzalez, M.A.; Estruch, R.; Salas-Salvado, J.; Corella, D.; Covas, M.I.; Aros, F.; Romaguera, D.; Gomez-Gracia, E.; Lapetra, J.; et al. Mediterranean dietary pattern and depression: The PREDIMED randomized trial. *BMC Med.* 2013, 11, 208.
55. Ros, E. *Nuts and CVD. Br. J. Nutr*. 2015, 113 (Suppl. 2), S111–S120.
56. Asadi-Shekaari, M.; Kalantaripour, T.P.; Nejad, F.A.; Namazian, E.; Eslami, A. *Avicenna J. Med. Biotechnol*. 2012, 4, 155–158.
57. Jiang, R.; Manson, J.E.; Stampfer, M.J.; Li.u, S.; Willett, W.C.; Hu, F.B. JAMA 2002, 288, 2554–2560.
58. Pan, A.; Sun, Q.; Manson, J.E.; Willett, W.C.; Hu, F.B. *J. Nutr*. 2013, 143, 512–518.
59. Ma, Y.; Njike, V.Y.; Millet, J.; Dutta, S.; Doughty, K.; Treu, J.A.; Katz, D.L. *Diabetes Care* 2010, 33, 227–232.
60. Banel, D.K.; Hu, F.B. *Am. J. Clin. Nutr*. 2009, 90, 56–63.
61. Kris-Etherton, P.M. *J. Nutr*. 2014, 144, 547S–554S.
62. Sabate, J.; Fraser, G.E.; Burke, K.; Knutsen, S.F.; Bennett, H.; Lindsted, K.D. *N. Engl. J. Med.* 1993, 328, 603–607.
63. Ndanuko, R.N.; Tapsell, L.C.; Charlton, K.E.; Neale, E.P.; Batterham, M.J. *J. Acad. Nutr. Diet*. 2017, 117, 228–239.
64. Guasch-Ferré, M.; Li, J.; Hu, F.B.; Salas-Salvadó, J.; Tobias, D.K. *Am. J. Clin. Nutr.* 2018, 108, 174–187.
65. Estruch, R.; Ros, E.; Salas-Salvadó, J.; Covas, M.I.; Corella, D.; Arós, F.; Gómez-Gracia, E.; Ruiz-Gutiérrez, V.; Fiol, M.; Lapetra, J.; et al. *N. Engl. J. Med.* 2018, 378, e34.
66. Tindall, A.M.; Petersen, K.S.; Skulas-Ray, A.C.; Richter, C.K.; Proctor, D.N.; Kris-Etherton, P.M. *J. Am. Heart Assoc*. 2019, 8, e011512.

67. Neale, E.P.; Tapsell, L.C.; Martin, A.; Batterham, M.J.; Wibisono, C.; Probst, Y.C. *Food Nutr. Res.* 2017, 61, 1344522.
68. Salas-Salvado, J.; Casas-Agustench, P.; Salas-Huetos, A. *Nutr Metab Cardiovasc Dis*. 2011;21(Suppl 1):S1-S6.
69. Engelhart, M.J.; Geerlings, M.I.; Ruitenberg, A.; van Swieten, J.C.; Hofman, A.; Witteman, J.C., et al. *JAMA*. 2002;287(24): 3223–3229.
70. Commenges, D.; Scotet, V.; Renaud, S.; Jacqmin-Gadda, H.; Barberger-Gateau, P.; Dartigues, J.F. *Eur J Epidemiol.* 2000;16(4): 357–363.
71. Morris, M.C.; Evans, D.A.; Tangney, C.C.; Bienias, J.L.; Wilson, R.S.; Aggarwal, N.T.; et al. *Am J Clin Nutr.* 2005;81(2):508–514.
72. Gu, Y.; Nieves, J.W.; Stern, Y.; Luchsinger, J.A.; Scarmeas, N. *Arch Neurol.* 2010;67(6):699–706.

10 A Molecular Insight into Plant Bioactive Congeners as Neuroprotective Therapeutics

Swarupananda Mukherjee, Dipanjan Karati, Mainak Chakraborty, Nilanjan Sarkar, Souvik Roy, and Sajal Kumar Jha

10.1 INTRODUCTION

With numerous risk factors contributing to the trend, the development of mental disorders and neurodegenerative diseases are the key challenges worldwide that modern society must overcome [1]. Targeting modifiable lifestyle behaviours, such as nutrition, has therefore been acknowledged as a crucial tactic to lessen this burden [2]. Approximately one new dementia case is diagnosed every 3.2 seconds, or 9.9 million new cases are reported yearly [3]. In addition to the hardship brought on by this debilitating neurological disease, healthcare costs are rising rapidly over time [4]. Alzheimer's disease (AD) is the senile decay of neurons. Two crucial characteristics, extracellular beta-amyloid deposition and intracellular tau hyperphosphorylation, are the hallmarks of the pathophysiology of AD illness [5–7]. Due to the complex nature of NDDs, new drug-based psychotropic therapies have had a modest success rate. New products that can target multiple receptors and improve behavioural capacities on their own or in conjunction with conventional treatments are thus becoming more and more necessary. Both industrialised and developing nations have seen significant growth in the usage of herbal products with a conventional focus in recent years. Scientists and medical professionals have created an arsenal of medicinal plants worldwide that may be able to treat the symptoms of brain dysfunctions. The importance of phytocompounds in reducing the symptoms of NDDs is highlighted in several articles. However, this article's main goal is to review recent studies critically.

10.2 BRIEF ATTENTION ON NDDS

Disorders of the nervous system known as neurodegenerative diseases (NDDs) are often distinguished by the loss of specific neurons. Studies showed that altered physicochemical qualities of proteins result in their accumulation in the human brain, where they promote neuronal degeneration [8]. NDDs are incurable to date and are significantly increasing the mortality and morbidity rates in developed countries. As the information piles up, the same scenario is expected in developing countries [9–10].

DOI: 10.1201/9781003300502-10

10.2.1 Alzheimer's Disease (AD)

The most common type of dementia is AD, a neuron-degenerative condition with an unidentified origin. The glutamate receptor subtype of the N-methyl-D-aspartate (NMDA) receptor. To regulate synaptic plasticity and brain activity, NMDA receptors are essential. The NMDA receptor, one of the inotropic glutamate receptors, gets its name from the specific way N-methyl-D-aspartate binds to this receptor (ion channel) [11–12]. The brain's cells are affected due to AD resulting in lessening intellectual functioning. Memory loss, senile dementia, the development of intra-neuronal neurofibrillary tangles, and the deposition of beta-amyloid protein in the form of amyloid plaques in the cerebral parenchyma are the cascade consequences of AD [13–20]. Microglia, resident myeloid cells in the central nervous system [21], play a significant role in the pathogenesis of AD and have become a potential therapeutic target for the disease's prevention and treatment. The pathophysiology of Alzheimer's disease has been linked to microglial provocation [22]. In the mature brain, microglia are involved in developing cognitive skills, including learning and intelligence [23–24]. By eliminating damaged cells and improperly folded proteins from the body, microglia contribute to the maintenance of homeostasis [25–30]. It was estimated that dementia would cost the world $1 trillion in 2018 and $2 trillion in 2030. Hence, understanding the pathophysiology of AD and developing efficient treatments are so necessary. The pathogenesis of AD is significantly influenced by TMP21. It plays a variety of significant roles in this illness. Additionally, AD seems to dysregulate it. This feature is highly important to highlight the connection between this protein and the aetiology of this illness [31–32].

Due to their positive impacts, bioactive dietary components can play a significant role in of AD therapy and prevention. However, there is currently inadequate information about their ideal dosages, bioavailability, variations in chemical structures, and potential interactions with other dietary supplement components. Although additional investigation is required, observational studies demonstrate that intake consuming food products containing these ingredients has undeniable advantages.

10.2.2 Parkinson's Disease (PD)

It is characterized by the loss of dopaminergic neurons in the substantia nigra (SN) [33]. Slow movement, shaking, rigidity, trouble walking, and behavioural issues are some of the disease's initial symptoms [34]. It is a condition with a late onset that affects 1–2% of adults over 60 [35]. The aberrant production of Lewy bodies is one of the unique neuropathological abnormalities in the brain. However, noradrenergic, adrenergic, glutamatergic, cholinergic, and GABAergic nerve cells also exhibit the same cytoskeleton damage as dopaminergic nigrostriatal neurons degenerate and develop Lewy bodies [36]. Progressive and targeted degeneration of dopaminergic neurons occurs in the SN [37].

10.2.2.1 Neuro-inflammation

A variety of cytokines, chemokines, reactive oxygen species, and secondary messengers are produced as part of an inflammatory reaction within the brain or spinal cord. One of the most likely paths leading to a broad range of psychiatric illnesses has been identified as this. It might also be a significant factor in the body's resistance to the current therapies for NPDs [38]. The cells responsible for halting brain apoptosis are activated because they exhibit increased phagocytic activity. Additionally, astrocytes have innate immune receptors that can be activated and lead to the production of cytokines and chemokines. The permeability of the blood-brain barrier (BBB) can be altered by creating various inflammatory mediators by activated glial cells, which then permit peripheral lymphocytes to be drawn in and activated in the CNS, resolving the initial injury and restoring homeostasis [39]. IL-1-3, 5-9, 11-18, interferons (IFN), and tumour necrosis factor (TNF)

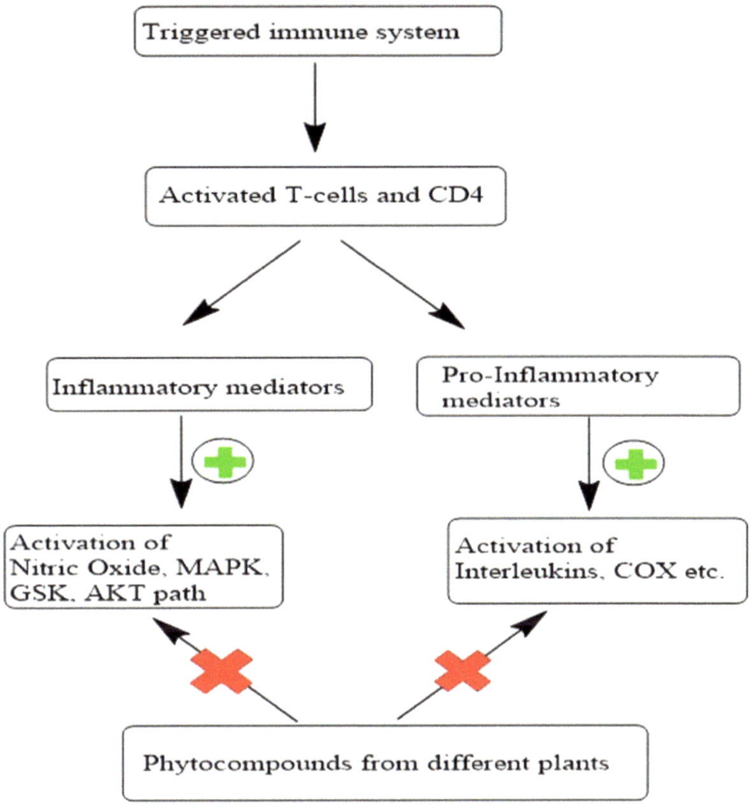

FIGURE 10.1 Mechanism of phytocompounds against neuro-inflammation.

are few examples of pro-inflammatory cytokines which are produced to help regulate inflammatory processes. Cognitive dysfunction, anxiety, and low mood have all been linked to certain inflammatory cytokines like TNF- and IL-6. Numerous phytocompounds with appealing biological features can reduce neuro-inflammation in various ways (Figure 10.1).

10.2.2.2 Neuropsychiatric disorders

Neuropsychiatric disorders like mania, major depressive syndrome, and bipolar disorder are thought to have complex pathogenesis, affecting 4.4% of the earth's total population (322 million individuals) [40–41]. Numerous symptoms, including neuro-inflammation, the existence of emotions of sadness, emptiness, anhedonia (lack of pleasure), and disrupted sleep, characterize these illnesses. An important pathogenic mechanism leading to NPDs may involve neuroinflammation, brought on by the interplay of immune system genetic predisposition and environmental variables such as hypoxia and infections [42]. Since elevated levels of inflammatory indicators have been found in patients with a variety of psychiatric issues, researchers are now examining the specific involvement of inflammatory pathways in the pathophysiology of neuropsychiatric illnesses [43–47].

10.2.2.2.1 Phytomedicines in the Treatment of Brain Disorders

A diverse range of natural bioactive substances in plants is known as phytochemicals. In neurological illnesses, many bioactive congeners, such as flavonoids, alkaloids, terpenoids, and phenols, operate as protective agents [48]. Treating neurological symptoms with phytochemicals has been widely employed, but its underlying mechanism of action is still unknown. One of the mechanisms

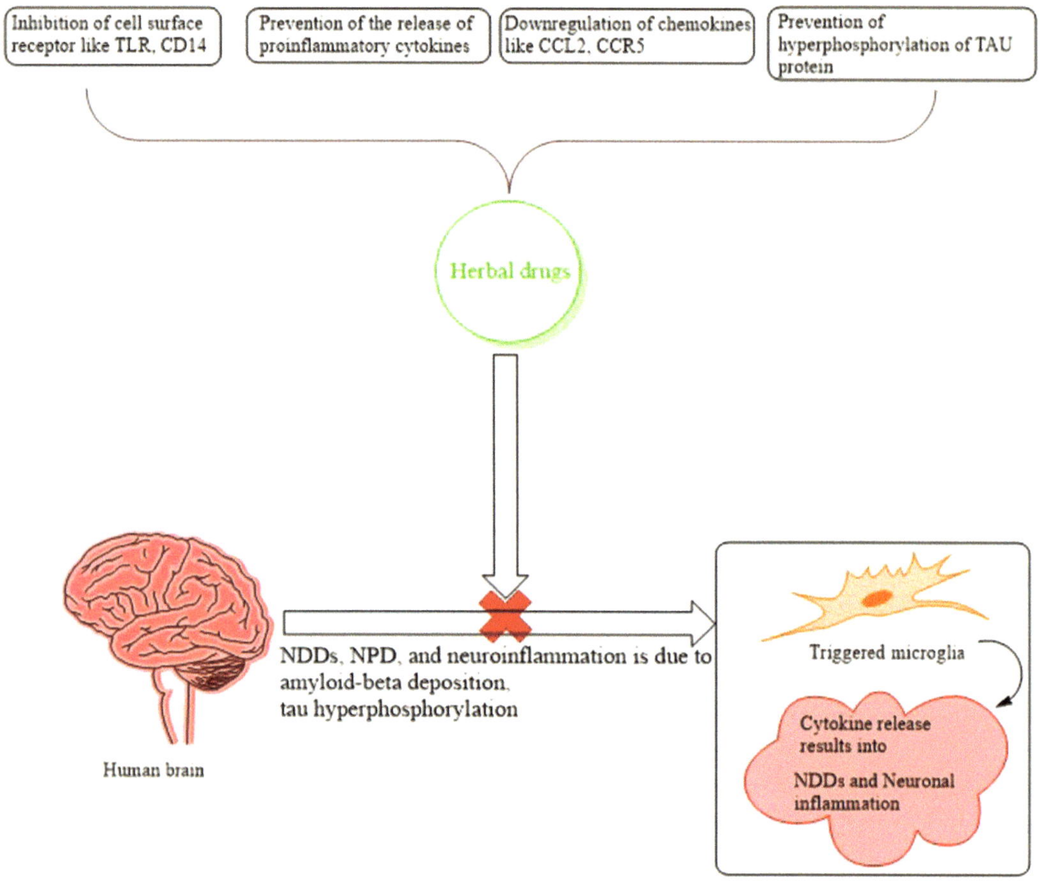

FIGURE 10.2 Mechanism of medicinal herbs and phytochemicals in the inhibition of NDDs, and neuro-inflammation.

by which phytomedicines work is that they may be effective as anti-oxidants and anti-inflammatory agents [49]. Numerous phytochemicals can change the excitability of neurons by blocking or activating certain ion channels or receptors [50-51]. By concentrating on the chemical interactions of these compounds with distinct cellular targets, we have explored the potential effectiveness of bioactive scaffolds as neuroprotective agents against NDDs in this review. The rationale behind administering herbal drugs as a curative for brain neurovegetative disorders and inflammation is elucidated in Figure 10.2.

10.2.2.2.2 Role of Phytotherapy in the Management of NDDs and NPD-

Historical records show that herbal treatments for NDDs were used as early as 6000 B. C. in Ayurvedic medicine in India, China, Africa, and pre-Columbian America, particularly with the Incas and the Aztecs [52]. Traditional herbal treatments are still very significant in the management of NDDs today. The Food and Drug Administration (FDA) views herbal treatments as dietary supplements rather than prescription medications. The majority of civilizations have a widespread usage of complementary and alternative medicine (CAM) for NDDs. The National Institutes of Health National Centre of Complementary and Alternative Medicine (NIH – NCCAM) defines CAM therapies as healthcare and medical practices that are not integral to conventional medicine as practised in the West [53]. Neuronal illnesses are becoming more common due to a paradigm shift in people's dietary preferences, types of employment, and lifestyles. Due to the rise in NDDs, dietary management is urgently needed to treat the symptoms connected to the former. However, the treatment plan and preventative strategy would mostly be tailored to the NDDs, taking into account the rate of progression, comorbidities, and personal traits of the patient. Consuming medicinal plants, however, may be able to prevent neuronal damage and provide the nerves with vital nutrients, while different bioactive may be able to block the disease's signalling pathways. Some prominent medicinal herbs employed for treating the latter have been elucidated below. Various phytochemicals effective in NDDs, NPDs, and neuro-inflammation have been explained in Figure 10.3.

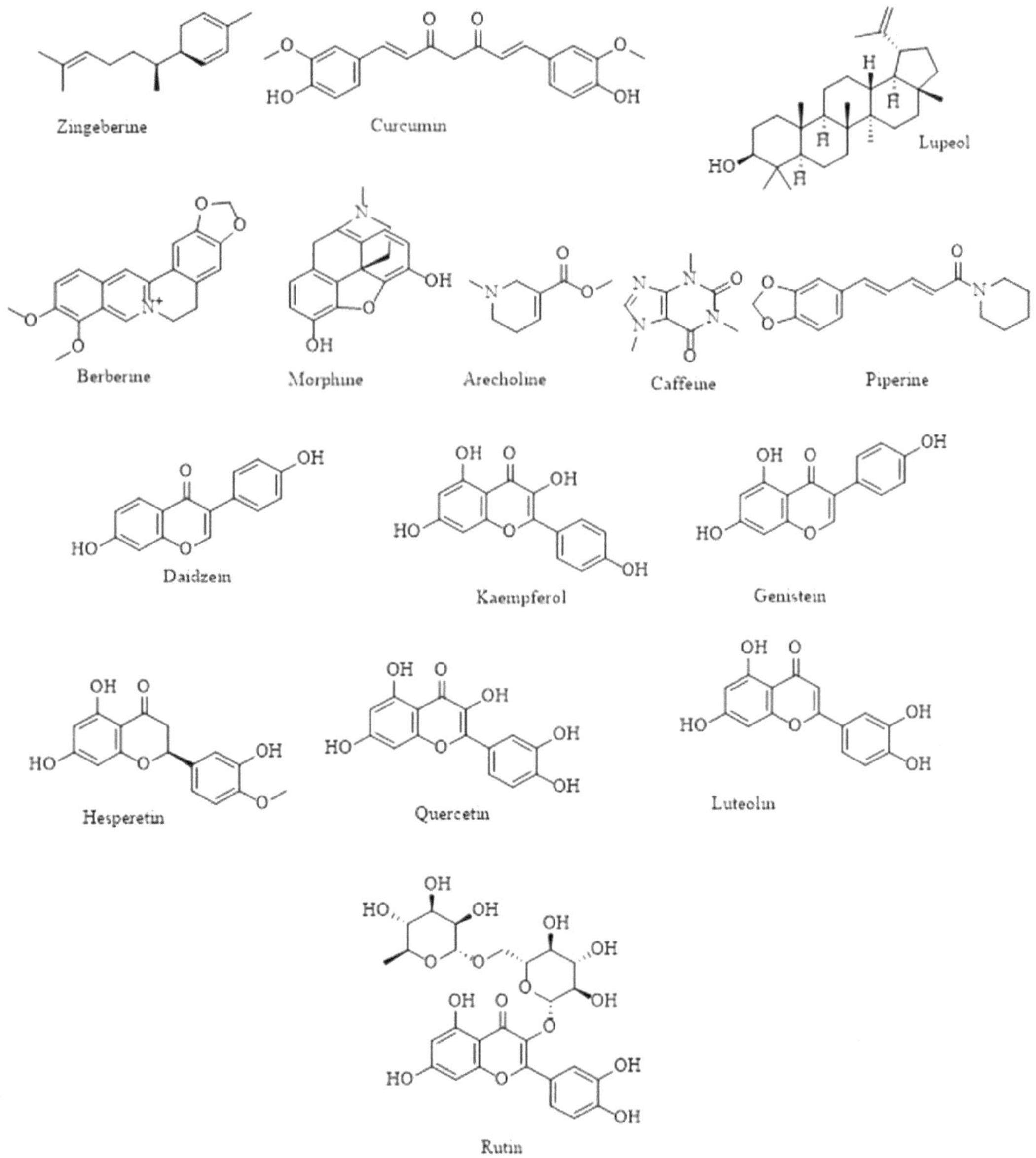

FIGURE 10.3 Chemical structures of phytochemicals.

10.2.2.2.2.1 Ginger Essential neuroprotectors include ginger and its constituent parts. Uncertainty exists regarding the precise mechanism of the action of ginger in this circumstance. However, due to its phenolic and flavonoid components, ginger is thought to have neuroprotective properties. In a crucial study, it was discovered that 6-shogaol suppresses microglia, which is beneficial for neuroprotection after acute global ischaemia [54]. Another study that supports ginger's function as a neuro-protectant demonstrates neuroprotective effects by enhancing brain antioxidant defences and restoring normal MDA levels in diabetic rats [55].

10.2.2.2.2.2 Curcumin Major depression is characterised by a persistently depressed mood, loss of interest in once-enjoyable activities, weight gain or loss, sleep disturbances (including insomnia and hypersomnia), agitation or slowing of motor function, exhaustion or a lack of energy, and an increased desire to commit suicide. About 15 to 20 percent of the world's population is affected when a new outbreak starts. Even though antidepressants are widely available, only 20 to 30 percent of those who experience depression can be successfully treated. Antidepressants have a long list of side effects as well as interactions with many other things, such as food, drink, and other medications. Therefore, it is critical to find novel pharmacological therapies for the treatment of severe depression that are both efficient and secure. Numerous studies done on animals have revealed that curcumin has antidepressant properties. When given intravenously at doses ranging from 10 to 80 mg/kg, curcumin showed an anti-immobility effect. The effect lasted for about 6 minutes. After receiving treatment for 90 minutes, the anti-immobility effect peaked. Reserpine-induced behavioural depression in rats might be reversed by curcumin when given in doses of 40 to 80 mg/kg. Since then, several studies have shown that curcumin increases the anti-immobility efficacy of the MAO inhibitors tranylcypromine (5 mg/kg, IV) and selegiline (5 mg/kg, IV). These two treatments are given intravenously. The results of this study suggest that the antidepressant properties of curcumin may be connected to the monoamine oxidase enzyme. Past studies have shown that curcumin has the ability to reduce the activity of both the MAO-A and MAO-B enzymes. It is crucial to remember that monoamine oxidase is the enzyme that degrades norepinephrine, serotonin, and dopamine. Curcumin increases the concentration of the neurotransmitter at the synapse, which prolongs the activity of the neurotransmitter by inhibiting the MAO enzymes [56].

Curcumin can also be used in the treatment of AD. The mechanistic path has been noted in Figure 10.4.

10.2.2.2.2.3 Lupeol Acetylcholinesterase inhibitors, which raise acetylcholine (Ach) levels in the brain, are one of the most important treatments for AD, a neurodegenerative condition that causes dementia and progresses over time. Rats were used to evaluate the ability of the bark's aqueous extract of the bark of *Ficus racemose* to boost Ach levels and anti-dementia activity. This extract (which contains flavonoids, lupeol, etc.) also has anti-inflammatory, antioxidant, and occasionally anticholinesterase effects. It may display a variety of the impact that could help treat AD, such as neuroprotection related to antioxidant and anti-inflammatory qualities. As with *Ficus hispida* extract, it might increase levels of Ach [57].

10.2.2.2.2.4 Berberine Berberine (BBR) belongs to the family of isoquinoline alkaloids [58]. Studies have indicated that BBR may have clinical implications for AD since it may reduce the formation and deposition of Aβ plaque [59]. The BACE-1 is the APP cleaving enzyme that initiates the Aβ production [60]. BBR lowered BACE-1 activity and prevented hippocampus neurodegeneration, improving behavioural impairment significantly [61]. It also has AChE and monoamine oxidase (MAO) inhibitory properties since both are known to play a role in AD progression. Recent evidence from additional literature shows that BBR inhibits BACE-1 expression and attenuates Aβ plaque deposition [62]. BBR improves motor stability and synchronization by preventing dopaminergic

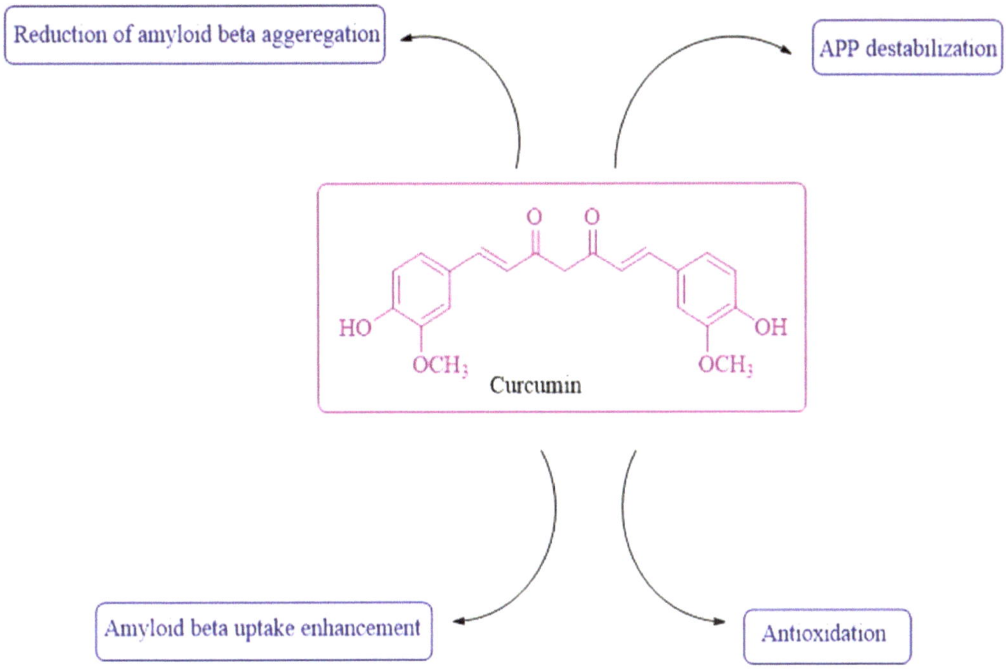

FIGURE 10.4 Anti-AD properties of curcumin.

neurons from suffering neuronal damage. Reducing apoptosis and enhancing neurogenesis in the hippocampus dentate gyrus also enhances short-term memory [63].

10.2.2.2.2.5 Morphine An isoquinoline alkaloid, morphine is used to lessen moderately severe excruciating conditions. It has potent narcotic and analgesic properties. Morphine mediates its analgesic activity through the μ-opioid receptor (MOR) [64–65]. It was discovered through various experimental models that morphine can be beneficial in preventing neural system damage [66]. By binding to MOR in the CNS, morphine plays a vital role in treating AD by raising GABA levels at brain synapses and protecting against oxidative stress-mediated neurotoxicity [67]. In primary neuronal cultures made from the neurons of rats and humans, morphine guards against the injurious effects of intracellular Aβ (iAβ).

10.2.2.2.2.6 Arecoline Arecoline is a pyridine alkaloid from the family Arecaceae that is primarily present in the fruit of the *Areca catechu* L. palm tree [68]. It has been hypothesised that oligodendrocytes (OLs) significantly contribute to the aetiology of schizophrenia [69]. By specifically targeting the OLs, arecoline effectively combats schizophrenia and also inhibits the demyelination of white matter. By promoting oligodendrocyte precursor cells (OPC) development and dephosphorylating activated protein kinase (AMPK), it improves social and cognitive function and prevents myelin degradation in the cortex [70]. Arecoline increases body excitability by stimulating muscarinic receptors (M receptors), which also helps to increase memory and learning capacity. This is how it may be effective against schizophrenia [68].

10.2.2.2.2.7 Caffeine Caffeine is a methylxanthine derivative, which is isolated from the *Coffea arabica* plant. Caffeine reduces the Aβ deposition in the cortex and hippocampus and suppresses the level of presinilin-1 (PS1), β- secretase [71]. Caffeine may lessen neurotoxicity in test models of AD because it can function as an adenosine receptor antagonist. An axonal protein called tau stabilises microtubules in a normal state, but when tau is misfolded, aggregates, or is abnormally phosphorylated, neuronal cells die and a neurofibrillary tangle forms [72]. Caffeine can effectively decrease the amount of proteolytic tau fragments and tau phosphorylation in the hippocampus [73].

In Parkinsonism, caffeine increases locomotor response as an adenosine A2 receptor antagonist. PD's neuro-inflammatory development has consistently been linked to increased activated microglial cells. Microglial cells that are triggered appear to have a negative impact on dopaminergic neurons [74]. According to certain reports, caffeine has the power to affect dopaminergic neurotransmission and reduce the mortality of neuronal cells. Additionally, it mediates dopamine receptor-mediated behavioural responses, including cognitive and locomotor processes [75]. In order to counteract adenosine's inhibitory effect on brain dopaminergic transmission, it instructs the downregulation of adenosine A2 receptors [76].

10.2.2.2.2.8 Daidzein Daidzein is a flavonoid that exclusively occurs naturally in legumes and soybeans and is a member of the isoflavones class [77]. Daidzein possesses diversified biological effects in various biological systems. It may be able to serve as an agent to demonstrate the therapeutic efficacy of flavonoids against several health issues [78], including improvement in blood cholesterol level, osteoporosis reduction [79], attenuating the risk of certain hormone-related cancer, and coronary artery diseases [80]. Additionally, it effectively reveals its property as an oestrogen agonist to further its effect as a neuroprotective drug [77]. Its structural resemblance to oestrogen gives it the capacity to bind with ERs in the brain. This explains how DNA binding and ER-dependent activation of oestrogen receptor promoters occur in various cell types [81]. Pro-inflammatory factors and microglia activation are thought to be the key factors in the neuronal cell loss that occurs in conjunction with Parkinson's disease [82]. As a result, it is known that the presence of neuroprotective qualities in PD is connected with the suppression of both factors. Daidzein has the ability to reduce the number of inflammatory mediators released by lipopolysaccharide-induced BV-2 microglial cells (LPS). One of the most prevalent hallmarks of neurodegenerative disorders like Parkinson's and Alzheimer's disease is an excess of free radicals followed by oxidative stress. It causes the body to produce ROS and NO, which have been found to impact the body's biosystem and the structure and function of brain cells.

Consequently, it causes various NDDs, such as AD and PD [83]. Daidzein reduces NO and ROS generation, which is related to oxidative stress [84]. As a result, an antioxidant agent that effectively neutralises ROS may have therapeutic potential for PD [83].

10.2.2.2.2.9 Kaempferol One of the most common dietary flavonoids and a phytoestrogen is kaempferol (3,4,5,7, -tetrahydroxyflavone). Tea, broccoli, apples, beans, strawberries, grapefruits, tea, and other foods typically contain it [85-86]. The most typical pathogenic sign in the emergence of NDDs is lipid peroxidation. It results in the development of oxidative damage brought on by the production of ROS. It was discovered that kaempferol, at a level of 30 M, prevented rotenone-induced acute toxicity from damaging the brain [87]. Similar to lipid peroxidation, monoamine oxidase-A (MAO-A) encourages the production of ROS, which results in the death of neuronal cells [88-89].

10.2.2.2.2.10 Quercetin Quercetin (3,30,40,5,7-pentahydroxylflavone) is a common flavonoid found in red wines, berries, apples, onions, tea, and tea leaves. It is important to highlight that quercetin enhances the neuroprotective benefits by reducing memory loss and neuronal cell death [91]. Additionally, quercetin reduces the amount of BACE-1, which mediates the cleavage of APP, and attenuates Aβ aggregation [92]. Additionally, quercetin strongly shields AD-related neuronal cells from neurotoxicity brought on by oxidative stress [93].

10.2.2.2.2.11 Genistein A major isoflavone found in soybeans called genistein (Gen) has many properties that are good for human health. It can mimic the pharmaceutical effects of oestrogen because of its structural resemblance to the endogenous steroid oestrogen [94].

Gen may be able to protect dopaminergic neurons against the neurotoxicity caused by lipopolysaccharide (LPS) in a dose-dependent way. In microglia and mesencephalic neuron-glia preparations, it prevents the formation of NO, TNF-α, and superoxide. Additionally, the primary immune cells in the brain, called microglia, are enthusiastically activated in response to any infection or damage, releasing pro-inflammatory substances like NO and superoxide that may interact with proteins to change their activities and ultimately result in cell death. Gen can reduce the generation and build-up of superoxide and NO, providing its neuroprotective effectiveness to dopaminergic neurons and shielding them from a post-injury reaction [95-96].

The main phytoestrogen in soybeans, Gen, successfully imitates the pharmacological effects of oestrogen [94]. The potential for oestrogen to prevent A-induced neuronal cell death is encouraging [97]. Because of its propensity to serve as an agonist for oestrogen receptors (ERs), which mediate the protective cascade against Aβ-induced damage, Gen has the potential to be neutrally neuroprotective [98]. Additionally, it lessens the critical clinical result of memory impairment in AD patients by preserving the brain's neural network. As a result, oestrogen influences the growth of memory and learning in several brain regions (hippocampus, neo-cortex, nuclei of the basal forebrain). Gen could be a helpful mediator in treating AD [99]. Last but not least, despite reports that Gen possesses neuroprotective properties, there have not yet been any clinical investigations on its use as a medicinal agent [100].

10.2.2.2.2.12 Hesperetin Hesperetin is a flavanone-class flavonoid that is present in citrus fruits. It is produced by hydrolyzing the aglycone hesperidin [101]. A dose of 97.2 μM hesperetin defends against A25-35-induced neuronal injury [102]. It also holds the capacity to ameliorate the Aβ impaired sugar uptake temperately by impeding autophagy. Hesperetin, which is highly effective at reducing neuronal autophagy, is taken in doses between 1 and 20 μM [103]. Importantly, oxidative damage brought on by lipid peroxidation is another aspect of AD pathogenesis that plays a role, along with Aβ aggregation. Hesperetin was shown to have a strong inhibitory property on the fat peroxidation that could lead to oxidative damage [102-103].

10.2.2.2.2.13 Luteolin The flavone category of flavonoids, which is usually distributed in the plant kingdom, includes luteolin (30,40,5,7-tetrahydroxyflavone) [104]. The most prevalent neurodegenerative condition that causes senile dementia is AD. AD is particularly responsible for cognitive impairment. According to reports, luteolin can prevent cognitive impairment in cerebral hypo-perfused rats when given in doses of 150 and 450 mg/kg [105]. The production of Aβ plaques is the pathophysiology of AD's defining feature. Lutein treatment of primary neuronal cells from SweAPP-overexpressing mice revealed that it momentarily reduced the Aβ generation. This was done to determine the Aβ lowering potential of luteolin [106]. The pathway behind the lessening of Aβ generation may encompass the GSK-3α isoform selective deactivation that augments the p-PS1 amount, which is the γ-secretase complex catalytic core [107].

According to an in vitro luteolin study, luteolin (5 μM) therapy may prevent LPS-induced dopaminergic neuronal degeneration by preventing microglia activation [108]. However, very little research has been done to examine the luteolin's advantageous effects on CNS.

10.2.2.2.2.14 Piperine Long pepper (*Piper longum*) and black pepper both contain the main alkaloid piperine (PIP) (*Piper nigrum*). It is a member of the Piperaceae family [109]. PIP enhances cognitive improvement by inhibiting AChE and -secretase enzymes. PIP has been thought to traverse the BBB [110]. Piper nigum L. methanolic extract was found to be helpful in improving cognition and reducing oxidative stress in the hippocampus of Aβ1-42 rat models of AD [49].

According to studies, PIP effectively treats Parkinson's disease (PD) by restoring balance and coordination in Parkinsonian rats produced by 6-OHDA. Additionally, it can reduce the inflammatory markers interleukin-1 (IL-1) and tumour necrosis factor (TNF-α) in 6-OHDA-induced Parkinson's rats [111].

By increasing the levels of serotonin and GABA in the cortical and hippocampus regions, PIP modifies the neurotransmitter systems (serotonin, GABA, and norepinephrine) and prevents tonic-clonic seizures. GABA receptor dysfunctions have been suggested to be the underlying cause of epilepsy [112]. Because PIP can interfere with the sodium/calcium channel, electrophysiological investigations have shed light on the fundamental mechanism by which it acts as an anti-convulsant [113]. Unfortunately, there is not enough data to demarcate how effective PIP is as an anti-convulsant. Studies must be conducted to determine its potential for therapeutic utility in clinical medicine.

10.2.2.2.2.15 Phenolic Compounds (anthocyanin-rich cherry juice, Rutin) Studies on neurodegenerative illnesses have looked at the health advantages of phenolic chemicals. Epigallocatechin gallate (EGCG) was demonstrated to be a potent inhibitor of tau aggregation and toxicity, which may help slow the progression of AD, which is characterized by the accumulation of amyloid-β (Aβ) and tau aggregates [114-115]. A clinical study has shown that older persons with Alzheimer's had improved verbal fluency, short-term memory, and long-term memory when consuming cherry juice rich in anthocyanins. Additionally, these patients' systolic and diastolic blood pressure levels were lower [116]. Some phenolic compounds have been shown to potentially be beneficial as treatment agents for Parkinson's disease, characterized by dopaminergic neuronal loss in the substantia nigra that is linked to oxidative stress, neuro-inflammation, and apoptosis. In a rotenone-induced rat model of Parkinson's, it was observed that the flavone apigenin reduced both α-synuclein build-up, a protein linked to sporadic and inherited forms of the illness, and motor impairments. Additionally, apigenin prevented neuronal death by controlling tyrosine hydroxylase, increasing dopamine production, and upregulating the expression of the D2 receptor [117]. Finally, it was observed that in rat pheochromocytoma PC12 cells, isoquercitrin inhibited Park5 and Park7 expression, whereas rutin suppressed the expression of genes linked to dopaminergic neural cell demise, such as Park5, Park7, and Casp3 [118].

10.2.2.2.2.16 L-Theanine Among the amino acids in green tea, theanine, also known as N-ethyl-L glutamine, is one of the healthiest. Scientists have been studying the pharmacology of L-theanine since it was recognized that it could have a calming effect. Studies on the neurochemistry of animals have revealed, for instance, that L-theanine raises the levels of serotonin, dopamine, and GABA in the brain and has a micromolar affinity for NMDA and AMPA receptors. This chemical's neuroprotective effects in animal models may be partially explained by its antagonistic effects on group 1 metabotropic glutamate receptors. Animal studies have shown that memory and learning can both be improved. The neuropharmacology of L-theanine states that more studies in both animals and people are needed to determine whether it possesses any neuroprotective and cognitive-improving properties [119].

10.2.2.2.2.17 Coumarins The phenolic chemicals known as coumarins are structural derivatives of o-coumaric acid and are produced through the shikimic acid pathway [120]. As noted in several research papers, coumarins represent another significant class of cholinesterase inhibitors [121–124].

The majority of the anticholinesterase action of coumarins depends on their ability to bind to the enzyme, and this activity is significantly enhanced in combination with several cholinesterase inhibitory medicines. Additionally, coumarins are excellent candidates for use as precursors in manufacturing medications due to their highly changeable structural properties. Given the prevalence of cholinesterase medications with coumarin nuclei, coumarin is a top candidate as a pharmacophore for cholinesterase inhibitors [125].

10.2.2.2.3 Plants in Memory Enhancement There are a number of plants reported in various literature and in Ayurveda used for prophylaxis as a supplement or in the treatment of memory loss to enhance the function of the brain [126]. The details about it are noted in the following Figure 10.5.

Various phytochemicals effective in neurodegenerative disorders have been elucidated in Figure 10.4, while their protective actions have been demonstrated in Table 10.1 [127–137]. Additional research is required to comprehend the therapeutic effectiveness of these bioactive phytomedicines and explore new potential botanicals.

Plant	Family	Active constituents
Ginkgo biloba	Ginkgoaceae	Bilobalide and Ginkgolides
Bacopa monniera	Scrophulariaceae	Bacoside A1. bacoside A2. bacoside A3
Acorus calamus	Araceae	Xanthone glycoside and triterpenoid saponins
Rhodiola rosea	Crassulaceae	Rosine, rosarin, salidroside
Celastrus paniculatus	Celastraceous	Pristimerin alkaloid
Withania somnifera	Solanaceae	Withanolides
Oryza sativa	Poaceae	Stigmastanol-3beta-p glyceroxydihydrocoumaroate
Prunus dulcis	Rosaceae	Kaempferol
Momordica charantia	Cucurbitaceae	Charantin Charine
Viscum album	Santalaceae	ß-amyrin and ß-amyrinacetate
Pistacia lentiscus	Anacardiaceae	Terpineol.ß-phellandrene

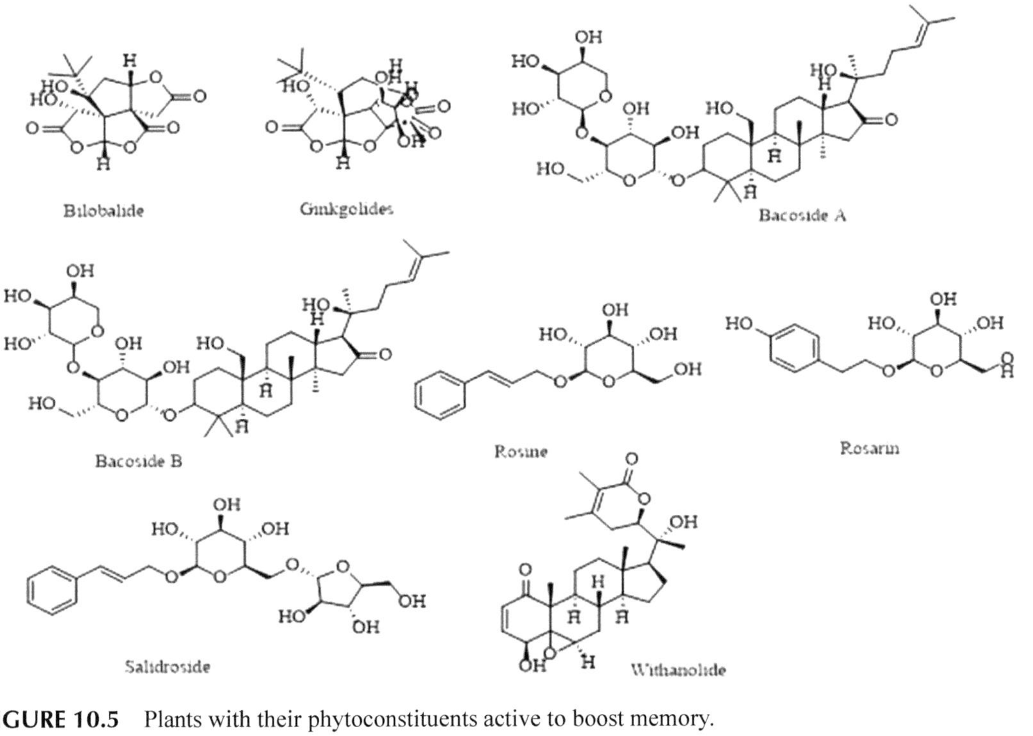

FIGURE 10.5 Plants with their phytoconstituents active to boost memory.

TABLE 10.1
Summarized beneficial effects of plant secondary metabolites in the pharmacotherapy of neurodegenerative disorders

Phytocompounds	Plant source	Mechanism of action	References
Acacetin	*Chrysanthemi indici*	↓ level of ROS, MAPK, p38	[138]
Aegeline	*Aegle marmelos*	Prevented α-synuclein-induced apoptosis	[139]
Andrographolide	*Andrographis paniculata*	↓PI3K/Akt/p38 MAPK	[140]
Apigenin	common constituent in plants	↓β-amyloid neurotoxicity	[141]
Berberine	*Berberis* genus	↓TNF-α, ↓Aβ plaque,	[142]
Carvacrol	*Cyperus rotundus*	↓antiplatelet aggregation	[143]
Curcumin	*Curcuma longa*	↓Aβ aggregation, ↓NF-κB,	[144]
Ginsenoside	*Panax ginseng*	↓excitotoxicity, ↓Ca2+ influx	[145]
Isoquercitrin	Common in plants	↓amyloid plaque, ↓tau hyperphosphorylation	[146]
Myricetin	Common constituent in plants	↓MAO	[147]
Quercetin	Tea, citrus	↓ROS, ↓Aβ aggregation	[148]
Rutin	Abundant in Citrus fruits	↓pro-inflammatory cytokines, ↓ROS	[149]
Withanone	*Withania somnifera*	Protect neurons and glial cells	[150]
Naringin	*Citrus* sp.	↓AChE, ↓cognitive deficit,	[151]
Genistein	Glycine max	↓NF-κB, ↓Aβ toxicity	[152-153]

10.3 CONCLUSION AND OUTLOOK

Numerous illnesses caused by persistent degeneration and deterioration of the CNS neurons are together referred to as 'neurodegenerative diseases'. Although there are numerous of these conditions, AD is the most common. AD patients exhibit progressive damage of cholinergic synapses in a definite portion of the brain along with a reduction in the neurotransmitter ACh, which is a key factor in the onset of dementia. Therefore, drugs that restore the amount of acetylcholine by blocking both of the foremost types of cholinesterase, AChE and BChE, could be used to treat AD and other types of dementia. In addition to successfully treating Alzheimer's disease, vascular dementia, Parkinson's disease, dementia with Lewy bodies and cognitive problems related to multiple sclerosis and Down syndrome, natural products may also be used to treat various types of dementia.

This review summarizes data on the therapeutic potential of small natural molecules from medicinal herbs, such as terpenoids, coumarins, and phenolic compounds, which have drawn particular attention because of their unique mechanisms of action, benefits of low toxicity and high efficacy in the treatment of neurodegenerative disorders. When given to patients with AD or other neurological illnesses, some of the natural substances mentioned in this paper have demonstrated numerous significant positive benefits. This has sparked research to determine the biological processes underlying these defences. These efforts have led to the identification of such mechanisms, which comprise, among others, (1) antioxidant capabilities, (2) radical scavenging properties reducing ROS content, (3) ability to interfere with the fibrillogenic process of amyloid peptides, such as those involved in AD, PD, neuro-inflammation, (4) inhibition of the activity of AChE, as well as to the development of synthetic methods for the synthesis of target natural flavonoids and the preparation of libraries of derivatives. To demonstrate their neuroprotective benefits, several natural with diverse modes of action are preferred. Additionally, the ability of these constituents to pass the blood-brain barrier is a significant determinant of the neuroprotective impact of these substances. To increase the neuroprotective action of natural products and their bioactive compounds for the prevention and treatment of neurodegenerative disorders, new approaches and tactics must be developed, such as the use of nanotechnology in the distribution of natural goods and compounds. Together, these findings support the ability of phytocompounds and their derivatives to operate as effective treatments for neurological disorders, which should inspire researchers to use this information to develop sensible prospective treatments for these maladies. We anticipate that the phytoconstituents included in this review will act as encouraging tools and chemical scaffolds for discovering fresh, powerful therapeutic leads for treating and alleviating NDDs and NPDs.

REFERENCES

1. GBD 2019 Diseases and Injuries Collaborators. 2020. *Lancet* 396:1204–1222.
2. GBD 2019 Diseases and Injuries Collaborators. 2020. *Lancet* 396: 1223–1249.
3. A.s.D. International. Dementia statistics. www.alz.co.uk/research/statistics (accessed December 9, 2017.
4. Wimo, A., Jonsson, L., Bond, J., Prince, M., Winblad, B. 2013. *Alzheimer's & Dementia* 9:1–11.
5. Cummings, J.L. Morstorf, T., Zhong. K. 2014. *Alzheimer's Research & Therapy* 6:37.
6. Meur, S., Mukherjee, S., Roy, S., Karati, D. 2024. *Mol Neurobiol.* 1–15.
7. Karati, D., Shaw, T.K., 2022. *RJPT* 15(8):3790–3795.
8. Kovacs, G.G. 2014. *Cit EMJ Neurol.* 1:78–86.
9. Van Assche, R., Temmerman, L., Dias, D.A., Boughton, B., Boonen, K., Braeckman, B.P. et al. 2015. *Metabolomics* 11(2):477–86.
10. Hussain, G., Shahzad, A., Anwar, H., Sohail, M.U., Baig, S.M., Shabbir, A., et al. 2017. *Pakistan J Neurol Sci.* 12(3).
11. Banerjee, A., Schepmann, D., Kohler, J., Würthwein, E.U., Wünsch, B.U. 2010. *Bioorganic & Medicinal Chemistry* 18:7855–7867.
12. Tamilselvan, M., Tamilanban, T., Chitra, V. 2020. *Research J. Pharm. and Tech.* 13(6):3021–3027.
13. Zhang, Y., Li, P., Feng, J., Wu, M. 2016. *Neurol Sci.* 37:1039–1047.
14. Surabhi, Singh B. K. 2019. *Int J Pharma Sci Res* 10(3):993–1000.
15. Bhushan, I., Kour, M., Kour, G., Gupta, S., Sharma, S., Yadav, A. 2018. *Ann Biotechnol* 1(1):1002.
16. Herholz, K., Ebmeier, K. 2011. *Lancet Neurol* 10:667–70.
17. Kabir, T., Sufian, M.A., Uddin, M.S., Begum, M.M., Akhter, S., Islam, A., et al. 2019. *Curr Pharm Des* 25(33):3507–14.
18. Uddin, M.S., Mamun, A.A., Takeda, S., Sarwar, M.S., Begum, M.M. 2019. *Psychogeriatrics* 19(2):87–94.
19. Velraj, M., Lavaniya, N. 2018. *Research J. Pharm. and Tech.* 11(6):2695–2700.
20. Dhinakaran, S., Tamilanban. T., Chitra, V. 2019. *Research J. Pharm. and Tech.* 12(6):3073–3077.
21. Heneka, M.T., Carson, M.J., El Khoury, J., *et al.* 2015. *Lancet Neurol.* 14(4):388–405.
22. Pierre, W.C., Smith, P.L., Londono, I., Chemtob, S., Mallard, C., Lodygensky, G.A. 2017. *Brain Behav Immun.* 59:333–345.
23. Parkhurst, C.N., Yang, G., Ninan, I., Savas, J.N., Yates, J.R., Lafaille, J.J., Hempstead, B.L., Littman, D.R. *et al.* 2013. *Cell* 155(7):1596–1609.
24. Yao, K., Zu, H.B. 2019. *Infllammo Pharmacol* 28:95–110
25. Hristovska, I., Pascual, O. 2015. *Front Integr Neuroscience.* 9:73.
26. Paolicelli, R.C., Bolasco, G., Pagani, F., Maggi, L., Scianni, M., Panzanelli, P., Giustetto, M., *et al.* 2011. *Science* 333(6048):1456–1458.
27. Sierra, A., Encinas, J.M., Deudero, J.J.P., Chancey, J.H., Enikolopov, G. *et al.* 2010. *Cell Stem Cell* 7(4):483–495.
28. Stevens, B., Allen, N.J., Vazquez, L.E., Howell, G.R., Christopherson, K.S., Nouri, N., *et al.* 2007. *Cell* 31(6): 1164–1178.
29. Tremblay, M.È., Lowery, R.L., Majewska, A.K. 2010. *PLoS Biol.* 8(11):e1000527.
30. Wake, H., Moorhouse, A.J., Jinno, S., Kohsaka, S., Nabekura, J. 2009. *J Neuroscience* 29(13):3974–3980.
31. Chen, F., Hasegawa, H., Schmitt-Ulms, G., Kawarai, T., Bohm, C., Katayama, T., et al. 2006. *Nature* 440:1208–1212.
32. Zhang, X., Wu, Y., Cai, F., Song, W. 2019. *Mol. Brain.* 12:39.

33. Klemann, C.J.H.M., Martens, G.J.M., Sharma, M., Martens, M.B., Isacson, O., Gasser, T., Visser, J.E., Poelmans, G. 2017. *NPJ Park. Dis*. 3:14.
34. Sveinbjornsdottir, S. 2016. *J. Neurochem* 1:318–324.
35. Sowell, R.A., Owen, J.B., Butterfield, D.A. 2009. *Ageing Res. Rev*. 8:1–17.
36. Braak, H., Braak, E. 2000. *J. Neurol.* 247:II3–II10.
37. Shulman, J.M., De Jager, P.L., Feany, M.B. 2011. *Annu. Rev. Pathol.* 6:193–222.
38. Al-Asmari, A.K., Khan, M.W. 2014. *Human & Experimental Toxicology.* 33(2):115–122.
39. Kumar, V. 2019. *J Neuroimmunol.* 332:16–30.
40. Northoff, G., Stanghellini, G., 2016. *Front Human Neurosci.* 10:172.
41. Tayab, M.A., Islam, M.N., Chowdhury, K.A., Tasnim, F.M. 2022. *Biomed. Pharmacother.* 147:112668.
42. Bechter, K., Reiber, H., Herzog, S., Fuchs, D., Tumani, H., Maxeiner, H.G. 2010. *J Psych Res.* 44(5):321–330.
43. Benros, M.E., Nielsen, P.R., Nordentoft, M., Eaton, W.W., Dalton, S.O., Mortensen, P.B. 2011. *Am J Psych.* 168(12):1303–1310.
44. Henje, B.E., Lekander, M., Ingvar, M., et al. 2012. *J. Affect Disord.* 136(3):716–723.
45. Pfau, M.L., Russo. S.J. 2016. *Trends Neurosci.* 39(6):353–355.
46. Kaletsky, R.L., Francica, J.R., Agrawal-Gamse, C., & Bates, P. 2009. *Proc Nat Acad Sci.* 106(8), 2886–2891.
47. Ivanovska, M., Abdi, Z., Murdjeva, M. et al. 2020. *Pharmaceuticals* 13:230.
48. Liu, R.H. 2003. *Am. J. Clin. Nutr.* 78:3–6.
49. Uriarte Pueyo, I.; Calvo, M.I. 2009. *Pharmacogn. Mag.* 5:287–290.
50. Quansah, E., Karikari, T.K., Asase, A., Safo, D.O., Kukuia, K.K.E., et al. 2018. *Evid-Based Complementary Altern Med* 2018:1–14.
51. Calvo, M.I., Cavero, R.Y. 2015. *J Ethnopharmacol.* 169:263–268.
52. Adams, M., Gmünder, F., Hamburger, M. 2007. *J. Ethnopharmacol.* 113:363–381.
53. Nahin, R.L., Barnes, P.M., Stussman, B.J., Bloom, B. 2009. *Natl. Health Stat. Rep.* 18:1–14.
54. Ha, K., Moon, E., Ju, M.S., Kim, D.H., Ryu, J.H., Oh M.S. et al., 2012. *Neuropharmacology* 63(2):211–23.
55. Shanmugam, K. R., Mallikarjuna, K., Kesireddy, N., Reddy, K.S. 2011. *Food Chem Toxicol.* 49(4):893–7.
56. Singh, Y., Paswan, S.K., Kumar, R., Otia, M.K., Acharya, S. et al., 2022. *J Res Appl Sci Biotechnol.* 1(2):10–24.
57. Pahari, N., Majumdar, S., Karati, D., Mazumder, R. 2022. *Pharmacol Res Mod Chin Med* 4:100137.
58. Schmitt, F., Hussain, G., Dupuis, L., Loeffler, J.P., Henriques, A. 2014. *Front Cell Neurosci.* 8:25.
59. Yan, R., Vassar, R. 2014. *Lancet Neurol.* 13:319–29.
60. Panahi, N., Mahmoudian, M., Mortazavi, P., Hashjin, G.S. 2013. *Arch Med Sci.* 9(1):146–50.
61. Mathew, B., Suresh, J., Mathew, G.E., Parasuraman, R., Abdulla, N. 2014. *Cent Nerv Syst Agents Med Chem.* 14(1):28–33.
62. Pohanka, M. 2014. *Int J Mol Sci.* 15:9809–25.
63. Jin, Y., Khadka, D.B., Cho, W.J. 2016. *Expert Opin Ther Pat.* 26(2):229–43.
64. Pang, B., Zhao, L., Zhou, Q., Zhao, T., Wang, H., Gu, C. et al. 2015. *Int J Endocrinol.* 2015:905749.
65. Kaur, R., Arora, S. 2015. *J Crit Rev.* 2(3):1–8.
66. Ye D, Bu H, Guo G, Shu B, Wang W, Guan X, et al. 2014. *J Mol Neurosci.* 53(4):571–9.
67. Cui, J., Wang, Y., Dong, Q., Wu, S., Xiao, X., Hu, J. et al. 2011. *J Neurosci.* 31(45):16227–40.
68. Grizzell, J.A., Patel, S., Barreto, G.E., Echeverria, V. 2017. *Prog Neuro-Psychopharmacol Biol Psych.* 78:75–81.
69. Coppola, M., Mondola, R. 2012. *Nordic J Psych.* 66:73–78.
70. Wang, Y., Reis, C., Applegate, R., Stier, G., Martin, R., Zhang, J.H. 2015. *Exp Neurol.* 272:26–40.
71. McIntire, L.K., McKinley, R.A., Goodyear, C., Nelson, J. 2014. *Brain Stimul.* 7(4):499–507.
72. Arendash, G.W., Mori, T., Cao, C., Mamcarz, M., Runfeldt, M., Dickson, A. et al. 2009. *J Alzheimer's Dis.* 17(3):661–80.
73. Collin, L., Bohrmann, B., Göpfert, U., Oroszlan-Szovik, K., Ozmen, L., Grüninger, F. 2014. *Brain.* 137(10):2834–46.

74. Ritchie, K., Carrière, I., de Mendonça, A., Portet, F., Dartigues, J.F., Rouaud, O. et al. 2017. *Neurology*. 23:272–90.
75. More, S.V., Kumar, H., Kim, I.S., Song, S.Y., Choi, D.K. 2013. *MediatInflamm.* 2013:1–12.
76. Rivera-oliver M, Manuel D, *Díaz-ríos M. Life Sci.* 2014;
77. Jung, W., Yu, O., Lau, S.-M.C., O'Keefe, D.P., Odell, J., Fader, G., McGonigle, B. 2000. *Nat. Biotechnol.* 18:208–212.
78. Choi, R.C.Y., Zhu, J.T.T., Yung, A.W.Y., Lee, P.S.C., Xu, S.L., Guo, A.J.Y., Zhu, K.Y., Dong, T.T.X., Tsim, K.W.K. 2013. *Hindawi* 2013:635694.
79. Civitelli, R. 1997. *Calcif. Tissue Int.* 61:12–14.
80. Peterson, G., Barnes, S. 1991. *Biochem. Biophys. Res. Commun.* 179:661–667.
81. Schreihofer, D.A. 2005. *Mol. Cell. Endocrinol.* 231:13–22.
82. Gao, H.M., Jiang, J., Wilson, B., Zhang, W., Hong, J.S., Liu, B. 2002. *J. Neurochem.* 81:1285–1297.
83. Chinta, S.J., Ganesan, A., Reis-Rodrigues, P., Lithgow, G.J., Andersen, J.K. 2013. *Neurotox. Res.* 23:145–153.
84. Jiang, T., Sun, Q., Chen, S. 2016. *Prog. Neurobiol.* 147:1–19.
85. Neumann, J., Sauerzweig, S., Ronicke, R., Gunzer, F., Dinkel, K., Ullrich, O., Gunzer, M., Reymann, K.G. 2008. *J. Neurosci.* 28:5965–5975.
86. Bhathena, S.J., Velasquez, M.T. 2002. *Am. J. Clin. Nutr.* 76:1191–1201.
87. Filomeni, G., Graziani, I., de Zio, D., Dini, L., Centonze, D., Rotilio, G., Ciriolo, M.R. 2012. *Neurobiol. Aging* 33:767–785.
88. Watanabe, Y., Himeda, T., Araki, T. 2005. *Med. Sci. Monit* 11:17–23.
89. Olanow, C.W., Obeso, J.A., Stocchi, F. 2006. *Lancet Neurol* 5:677–687.
90. Chen, C., Zhou, J., Ji, C. 2010. Quercetin: A potential drug to reverse multidrug resistance. *Life Sci* 87: 333–338.
91. Pu, F., Mishima, K., Irie, K., Motohashi, K., Tanaka, Y., Orito, K., Egawa, T., Kitamura, Y., Egashira, N., Iwasaki, K., et al. 2007. *J. Pharmacol. Sci.* 104:329–334.
92. Sabogal-Guáqueta, A.M.; Muñoz-Manco, J.I.; Ramírez-Pineda, J.R.; Lamprea-Rodriguez, M.; Osorio, E.; Cardona-Gómez, G.P. 2015. *Neuropharmacology* 93:134–145.
93. Heo, H.J., Lee, C.Y. 2004. *J. Agric. Food Chem* 52:7514–7517.
94. Vegeto, E., Bonincontro, C., Pollio, G., Sala, A., Viappiani, S., Nardi, F., Brusadelli, A., Viviani, B., Ciana, P., Maggi, A. 2001. *J. Neurosci.* 21:1809–1818.
95. Sonee, M., Sum, T., Wang, C., Mukherjee, S.K. 2004. *Neurotoxicology* 25:885–891.
96. Gao, H.M., Jiang, J., Wilson, B., Zhang, W., Hong, J.S., Liu, B. 2002. *J. Neurochem.* 81:1285–1297.
97. Hardy, J. 2002. *Science* 297:353–356.
98. Kim, H., Bang, O.Y., Jung, M.W., Ha, S.D., Hong, H.S., Huh, K., Kim, S.U., Mook-Jung, I. 2001. *Neurosci. Lett.* 302:58–62.
99. Bang, O.Y., Hong, H.S., Kim, D.H., Kim, H., Boo, J.H., Huh, K., Mook-Jung, I. 2004. *Neurobiol. Dis.* 16:21–28.
100. Devi, K.P., Shanmuganathan, B., Manayi, A., Nabavi, S.F., Nabavi, S.M. 2016. *Mol. Neurobiol.* 1–14.
101. Kim, J.Y., Jung, K.J., Choi, J.S., Chung, H.Y. 2004. *Free Radic. Res.* 38:761–769.
102. Cho, J. 2006.*Arch. Pharm. Res.* 29:699–706.
103. Huang, S.M., Tsai, S.Y., Lin, J.A., Wu, C.H., Yen, G.C. 2012. *Mol. Nutr. Food Res.* 56:601–609.
104. Harborne, J.B., Williams, C.A. 2000. *Phytochemistry* 55:481–504.
105. Fu, X., Zhang, J., Guo, L., Xu, Y., Sun, L., Wang, S., Feng, Y., Gou, L., Zhang, L., Liu, Y. 2014. *Pharmacol. Biochem. Behav.* 126:122–130.
106. Rezai-Zadeh, K., Douglas Shytle, R., Bai, Y., Tian, J., Hou, H., Mori, T., Zeng, J., Obregon, D., Town, T.; Tan, J. 2009. *J. Cell. Mol. Med.* 13:574–588.
107. Wang, H., Wang, H., Cheng, H., Che, Z. 2016. *Mol. Med. Rep.* 13:4215–4220.
108. Chen, H.Q., Jin, Z.Y., Wang, X.J., Xu, X.M., Deng, L., Zhao, J.W. 2008. *Neurosci. Lett.* 448:175–179.
109. Ghavami, S., Shojaei, S., Yeganeh, B., Ande, S.R., Jangamreddy, J.R., Mehrpour, M., et al. 2014. *Prog Neurobiol.* 112:24–49.
110. Hirose S. 2014. *Prog brain Res.* 213:55–85.
111. Rinwa, P., Kumar, A. 2017. *Arch Pharm Res.* 40(10):1166–75.
112. Goswami, P.K., Samant, M. 2014. *Int J Pharm Pharm Sci.* 6(5):4–7.

113. Mikkilineni, S., Cantuti-Castelvetri, I., Cahill, C.M., Balliedier, A., Greig, N.H., Rogers, J.T. 2012. *Parkinson's Dis.* 2012:5–8.
114. Wobst, H.J., Sharma, A., Diamond, M.I., Wanker, E.E., Bieschke, J. 2015. *FEBS Lett.* 589:77–83.
115. Cosme, P., Rodríguez, A.B., Espino, J., Garrido, M. 2020. *Antioxidants* 9:1263.
116. Kent, K., Charlton, K., Roodenrys, S., Batterham, M., Potter, J., Traynor, V., Gilbert, H., Morgan, O., Richards, R. 2017. *Eur. J. Nutr.* 56:333–341.
117. Anusha, C., Sumathi, T., Joseph, L.D. 2017. *Chem. Biol. Interact.* 269:67–79.
118. Magalingam, K.B., Radhakrishnan, A.; Ramdas, P.; Haleagrahara, N. 2014. *J. Mol. Neurosci.* 55:609–617.
119. Kim T. I., Lee Y. K., Park S. G., et al. 2009. *Free Radical Biology and Medicine* 47(11):1601–1610.
120. Matos, M.J.; Santana, L.; Uriarte, E.; Abreu, O.A.; Molina, E.; Yordi, E.G. 2015. InTech: *London* 113–140.
121. Orhan, I.E., Senol, F.S., Shekfeh, S., Skalicka-Wozniak, K., Banoglu, E. 2019. *Food Chem. Toxicol.* 109:970–974.
122. Athar, A.M., Hussain, G., Aziz-Ur-Rehman, Siddiqui, S.Z., Ahmad, V.U. 2017. *Asian J. Chem.* 29:758–762.
123. Dastan, D., Validi, S., Ebadi, A. 2020. *Struct. Chem.* 31:965–973.
124. Heo, J.H., Eom, B.H., Ryu, H.W., Kang, M.G., Park, J.E., Kim, D.Y, Kim, J.-H., Park, D., Oh, S.R., Kim, H. 2020. *Sci. Rep.* 10:21695.
125. de Souza, L.G., Renna, M.N., Figueroa-Villar, J.D. 2016. *Chem. Biol. Interact.* 254:11–23.
126. Gudalwar, B.R., Panchale, W.A., Manwar, J.V., Nimbalwar, M.G., Badukale, N.A., Bakal, R.L. 2021. *GSC Adv Res Rev.* 08(02):019–029.
127. Sharifi-Rad, M., Lankatillake, C., Dias, D.A., Docea, A.O., Mahomoodally, M.F., Lobine, D. 2020. *J. Clin. Med.* 9:1061.
128. Chen, X., Drew, J., Berney, W., Lei, W. 2021. *Cells* 10:1309.
129. Pannangrong, W., Wattanathorn, J., Muchimapura, S., Tiamkao, S., Tong-Un, T. 2011. *J. Med. Food* 14:688–694.
130. Kulkarni, K.S., Kasture, S.B., Mengi, S.A. 2010. *Indian J. Pharm.* 42:168–173.
131. Haider, S., Batool, Z., Haleem, D.J. 2012. *Nutr. Hosp.* 27:2109–2115.
132. Batool, Z., Agha, F., Ahmad, S., Liaquat, L., Tabassum, S., Khaliq, S., Anis, L., Sajid, I., Emad, S., Perveen, T. et al. 2017. *Pak. J. Pharm. Sci.* 30:273–279.
133. Batool, Z., Sadir, S., Liaquat, L., Tabassum, S., Madiha, S., Rafiq, S., Tariq, S., Batool, T.S., Saleem, S., Naqvi, F. et al. 2016. *Brain Res. Bull.* 120:63–74.
134. Huang, H.J., Chen, S.L., Chang, Y.T., Chyuan, J.H., Hsieh-Li, H.M. 2018. *Nutrients* 10:1888.
135. Sepehri, H., Hojati, A., Safari, R. 2019. *J. Exp. Pharm.* 11:115–119.
136. Szurpnicka, A., Kowalczuk, A., Szterk, A. 2020. *Arch. Pharmacal. Res.* 43:593–629.
137. Ammari, M., Othman, H., Hajri, A., Sakly, M., Abdelmelek, H. 2018. *Brain Res. Bull.* 140:140–147.
138. Kim, S.M., Park, Y.J., Shin, M.S., Kim, H.R., Kim, M.J., Lee, S.H., Yun, S.P., Kwon, S.H. 2017. *Bioorganic. Med. Chem. Lett.* 27:5207–5212.
139. Derf, A., Sharma, A., Bharate, S.B., Chaudhuri, B. 2018. *Bioorg. Med. Chem. Lett.* 29:454–460.
140. Yang, C.H., Yen, T.L., Hsu, C.Y., Thomas, P.A., Sheu, J.R., Jayakumar, T. 2017. *Int. J. Mol. Sci.* 18:1638.
141. Santos, G., Giraldez-Alvarez, L.D., Avila-Rodriguez, M., Capani, F., Galembeck, E., Neto, A.G., Barreto, G.E., Andrade, B. 2016. *Front. Aging Neurosci.* 8:97.
142. Kaufmann, D., Kaur Dogra, A., Tahrani, A., Herrmann, F., Wink, M. 2016. *Molecules* 21:1161.
143. Pirzada, A.M., Ali, H.H., Naeem, M., Latif, M., Bukhari, A.H., Tanveer, A. 2015. *J. Ethnopharmacol.* 174:540–560.
144. Seo, E.J.; Fischer, N.; Efferth, T. 2018. *Pharmacol. Res.* 129:262–273.
145. Nabavi, S.F., Sureda, A., Habtemariam, S., Nabavi, S.M. 2015. *J. Ginseng Res.* 39:299–303.
146. Carmona, V.; Martin-Aragon, S.; Goldberg, J.; Schubert, D.; Bermejo-Bescos, P. Nutr. Neurosci. 2018, 1–16.
147. Larit, F., Elokely, K.M., Chaurasiya, N.D., Benyahia, S.; Nael, M.A., Leon, F., Abu-Darwish, M.S., Efferth, T., Wang, Y.H., Belouahem-Abed, D. et al. 2018. *Phytomed. Int. J. Phytother. Phytopharm.* 40:27–36.

148. Velmurugan, B.K., Rathinasamy, B., Lohanathan, B.P., Thiyagarajan, V., Weng, C.F. 2018. *Molecules* 23:2485.
149. Sivanantham, B., Krishnan, U., Rajendiran, V. 2018. *Biomed. Pharm.* 108:15–26.
150. Dar, N.J., Hamid, A., Ahmad, M. 2015. *Cell. Mol. Life Sci.* 72:4445–4460.
151. Costa, S.L., Silva, V.D., Dos Santos Souza, C., Santos, C.C., Paris, I., Munoz, P., Segura-Aguilar, J. 2016. *Neurotox. Res.* 30:41–52.
152. Dey, A., Bhattacharya, R., Mukherjee, A., Pandey, D.K. 2017. *Biotechnol. Adv.* 35:178–216.
153. Rahman, M.M., Rahaman, M.S., Islam, M.R., Rahman, F., Mithi, F. M. et al., 2022. *Molecules* 27:233.

11 Plant Phenols and Their Potential Modulation of Neurochemistry

Rahul Kumar Sharma, Shilpa Thakur, Navidha Aggarwal, and Hitesh Chopra

11.1 INTRODUCTION

Flora and fauna are the gifts of nature. Many plants have been used to treat many ailments for many decades. The plants have a pharmacologically active substance that is responsible for many biological activities [1–3]. They are alkaloids, terpenoids, phenols and phenolic glycosides, coumarins and their glycosides, anthraquinones and their glycosides, flavones and flavonoid glycosides or heterosides, mucilage and gums, tannins, volatile oils, saponins, cardioactive glycosides, cyanogenic glycosides, etc. The majority of plants' secondary metabolites are polyphenols, also referred to as polyhydroxyphenols. They are part of a structural class of organic compounds that are distinguished by the presence of many phenols' structural units. The distinct physical, chemical, and biological (metabolic, poisonous, medicinal, etc.) characteristics of certain members of the class are a result of these phenol structures. Its chemical structure, produced by joining an aromatic phenyl ring to a hydroxyl (-OH) group, is known as phenol. The majority of fruits, vegetables, cereals, and beverages contain polyphenols, which are naturally occurring chemicals. Up to 200–300 mg of polyphenols is present in every 100 g of dry mass of fruits like berries, grapes, apples, pears, cherries [4]. Quercetin and catechin are two instances of free monomers of polyphenols that can be present in food, along with oligomers (procyanidins). They often exist as free aglycones or are associated with saccharides as glycosides [5]. Strong evidence came from epidemiological studies and related meta-analyses suggesting long-term intake of diets high in plant polyphenols provides some protection against the onset of malignancies, cardiovascular illnesses, diabetes, osteoporosis, and neurological diseases [6–13]. Dietary flavonoids have been postulated to provide health benefits through antioxidant pathways since oxidative stress has been thought to play a part in the pathophysiology of these diseases, in addition to other variables. Flavonoids exhibit a significant antioxidant effect in in vitro tests of redox temperingability [14–18]. In vitro, polyphenols exert powerful antioxidant effects via a variety of mechanisms, including radical scavenging, chelation of metal ions (Fe, Cu, and others), and regulation of antioxidant enzyme activity. By giving a hydrogen atom from their hydroxyl groups to radicals, phenolic -OH groups' position and number affect their capability to scavenge, which leads to the removal of radical moiety. The phenoxy radical that can produce stable compounds and end radical reactions by reacting with another radical is created during this reaction.

When consumed in food, polyphenols are present as esters, glycosides, or polymers that the intestines are unable to absorb. The native polyphenolic chemical molecules are digested by microbial enzymes in the colon, changed to derivatives by methylation, sulfation, and glucuronidation, and then absorbed in the colon and transported through the blood to numerous tissues and organs, including the brain. Moreover, the bioavailability of phenolic substances can range from extremely

DOI: 10.1201/9781003300502-11

low to incredibly high due to their various susceptibilities to colon enzyme metabolism [19,20]. In comparison to other low-molecular-weight antioxidants like vitamins C and E and uric acid, polyphenolic substances have lower direct antioxidant action because of their low bioavailability and in vivo conversion to various derivatives.

Polyphenols can participate in the modulation of various signalling pathways, thus influencing the fate of cells [21,22], including nerve cells, via influencing neuronal survival, regeneration, development, or death. The neuroprotective properties of polyphenols are responsible for the stimulation of NO production in the endothelium and the suppression of platelet aggregation, both of which result in an increase in cerebral blood flow.

Polyphenols might modify brain function at three locations:

(i) By increasing cerebral blood flow outside the central nervous system (CNS) or by altering the signalling pathways that connect the brain to peripheral organs.
(ii) By modifying the influx/efflux mechanisms of many biomolecules at the blood-brain barrier, including multi-drug resistance proteins.
(iii) Within the CNS, via altering the activity of neurons and glial cells.

The neuroprotective effects of polyphenols, in addition to their antioxidant properties, have been demonstrated to be mediated by modulation of cellular signalling pathways involved in cognitive processes like synaptic plasticity, particularly those that involve CREB (cAMP-response element-binding protein) signalling. This transcription factor, known as CREB, is connected to genes that express brain-derived neurotrophic factor (BDNF). Many researchers suggested that CREB activity can affect memory development and, conversely, that higher CREB activity can accelerate memory formation, highlighting the crucial role that CREB plays in brain function [23,24]. By activating BDNF after inducing CREB, polyphenols can potentially modify these signalling pathways [25].

11.2 POLYPHENOLSSTRUCTURE AND CLASSIFICATION

More than 8,000 polyphenolic chemical compounds can be found in a variety of plant species. All plant phenolic compounds begin with phenylalanine or a nearby precursor called shikimic acid. Although sugar (polysaccharide or monosaccharide) and an aromatic carbon can also be linked directly, they are most frequently found in conjugated forms with one or more sugar residues attached to hydroxyl groups. Polyphenols can be categorised into different classes based on the number of phenol rings they contain, as well as the structural elements that hold these rings together (See Figure 11.1).

11.3 NEURODEGENERATIVE DISORDERS

Neurodegeneration is the term used to describe the gradual loss of a neuron's structure and/or function, as well as their death. This deficit may be hereditary and assessed at birth or shortly after birth, or it may even be caused by an injury [26,27]. There are thought to be a few hundred different neurodegenerative disorders, and many of them appear to share clinical and pathological characteristics. Longer life lengths are one of the greatest issues facing medical science, as neurodegenerative illnesses are a major source of mortality and disability. Although the causes of neurodegeneration differ, they can be broken down into five categories of mechanisms: excitotoxicity, apoptosis, amyloid cascade, oxidative stress, and sodium nitrite-mediated neurodegeneration. Superoxide, hydroxyl radicals, and hydrogen peroxide are examples of reactive oxygen species (ROS) that are produced endogenously and play helpful roles in cellular processes. However, they can take part in detrimental processes, including neurodegeneration. Neurodegenerative illnesses like

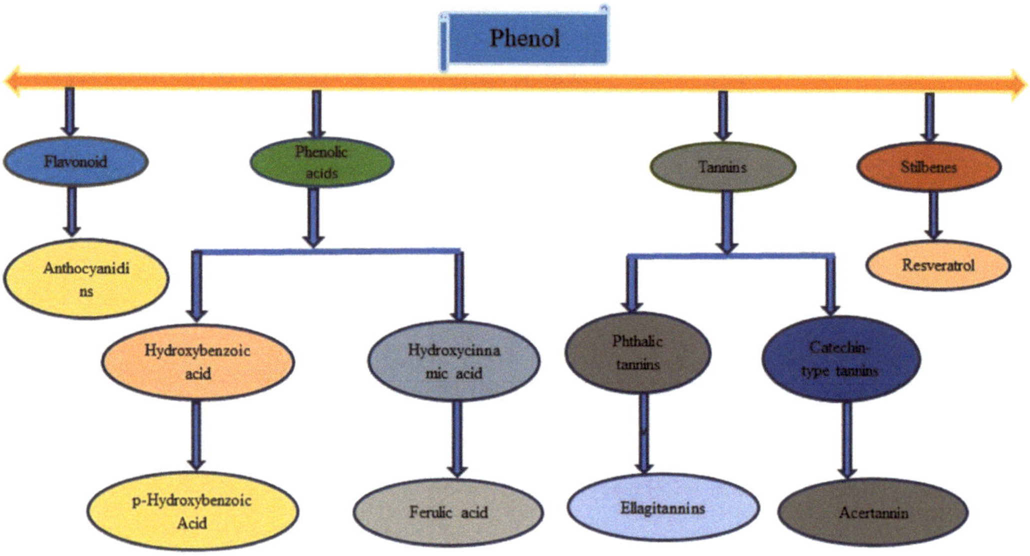

FIGURE 11.1 Types of Phenol.

amyotrophic lateral sclerosis, Alzheimer's, Huntington's, and Parkinson's diseases have received a lot of attention, and it has been established that these illnesses are closely related to mitochondrial malfunction brought on by ROS. Neuroprotection refers to any medical method, including drugs and treatments, that stop neurons from dying. Limiting neuronal dysfunction following damage is the aim of neuroprotection, which also aims to preserve the integrity of any cellular interactions that might occur in the brain and preserve neural function. Antioxidants are useful in reducing or avoiding certain illnesses because they neutralise free radicals[28–30].

11.4 ROLE OF PLANT PHENOL ON NEUROCHEMISTRY

Plants have been used to treat a variety of human diseases since antiquity. India is one of the world's top 12 mega-diversity countries due to the large number of species there. Meanwhile, neurological illnesses are frequently not regarded as common diseases, and many persons with psychological disorders, like epilepsy, are profoundly impacted by stigma and discrimination connected to health. The majority of the population relies on traditional medicine for basic health care. Others, such as meningitis, stroke, Parkinson's, and Alzheimer's. Speaking, breathing, movement, mood, and memory are all impacted by nervous system problems. The brain and spinal cord are impacted by neurological conditions. By attaching to different receptors, polyphenols in cerebral tissue can produce anti-inflammatory effects. Various kinases, including the nuclear factor-kappaB (resveratrol and epigallocatechin gallate) and protein kinase C (PKC) pathways, phosphoinositide-3-(PI3) kinase (curcumin), mitogen-activated protein (MAP) kinase, and phosphoinositide-3-(PI3) kinase, can have their active states modified by flavonoid-induced receptor (resveratrol). Thus, polyphenols can affect apoptosis and differentiation, inflammatory response, learning and memory, and the prevention of amyloid plaque formation. They can also affect cell survival (by inhibiting apoptosis). By preventing the inflammation that macrophages cause and lowering the production of proinflammatory cytokines (IL-1 beta, TNF-alpha), polyphenols have the potential to influence the activation of glial cells in the brain, which are where macrophages dwell. By encouraging NO production in the endothelium and inhibiting platelet aggregation, polyphenols can be credited with improving cerebral blood flow and providing neuroprotection [31].

TABLE 11.1
Phenolic plant sources and their medicinal effects

Plant	Family	Chemicals	Plant part	Pharmacologically use	References
Rauwolfia serpentina	*Apocyanaceae*	Ajmaline, ajmalicine, reserpine, serpentine	Root	Insomnia, mental disorders, aggressive behaviour, schizophrenia, epilepsy, psychosis	[32]
Aegle marmelos	*Rutaceae*	Carotenoids, tannic acid, ageline, aegelenine, mermelosinmarmesinin, ascorbic acid, eugenol, skimmianine and saponin	Leaves	Fatigue, anxiety, depression	[33]
Rosmarinus officinalis	*Lamiaceae*	Rosemary diterpenes, p-cymene, linalool, eucalyptol, beta-pinene	Whole plant	Depression, anxiety	[34]
Withaniasomnifera	*Solanaceae*	Somniferinine, withananine, Withanine, somniferine, somnine	Roots, stems, and leaves	Nervous disorders, arthritis, rheumatism, amnesia, insomnia, stress and anxiety, paralysis, Alzheimer's disease	[35]
Bacopa monnieri	*Plantaginaceae*	Alkaloids, brahmine, herpestine, saponins, β-sitosterol, and stigmasterols	Whole plant, leaf	epilepsy, schizophrenia, and dissociative disorder	[36]
Datura metel	*Solanaceae*	Hyoscyamine, norhyoscyamine, norscopolamine	Leaf, seed	Increase the heart rate, induce relaxation and motor inhibition in smooth muscles, decrease secretions and induce dilation of the pupils of the eyes. Rheumatism	[37]
Centella asiatica	*Apiaceae*	Triterpenoids, volatile fatty acids, glycosides, flavonoids, alkaloids, and tannins,	Whole plant, leaf, stem	psychosis, epilepsy and insomnia, mental disorders, memory	[38]
Ginkgo biloba	*Ginkgoaceae*	Trilactonic diterpenes (ginkgolide A-C, ginkgolide J-M), trilactonic sesquiterpene (Bilobalide), and flavanoids	Leaf	Alzheimer's disease	[39]
Turmeric	*Zingiberaceae*	Curcumin, curcuminoids, turmerones, sabinene, cineol	Rhizome	Alzheimer's disease, arthritis, bone fracture, depression, hypertension	[40]

(*continued*)

TABLE 11.1 (Continued)
Phenolic plant sources and their medicinal effects

Plant	Family	Chemicals	Plant part	Pharmacologically use	References
Embilica officinalis	*Euphorbiaceae*	Gallic acid, catechol, ascorbic acid, ellagic acid, rutin, quercetin	Fruit	memory loss, mental fatigue, anxiety, depression, migraine, insomnia, hyperactivity disorder	[41]
Acorus calamus	*Acoraceae*	α-asarone, calamine, β-pinene, camphene, isoeugenol, camphor, calamol, azulene	Whole plant, rhizome	Epilepsy, mental problems, psychosis, dissociative disorder	[42]
Terminalia chebula	*Combretaceae*	chebulinic acid, chebulic acid, gallic acid, coumarin, punicalagin, corilagin, ellagic acid, chebulagic acid	Fruit	Schizophrenia and mental retardation	[43]
Panax Ginseng	*Araliaceae*	Ginsenoside	Root	Parkinson's disease, Alzheimer's disease, cancer, antianxiety, antidepressant, hypertension and diabetes	[44]
Cassia fistula	*Fabaceae*	Glycosides, flavonoids, proanthocyanides, polyphenol, terpenoids	Leaves, fruit	anti-tumour, antidepressant, anxiety, antidiabetic, paralysis, rheumatism and gout	[45]
Zingiber officinale	*Zingiberaceae*	Gingerols, shogaols	Rhizome	Rheumatism, migraine, muscular strength, schizophrenia	[46]

11.5 ROLE OF PLANT PHENOLS IN NEURODEGENERATIVE DISORDERS

Mental health issues in children, adolescents, and adults are usually caused by neurodegenerative illnesses such as schizophrenia, ADHD, and major depressive disorder. The term 'oxidative stress' describes a condition in which the production of free radicals and/or reactive oxygen species/reactive nitrogen species (ROS/RNS) outweighs the antioxidant defence, favouring ROS/RNS, resulting in oxidative damage to lipids, proteins, and DNA, and eventually cell and organ dysfunction. The brain is a lipid-rich organ, which can contribute to its susceptibility to oxidative damage. Compared to other organs of the body, the CNS's cells are more vulnerable to ROS' harmful effects. Additionally, compared to other cells, the brain has lower levels of catalase, glutathione peroxidase, and superoxide dismutase (enzymes that break down hydrogen peroxide or organic peroxides), as well as a higher concentration of iron ions and ascorbic acid, which together create the ideal environment for the formation of the extremely toxic hydroxyl radical. The brain is more vulnerable to oxidative and peroxidative damage to biomolecules as a result of these facts. By altering particular cellular signalling pathways involved in cognitive processes like synaptic plasticity, polyphenols can exhibit their neuroprotective effects. Polyphenols can directly modulate these signalling pathways by induction of CREB [47].

11.5.1 Alzheimer's Disease

Alzheimer's disease (AD) is a prevalent form of dementia that is caused by ageing as opposed to vascular dementia linked to brain infarction. A loss of neurons, neurofibrillary tangles, and amyloid plaques are all symptoms of AD (particularly cholinergic neurons of the basal forebrain) [48–50]. These anatomical defects can be seen in the brains of AD patients. Loss of neurons that release acetylcholine: The nucleus basalis, a significant hub of neurons that release Ach, is located below the globus pallidus. The dispersion of these neurons' axons, which are characteristic of Alzheimer's disease, is widespread across the limbic and cerebral cortex. β-Amyloid plaques, aberrant protein clusters that are deposited outside of neurons, and neurofibrillary tangles, abnormal filament bundles inside of neurons in affected brain regions. This tau protein, which has undergone hyperphosphorylation, makes up the filaments. These anomalies, along with others, play a role in a variety of neurotoxic processes, such as oxidative stress and inflammation, dysfunctional synapses, neurotransmitter depletion, and, ultimately, cell death. The loss of cortical and subcortical neurons and severe atrophy of the cerebral cortex are hallmarks of AD. Numerous neuronal systems employ glutamate as a neurotransmitter, and it is thought that glutamate mediates the majority of excitatory synaptic transmission in mammalian brains. Even while glutamate is necessary for healthy brain function, too much of it can cause excitotoxicity, which is the death of cells [51]. Glutamate receptors, especially those of the N-Methyl-D- Aspartate (NMDA) type, mediate the damaging effects of glutamate. The complicated character of polyphenols with regard to their bioactivity and bioavailability in the brain hinders their progression in the treatment of neurological problems. Dietary intake of polyphenols has been designed to improve cognitive abilities while lowering free radical controlling signalling pathways and reducing the risk of AD. Curcumin in AD is that it can cohere to amyloid beta (Aβ) by defeating nuclear factor kappa B (NF-κβ), granting this polyphenol the potential to less AD pathological process [52]. The predominance of curcumin's chemical makeup makes it a lipophilic substance that easily crosses the blood-brain barrier, forms plaques, and inhibits the amyloid peptide's aggregation and disaggregation models. Increasing evidence indicates that the PI3K/Akt/GSK-3 flagging pathway is altered in AD brain [53,54].

11.5.2 Parkinson's Disease

Parkinson's disease is a progressive, age-related movement disorder whose neuropathology is characterized by deterioration of the afferent pigmented neurons of the substantia nigra [55]. Dopaminergic neurons in the substantia nigra pars compacta deteriorate as a result of the motor and cognitive impairment that underlies PD (SNpc). Due to the persistent accumulation of protein inclusions containing ubiquitin in the cytoplasm of the chosen neuron, which results in their death, trademark lesions known as Lewy bodies are produced [56,57]. Tremor, stiffness, slowness, decreased balance, and eventually a shuffling stride are among the primary symptoms. Anxiety, depression, and dementia are a few subsequent symptoms.

In a rat model of PD, resveratrol has been demonstrated to lessen oxidative stress, lipid peroxidation, and protein carbonyl (PC), as well as to decrease cyclooxygenase-2 (COX-2) and TNF-alpha encoded mRNA levels in the substantia nigra [58]. Oxyresveratrol has been shown to lessen brain damage in SH-SY5Y cells by increasing SIRT1 levels and decreasing the expression of caspase-3, c-Jun N-terminal kinase (JNK), and c-Jun transcription factors. Similar to oxyresveratrol, ferulic acid has shown that inhibiting the JNK pathway has neuroprotective effects [59].

11.5.3 Huntington's Disease

Huntington's disease (HD) is a hereditary (autosomal dominant) disorder characterised by progressive brain deterioration that begins in adulthood and progresses quickly till death [60–62].

HD's foundation is made up of the enlarged polyglutamine region of the huntingtin protein and the AG triplet nucleotide repetitions. It results in severe motor symptoms such as uncontrollable writhing motions, which are quite incapacitating. Balance is hampered, movements become more pronounced, and dysarthria and dysphasia start to occur. The striatum (caudate/putamen) of the brain exhibits substantial neuronal loss in HD. The caudate nucleus tail is first affected by the atrophy of these structures, which then affects the anterior region from the mediodorsal to the ventrolateral. The emergence of chorea, the most obvious clinical symptom of HD, also appears to be mediated by selective vulnerability. The typical age of onset is between 35 and 40 years old, but the age ranges from 2 years old to the middle of one's 80s. Even though both parents contribute equally to the defect's inheritance, the father accounts for more than 80% of cases when symptoms appear before age 20. Particularly if one parent possesses an intermediate gene, the disease can develop in unaffected parents.

Polyphenols have therapeutic significance because they provide a variety of health benefits, such as anti-inflammatory, anti-cancer, and anti-ageing properties. In a transgenic mouse model of HD, resveratrol has been reported to have beneficial benefits via SIRT1 activation of the PGC-1alpha (peroxisome proliferative activated receptor, gamma) signalling pathway [63]. Hesperidin and naringenin, abundant in citrus fruits, induced neuroprotection in rats, possibly via nitric oxide synthase (NOS) inhibition [64]. Curcumin has been demonstrated to alter a number of transgene-dependent metrics and reduce Huntington aggregates, suggesting a potential therapeutic effect in HD [65]. According to the available preclinical evidence, polyphenols provide substantial neuroprotection through genetic and immunological regulation, offering hope for the clinical delay or prevention of neurological illnesses like PD and HD.

Stroke is the second leading cause of mortality worldwide and the fifth major cause of death in the United States, which is brought on by an abrupt obstruction of the CNS's blood supply. They are frequently brought on by an embolus, thrombus, or hypoperfusion. There may be sufficient collateral circulation if an obstruction develops within or close to the cerebral arterial circle. When a thrombotic event occurs, the blood vessel that supplies blood to the brain becomes dysfunctional due to atherosclerosis, arterial dissection, fibromuscular dysplasia, or an inflammatory disorder. When an embolic event occurs, bodily debris obstructs blood flow through the afflicted vessel.

Neuroprotective activity by down regulation of the matrix metalloproteinase (MMP) in the mice model of the cerebral ischaemia has been exhibited by green tea polyphenol, EGCG [66]. Green tea polyphenols have been discovered to play a role in the defence of neurons against hypoxia-induced ischaemic injury by regulating the inflammatory cascade and reducing trans-membrane potential degeneration [67]. To attenuate ischaemic injury, quercetin played a crucial role by controlling the peroxidation of lipids in neurons and the acid-sensing of ion channels that led to dysregulation [68]. Quercetin has decreased blood-brain barrier interruption and lowered MMP-9 levels in tests of cerebral ischaemia by using the same antioxidant analyses as green tea polyphenols [69].

Serotonin abnormalities, abnormal serotonin receptors, and depression are thought to be causally related, but dopamine, adrenaline, and glutamate are also thought to play a role [70,71]. Enzymes involved in the breakdown of neurotransmitters, such as monoamine oxidase (MAO), and the synthesis of their amino acid precursor tryptophan by the enzyme tryptophan hydroxylase both affect neurotransmitter metabolism [72]. Pro-inflammatory cytokines like tumour necrosis factor (TNF-alpha), IL-6, and C-reactive protein (CRP) are known to reliably indicate elevated inflammation in depressed people when their blood levels of these cytokines are elevated. These reliable markers of nonspecific inflammation showed a positive correlation with the severity of depressive symptoms as well as several co-morbid conditions like fatigue, poor sleep, and cognitive impairment [73,74].

When curcumin and the alkaloid piperine were taken together, MAO activity was inhibited, and levels of serotonin and dopamine were raised [75]. It is thought that curcumin's ability to inhibit lipid peroxidation and neutralise ROS and RNS is what allows it to easily cross the blood-brain barrier and induce neuroprotection [76,77]. Currently, EGCG is recognised for its potent anti-inflammatory and antioxidant activities as well as its capacity to reduce stress and depression. In a mouse study, higher levels of BDNF or lower levels of serum corticosterone and adrenocorticotropic hormone were found in response to long-term administration of green tea polyphenols or a forced swimming test, respectively [78].

Schizophrenia is a severe mental illness that manifests as distorted behaviour and abnormal thinking processes. Psychological and social processes, as well as genetic and early environmental influences, seem to be significant contributing variables. Although it is believed that the disease primarily affects thinking, it frequently exacerbates long-term issues with behaviour and emotions. The pathophysiology of schizophrenia is thought to be related to increased oxidative stress [79]. The brain has a very high metabolic rate of oxidation. The brain's high concentrations of the neurotransmitters dopamine, adrenaline, and noradrenaline can also cause auto-oxidation, which can result in a significant amount of hydrogen peroxide being produced. The superoxide anion radical can also be overproduced by neuronal mitochondria. Mitochondria become harmed and dysfunctional due to insufficient Mn-superoxide dismutase (Mn SOD) activity and low levels of glutathione (GSH), the main free radical scavenger in the brain [80]. In the prefrontal cortex of rodents and humans, glutathione and redox regulation play a crucial role in myelination processes and white matter maturation, a mechanism that may be compromised in schizophrenia.

Schizophrenia patients may benefit from a green tea extract containing the main polyphenol epigallocatechin-3-gallate (EGCG) [81]. Epicatechin, a key flavonoid found in green tea, has been shown in an ex vivo study to lessen the lipid peroxidation caused by haloperidol in human plasma [82]. The blood-brain barrier is crossed by EGb-761, a Ginkgo biloba extract that shields the brain from the harmful effects of oxidative stress due to its predominantly lipophilic components [83].

11.6 CONCLUSION

Dietary polyphenols may have several beneficial benefits on the brain, including protection of neurons from damage caused by neurotoxins, reduction of neuroinflammation, and enhancement of memory, learning, and cognitive abilities. Despite the fact that many of the mechanisms underlying their beneficial effects have yet to be elucidated, it has become clear that they, in part, involve reductions in oxidative/inflammatory stress signalling and increases in protective signalling, and may also involve hormetic effects to protect neurons against oxidative and inflammatory stressors. Protective against age-related diseases, dietary polyphenols are overwhelmingly chemically reactive, with the vast majority being electrophilic. Because of their unique chemical composition, these molecules are able to influence a cascade of transcription factors, leading to the activation of genes involved in both the first and second stages of the metabolic pathway (phase I and phase II, respectively). However, much of the information we have on their bioactivity comes from in vitro or in vivo research that lasted just a few weeks and employed doses that were not relevant to human nutrition. It is still unclear how much polyphenol effects contribute in vivo and at physiologically relevant concentrations to the advantages of consuming a diet rich in fruits and vegetables, despite the fact that the weight of evidence suggests that they do. Whether or if this class of chemicals is associated with health advantages has to be shown, as does the positive impact these compounds have on slowly progressing neurodegenerative illnesses. The intake of polyphenol-rich foods throughout life has the potential to limit neurodegeneration and prevent or repair age-dependent deteriorations in cognitive function due to their various biological activities. However, clinical translation of the therapeutic and pharmacological potential of these natural chemicals is still lacking. In addition, random clinical trial effectiveness is required to back up the generally consistent epidemiological and mechanistic findings. Despite the lack of efficacy data and the ambiguity of their effects in vivo, studies on the absorption and metabolism of different polyphenols in humans have shown that the majority of polyphenols share common pathways for their metabolism, most notably via their bacterial metabolism in the large intestine. Therefore, studies of smaller polar polyphenols for brain bioavailability and bioactivity should be prioritized in the development of dietary polyphenols for applications in neurodegenerative illnesses. Until thorough randomized controlled clinical studies are conducted to ascertain experimentally if polyphenols and/or their metabolites have effectiveness in people with dementia and other neurodegenerative disorders, the task ahead is to proceed carefully.

REFERENCES

1. R. Goyal, G. Mittal, S. Khurana, N. Malik, V. Kumar, A. Soni, H. Chopra, and M.A. Kamal. "Insights on Quercetin Therapeutic Potential for Neurodegenerative Diseases and Its Nano-technological Perspectives." Current Pharmaceutical Biotechnology 24, no. 10 (2023): 1110–1123. https://doi.org/10.2174/1389201025666230830125410.
2. A.V. Kaaviya, P. Hridya, D. Prasanth, D. Abernaa, P.S. Harisankaran, C.R. Hari Sankar, N.S. Rajan, S. Karthik, J. Adinan, K.S. Abhijith, R. Krishnan, S. Akash, R. Amin, S. Chakraborty, H. Chopra, A. Dey, A.K. Sharma, M. Alagawany, K. Dhama, and D. Chandran. "Cinnamon as a Potential Feed Additive: Beneficial Effects on Poultry Health and Production Performances – An Update." Journal of Experimental Biology and Agricultural Sciences 11, no. 3 (2023): 444–461. https://doi.org/10.18006/2023.11(3).444.461.
3. C.R. Hari Sankar, N.S. Rajan, Raida, V.K. Sreya, S. Suresh, P.S. Harisankaran, P. Sheela, M. Pran, R. Priya, M.I. Yatoo, H. Chopra, T. Bin Emran, A. Dey, K. Dhama, and D. Chandran. "Potential Effects of Essential Oils in Safeguarding the Health and Enhancing Production Performance of Livestock Animals: The Current Scientific Understanding." Journal of Experimental Biology and Agricultural Sciences 10, no. 6 (2022): 1222–1240. https://doi.org/10.18006/2022.10(6).1222.1240.
4. Scalbert, C. Manach, C. Morand, C. Rémésy, and L. Jiménez. "Dietary Polyphenols and the Prevention of Diseases." Critical Reviews in Food Science and Nutrition 45, no. 4 (2005): 287–306. https://doi.org/10.1080/1040869059096.
5. V. Křen. "Glycoside vs. Aglycon: The Role of Glycosidic Residue in Biological Activity." Glycoscience (2008): 2589–2644. https://doi.org/10.1007/978-3-540-30429-6_67.
6. Z. Yan, Y. Zhong, Y. Duan, Q. Chen, and F. Li. "Antioxidant Mechanism of Tea Polyphenols and Its Impact on Health Benefits." Animal Nutrition 6 (2020): 62–73. https://doi.org/10.1016/j.aninu.2020.01.001.
7. L. Kan, T. Oliviero, R. Verkerk, V. Fogliano, and E. Capuano. "Interaction of Bread and Berry Polyphenols Affects Starch Digestibility and Polyphenols Bio-accessibility." Journal of Functional Foods 68 (2020): 103924. https://doi.org/10.1016/j.jff.2020.103924.
8. C. Del Bo, S. Bernardi, M. Marino, M. Porrini, M. Tucci, S. Guglielmetti, A. Cherubini, B. Carrieri, B. Kirkup, P. Kroon, R. Zamora-Ros, N.H. Liberona, C. Andres-Lacueva, and P. Riso. "Systematic Review on Polyphenol Intake and Health Outcomes: Is There Sufficient Evidence to Define a Health-Promoting Polyphenol-Rich Dietary Pattern?" Nutrients 11, no. 6 (2019): 1355. https://doi.org/10.3390/nu11061355.
9. H. Shakoor, J. Feehan, V. Apostolopoulos, C. Platat, A.S. Al Dhaheri, H.I. Ali, L.C. Ismail, M. Bosevski, and L. Stojanovska. "Immunomodulatory Effects of Dietary Polyphenols." Nutrients 13, no. 3 (2021): 728. https://doi.org/10.3390/nu13030728.
10. Y.A. Kim, J.B. Keogh, and P.M. Clifton. "Polyphenols and Glycémie Control." Nutrients 8, no. 1 (2016): 17. https://doi.org/10.3390/nu8010017.
11. M. Abbas, F. Saeed, F.M. Anjum, M. Afzaal, T. Tufail, M.S. Bashir, A. Ishtiaq, S. Hussain, and H.A.R. Suleria. "Natural Polyphenols: An Overview." International Journal of Food Properties 20, no. 2 (2017): 247–284. https://doi.org/10.1080/10942912.2016.1220393.
12. C. Manach, A. Scalbert, C. Morand, C. Rémésy, and L. Jiménez. "Polyphenols: Food Sources and Bioavailability." American Journal of Clinical Nutrition 79, no. 5 (2004): 727–747. https://doi.org/10.1093/ajcn/79.5.727.
13. C. Di Lorenzo, F. Colombo, S. Biella, C. Stockley, and P. Restani. "Polyphenols and Human Health: The Role of Bioavailability." Nutrients 13, no. 1 (2021): 273. https://doi.org/10.3390/nu13010273.
14. Kerimi and G. Williamson. "Differential Impact of Flavonoids on Redox Modulation, Bioenergetics, and Cell Signaling in Normal and Tumor Cells: A Comprehensive Review." Antioxidants & Redox Signaling 29, no. 7 (2018): 641–667. https://doi.org/10.1089/ars.2017.7086.

15. S.T. Raman, A.K.P. Ganeshan, C. Chen, C. Jin, S.H. Li, H.J. Chen, and Z. Gui. "In Vitro and In Vivo Antioxidant Activity of Flavonoid Extracted from Mulberry Fruit (Morus alba L.)." Pharmacognosy Magazine 12, no. 45 (2016): 282–290. https://doi.org/10.4103/0973-1296.177910.
16. R.A. Khan, M.R. Khan, S. Sahreen, and M. Ahmed. "Assessment of Flavonoids Contents and In Vitro Antioxidant Activity of Launaea procumbens." Chemistry Central Journal 6 (2012): 1–11. https://doi.org/10.1186/1752-153X-6-43/TABLES/4.
17. C. Sarkar, P. Chaudhary, S. Jamaddar, P. Janmeda, M. Mondal, M.S. Mubarak, and M.T. Islam. "Redox Activity of Flavonoids: Impact on Human Health, Therapeutics, and Chemical Safety." Chemical Research in Toxicology 35, no. 4 (2022): 549–572. https://doi.org/10.1021/acs.chemrestox.1c00348.
18. H. Arif, A. Sohail, M. Farhan, A.A. Rehman, A. Ahmad, and S.M. Hadi. "Flavonoids-Induced Redox Cycling of Copper Ions Leads to Generation of Reactive Oxygen Species: A Potential Role in Cancer Chemoprevention." International Journal of Biological Macromolecules 106 (2018): 208–214. https://doi.org/10.1016/j.ijbiomac.2017.08.049.
19. P. Cosme, A.B. Rodríguez, J. Espino, and M. Garrido. "Plant Phenolics: Bioavailability as a Key Determinant of Their Potential Health-Promoting Applications." Antioxidants 9, no. 12 (2020): 1263. https://doi.org/10.3390/antiox9121263.
20. L.F. Călinoiu and D.C. Vodnar. "Whole Grains and Phenolic Acids: A Review on Bioactivity, Functionality, Health Benefits and Bioavailability." Nutrients 10, no. 2 (2018): 204. https://doi.org/10.3390/nu10020204.
21. L.R. Sun, W. Zhou, H.M. Zhang, Q.S. Guo, W. Yang, B.J. Li, Z.H. Sun, S.H. Gao, and R.J. Cui. "Modulation of Multiple Signaling Pathways of the Plant-Derived Natural Products in Cancer." Frontiers in Oncology 9 (2019): 1153. https://doi.org/10.3389/FONC.2019.01153/XML/NLM.
22. M.H. Cháirez-Ramírez, K.G. de la Cruz-López, and A. García-Carrancá. "Polyphenols as Antitumor Agents Targeting Key Players in Cancer-Driving Signaling Pathways." Frontiers in Pharmacology 12 (2021). https://doi.org/10.3389/fphar.2021.710304.
23. N. Sen. "ER Stress, CREB, and Memory: A Tangled Emerging Link in Disease." Neuroscientist 25 (2019). https://doi.org/10.1177/1073858418816611.
24. S. Ortega-Martínez. "A New Perspective on the Role of the CREB Family of Transcription Factors in Memory Consolidation via Adult Hippocampal Neurogenesis." Frontiers in Molecular Neuroscience 8 (2015). https://doi.org/10.3389/fnmol.2015.00046.
25. F. Moosavi, R. Hosseini, L. Saso, and O. Firuzi. "Modulation of Neurotrophic Signaling Pathways by Polyphenols." Drug Design, Development and Therapy 10 (2016): 23. https://doi.org/10.2147/DDDT.S96936.
26. F. Islam, S. Bepary, M.H. Nafady, R. Islam, T. Bin Emran, S. Sultana, A. Huq, S. Mitra, H. Chopra, R. Sharma, and S.H. Sweilam. "Polyphenols Targeting Oxidative Stress in Spinal Cord Injury: Current Status and Future Vision." 2022.
27. T. Bhattacharya, G.A.B.E. Soares, H. Chopra, M.M. Rahman, Z. Hasan, S.S. Swain, and S. Cavalu. "Applications of Phyto-Nanotechnology for the Treatment of Neurodegenerative Disorders." Materials 15 (2022): 804. https://doi.org/10.3390/MA15030804.
28. H. Chopra, A.K. Mishra, A.A. Baig, T.K. Mohanta, Y.K. Mohanta, and K.-H. Baek. "Narrative Review: Bioactive Potential of Various Mushrooms as the Treasure of Versatile Therapeutic Natural Product." Journal of Fungi 7 (2021): 728. https://doi.org/10.3390/JOF7090728.
29. T. Bhattacharya, G.A.B.E. Soares, H. Chopra, M.M. Rahman, Z. Hasan, S.S. Swain, and S. Cavalu. "Applications of Phyto-Nanotechnology for the Treatment of Neurodegenerative Disorders." Materials 15 (2022): 804. https://doi.org/10.3390/MA15030804.
30. R.K. Singla, C.S. Sai, H. Chopra, S. Behzad, H. Bansal, R. Goyal, R.K. Gautam, C. Tsagkaris, S. Joon, S. Singla, and B. Shen. "Natural Products for the Management of Castration-Resistant Prostate Cancer: Special Focus on Nanoparticles-Based Studies." Frontiers in Cell and Developmental Biology 9 (2021). https://doi.org/10.3389/FCELL.2021.745177.
31. S.A. Abdelsalam, K. Renu, H.A. Zahra, B.M. Abdallah, E.M. Ali, V.P. Veeraraghavan, K. Sivalingam, L. Ronsard, R. Ben Ammar, D.S. Vidya, P. Karuppaiya, S.Y. Al-Ramadan, and P. Rajendran. "Polyphenols Mediate Neuroprotection in Cerebral Ischemic Stroke—An Update." Nutrients 15, no. 5 (2023). https://doi.org/10.3390/nu15051107.

32. B. Jayaprakasam, K. Padmanabhan, and M.G. Nair. "Withanamides in Withania somnifera Fruit Protect PC-12 Cells from Beta-Amyloid Responsible for Alzheimer's Disease." Phytotherapy Research 24, no. 6 (2010): 859–863. https://doi.org/10.1002/PTR.3033.
33. S. Rahman and R. Parvin. "Therapeutic Potential of Aegle marmelos (L.)—An Overview." Asian Pacific Journal of Tropical Disease 4, no. 6 (2014). https://doi.org/10.1016/S2222-1808(14)60318-2.
34. K. Sasaki, A. El Omri, S. Kondo, J. Han, and H. Isoda. "Rosmarinus officinalis Polyphenols Produce Anti-Depressant-Like Effect through Monoaminergic and Cholinergic Functions Modulation." Behavioral Brain Research 238 (2013): 183–190. https://doi.org/10.1016/j.bbr.2012.10.010.
35. H.K. Afewerky, A.E. Ayodeji, B.B. Tiamiyu, J.I. Orege, E.S. Okeke, A.O. Oyejobi, P.N.N. Bate, S.B. Adeyemi. "Critical Review of the Withania somnifera (L.) Dunal: Ethnobotany, Pharmacological Efficacy, and Commercialization Significance in Africa." Bulletin of the National Research Centre 45 (2021). https://doi.org/10.1186/s42269-021-00635-6.
36. D. Mathur, K. Goyal, V. Koul, and A. Anand. "The Molecular Links of Re-Emerging Therapy: A Review of Evidence of Brahmi (Bacopa monniera)." Frontiers in Pharmacology 7 (2016). https://doi.org/10.3389/fphar.2016.00044.
37. P. Soni, A.A. Siddiqui, J. Dwivedi, and V. Soni. "Pharmacological Properties of Datura stramonium L. as a Potential Medicinal Tree: An Overview." Asian Pacific Journal of Tropical Biomedicine 2, no. 5 (2012). https://doi.org/10.1016/S2221-1691(13)60014-3.
38. K.J. Gohil, J.A. Patel, and A.K. Gajjar. "Pharmacological Review on Centella asiatica: A Potential Herbal Cure-All." Indian Journal of Pharmaceutical Sciences 72, no. 5 (2010): 546–556. https://doi.org/10.4103/0250-474X.78519.
39. A. Nowak, K. Kojder, J. Zielonka-Brzezicka, J. Wróbel, M. Bosiacki, M. Fabiańska, M. Wróbel, J. Sołek-Pastuszka, and A. Klimowicz. "The Use of Ginkgo Biloba L. as a Neuroprotective Agent in Alzheimer's Disease." Frontiers in Pharmacology 12 (2021). https://doi.org/10.3389/fphar.2021.775034.
40. C.A.C. Araújo and L.L. Leon. "Biological Activities of Curcuma longa L." Memórias do Instituto Oswaldo Cruz 96, no. 5 (2001): 723–728. https://doi.org/10.1590/S0074-02762001000500026.
41. B.C. Variya, A.K. Bakrania, and S.S. Patel. "Emblica officinalis (Amla): A Review for Its Phytochemistry, Ethnomedicinal Uses and Medicinal Potentials with Respect to Molecular Mechanisms." Pharmacological Research 111 (2016). https://doi.org/10.1016/j.phrs.2016.06.013.
42. V. Sharma, I. Singh, and P. Chaudhary. "Acorus calamus (The Healing Plant): A Review on Its Medicinal Potential, Micropropagation and Conservation." Natural Product Research 28 (2014). https://doi.org/10.1080/14786419.2014.915827.
43. M. Nigam, A.P. Mishra, A. Adhikari-Devkota, A.I. Dirar, M.M. Hassan, A. Adhikari, T. Belwal, and H.P. Devkota. "Fruits of Terminalia chebula Retz.: A Review on Traditional Uses, Bioactive Chemical Constituents and Pharmacological Activities." Phytotherapy Research 34 (2020). https://doi.org/10.1002/ptr.6702.
44. Z.A. Ratan, M.F. Haidere, Y.H. Hong, S.H. Park, J.O. Lee, J. Lee, and J.Y. Cho. "Pharmacological Potential of Ginseng and Its Major Component Ginsenosides." Journal of Ginseng Research 45 (2021). https://doi.org/10.1016/j.jgr.2020.02.004.
45. R.W. Mwangi, J.M. Macharia, I.N. Wagara, and R.L. Bence. "The Medicinal Properties of Cassia fistula L.: A Review." Biomedical Pharmacotherapy 144 (2021). https://doi.org/10.1016/j.biopha.2021.112240.
46. Q.Q. Mao, X.Y. Xu, S.Y. Cao, R.Y. Gan, H. Corke, T. Beta, and H. Bin Li. "Bioactive Compounds and Bioactivities of Ginger (Zingiber officinale Roscoe)." Foods 8 (2019). https://doi.org/10.3390/foods8060185.
47. F. Di Meo, A. Valentino, O. Petillo, G. Peluso, S. Filosa, and S. Crispi. "Bioactive Polyphenols and Neuromodulation: Molecular Mechanisms in Neurodegeneration." International Journal of Molecular Sciences 21 (2020). https://doi.org/10.3390/ijms21072564.
48. H.M. Soliman, G.A. Ghonaim, S.M. Gharib, H. Chopra, A.K. Farag, M.H. Hassanin, A. Nagah, M. Emad-Eldin, N.E. Hashem, G. Yahya, S.E. Emam, A.E.A. Hassan, and M.S. Attia. "Exosomes in Alzheimer's Disease: From Being Pathological Players to Potential Diagnostics and Therapeutics." International Journal of Molecular Sciences 22 (2021): 10794. https://doi.org/10.3390/IJMS221910794.

49. T.B. Emran and H. Chopra. "Autophagy and Alzheimer's Disease: How Far Science Has to Be Progressed? – Correspondence." Annals of Medicine and Surgery 85 (2023): 323–325. https://doi.org/10.1097/MS9.0000000000000156.
50. D.I. Boiko, A.D. Shkodina, M.M. Hasan, M. Bardhan, S.K. Kazmi, H. Chopra, P. Bhutra, A.A. Baig, and A.M. Skrypnikov. "Melatonergic Receptors (Mt1/Mt2) as a Potential Additional Target of Novel Drugs for Depression." Neurochemical Research (2022). https://doi.org/10.1007/s11064-022-03646-5.
51. Y. Zhou and N.C. Danbolt. "Glutamate as a Neurotransmitter in the Healthy Brain." Journal of Neural Transmission 121 (2014): 799. https://doi.org/10.1007/S00702-014-1180-8.
52. S. Mishra and K. Palanivelu. "The Effect of Curcumin (Turmeric) on Alzheimer's Disease: An Overview." Annals of Indian Academy of Neurology 11 (2008): 13. https://doi.org/10.4103/0972-2327.40220.
53. P.H. Reddy. "Amyloid Beta-Induced Glycogen Synthase Kinase 3β Phosphorylated VDAC1 in Alzheimer's Disease: Implications for Synaptic Dysfunction and Neuronal Damage." Biochimica et Biophysica Acta 1832 (2013): 1913–1921. https://doi.org/10.1016/J.BBADIS.2013.06.012.
54. C. Hooper, R. Killick, and S. Lovestone. "The GSK3 Hypothesis of Alzheimer's Disease." Journal of Neurochemistry 104 (2008): 1433. https://doi.org/10.1111/J.1471-4159.2007.05194.X.
55. D.W. Dickson. "Parkinson's Disease and Parkinsonism: Neuropathology." Cold Spring Harbor Perspectives in Medicine 2 (2012). https://doi.org/10.1101/CSHPERSPECT.A009258.
56. E. Matar, J.M. Shine, G.M. Halliday, and S.J.G. Lewis. "Cognitive Fluctuations in Lewy Body Dementia: Towards a Pathophysiological Framework." Brain 143 (2020). https://doi.org/10.1093/brain/awz311.
57. J. Schumacher, L.R. Peraza, M. Firbank, A.J. Thomas, M. Kaiser, P. Gallagher, J.T. O'Brien, A.M. Blamire, and J.P. Taylor. "Dysfunctional Brain Dynamics and Their Origin in Lewy Body Dementia." Brain 142 (2019). https://doi.org/10.1093/brain/awz069.
58. M.M. Khan, A. Ahmad, T. Ishrat, M.B. Khan, M.N. Hoda, G. Khuwaja, S.S. Raza, A. Khan, H. Javed, K. Vaibhav, and F. Islam. "Resveratrol Attenuates 6-Hydroxydopamine-Induced Oxidative Damage and Dopamine Depletion in Rat Model of Parkinson's Disease." Brain Research 1328 (2010): 182–192. https://doi.org/10.1016/j.brainres.2010.02.031.
59. H.Y. Kim and S.M. Lee. "Ferulic Acid Attenuates Ischemia/Reperfusion-Induced Hepatocyte Apoptosis via Inhibition of JNK Activation." European Journal of Pharmaceutical Sciences 45 (2012): 654–661. https://doi.org/10.1016/j.ejps.2012.01.010.
60. D.G. Anderson, A. Ferreira-Correia, F.B. Rodrigues, N.A. Aziz, J. Carr, E.J. Wild, R.L. Margolis, and A. Krause. "Comparison of the Huntington's Disease Like 2 and Huntington's Disease Clinical Phenotypes." Movement Disorders Clinical Practice 6 (2019). https://doi.org/10.1002/mdc3.12742.
61. J.S. Snowden. "The Neuropsychology of Huntington's Disease." Archives of Clinical Neuropsychology 32 (2017). https://doi.org/10.1093/arclin/acx086.
62. P. McColgan and S.J. Tabrizi. "Huntington's Disease: A Clinical Review." European Journal of Neurology 25 (2018). https://doi.org/10.1111/ene.13413.
63. D.J. Ho, N.Y. Calingasan, E. Wille, M. Dumont, and M.F. Beal. "Resveratrol Protects Against Peripheral Deficits in a Mouse Model of Huntington's Disease." Experimental Neurology 225 (2010): 93–99. https://doi.org/10.1016/j.expneurol.2010.05.006.
64. P. Kumar and A. Kumar. "Protective Effect of Hesperidin and Naringin Against 3-Nitropropionic Acid Induced Huntington's Like Symptoms in Rats: Possible Role of Nitric Oxide." Behavioural Brain Research 206 (2010): 216–223. https://doi.org/10.1016/j.bbr.2009.08.028.
65. M.A. Hickey, C. Zhu, V. Medvedeva, R.P. Lerner, S. Patassini, N.R. Franich, P. Maiti, S.A. Frautschy, S. Zeitlin, M.S. Levine, and M.F. Chesselet. "Improvement of Neuropathology and Transcriptional Deficits in CAG 140 Knock-In Mice Supports a Beneficial Effect of Dietary Curcumin in Huntington's Disease." Molecular Neurodegeneration 7 (2012). https://doi.org/10.1186/1750-1326-7-12.
66. J.W. Park, J.S. Hong, K.S. Lee, H.Y. Kim, J.J. Lee, S.R. Lee. "Green Tea Polyphenol (−)-Epigallocatechin Gallate Reduces Matrix Metalloproteinase-9 Activity Following Transient Focal Cerebral Ischemia." Journal of Nutritional Biochemistry 21 (2010): 700–708. https://doi.org/10.1016/j.jnutbio.2009.08.009.

67. K.S. Panickar, M.M. Polansky, and R.A. Anderson. "Green Tea Polyphenols Attenuate Glial Swelling and Mitochondrial Dysfunction Following Oxygen-Glucose Deprivation in Cultures." Nutritional Neuroscience 12 (2009): 212–222. https://doi.org/10.1179/147683009X423300.
68. Y. Yao, D.D. Han, T. Zhang, and Z. Yang. "Quercetin Improves Cognitive Deficits in Rats with Chronic Cerebral Ischemia and Inhibits Voltage-Dependent Sodium Channels in Hippocampal CA1 Pyramidal Neurons." Phytotherapy Research 24 (2010): 546–552. https://doi.org/10.1002/ptr.2902.
69. J.K. Lee, H.J. Kwak, M.S. Piao, J.W. Jang, S.H. Kim, and H.S. Kim. "Quercetin Reduces the Elevated Matrix Metalloproteinases-9 Level and Improves Functional Outcome After Cerebral Focal Ischemia in Rats." Acta Neurochirurgica 153 (2011): 1321–1329. https://doi.org/10.1007/S00701-010-0889-X.
70. M.J. Owens. "Selectivity of Antidepressants: From the Monoamine Hypothesis of Depression to the SSRI Revolution and Beyond." Journal of Clinical Psychiatry 65 (2004): 5–11.
71. V. Maletic, M. Robinson, T. Oakes, S. Iyengar, S.G. Ball, and J. Russell. "Neurobiology of Depression: An Integrated View of Key Findings." International Journal of Clinical Practice 61 (2007): 671–688. https://doi.org/10.1111/j.1742-1241.2007.01602.x.
72. S. Matthes, V. Mosienko, S. Bashammakh, N. Alenina, and M. Bader. "Tryptophan Hydroxylase as Novel Target for the Treatment of Depressive Disorders." Pharmacology 85 (2010): 133–139. https://doi.org/10.1159/000279322.
73. C.A. Meyers, M. Albitar, and E. Estey. "Cognitive Impairment, Fatigue, and Cytokine Levels in Patients with Acute Myelogenous Leukemia or Myelodysplastic Syndrome." Cancer 104 (2005): 1555–1563. https://doi.org/10.1002/cncr.21234.
74. S.J. Motivala, A. Sarfatti, L. Olmos, and M.R. Irwin. "Inflammatory Markers and Sleep Disturbance in Major Depression." Psychosomatic Medicine 67 (2005): 75–82. https://doi.org/10.1097/01.psy.0000149259.72488.09.
75. S.K. Kulkarni, M.K. Bhutani, and M. Bishnoi. "Antidepressant Activity of Curcumin: Involvement of Serotonin and Dopamine System." Psychopharmacology (Berlin) 201 (2008): 435–442. https://doi.org/10.1007/s00213-008-1300-y.
76. E.B. Kurutas. "The Importance of Antioxidants Which Play the Role in Cellular Response Against Oxidative/Nitrosative Stress: Current State." Nutrition Journal 15 (2016). https://doi.org/10.1186/s12937-016-0186-5.
77. C.A. Cobb and M.P. Cole. "Oxidative and Nitrative Stress in Neurodegeneration." Neurobiology of Disease 84 (2015): 4–14. https://doi.org/10.1016/J.NBD.2015.04.020.
78. W.L. Zhu, H.S. Shi, Y.M. Wei, S.J. Wang, C.Y. Sun, Z.B. Ding, and L. Lu. "Green Tea Polyphenols Produce Antidepressant-Like Effects in Adult Mice." Pharmacological Research 65 (2012): 74–80. https://doi.org/10.1016/j.phrs.2011.09.007.
79. B.K.Y. Bitanihirwe and T.U.W. Woo. "Oxidative Stress in Schizophrenia: An Integrated Approach." Neuroscience & Biobehavioral Reviews 35 (2011): 878–893. https://doi.org/10.1016/J.NEUBIOREV.2010.10.008.
80. S. Madireddy and S. Madireddy. "Therapeutic Interventions to Mitigate Mitochondrial Dysfunction and Oxidative Stress–Induced Damage in Patients with Bipolar Disorder." International Journal of Molecular Sciences 23 (2022). https://doi.org/10.3390/ijms23031844.
81. J.M. Loftis, C.J. Wilhelm, and M. Huckans. "Effect of Epigallocatechin Gallate Supplementation in Schizophrenia and Bipolar Disorder: An 8-Week, Randomized, Double-Blind, Placebo-Controlled Study." Therapeutic Advances in Psychopharmacology 3 (2013): 185–192. https://doi.org/10.1177/2045125312464103.
82. A. Dietrich-Muszalska, B. Kontek, B. Olas, and J. Rabe-Jabłońska. "Epicatechin Inhibits Human Plasma Lipid Peroxidation Caused by Haloperidol in Vitro." Neurochemical Research 37 (2012): 1290–1296. https://doi.org/10.1007/s11064-011-0642-8.
83. W.F. Zhang, Y.L. Tan, X.Y. Zhang, R.C.K. Chan, H.R. Wu, and D.F. Zhou. "Extract of Ginkgo biloba Treatment for Tardive Dyskinesia in Schizophrenia: A Randomized, Double-Blind, Placebo-Controlled Trial." Journal of Clinical Psychiatry 72 (2011). https://doi.org/10.4088/JCP.09m05125yel.

12 Plant Phytochemicals as Potential Therapeutics in Neurological Disorders

Pawan Prabhakar, Ankit Kumar, and Mamoni Banerjee

12.1 INTRODUCTION

According to medical terminology, neurological disorders affect the spinal cord, brain, and body's nerves and are caused by structural, metabolic, or electrical abnormalities in the brain, spinal cord, or other nerves in the body. There are several symptoms of neurological disorder, including paralysis, weak muscles, unstable coordination, loss of sensation, convulsions, confusion, pain, and altered levels of consciousness [1, 2, 3]. The particular causes of neurological issues can vary e.g., starvation, brain damage, spinal cord injury, nerve injury, inherited abnormalities or malformations, infections, lifestyle choices, or environmental health problems [4]. There are several known other neurological illnesses, some of which are rare, but others are not [5]. More than 600 different neurologic illnesses and numerous conditions fall under the category of neurological disability, including epilepsy, learning difficulties, neuromuscular diseases, autism, attention deficit disorder (ADD), brain tumours, and cerebral palsy [6]. Among the various identified diseases that come under central nervous system (CNS) disorder include Alzheimer's and Parkinson's, mental illnesses including anxiety, depression, psychosis, epilepsy, multiple sclerosis, neuropathic pain, autism, and many more [7]. Only Alzheimer's disease, or AD itself, accounts for 80% of occurrences of dementia in senior persons, resulting in memory loss being one of the prime symptoms. It can be advanced to a stage where it manifests as agitation, apathy, aggression, hallucinations, false beliefs, and cognitive dysfunction. In the end, severe AD patients pass away from losing basic physiological functions and complications from infections [8]. Alzheimer's disease results in a significant loss of neurons, creating disorganized plaques and neurofibrillary masses. The principal component of senile plaques that plays a major role in the progression of Alzheimer's disease (AD) is a beta-amyloid peptide (Aβ). This Aβ peptide plaque and neurofibrillary masses are connected with acetylcholinesterase enzyme (AChE). Through generating energy loss, oxidative stress, inflammation, and apoptosis, the accumulation of Aβ results in destruction of neural cells. The neurological impairments result from injury to the nervous system, regardless of the source and how much communication, vision, hearing, mobility, and cognition are affected depending on, and, where the injury occurs. According to studies, neurological conditions were the second-largest cause of death (9 million) and the main cause of disability-adjusted life years (DALYs), 280 million worldwide [9]. In the upcoming ten years, low and middle-income countries (LMICs) are anticipated to exponentially increase in the neurological illness burden [10]. Even though neurological illnesses have a large negative impact on patients and society, our understanding of their epidemiology, including the variations in disease frequency over space and time. The related risk factors and outcomes are still inadequate, especially in LMICs [11]. Due to physical, cognitive, and psychological impairments, patients with neurological illnesses frequently require extensive social and financial assistance. Despite the high incidence of disability, services and resources are

DOI: 10.1201/9781003300502-12

increasingly considered disproportionately limited, particularly in LMICs, with less information on effective efforts to enhance neurological care in these settings [12]. No neurological disorder condition is now curable, and the available medicines merely treat the symptoms or slow the disease's course. Since the World Health Organization has predicted that neurodegenerative illnesses affecting motor function will become the second most common cause of mortality in the next 20 years hence, there is an urgent need for innovative treatment and medications for this type of disease [13]. Modern drug therapy may also lead to the creation of new illnesses and adverse effects, which are recognised along with improved health outcomes [14, 15, 16, 17]. Additionally, an overdose of specific medicines like antibiotics causes gastrointestinal dysbiosis, which also raises the chance of developing various illnesses, including neurological and psychiatric abnormalities [18].

A growing trend in drug discovery focused on the drugs of natural origin in which methods for identifying small molecule leads are used as novel therapeutic in neurological disorders. These natural drugs have few or no side effects and are suitably compatible with human neural physiology [19, 20]. About 80% of people in underdeveloped nations primarily rely on plant-based traditional medicine for their primary healthcare needs [21]. Traditional remedies are typically prepared using various sections of medicinal plants rather than the entire plant [22, 23]. Due to the close connection between natural products and the human brain, modern medicine has been now developed as conjugate with natural products for better efficacy in treating CNS problems. According to a recent study, 400 clinically approved CNS medications are linked to 20 natural product scaffolds, and 84% of approved drugs for the treatment of CNS illnesses are thought to be natural products or natural product-inspired [24, 25]. Many natural medication molecules with effective therapeutic efficacy for neurological illnesses are still unknown. Therefore, ongoing efforts are made to pinpoint their molecular makeup, pharmacology, and pharmacokinetic activities. Natural drug compounds, also known as phytochemicals or plant's secondary metabolites, are produced under stress conditions to safeguard the plants. In viewing their wide acceptability and popularity among researchers and common people, especially for chronic and neurological disorders, they are promising areas of novel therapeutics. Therefore, this chapter focuses on detailed discussions of phytochemicals, their classes and extraction techniques, functions in neurological disorders, and future perspectives.

12.2 PHYTOCHEMICALS AND THEIR CLASSIFICATION

12.2.1 Alkaloids

An alkaloid belongs to a group of naturally occurring organic nitrogen-containing substances that are widely distributed across the plant kingdom. They are found in all higher and lower categories of plants and even in microbes like fungi and bacteria having low molecular weight [26]. Over 18,000 alkaloids, including their methylated, glycosylated, and hydroxylated variations, have been found. Based on their chemical makeup, alkaloids are frequently divided into two groups: proto-alkaloids, which contain nitrogen in a side chain, and heterocyclic or typical alkaloids, which have nitrogen in the heterocycle. Regarding structure, alkaloids are divided into pyridine, indole, isoquinoline, benzylisoquinoline, piperidine, indolizidine, histamine, tropane, pyrrolidine, and pyrrolizidine [27, 28]. The alkaloid biosynthesis typically involves the accumulation of an amine precursor, the accumulation of an aldehyde precursor, and a mannich-like reaction that leads to scaffold-formation [29]. Alkaloids were examined for their propensity to establish hydrogen bonds with enzymes, receptors, and proteins due to protons receiving N-atoms and one or more protons donating amine H-atoms [30]. Many plant families, like the Solanaceae (nightshades), Papaveraceae (poppies family), Ranunculaceae (buttercups), and Amaryllidaceae (amaryllis), are primarily rich in various types of alkaloids [31]. Alkaloids are useful in the treatment of a variety of ailments, including malaria, diabetes, cancer, and heart dysfunction. They are also potential antiproliferative, antibacterial, and antioxidant agents that can be exploited in manufacturing pharmaceuticals [32, 33]. The molecular structure of some common alkaloids derived from different plant species is given in Figure 12.1.

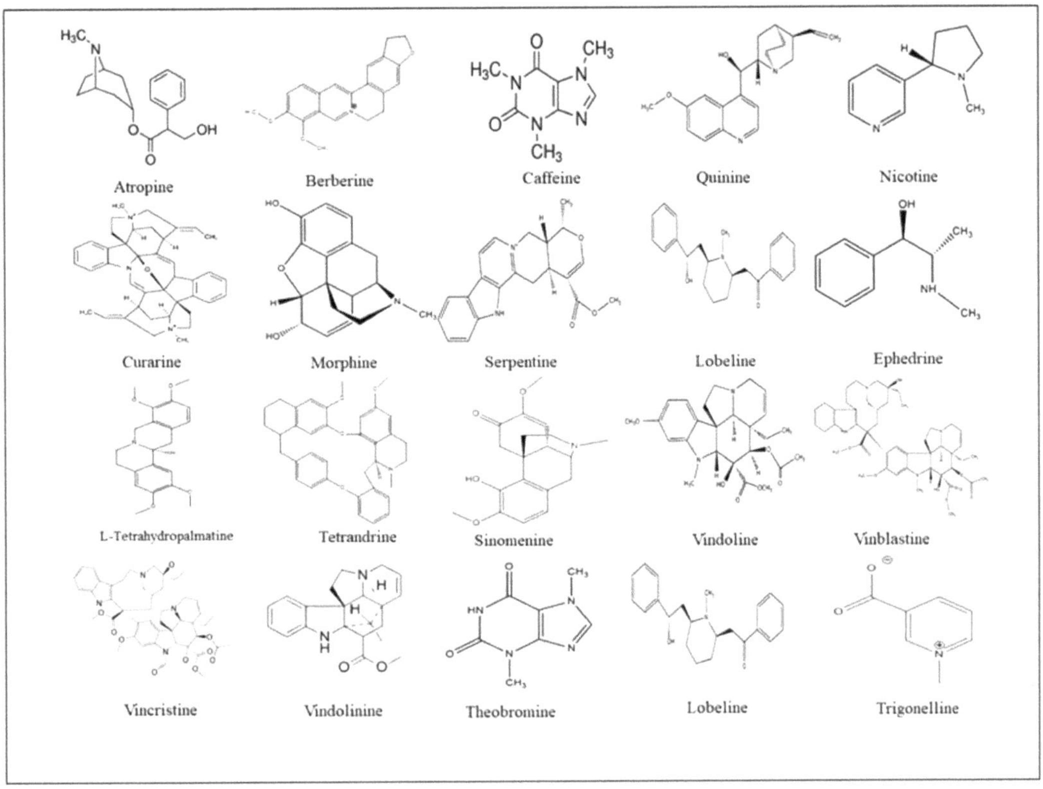

FIGURE 12.1 Molecular structure of some common alkaloids.

12.2.2 Phenolics

Phenolics, which include one or more aromatic rings with varying hydroxyl groups attached, are the most prevalent secondary metabolites in the plant kingdom. Due to their ability to transfer electrons and hydrogen atoms to free radicals, which stops oxidation, they are excellent antioxidants and are found in various forms, from basic to complex phenolic structures [34]. When a hydroxy group is replaced in an aromatic ring's 'ortho', 'meta', or 'para' locations, the parent structure is created. L-phenylalanine and/or l-tyrosine serve as the building blocks for the biosynthesis of phenolic acids in plants, primarily via the shikimate/phenylpropanoid pathway. Furthermore, deamination, hydroxylation, and methylation are three mechanisms contributing to the production of phenolic acids. The biosynthesis of the amino acids occurs when phenylalanine and tyrosine are deaminated to produce cinnamic acid and p-coumaric acid, respectively. Cinnamic acid (C6–C3) becomes benzoic acid when the ethyl side chain is removed (C6–C1) [35]. All phenolic acids are structurally hydroxylated benzoic acid or cinnamic acid derivatives. They can be insoluble cell wall structural elements linked to xylan, pectin, and lignin, or they can be conjugated with sugars to create depsides or simple glycosides [36]. Resorcinol (1,3-dihydroxybenzene), hydroquinone (1,4-dihydroxybenzene), and catechol (1,2-dihydroxybenzene) are examples of simple phenolic compounds. These complex phenolic compounds include luteolin (3',4',5,7-tetrahydroxyflavone), kaempferol (3 3, 4',5,7-tetrahydroxyflavone), and naringenin (5,7,4'-trihydroxyflavanone) [37, 38]. Phenolic compounds can scavenge free radicals, donate electrons or hydrogen atoms, or bind metal cations. The ability of phenolic compounds to inactivate free radicals is conferred by their molecular structures, specifically the quantity and locations of their hydroxyl groups. The types of substitutions they undergo on their aromatic rings are known as the structure-activity relationship [39, 40]. The phenolic acids, flavonoids, and tannins are considered to be the primary dietary phenolic components among the several types of phenolic compounds, and their antioxidant potential is strongly correlated. Phenolic compounds have anti-inflammatory, antitumor, antimicrobial, antidiabetic, and neuroprotective properties [41, 42, 43, 44, 45, 46]. Fruits contain several different phenolic chemicals; among them, the five most common phenolic acids are stilbenes, lignans, flavonoids, and tannins or proantho cyanidins (Figure 12.2). These groups of phenolic compounds are the most prevalent and are widely consumed as dietary antioxidants by people worldwide [47]. Furthermore, the phenolic acids (benzoic and cinnamic acid derivatives) and flavonoids are the most

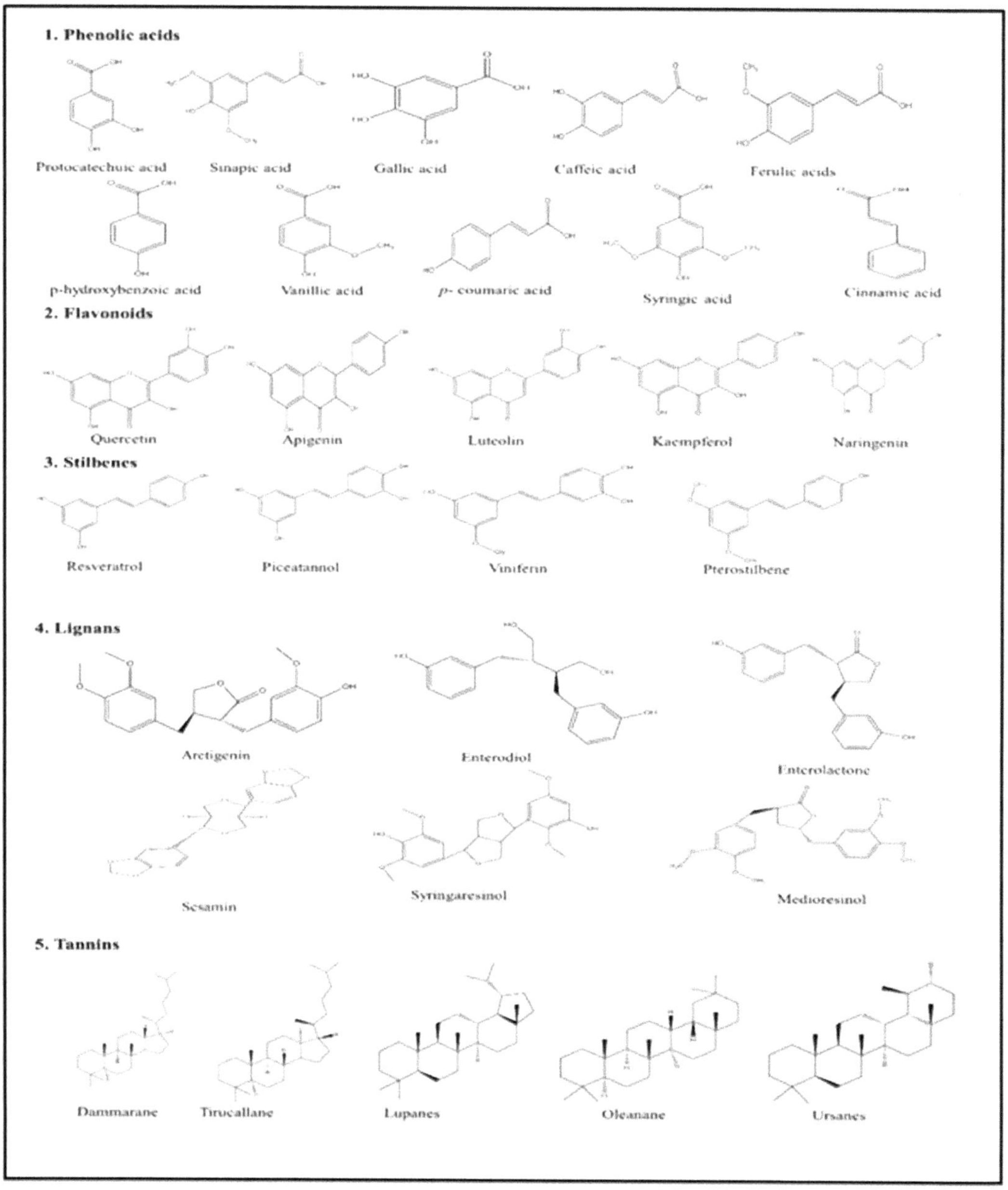

FIGURE 12.2 Phenolics and their sub-classes: Phenolic acids, flavonoids, stilbenes, lignans, and tannins.

prevalent phenolic compounds in the diet, making up 60% and 30%, respectively, of all dietary phenolic compounds [48].

12.2.3 Terpenoids

Terpenoids are another essential and most prominent secondary metabolite in the plant kingdom and dominate among all classes of secondary metabolites in plants. Five isoprene units make up each one, which is then put together in countless different ways with distinct functional groups joined in various locations to offer a range of functions [49]. Green flowering plants, in particular, synthesize large amounts of terpenoids to promote plant growth and development, environmental stimulation, and other physiological processes. Based on the number of isoprene units, terpenoids are divided as monoterpene (C10) < sesquiterpene (C15) < diterpene (C20) < triterpene (C30) < tetraterpene (C40) < polyterpene (C > 40) [50]. It has been estimated that the plant kingdom has more than 50,000 terpenoids [51].

Mevalonic acid is an intermediary in the most well-known mevalonate pathway, from which terpenoids are made. Acetyl-CoA serves as the initial primary metabolite in the production of terpenoids, and it undergoes a series of processes to condense into three molecules of mevalonic acid. Terpenes are essential components of natural oils, secondary metabolites produced to fight infections or secreted in times of stress. These are from various fragrant plants typically found in mild to warm climates, such as the Mediterranean and tropical countries, where they make up a sizable portion of the traditional pharmacopoeias. They dissolve in natural solvents with a generally lower thickness than water, are lipid-dissolvable, fluidic in nature, clear, and rarely coloured [52]. Terpenoids have a wide range of physical and biological qualities, including memory boosters [53], antibacterial [54], antifungal [55], antiviral [56], anti-hyperglycemic [57], anti-inflammatory [58], and cancer chemopreventive actions [59]. The molecular structure of some bioactive terpenoids is given in Figure 12.3.

12.2.4 Saponins

The broad family of amphiphilic glycosides of steroids and triterpenes known as saponins are present in many plants and particular marine creatures [60]. Saponins have a variety of biological and pharmacological activities. They are the main active ingredients in many traditional medicines because they express various structural variations on sugar chains and aglycones. The aglycone unit in saponins is connected to one or more carbohydrate chains. A triterpene unit, which is more prevalent, or a sterol unit, makes up the aglycone or sapogenin unit. The carbohydrate side chain is frequently joined to sapogenin's third carbon in both steroid and triterpenoid saponins [61, 62]. The pharmaceutical, cosmetic, and food industries are all interested in the biological activity that the saponins and their biosynthetic intermediates exhibit. Although their importance in industrial applications has long been understood, little is known about their function in plants. The importance of saponins and their biosynthetic intermediates in plant growth and development has been proven by recent research on altering native pathway flow in saponin production [63].

12.2.5 Phytosterols

The term 'phytosterol' refers to all plant sterols and their saturated forms (stanols). They can be found in foods such as unprocessed plant oils, legumes, nuts, and seeds. Due to their structural resemblance to cholesterol, phytosterols are sometimes referred to as 'plant cholesterol'. They differ from cholesterol by having an additional double bond at the C-22 position (brassicasterol or stigmasterol), an additional methyl group (campesterol), an additional ethyl group (sitosterol), or

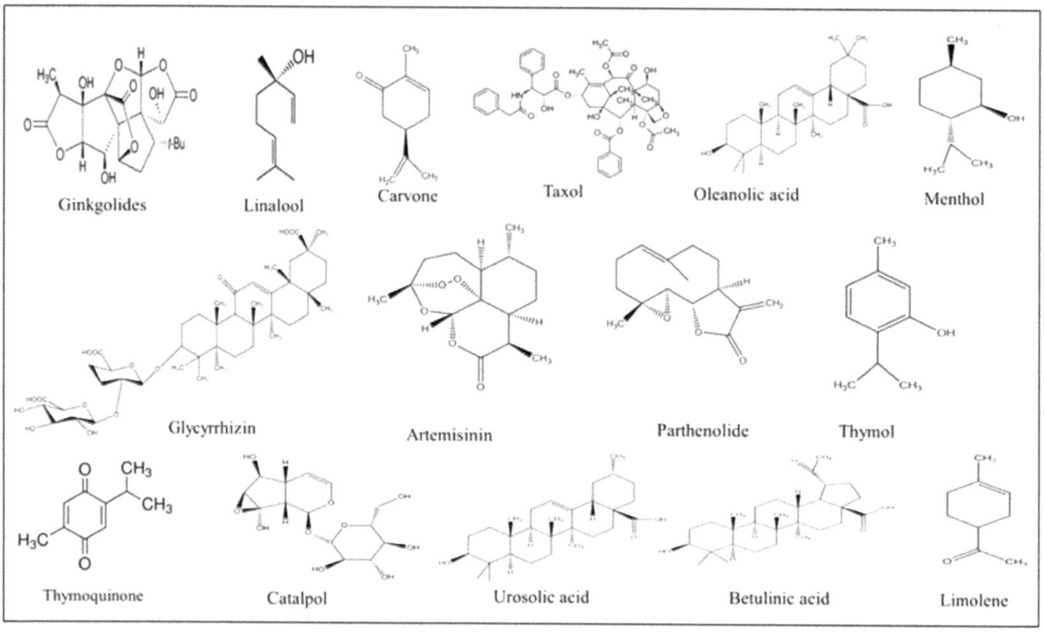

FIGURE 12.3 Basic molecular structure and derivatives of terpenoids.

both at the C-24 position. Plant cell walls contain phytosterols, which are structurally like human cholesterol [64]. Plants produce more than 250 phytosterols, and each plant species has a different phytochemical profile of phytosterols. The three phytosterols that are most prevalent in our diet are campesterol, stigmasterol, and sitosterol. The C-5 and C-6 carbon positions in phytosterols' skeletons are filled with saturated bonds, whereas the C-17 carbon atom has an aliphatic side chain and a functional group called hydroxyl. These compounds were initially created as fatty acids, cinnamic acids, and non-esterified and esterified glycosides. Phytosterols can pass the blood-brain barrier (BBB) and build up in the brain. As a result, they may play a crucial part in altering several brain circuits related to neurodegeneration [65]. Some of the derivatives of phytosterols are given in Figure 12.4.

12.2.6 Cardiac Glycoside

These secondary metabolites of bioactive phytocompounds have a steroid backbone connected by a lactone ring at carbon position C-17b and a sugar moiety and sugar unit at carbon position 3-b (Figure 12.5). The varied class of naturally occurring substances known as cardiac glycosides is well-known for its capacity to bind to and block the salt pump. Aldehyde, hydroxyl, and methyl functional groups are attached to the steroid nucleus and are responsible for their biological activity [66]. Cardiac glycosides and their derivatives have long been utilized to treat atrial arrhythmia and heart failure (Figure 12.5) [67]. Recent research describes additional sodium pump signalling pathways and links cardiac glycosides to control numerous critical physiological functions, including gene expression profiling, apoptosis, proliferation, and cell-cell communication [68].

12.3 PHYTOCHEMICAL EXTRACTION TECHNIQUES

Alkaloids, flavonoids, terpenes, saponins, steroids, and glycosides are the secondary metabolites extracted from medicinal plants and herbs to separate them from their matrix, inert or inactive

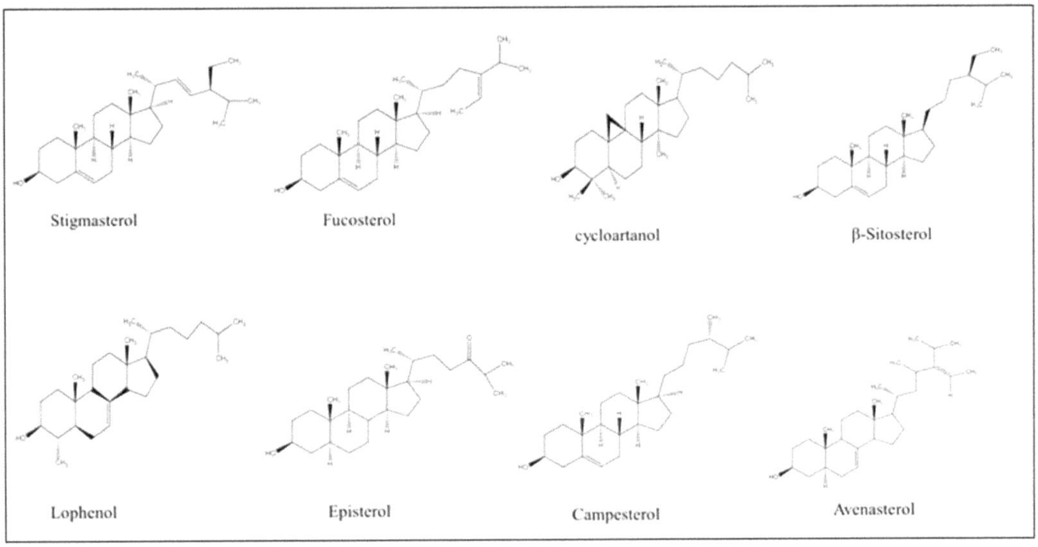

FIGURE 12.4 Phytosterols and their derivatives.

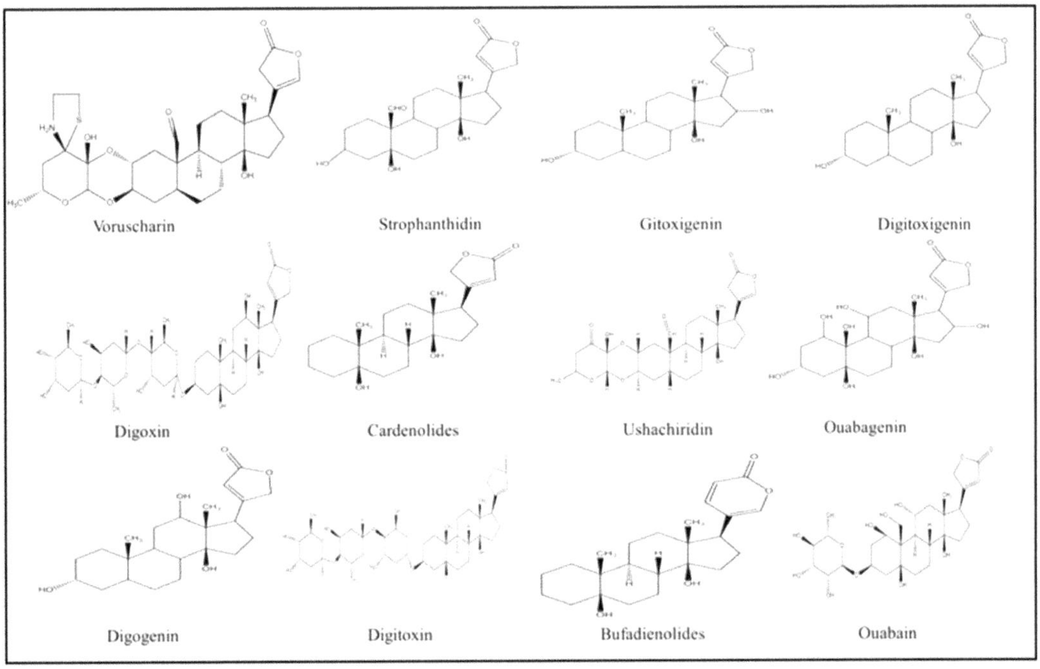

FIGURE 12.5 Major cardiac glycosides.

material is called extraction [69]. Several methods are used to extract these metabolites, including maceration, infusion, decoction, percolation, digestion, Soxhlet extraction, ultrasound-assisted extraction, microwave-assisted, hydrodistillation, and enzyme-assisted extraction [70]. Using thin-layer chromatography (TLC), high-performance liquid chromatography (HPLC), and gas chromatography (GC), the secondary metabolites were also isolated from one another, identified, and purified. The plant material, the solvent employed, its pH, the temperature, and the solvent-to-sample ratio all play a role in determining the optimum extraction procedure [71, 72]. Thus, the commonly used conventional and modern methods in the extraction of medicinal plants are:

12.3.1 Conventional Methods for Extraction

Maceration: Maceration is an ancient technique of extraction in which a finely coarsen plant material such as leaves, stem bark and root taken in a container. The solvent or menstruum is poured into it which completely dissolves the powdered material. The solution is kept for few days and agitated regularly on some time intervals to speed up the extraction process. Filtration or decantation is used at the extraction end to separate the micelle from the marc (unwanted or insoluble substances). In an oven or on top of a water bath, the micelle is evaporated to separate it from the menstruum. This approach is practical and appropriate for phytochemicals that are thermolabile in nature [73].

Infusion: This is a method of extraction similar to maceration, where the drug substance is reduced to a powder form and put into a clean container. The drug material is covered with the hot or cold extraction solvent, which is then left to soak for a brief time. This technique works well for extracting easily soluble bioactive components. It is also a suitable method for making fresh extract before usage. Depending on the intended purpose, the solvent-to-sample ratio is typically 4:1 or 16:1 [70, 69].

Digestion: With this extraction technique, the extraction procedure uses a modest amount of heat. Powdered material is added to a clean container after the extraction solvent. At a temperature of around 50°C, the mixture is put over a water bath or in an oven. Throughout the extraction procedure, heat was used to reduce the solvent's viscosity and improve the removal of secondary metabolites. Plant materials that are easily soluble lend themselves to this approach well [74].

Decoction: This procedure uses the given amount of water as a solvent for continuous hot extraction. Plant material that has been dried, processed and powdered is added to a spotless container. After that, water is added and mixed. The extraction is then accelerated by using heat throughout the process. The procedure just takes a few minutes typically around 15 minutes to complete. Typically, the ratio of solvent to crude medication is 4:1 or 16:1. It is employed to extract plant material that is both heat- and water-soluble [75, 69].

Percolation: The tool used in this procedure is a percolator, a thin glass jar with openings on both ends. In a clean container, dried, ground, and finely powdered plant material is soaked with the extraction solvent. Filtration is used to separate the extract, followed by decantation. After the marc has been expressed, the remaining solvent is added to get the necessary volume.

Soxhlet extraction: The glass-based device is known as a Soxhlet extractor which is a round-bottomed flask, extraction chamber, syphon tube, and condenser at the top make up this apparatus. The porous bag (thimble), which is made of clean cloth or sturdy filter paper, is filled with dried, pulverised, and finely powdered plant material, and then it is firmly closed. The bottom flask is filled with the extraction solvent before the thimble is placed within the extraction chamber. The solvent is then heated from the bottom flask, evaporates via the condenser, condenses, flows downward to the extraction chamber, and by coming into touch with the drug, extracts it. As a result, the solvent and the extracted plant material flow back to the flask when the solvent level in the extraction chamber reaches the top of the syphon. The entire procedure is repeated up until the point at which a solvent running out of the extraction chamber leaves no trace of any residual medication. Plant materials with insoluble contaminants, as well as plant materials that are only partially soluble in the selected solvent, can be used in this approach. For plant materials that are thermolabile, it is an inadequate approach. A tiny amount of solvent can be used to extract a lot of medication. Plant materials that can withstand heat can also use it. There is no need for filtering, and a lot of heat might be used, this technique is inappropriate for materials that are thermolabile and regular shaking is not achievable [76, 70].

12.3.2 Modern Methods for Extraction

Microwave-assisted extraction: The method employs the displacement of charged ions found in the solvent and drug material as well as the dipole rotation and ionic transfer mechanisms. This technique works well for extracting flavonoids. Electromagnetic radiation with frequencies between 300 MHz and 300 GHz and wavelengths between 1 cm and 1 m is used in this process. Energy produced by the microwaves started from 300 W and ranged up to 1 kW. The process exposes an object to microwave radiation, which has the potential to absorb electromagnetic energy and convert it to heat. The heat created as a result makes it easier for the solvent to enter the drug matrix. The microwave extraction assembly comprises four primary elements: a magnetron/microwave generator for microwave production, a Wave Guide for directing microwaves from the source to the microwave cavity, an applicator for holding the sample holder containing the sample, and a circulator for regulating the forward propagation of microwaves. Dipole rotation and ion migration that boost solvent penetration and aid in the extraction process happen when the polar solvent is utilized (Figure 12.6).

Nonpolar solvents are not recommended for this procedure since they will reduce the amount of heat produced by the microwave radiation when employed. Microwave-assisted extraction provides certain benefits, such as cutting down on solvent use and extraction time while improving results. Many flavonoids and phenolic chemicals can be used in this approach as the high temperature involved, substances like tannins and anthocyanins may be destroyed [79, 78, 77].

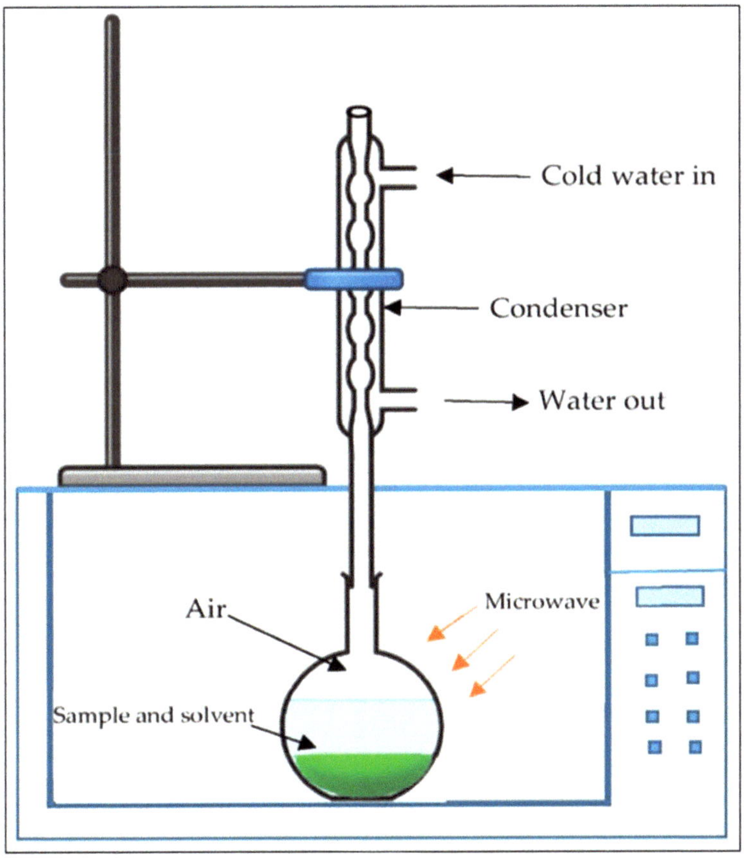

FIGURE 12.6 Microwave-assisted extraction system [77]. Copyright ©Saifullah et al. (2021), Processes, 2021, 9(12), 2212; https://doi.org/10.3390/pr9122212.

Ultrasound-assisted extraction. In this method, sound energy is used at a very high frequency of more than 20 kHz to damage plant cell walls and increase the drug's surface area for solvent penetration. Secondary metabolites will consequently be released. In this procedure, plant material must first be dried before being ground into a fine powder and carefully sieved. The prepared sample is then placed into the ultrasonic extractor along with a suitable solvent for extraction. By lowering the amount of heat needed, the high sound energy speeds up the extraction process. Advantages. Small samples can be extracted using ultrasound; it maximizes yield while cutting down on extraction time and solvent use. It is challenging to replicate this procedure, because using much energy may cause the phytochemical to deteriorate by generating free radicals [82, 80, 81].

12.3.2.1 Supercritical Fluid Extraction

When separate liquid and gas phases do not exist, any material above its critical temperature and pressure is referred to as a supercritical fluid (SCF). SCFs can diffuse through solid matrices like gas and then dissolve substances like typical liquids. The density-dependent selectivity of supercritical fluid extraction (SFE) results from the fact that many SFEs are carried out very near the fluid's critical point, where slight changes in pressure or temperature substantially impact the fluid's density [83]. In their sub- and supercritical forms, ethanol, water, butane, propane, and carbon dioxide have all been employed for extraction. Because it is non-toxic, non-flammable, and non-polluting, as well as being readily available, carbon dioxide is recommended for utilisation. Due to the fact that CO_2 is a gas at standard temperature and pressure and is therefore readily eliminated after the extraction, using pressurized CO_2 ensures that there are no chemical leftovers in the extract [84].

Technically, several methods were employed in the extraction of medicinal herbs, while some more recent techniques are still developing, and those already in use are changing. The selection of an effective extraction method is crucial and, in certain situations, relies on the intended application of an extract [85]. Heat-stable plant material is extracted using Soxhlet or microwave-assisted extraction, whereas plant materials that are not heat-stable are extracted using maceration or percolation. If the solvent of extraction is water, maceration is a suitable method, but for volatile solvents, percolation and Soxhlet extraction are more appropriate. Cheap drugs are extracted using maceration, whereas costly drugs are preferably extracted using percolation. Maceration is suitable for plant material requiring long exposure to the menstruum, whereas microwave- or ultrasound-assisted extraction techniques are used for a shorter duration. Large-volume products such as tinctures are prepared by maceration, whereas concentrated products are produced by percolation or Soxhlet extraction. Extracts intended for consumption by humans are usually prepared by maceration, whereas products intended for experimental testing are prepared using other methods in addition to maceration [86, 87, 88].

12.4 PROMISING PHYTOCHEMICALS FOR PROMINENT DRUG TARGET IN NEUROLOGICAL DISORDERS

12.4.1 Phytochemicals as Acetylcholinesterase (AChE) and Butyrylcholinesterase (BChE) Inhibitors in Alzheimer's Disease (AD)

Inhibiting the activity of the cholinesterase enzymes (AChE and BuChE) is the principal purpose of cholinesterase inhibitors. This contributes to raising the ACh levels in the brain, which enhances central cholinergic activity. The cholinergic synapses are the primary location of the acetylcholinesterase (AChE) and Butrylcholinterase (AChE) enzymes, and as AChE levels fall and BChE level rises, the progression of Alzheimer's disease advances. BChE has less impact in controlling acetylcholine levels compared to AChE but is still a crucial part of the pathogenesis of AD. Therefore, the inhibition of cholinesterase enzymes is one of the prime treatments for the suppression or delaying of Alzheimer's disease progression, as shown in Figure 12.7 [89, 90].

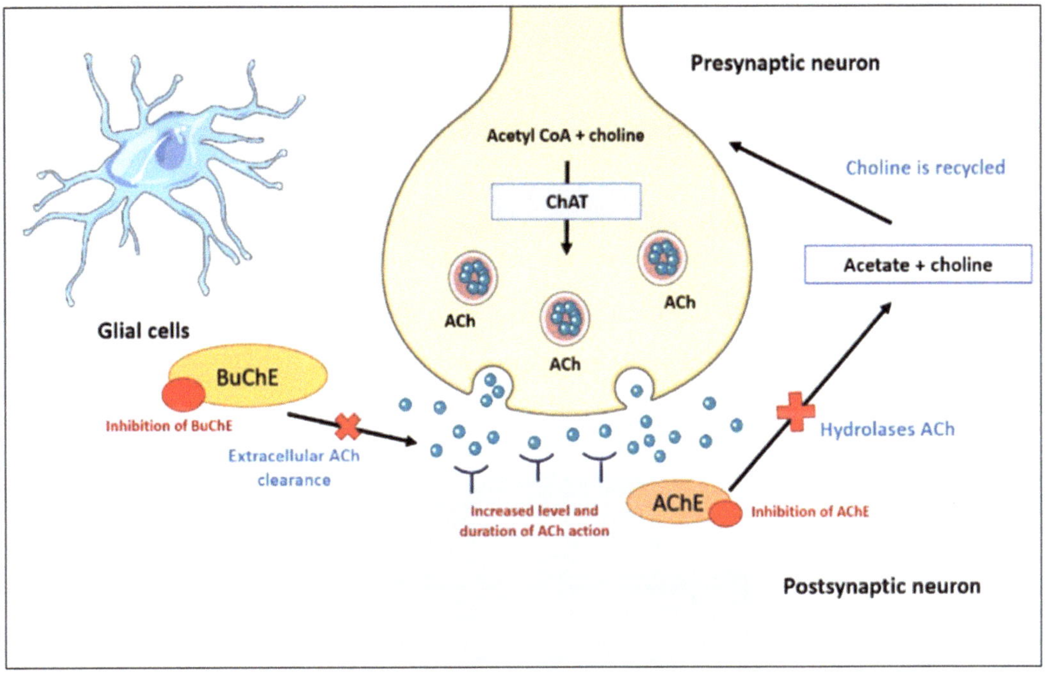

FIGURE 12.7 An overview of the cholinesterase hypothesis and inhibition of AChE and BuChE [90]. *Copyright*©Lai Shi Min *et al. (2022), Biology 2022, 11(2), 307.* https://doi.org/10.3390/biology11020307.

Bioactive phytocompounds group alkaloids have been extensively studied for their cholinesterase inhibition activity, which leads to the suppression of Alzheimer's disease. A six-membered aromatic ring fused to a five-membered pyrrole ring, which contains nitrogen, makes up the biggest category of heterogeneous secondary metabolites known as indole alkaloids that inhibit the activity of AChE. Berberine, a benzylisoquinoline alkaloid isolated from *Berberis darwinii*, was found to inhibit the activity of acetylcholinesterase significantly and suggested to be a potent AChE inhibitor [91]. Indole alkaloids like rescinnamine, obtained from *Rauvolfia reflexa* and harmane, 3-Hydroxylated harmine, harmol, harmine, 1-Hydroxy-7-methoxy-carboline, and harmaline, obtained from *Peganum harmala* were found to be potent cholinesterase inhibitors with having the IC50 1.21 μM to higher than10 μM [92, 93]. Eserine, another alkaloid obtained from *Physostigma venenosum* Balf., is also found to be a potent cholinesterase inhibitor but more specific to AChE [94]. The alkaloids galantamine, caranine, N-demethylgalanthamine, and lycoramine obtained from the leaves, roots, and bulbs of *Zephyranthes, Habranthus*, and *Crinum* have shown their inhibitory activity for acetylcholinesterase enzyme [95]. Quercetin, one of the most prominent flavonoids, and its glycosylated conjugate, rutin, have been reported for the cholinesterase inhibition, suppression of Fe^{2+}-induced lipid peroxidation and radical scavenging activities in rat brain homogenates in the concentration range 0.06 to 0.6 mM. Quercetin has shown higher AChE and BChE inhibition activities than rutin. It has also shown a strong Fe^{2+} chelating ability and higher radical scavenging capacity than rutin. It has been suggested that the control of oxidative stress-induced neurodegeneration may be aided by the flavonoids' ability to inhibit cholinesterases and their antioxidative effects [96]. Other phenolic compounds like resveratrol, curcumin, gallocatechin, cinnamic acid, and caffeic acid were found to have significant inhibitory activity for the acetylcholinesterase enzyme [97, 98, 99, 100, 101, 102]. Terpenoids, the largest among the secondary metabolite classes, have also shown significant acetylcholinesterase inhibition. Abietane diterpenoids compounds, 12-O-demethylcrptojaponol and

6-hydroxydemethylcryptojaponol obtained from *Caryopteris mongolica* have shown inhibitory activities against AChE and BuChE in which 6-hydroxydemethylcryptojaponol inhibited both AChE and BuChE (AChE IC50 = 50.8 μM, BuChE IC50 = 70.1 μM) while 12-O-demethylcrptojaponol was specific to BuChE (IC50 = 7.7 μM) (102). In another study, terpenoids like arudiol, tanshinone, militrone, cryptotanshinone, 1,2-didehydromiltirone *Caryopteris mongolica*, *Perovskia atriplicifolia*, *Salvia Glutinosa,* and *Lycopodiastrum Casuarinoides* have demonstrated notable AChE and BuChE inhibition activities [104, 105]. The phytosterols did not seem to contribute much to the cholinesterase activity since the mature fronds' BChE inhibitory impact was >10 times less than that of the juvenile fronds. β-sitosterol alone, according to prior investigations, had no discernible inhibitory impact on either AChE or BuChE [106].

12.4.2 Phytochemicals as Nrf2 Activators in Autism Spectral Disorder (ASD)

The autism spectrum disorder is a complicated illness with behavioural side effects and comorbid conditions, including anxiety, epilepsy, gastrointestinal issues leading to gut dysbiosis, oxidative stress, mitochondrial dysfunction, and biochemical changes [107]. The physiological difficulties have the potential to exacerbate biochemical and behavioural issues. No therapeutic cure has been recognized to treat the disease, but signs like self-mutilation, animosity, monotonous and stereotyped behaviours, carelessness, hyperactivity, and resting disarranges can be treated utilizing medications [108]. Therefore, Immune dysregulation/inflammation, mitochondria, and oxidative dysfunction stress all impact the development of ASD [109]. Nrf2 (nuclear factor erythroid 2 related factors 2), a universal master transcription factor, controls the production of antioxidant enzymes and cytoprotective proteins via the antioxidant response elements (AREs) pathway. Cell survival and defence are the results of this transcriptional factor. Chemoprevention and defence against cardiovascular, renal, and pulmonary illnesses depend on Nrf2 stimulation of gene expression [110, 111].

One method of activating Nrf2 requires cysteine oxidation at particular Keap1 sites to break the homodimer's attachment to it. The target genes' antioxidant response element (ARE) sequence is bound by free Nrf2, which then enters the nucleus and forms a dimer with small musculoaponeurotic fibrosarcoma proteins (sMafs). Because oxidative stress, along with neuroinflammation and mitochondrial dysfunction, is one of the hallmarks of neurodegenerative pathologies. It is a target for altering Nrf2/ARE signalling and the transcriptional activity of certain genes to prevent or postpone the onset of age-related and hereditary neurogenerative disorders [112]. Numerous studies have identified oxidative stress and poor methylation in autism spectrum disorder (ASD), including lower levels of methylcobalamin (III) in the brain. Nuclear factor erythroid 2-related factor 2 (Nrf2) stimulates the expression of many antioxidant genes (MeCbl) [113]. When used as a neurotherapeutic, phytochemicals can help to reduce the oxidative stress that leads to mitochondrial dysfunction in ASD. Phenolic compounds such as quercetin reduced mitochondrial damage and increased the expression of genes linked to mitochondrial DNA copy number. Additionally, quercetin supported Nrf2 activation, and elevated heme oxygenase-1 gene expression leads to an effective therapeutic compound in autism [114]. Wang et al. (2017) reported naringenin flavanone in the *Citrus* genus that induced higher levels of high-energy phosphates, improved mitochondrial ANT transport activity, and increased mitochondrial membrane potential; all suggest that this compound could help with mitochondrial dysfunction [115]. Naringenin also reduced the rate of cell death and enhanced cell viability. These results showed that NAR activated the Nrf2/ARE signalling pathway in neurons, which reduced oxidative stress and improved mitochondrial dysfunction, which might be helpful in neurological disorders like ASD. Curcumin, an important flavonoid obtained from *Curcuma longa,* has demonstrated exceptional therapeutic effect in suppressing cytokine-induced inflammation and restoring mitochondrial dysfunction, hence acting *as a* neurotherapeutic to treat the neurobehavioral, biochemical, and molecular changes associated with

ASD [116]. Sulforaphane, a member of the isothiocyanate group, is produced when myrosinase catalyses glucoraphanin. Sulforaphane reduces oxidative stress and mitochondrial dysfunction linked to Nrf2 activation [117].

12.4.3 Role of Phytochemicals Targeting α-Synuclein Aggregation in Parkinson's Disease

Parkinson's disease (PD) is the second most prevalent chronic neurodegenerative illness that impairs both motor and cognitive function [118]. The standard treatment strategies for managing PD only effectively treat symptoms [119]. Numerous environmental and genetic risk factors have been linked to the development of Parkinson's disease, including ambient air pollution and pesticides, especially organochlorines [120, 121]. A presynaptic protein known as α-synuclein (α-synor αS) is a 140-residue presynaptic protein that controls the rate of releasing neurotransmitters in synaptic vesicles. α-synuclein is thought to be linked with the pathogenesis of Parkinson's disease in many ways. However, most preferably, it has been predicted cellular homeostasis disturbance and neuronal death through impacts on a variety of intracellular targets, including synaptic function, are caused by its abnormal soluble oligomeric conformations, known as protofibrils [122, 124]. Lewy bodies and synuclein aggregations are characteristics of both familial and sporadic types of PD. These aggregates go through several critical phases of aggregation, oligomerization, and fibrillation, leading to the pathogenesis of PD [127, 125, 126]. Thus, the progression of PD is linked with oligomerization and fibrillation of α-syn and can be a novel target in the process of drug development. Phytochemicals have been used to treat neurodegenerative disorders and have demonstrated significant promise in preventing amyloid aggregation. [128]. Zhu et al. (2013) demonstrated that the oxidized quercetin species like benzyfuranone, chalcantrione, and quercetinchinone with other derivatives have significantly inhibited the fibril formation activity of α-synuclein due to their high polarity and hydrophilicity than quercetin [128]. In an experiment, Li et al. (2017) demonstrated that Baicalein, a flavonoid, was involved in the reduction of α-syn in dopaminergic cell lines (SN4741) and reduced its transmission by facilitating α-syn polymerization to giant molecule, thus inhibiting its propagation in neurons [129]. Epigallocatechin gallate (epigallocatechin-3-gallate), a phenolic compound of the catechin sub-class, was found to inhibit the size and adverse effects due to oligomerization/fibrillation of α-Syn [130]. Polyphenol, brazilin isolated from red cedarwood trees, was found to inactivate α-syn isolated from the brain of a patient suffering from PD. Brazilin is also thought to detoxify the aggregation of α-syn and help stabilize amyloid fibril clusters. Fu et al. (2017) reported that diarylheptanoids like alpinin a and alpinin b isolated from *Alpinia officinarum* significantly inhibited the aggregation of α-Syn [131]. Ghanem et al. (2021) reported that the results from Th-S assays demonstrated six out of nine alkaloid compounds from Chinese herbs: cytisine, harmine, hupehenine, koumine, synephrine, and trigonelline were found to be potent inhibitors of α-syn at different concentrations [132]. In silico experiments, driminen, a terpenoid identified in *Abrus precatorius*, showed a higher binding affinity for α-syn (- 9kcal/mol) [133].

12.4.4 Phytochemicals Promote Γ-Amino Butyric Acid (GABA) in Hungtinton's Disease

A neurodegenerative condition called Huntington's disease (HD) or Huntington's chorea is characterized by increasing motor dysfunction, such as chorea and dystonia, as well as emotional problems, memory problems, and weight loss. Huntington's disease primarily affects the medium spiny neurons in the brain and striatum [134]. Along with changes in N-methyl-D-aspartate (NMDA) receptors, HD also results in the loss of enkephalin and gamma-aminobutyric acid (GABA) neurons from the basal ganglia. By attaching to gamma-aminobutyric acid receptors, which are found either pre- or postsynaptically, that controls neuronal function [135]. The mammalian central nervous system primarily uses the amino acid GABA for inhibitory purposes. The GABAergic system is made up of structures that are involved in the release or binding of GABA as a neurotransmitter. The GABAergic system has a role in the control of memory, alertness, anxiety, muscular tension, and epileptogenic activity [136]. Morphin, the oldest known alkaloid for pain-relieving action, is reported to bind with μ-opioid receptor (MOR) in CNS, resulting in high Gamma-aminobutyric acid content in the synapses [137]. They reported that zerumin A and labdane, diterpenes, and isolated from *Curcuma kwangsiensis* were found to have potent GABA receptor modulations [138]. The methanol extract of *Crataeva nurvala* leaves showed the presence of D-allose, and α-Tocopherol, which found to have GABArgic modulating effect in swiss albino mice as reported by [139]. Withanolide isolated from *Withania somnifera* and purnavin, an alkaloid isolated from *Boerhaavia diffusa* L., was reported to have practical modulatory activities in the GABAergic system [140].

12.4.5 Phytochemicals as Inhibitors of the Expression of NF-κB Signalling Pathway in Cerebral Palsy

The most prevalent motor impairment in children is cerebral palsy (CP), and to date, the exact cause of cerebral palsy is unknown to researchers. The term 'cerebral palsy' refers to a collection of long-term abnormalities of movement and posture development that impair daily activities and are thought to be caused by non-progressive problems of the fetal or newborn brain. In an animal model, anti-inflammatory medication can significantly lessen CP symptoms. This finding may have significance for the postnatal care of infants at risk of developing CP [141]. The most well-known transcription factor, nuclear factor kappa B (NF-κB), is ubiquitously expressed and controls the expression of several genes that code for proteins involved in the processes of immunity and inflammation; thus, a prominent pathological feature of neurodegenerative disorders [142]. Many bioactive phytocompounds have been identified as potential anti-inflammatory compounds, and the majority belong to alkaloids, phenolics, terpenoids, and others. Can this-6-one (CO) alkaloids have a variety of biological effects, including antimicrobial, anticancer, antifungal, and antiviral were found to have anti-inflammatory effects, isolated from *Ailanthus altissima.* The compounds were found to inhibit the phosphorylation of inhibitor kappa B as well as the transcriptional activation of nuclear factor kappa B (NF-κB) in lipopolysaccharide (LPS)-stimulated macrophages [143]. Gao et al. (2020) reported preskimmianine an alkaloid isolated from the root bark of *Dictamnus dasycarpus* had shown significant suppression of nuclear factor kappa B (NF-κB) in LPS induced-BV-2 microglial cells [144]. Resveratrol, apigenin, quercetin, and epicatechin polyphenols were found to be potent phytocompounds as they the suppress NF-κB signalling pathway in various diseases [145, 146, 147, 148], and thus might help treat neurological diseases like cerebral palsy.

12.4.6 Role of Phytochemicals as Interleukin-1β (IL-1β) Inhibitors in Epilepsy

Recurrent seizures, which are the hallmark of the brain condition epilepsy, are often defined as a sharp change in behaviour brought on by a momentary disturbance in the electrical activity of the brain [149]. Typically, the brain continuously generates tiny electrical impulses that have a consistent pattern. Neurotransmitters are the chemical messengers that carry these signals along neurons, the network of nerve cells in the brain, and throughout the whole body. In epilepsy, the brain's electrical rhythms have a tendency to become imbalanced, resulting in recurrent seizures. Despite the fact that epilepsy is widespread, almost 80% of epileptic patients reside in developing or low-income nations [150, 151]. The current experimental models suggest that epilepsy involves the inflammatory pathways in the brain for the origin and spreading of epileptic activities. Seizures cause an increase in IL-1, but they can also modify a person's vulnerability to stimuli that cause seizures. The intrinsic brain inflammation results from IL-1 receptor signalling activation in the glia and neurons. Thus, inhibiting Interleukin-1β (IL-1β) may be a valuable tactic for reducing epilepsy

[149]. Phenolic compounds like rutin, hesperidin, gallic acid, kaempferol, and curcumin were found to be effective therapeutics in targeting the suppression of Interleukin-1β (IL-1β), leading to a reduction in inflammatory activities [152, 153, 154]. Gabay et al. (2010) reported stigmasterol's potential anti-inflammatory activities, which have inhibited the interleukin-1β inflammatory pathway in newborn mouse chondrocytes and human osteoarthritis (OA) chondrocytes [155]. Alkaloids like berberine and hirustine were identified as potential inhibitors of inflammation induced by interleukin-1β; therefore, they might be helpful in targeted therapy for the treatment of epilepsy [156, 157]. Therefore, the use of phytoconstituents to treat epilepsy and the comorbidities that it is linked with gives support for the research and creation of new antiepileptic medicines derived from herbal remedies.

12.5 CONCLUSION

Neurological disorders are fatal, and they have a profound impact on the overall standard of living of the patients. These disorders are more often acknowledged as one of the conditions having the highest burden on sick people, their loved ones, and society. They are characterized by inflammation, aggregation of the protein, and oxidative stress. Phytochemicals comprise an enormous source of compounds and molecules that have a beneficial role in improving human health. In this chapter, we have discussed and reviewed various classes of phytochemicals, their extraction techniques, and their mode of action in the pathophysiology of neurological disorders. Therefore, in the future, phytochemicals may be utilized as a treatment for neurological diseases by altering the disease mechanism, but a profound investigation is needed to explore new compounds and their mechanism of action to stop disease progression.

CONTRIBUTION OF AUTHORS

Pawan Prabhakar has given the framework, designed, and written the manuscript. Ankit Kumar has designed, interpreted, and technically corrected the manuscript while Mamoni Banerjee has conceptualized, provided technical suggestions, and final corrections to the manuscript. All authors critically reviewed and approved the final version of the manuscript for submission.

CONFLICT OF RESEARCH INTERESTS

Authors declare no conflict of research interests

REFERENCES

1. Chandra, V., Pandav, R., Laxminarayan, R., Tanner, C., Manyam, B., Rajkumar, S., ... & Zhang, Z. X. (2006). Neurological disorders. *Disease Control Priorities in Developing Countries*, Chapter *32*, 627–643.
2. Thakur, K. T., Albanese, E., Giannakopoulos, P., Jette, N., Linde, M., Prince, M. J., Steiner, T. J., & Dua, T. (2016). Neurological disorders. In V. Patel (Eds.) et. al., *Mental, Neurological, and Substance Use Disorders: Disease Control Priorities*, Third Edition (Volume *4*). The International Bank for Reconstruction and Development / The World Bank.
3. Khalil, M., Teunissen, C. E., Otto, M., Piehl, F., Sormani, M. P., Gattringer, T., ... & Kuhle, J. (2018). Neurofilaments as biomarkers in neurological disorders. *Nature Reviews Neurology*, *14*(10), 577–589.
4. World Health Organization. (2006). *Neurological Disorders: Public Health Challenges*. World Health Organization
5. Malla, A., Joober, R., & Garcia, A. (2015). "Mental illness is like any other medical illness": a critical examination of the statement and its impact on patient care and society. *Journal of Psychiatry & Neuroscience*, *40*(3), 147.

6. Lima, A. A., Mridha, M. F., Das, S. C., Kabir, M. M., Islam, M. R., & Watanobe, Y. (2022). A comprehensive survey on the detection, classification, and challenges of neurological disorders. *Biology, 11*(3), 469.
7. Chen, X., & Pan, W. (2014). The treatment strategies for neurodegenerative diseases by integrative medicine. *Integrative Medicine International, 1*(4), 223–225.
8. Jabir, N. R., Khan, F. R., & Tabrez, S. (2018). Cholinesterase targeting by polyphenols: A therapeutic approach for the treatment of Alzheimer's disease. *CNS Neuroscience & Therapeutics, 24*(9), 753–762.
9. Lin, S. X., Curtis, M. A., & Sperry, J. (2020). Pyridine alkaloids with activity in the central nervous system. *Bioorganic & Medicinal Chemistry, 28*(24), 115820.
10. Winkler, A. S. (2020). The growing burden of neurological disorders in low-income and middle-income countries: Priorities for policy making. *The Lancet Neurology*, 19(3), 200–202.
11. Feigin, V. L., Nichols, E., Alam, T., Bannick, M. S., Beghi, E., Blake, N., & Fischer, F. (2019). Global, regional, and national burden of neurological disorders, 1990–2016: A systematic analysis for the Global Burden of Disease Study 2016. *The Lancet Neurology, 18*(5), 459-480.
12. Kruk, M. E., Gage, A. D., Arsenault, C., Jordan, K., Leslie, H. H., Roder-DeWan, S., & Pate, M. (2018). High-quality health systems in the Sustainable Development Goals era: Time for a revolution. *The Lancet Global Health, 6*(11), e1196–e1252.
13. Durães, F., Pinto, M., & Sousa, E. (2018). Old drugs as new treatments for neurodegenerative diseases. *Pharmaceuticals, 11*(2), 44.
14. Center, B. P. (2014). Drug-induced neurologic conditions. *US Pharm, 39*(1), 47–51.
15. Duma, S. R., & Fung, V. S. (2019). Drug-induced movement disorders. *Australian Prescriber, 42*(2), 56.
16. Han, S., Kim, S., Kim, H., Shin, H. W., Na, K. S., & Suh, H. S. (2019). Prevalence and incidence of Parkinson's disease and drug-induced parkinsonism in Korea. *BMC Public Health, 19*(1), 1–9.
17. Ghasemiyeh, P., Borhani-Haghighi, A., Karimzadeh, I., Mohammadi-Samani, S., Vazin, A., Safari, A., & Qureshi, A. I. (2020). Major neurologic adverse drug reactions, potential drug-drug interactions and pharmacokinetic aspects of drugs used in covid-19 patients with stroke: A narrative review. *Therapeutics and Clinical Risk Management, 16*, 595.
18. Champagne-Jorgensen, K., Kunze, W. A., Forsythe, P., Bienenstock, J., & Neufeld, K. A. M. (2019). Antibiotics and the nervous system: More than just the microbes? *Brain, Behavior, and Immunity, 77*, 7–15.
19. Velmurugan, B. K., Rathinasamy, B., Lohanathan, B. P., Thiyagarajan, V., & Weng, C. F. (2018). Neuroprotective role of phytochemicals. *Molecules, 23*(10), 2485.
20. Prabhakar, P., & Mamoni, B. (2021). Technical problems, regulatory and market challenges in bringing herbal drug into mainstream of modern medicinal practices. *Research Journal of Biotechnology, 16* (3): 212–222.
21. Ekor, M. (2014). The growing use of herbal medicines: issues relating to adverse reactions and challenges in monitoring safety. *Frontiers in Pharmacology, 4*, 177.
22. Deb, L., Laishram, S., Khumukcham, N., Ningthoukhongjam, D., Nameirakpam, S. S., Dey, A., ... & Ningthoukhongjam, T. R. (2015). Past, present and perspectives of Manipur traditional medicine: A major health care system available for rural population in the North-East India. *Journal of Ethnopharmacology, 169*, 387–400.
23. Che, C. T., George, V., Ijinu, T. P., Pushpangadan, P., & Andrae-Marobela, K. (2017). Traditional medicine. In *Pharmacognosy* (pp. 15–30). Academic Press.
24. Bharate, S. S., Mignani, S., & Vishwakarma, R. A. (2018). Why are the majority of active compounds in the CNS domain natural products? A critical analysis. *Journal of Medicinal Chemistry, 61*(23), 10345–10374.
25. Lin, S. X., Curtis, M. A., & Sperry, J. (2020). Pyridine alkaloids with activity in the central nervous system. *Bioorganic & Medicinal Chemistry, 28*(24), 115820.
26. Dey, A., Nanda, B., Nandy, S., Mukherjee, A., & Pandey, D. K. (2020). Implications of phytochemicals as disease-modifying agents against Huntington's disease (HD): Bioactivity, animal models and transgenics, synergism and structure-activity studies. *Studies in Natural Products Chemistry, 67*, 27–79.

27. Casciaro, B., Mangiardi, L., Cappiello, F., Romeo, I., Loffredo, M. R., Iazzetti, A., & Quaglio, D. (2020). Infections. *Molecules*, *25*(16), 3619.
28. Rasouli, H., Yarani, R., Pociot, F., & Popović-Djordjević, J. (2020). Anti-diabetic potential of plant alkaloids: Revisiting current findings and future perspectives. *Pharmacological Research*, 155, 104723.
29. Lichman, B. R. (2021). The scaffold-forming steps of plant alkaloid biosynthesis. *Natural Product Reports*, *38*(1), 103–129.
30. Aryal, B., Raut, B. K., Bhattarai, S., Bhandari, S., Tandan, P., Gyawali, K., & Parajuli, N. (2022). Potential therapeutic applications of plant-derived alkaloids against inflammatory and neurodegenerative diseases. *Evidence-Based Complementary and Alternative Medicine*, *2022*.
31. Cushnie, T. T., Cushnie, B., & Lamb, A. J. (2014). Alkaloids: An overview of their antibacterial, antibiotic-enhancing and antivirulence activities. *International Journal of Antimicrobial Agents*, 44(5), 377–386.
32. Ain, Q. U., Khan, H., Mubarak, M. S., & Pervaiz, A. (2016). Plant alkaloids as antiplatelet agent: Drugs of the future in the light of recent developments. *Frontiers in Pharmacology*, *7*, 292.
33. Thawabteh, A., Juma, S., Bader, M., Karaman, D., Scrano, L., Bufo, S. A., & Karaman, R. (2019). The biological activity of natural alkaloids against herbivores, cancerous cells and pathogens. *Toxins*, *11*(11), 656.
34. Minatel, I. O., Borges, C. V., Ferreira, M. I., Gomez, H. A. G., Chen, C. Y. O., & Lima, G. P. P. (2017). Phenolic compounds: Functional properties, impact of processing and bioavailability. *Phenolic Compd. Biol. Act*, *8*, 1–24.
35. Al Jitan, S., Alkhoori, S. A., & Yousef, L. F. (2018). Phenolic acids from plants: Extraction and application to human health. *Studies in Natural Products Chemistry*, 58, 389–417.
36. Shahidi, F., & Ambigaipalan, P. (2015). Phenolics and polyphenolics in foods, beverages and spices: Antioxidant activity and health effects–A review. *Journal of Functional Foods*, *18*, 820–897.
37. Giada, M. D. L. R. (2013). *Food Phenolic Compounds: Main Classes, Sources and Their Antioxidant Power. Oxidative stress and chronic degenerative diseases-A role for antioxidants*, *2013*, 87–112.
38. Gharaati Jahromi, S. (2019). *Extraction Techniques of Phenolic Compounds from Plants*. IntechOpen. doi: 10.5772/intechopen.84705
39. Mathew, S., Abraham, T. E., & Zakaria, Z. A. (2015). Reactivity of phenolic compounds towards free radicals under in vitro conditions. *Journal of Food Science and Technology*, *52*(9), 5790–5798.
40. Ambriz-Pérez, D. L., Leyva-López, N., Gutierrez-Grijalva, E. P., & Heredia, J. B. (2016). Phenolic compounds: Natural alternative in inflammation treatment. A Review. *Cogent Food & Agriculture*, *2*(1), 1131412.
41. Ambriz-Pérez, D. L., Leyva-López, N., Gutierrez-Grijalva, E. P., & Heredia, J. B. (2016). Phenolic compounds: Natural alternative in inflammation treatment. A Review. *Cogent Food & Agriculture*, *2*(1), 1131412.
42. Roleira, F. M., Tavares-da-Silva, E. J., Varela, C. L., Costa, S. C., Silva, T., Garrido, J., & Borges, F. (2015). Plant-derived and dietary phenolic antioxidants: Anticancer properties. *Food Chemistry*, *183*, 235–258.
43. Tungmunnithum, D., Thongboonyou, A., Pholboon, A., & Yangsabai, A. (2018). Flavonoids and other phenolic compounds from medicinal plants for pharmaceutical and medical aspects: An overview. *Medicines*, 5(3), 93.
44. Walsh, D. J., Livinghouse, T., Goeres, D. M., Mettler, M., & Stewart, P. S. (2019). Antimicrobial activity of naturally occurring phenols and derivatives against biofilm and planktonic bacteria. *Frontiers in Chemistry*, *7*, 653.
45. Prabhakar, P., & Banerjee, M. (2020). Antidiabetic phytochemicals: a comprehensive review on opportunities and challenges in targeted therapy for herbal drug development. *International Journal of Pharmaceutical Research*, 12, 1673–1696.
46. Caruso, G., Godos, J., Privitera, A., Lanza, G., Castellano, S., Chillemi, A., & Grosso, G. (2022). Phenolic acids and prevention of cognitive decline: Polyphenols with a neuroprotective role in cognitive disorders and Alzheimer's disease. *Nutrients*, *14*(4), 819.
47. Pandey, K. B., & Rizvi, S. I. (2009). Plant polyphenols as dietary antioxidants in human health and disease. *Oxidative Medicine and Cellular Longevity*, *2*(5), 270–278.

48. Shahidi, F., & Ambigaipalan, P. (2015). Phenolics and polyphenolics in foods, beverages and spices: Antioxidant activity and health effects–A review. *Journal of Functional Foods*, *18*, 820–897.
49. Singh, B., & Sharma, R. A. (2015). Plant terpenes: defense responses, phylogenetic analysis, regulation and clinical applications. *3 Biotech*, *5*(2), 129–151.
50. Perveen, S. (2018). Introductory Chapter: Terpenes and Terpenoids. In S. Perveen, & A. Al-Taweel (Eds.), *Terpenes and Terpenoids. IntechOpen*. https://doi.org/10.5772/intechopen.79683 .
51. Al Hoqani, U., Leon, R., & Purton, S. (2022). Over-expression of a cyanobacterial gene for 1-deoxy-D-xylulose-5-phosphate synthase in the chloroplast of Chlamydomonas reinhardtii perturbs chlorophyll: carotenoid ratios. *Journal of King Saud University-Science*, 102141.
52. Dash, D. K., Tyagi, C. K., Sahu, A. K., & Tripathi, V. (2022). Revisiting the Medicinal Value of Terpenes and Terpenoids. In V. S. Meena, H. P. Parewa, & S. K. Meena (Eds.), *Revisiting Plant Biostimulants. IntechOpen.* https://doi.org/10.5772/intechopen.102612.
53. Wojtunik-Kulesza, K. A., Targowska-Duda, K., Klimek, K., Ginalska, G., Jóźwiak, K., Waksmundzka-Hajnos, M., & Cieśla, Ł. (2017). Volatile terpenoids as potential drug leads in Alzheimer's disease. *Open Chemistry*, 15(1), 332–343.
54. Guimarães, A. C., Meireles, L. M., Lemos, M. F., Guimarães, M. C. C., Endringer, D. C., Fronza, M., & Scherer, R. (2019). Antibacterial activity of terpenes and terpenoids present in essential oils. *Molecules*, *24*(13), 2471.
55. Fernández, L. R., Butassi, E., Svetaz, L., Zacchino, S. A., Palermo, J. A., & Sánchez, M. (2014). Antifungal Terpenoids from Hyalis argentea var. latisquama. *Journal of Natural Products*, *77*(7), 1579–1585.
56. Yang, J., Fu, X., Liao, X., & Li, Y. (2020). Nrf2 activators as dietary phytochemicals against oxidative stress, inflammation, and mitochondrial dysfunction in autism spectrum disorders: a systematic review. *Frontiers in Psychiatry*, *11*, 561998.
57. Smitha Grace, S. R., Chandran, G., & Chauhan, J. B. (2019). Terpenoids: an activator of "fuel-sensing enzyme AMPK" with special emphasis on antidiabetic activity. *Plant and Human Health, Volume 2*, 227–244.
58. Prakash, V. E. D. (2017). Terpenoids as source of anti-inflammatory compounds. *Asian Journal of Pharmaceutical and Clinical Research*, *10*(3), 68–76.
59. Phulara, S. C., Rajput, V. S., Mazumdar, B., & Runthala, A. (2020). Metabolic and enzyme engineering for the microbial production of anticancer terpenoids. In *Essentials of Cancer Genomic, Computational Approaches and Precision Medicine* (pp. 237–259). Springer, Singapore.
60. Xu, P., & Yu, B. (2021). Chemical synthesis of saponins: An update. In *Advances in Carbohydrate Chemistry and Biochemistry* (Vol. 79, pp. 1–62). Academic Press.
61. Lu, Y. A. O., Jun, L. U., Juan, W. A. N. G., & Wen-Yuan, G. A. O. (2020). Advances in biosynthesis of triterpenoid saponins in medicinal plants. *Chinese Journal of Natural Medicines*, *18*(6), 417–424.
62. Juang, Y. P., & Liang, P. H. (2020). Biological and pharmacological effects of synthetic saponins. *Molecules*, *25*(21), 4974.
63. Moses, T., Papadopoulou, K. K., & Osbourn, A. (2014). Metabolic and functional diversity of saponins, biosynthetic intermediates and semi-synthetic derivatives. *Critical Reviews In Biochemistry and Molecular Biology*, *49*(6), 439–462.
64. Shahzad, N., Khan, W., Shadab, M. D., Ali, A., Saluja, S. S., Sharma, S., & Al-Ghamdi, S. S. (2017). Phytosterols as a natural anticancer agent: Current status and future perspective. *Biomedicine & Pharmacotherapy*, *88*, 786–794.
65. Sharma, N., Tan, M. A., & An, S. S. A. (2021). Phytosterols: Potential metabolic modulators in neurodegenerative diseases. *International Journal of Molecular Sciences*, 22(22), 12255.
66. Du, J., Jiang, L., Chen, F., Hu, H., & Zhou, M. (2021). Cardiac glycoside ouabain exerts anticancer activity via downregulation of STAT3. *Frontiers in Oncology*, *11*, 2565.
67. Prassas, I., & Diamandis, E. P. (2008). Novel therapeutic applications of cardiac glycosides. *Nature Reviews Drug Discovery*, *7*(11), 926–935.
68. Kumavath, R., Paul, S., Pavithran, H., Paul, M. K., Ghosh, P., Barh, D., & Azevedo, V. (2021). Emergence of cardiac glycosides as potential drugs: Current and future scope for cancer therapeutics. *Biomolecules*, *11*(9), 1275.
69. Abubakar, A. R., & Haque, M. (2020). Preparation of medicinal plants: Basic extraction and fractionation procedures for experimental purposes. *Journal of Pharmacy & Bioallied Sciences*, *12*(1), 1.

70. Azwanida, N. N. (2015). A review on the extraction methods used in medicinal plants, principle, strength and limitation. *Med Aromat Plants*, *4*(196), 2167–0412.
71. Sasidharan, S., Chen, Y., Saravanan, D., Sundram, K. M., & Latha, L. Y. (2011). Extraction, isolation and characterization of bioactive compounds from plants' extracts. *African Journal of Traditional, Complementary and Alternative Medicines*, *8*(1).
72. Ntalikwa, J. W. (2021). Solvent extraction of jatropha oil for biodiesel production: effects of solvent-to-solid ratio, particle size, type of solvent, extraction time, and temperature on oil yield. *Journal of Renewable Energy*, 2021.
73. Pandey, R. K., Shukla, S. S., Vyas, A., Jain, V., Jain, P., & Saraf, S. (2018). *Fingerprinting Analysis and Quality Control Methods of Herbal Medicines*. CRC Press.
74. Pandey, A., & Tripathi, S. (2014). Concept of standardization, extraction and pre phytochemical screening strategies for herbal drug. *Journal of Pharmacognosy and Phytochemistry*, *2*(5).
75. Suna, S., Tamer, C. E., & Özcan-Sinir, G. (2019). Trends and possibilities of the usage of medicinal herbal extracts in beverage production. In *Natural Beverages* (pp. 361–398). Academic Press.
76. Handa, S. S. (2008). An overview of extraction techniques for medicinal and aromatic plants. *Extraction Technologies for Medicinal and Aromatic Plants*, *1*, 21–40.
77. Saifullah, M., McCullum, R., & Vuong, Q. V. (2021). Optimization of microwave-assisted extraction of polyphenols from Lemon Myrtle: Comparison of modern and conventional extraction techniques based on bioactivity and total polyphenols in dry extracts. *Processes*, *9*(12), 2212.
78. Mirzadeh, M., Arianejad, M. R., & Khedmat, L. (2020). Antioxidant, antiradical, and antimicrobial activities of polysaccharides obtained by microwave-assisted extraction method: A review. *Carbohydrate Polymers*, *229*, 115421
79. Garcia-Vaquero, M., Ummat, V., Tiwari, B., & Rajauria, G. (2020). Exploring ultrasound, microwave, and ultrasound–microwave assisted extraction technologies to increase the extraction of bioactive compounds and antioxidants from brown macroalgae. *Marine Drugs*, *18*(3), 172.
80. Dzah, C. S., Duan, Y., Zhang, H., Wen, C., Zhang, J., Chen, G., & Ma, H. (2020). The effects of ultrasound-assisted extraction on yield, antioxidant, anticancer and antimicrobial activity of polyphenol extracts: A review. *Food Bioscience*, *35*, 100547.
81. Raj, G. B., & Dash, K. K. (2020). Ultrasound-assisted extraction of phytocompounds from dragon fruit peel: Optimization, kinetics and thermodynamic studies. *Ultrasonics Sonochemistry*, *68*, 105180.
82. Kumar, A., Rout, R. K., Rao, P. S., & Prabhakar, P. (2022). Optimization of pulsed mode sonication and in silico molecular docking study for antioxidant properties of mandarin (Citrus reticulata Blanco) peels. *Journal of Food Process Engineering*, e14111.
83. Singh, S., Verma, D. K., Thakur, M., Tripathy, S., Patel, A. R., Shah, N., & Aguilar, C. N. (2021). Supercritical fluid extraction (SCFE) as green extraction technology for high-value metabolites of algae, its potential trends in food and human health. *Food Research International*, *150*, 110746.
84. Ahangari, H., King, J. W., Ehsani, A., & Yousefi, M. (2021). Supercritical fluid extraction of seed oils–A short review of current trends. *Trends in Food Science & Technology*, *111*, 249–260.
85. Swami, H. S., Singh, K. S. P., Gennaro, L., & Dutt, R. D. (2008). Extraction technologies for medicinal and aromatic plants. *Trieste: United Nations Industrial Development Organization and the International Centre for Science and High Technology*, *200*, 266.
86. Doughari, J. H. (2012). *Phytochemicals: Extraction Methods, Basic Structures and Mode of Action as Potential Chemotherapeutic Agents* (pp. 1–33). Rijeka, Croatia: INTECH Open Access Publisher.
87. Altemimi, A., Lakhssassi, N., Baharlouei, A., Watson, D. G., & Lightfoot, D. A. (2017). Phytochemicals: Extraction, isolation, and identification of bioactive compounds from plant extracts. *Plants*, *6*(4), 42.
88. Ingle, K. P., Deshmukh, A. G., Padole, D. A., Dudhare, M. S., Moharil, M. P., & Khelurkar, V. C. (2017). Phytochemicals: Extraction methods, identification and detection of bioactive compounds from plant extracts. *Journal of Pharmacognosy and Phytochemistry*, *6*(1), 32–36.
89. Thakur, A. K., Kamboj, P., Goswami, K., & Ahuja, K. (2018). Pathophysiology and management of Alzheimer's disease: An overview. *Journal of Analytical & Pharmaceutical Research*, *7*(1).
90. Lai Shi Min, S., Liew, S. Y., Chear, N. J. Y., Goh, B. H., Tan, W. N., & Khaw, K. Y. (2022). Plant Terpenoids as the Promising Source of Cholinesterase Inhibitors for Anti-AD Therapy. *Biology*, *11*(2), 307.

91. Habtemariam, S. (2011). The therapeutic potential of Berberis darwinii stem-bark: quantification of berberine and in vitro evidence for Alzheimer's disease therapy. *Natural Product Communications*, *6*(8), 1934578X1100600809.
92. Fadaeinasab, M., Basiri, A., Kia, Y., Karimian, H., Ali, H. M., & Murugaiyah, V. (2015). New indole alkaloids from the bark of Rauvolfia reflexa and their cholinesterase inhibitory activity. *Cellular Physiology and Biochemistry*, *37*(5), 1997–2011.
93. Yang, Y., Cheng, X., Liu, W., Chou, G., Wang, Z., & Wang, C. (2015). Potent AChE and BChE inhibitors isolated from seeds of Peganum harmala Linn by a bioassay-guided fractionation. *Journal of Ethnopharmacology*, *168*, 279–286.
94. Konrath, E. L., Passos, C. D. S., Klein-Júnior, L. C., & Henriques, A. T. (2013). Alkaloids as a source of potential anticholinesterase inhibitors for the treatment of Alzheimer's disease. *Journal of Pharmacy and Pharmacology*, *65*(12), 1701–1725.
95. Stavrakov, G., Philipova, I., Lukarski, A., Atanasova, M., Zheleva, D., Zhivkova, Z. D., & Doytchinova, I. (2020). Galantamine-curcumin hybrids as dual-site binding acetylcholinesterase inhibitors. *Molecules*, *25*(15), 3341.
96. Ademosun, A. O., Oboh, G., Bello, F., & Ayeni, P. O. (2016). Antioxidative properties and effect of quercetin and its glycosylated form (Rutin) on acetylcholinesterase and butyrylcholinesterase activities. *Journal of Evidence-Based Complementary & Alternative Medicine*, *21*(4), NP11–NP17.
97. Chen, X., & Pan, W. (2014). The treatment strategies for neurodegenerative diseases by integrative medicine. *Integrative Medicine International*, *1*(4), 223–225.
98. Pan, L. F., Wang, X. B., Xie, S. S., Li, S. Y., & Kong, L. Y. (2014). Multitarget-directed resveratrol derivatives: anti-cholinesterases, anti-β-amyloid aggregation and monoamine oxidase inhibition properties against Alzheimer's disease. *MedChemComm*, *5*(5), 609–616.
99. Kalaycıoğlu, Z., Gazioğlu, I., & Erim, F. B. (2017). Comparison of antioxidant, anticholinesterase, and antidiabetic activities of three curcuminoids isolated from Curcuma longa L. *Natural Product Research*, *31*(24), 2914–2917.
100. Abdel Bar, F. M., Salkini, A. A., Amen, Y., & Sherif, A. E. (2022). Acetylcholinesterase inhibitors from Conocarpus lancifolius Engl.(Combretaceae). *Natural Product Research*, 1–6.
101. Ghafary, S., Ghobadian, R., Mahdavi, M., Nadri, H., Moradi, A., Akbarzadeh, T., & Amini, M. (2020). Design, synthesis, and evaluation of novel cinnamic acid-tryptamine hybrid for inhibition of acetylcholinesterase and butyrylcholinesterase. *DARU Journal of Pharmaceutical Sciences*, *28*(2), 463–477.
102. Aluko, R. E. (2021). Food-derived acetylcholinesterase inhibitors as potential agents against Alzheimer's Disease. *eFood*, *2*(2), 49–58.
103. Murata, T., Selenge, E., Oikawa, S., Ageishi, K., Batkhuu, J., Sasaki, K., & Yoshizaki, F. (2015). Cholinesterase-inhibitory diterpenoids and chemical constituents from aerial parts of Caryopteris mongolica. *Journal of Natural Medicines*, *69*(4), 471–478.
104. Senol, F. S., Ślusarczyk, S., Matkowski, A., Pérez-Garrido, A., Girón-Rodríguez, F., Cerón-Carrasco, J. P., ... & Orhan, I. E. (2017). Selective in vitro and in silico butyrylcholinesterase inhibitory activity of diterpenes and rosmarinic acid isolated from Perovskia atriplicifolia Benth. and Salvia glutinosa L. *Phytochemistry*, *133*, 33–44.
105. Liu, Y., Li, J., Li, D., Li, X. M., Li, D., Zhou, G., & Tan, G. S. (2019). Anti-cholinesterase activities of constituents isolated from Lycopodiastrum casuarinoides. *Fitoterapia*, *139*, 104366.
106. Chear, N. J. Y., Khaw, K. Y., Murugaiyah, V., & Lai, C. S. (2016). Cholinesterase inhibitory activity and chemical constituents of Stenochlaena palustris fronds at two different stages of maturity. *Journal of Food and Drug Analysis*, *24*(2), 358–366.
107. Fattorusso, A., Di Genova, L., Dell'Isola, G. B., Mencaroni, E., & Esposito, S. (2019). Autism spectrum disorders and the gut microbiota. *Nutrients*, *11*(3), 521.
108. Lord, C., Elsabbagh, M., Baird, G., & Veenstra-Vanderweele, J. (2018). Autism spectrum disorder. *The Lancet*, *392*(10146), 508–520.
109. Wang, W., Zhang, Y., Jin, W., Xing, Y., & Yang, A. (2018). Catechin weakens diabetic retinopathy by inhibiting the expression of NF-κB signaling pathway-mediated inflammatory factors. *Annals of Clinical & Laboratory Science*, *48*(5), 594–600.

110. Zhao, H., Eguchi, S., Alam, A., & Ma, D. (2017). The role of nuclear factor-erythroid 2 related factor 2 (Nrf-2) in the protection against lung injury. *American Journal of Physiology-Lung Cellular and Molecular Physiology*, *312*(2), L155–L162.
111. Saha, S., Buttari, B., Panieri, E., Profumo, E., & Saso, L. (2020). An overview of Nrf2 signaling pathway and its role in inflammation. *Molecules*, *25*(22), 5474.
112. Zgorzynska, E., Dziedzic, B., & Walczewska, A. (2021). An overview of the nrf2/are pathway and its role in neurodegenerative diseases. *International Journal of Molecular Sciences*, *22*(17), 9592.
113. Schrier, M. S., Zhang, Y., Trivedi, M. S., & Deth, R. C. (2022). Decreased cortical Nrf2 gene expression in autism and its relationship to thiol and cobalamin status. *Biochimie*, *192*, 1–12.
114. Alvarez-Arellano, L., Salazar-García, M., & Corona, J. C. (2020). Neuroprotective effects of quercetin in pediatric neurological diseases. *Molecules*, *25*(23), 5597.
115. Wang, K., Chen, Z., Huang, L., Meng, B., Zhou, X., Wen, X., & Ren, D. (2017). Naringenin reduces oxidative stress and improves mitochondrial dysfunction via activation of the Nrf2/ARE signaling pathway in neurons. *International Journal of Molecular Medicine*, *40*(5), 1582–1590.
116. Bhandari, R., & Kuhad, A. (2015). Neuropsychopharmacotherapeutic efficacy of curcumin in experimental paradigm of autism spectrum disorders. *Life Sciences*, *141*, 156–169.
117. Bhandari, R., Paliwal, J. K., & Kuhad, A. (2020). Dietary phytochemicals as neurotherapeutics for autism spectrum disorder: plausible mechanism and evidence. *Personalized Food Intervention and Therapy for Autism Spectrum Disorder Management*, 615–646.
118. DeMaagd, G., & Philip, A. (2015). Parkinson's disease and its management: part 1: disease entity, risk factors, pathophysiology, clinical presentation, and diagnosis. *Pharmacy and Therapeutics*, *40*(8), 504.
119. Oertel, W. H. (2017). Recent advances in treating Parkinson's disease. *F1000Research*, *6*.
120. Schneider Medeiros, M., P Reddy, S., P Socal, M., Schumacher-Schuh, A. F., & Mello Rieder, C. R. (2020). Occupational pesticide exposure and the risk of death in patients with Parkinson's disease: an observational study in southern Brazil. *Environmental Health*, *19*(1), 1–8.
121. Calderón-Garcidueñas, L. (2021). Parkinson disease and air pollution: does what we breathe matter?. *Nature Reviews Neurology*, *17*(8), 467–468.
122. de Oliveira, G. A., & Silva, J. L. (2019). Alpha-synuclein stepwise aggregation reveals features of an early onset mutation in Parkinson's disease. *Communications Biology*, *2*(1), 1–13.
123. Meade, R. M., Fairlie, D. P., & Mason, J. M. (2019). Alpha-synuclein structure and Parkinson's disease–lessons and emerging principles. *Molecular Neurodegeneration*, *14*(1), 1–14.
124. Meade, R. M., Fairlie, D. P., & Mason, J. M. (2019). Alpha-synuclein structure and Parkinson's disease–lessons and emerging principles. *Molecular Neurodegeneration*, *14*(1), 1–14.
125. Wang, Y., Shi, M., Chung, K. A., Zabetian, C. P., Leverenz, J. B., Berg, D., ... & Zhang, J. (2012). Phosphorylated α-synuclein in Parkinson's disease. *Science Translational Medicine*, *4*(121), 121ra20–121ra20.
126. Shahpiri, Z., Bahramsoltani, R., Farzaei, M. H., Farzaei, F., & Rahimi, R. (2016). Phytochemicals as future drugs for Parkinson's disease: a comprehensive review. *Reviews in the Neurosciences*, *27*(6), 651–668.
127. Lei, P., Ayton, S., & Bush, A. I. (2021). The essential elements of Alzheimer's disease. *Journal of Biological Chemistry*, *296*.
128. Zhu, M., Han, S., & Fink, A. L. (2013). Oxidized quercetin inhibits α-synuclein fibrillization. *Biochimica et Biophysica Acta (BBA)-General Subjects*, *1830*(4), 2872–2881.
129. Li, X., Zhang, G., Nie, Q., Wu, T., Jiao, L., Zheng, M., ... & Lin, X. (2017). Baicalein blocks α-synuclein secretion from SN4741 cells and facilitates α-synuclein polymerization to big complex. *Neuroscience Letters*, *655*, 109–114.
130. Jha, N. N., Kumar, R., Panigrahi, R., Navalkar, A., Ghosh, D., Sahay, S., & Maji, S. K. (2017). Comparison of α-synuclein fibril inhibition by four different amyloid inhibitors. *ACS Chemical Neuroscience*, *8*(12), 2722–2733.
131. Fu, G., Zhang, W., Du, D., Ng, Y. P., Ip, F. C., Tong, R., & Ip, N. Y. (2017). Diarylheptanoids from rhizomes of Alpinia officinarum inhibit aggregation of α-synuclein. *Journal of Agricultural and Food Chemistry*, *65*(31), 6608–6614.

132. Ghanem, S. S., Fayed, H. S., Zhu, Q., Lu, J. H., Vaikath, N. N., Ponraj, J., & El-Agnaf, O. M. (2021). Natural alkaloid compounds as inhibitors for alpha-synuclein seeded fibril formation and toxicity. *Molecules*, *26*(12), 3736.
133. Omoboyowa, D. A., Balogun, T. A., Omomule, O. M., & Saibu, O. A. (2021). Identification of terpenoids from Abrus precatorius against Parkinson's disease proteins using in silico approach. *Bioinformatics and Biology Insights*, *15*, 11779322211050757.
134. Mestre, T. A., & Sampaio, C. (2017). Huntington disease: linking pathogenesis to the development of experimental therapeutics. *Current Neurology and Neuroscience Reports*, *17*(2), 1–8.
135. Reilmann, R. (2013). Pharmacological treatment of chorea in Huntington's disease–good clinical practice versus evidence-based guideline. *Movement Disorders*, *28*(8), 1030–1033.
136. Hsu, Y. T., Chang, Y. G., & Chern, Y. (2018). Insights into GABAAergic system alteration in Huntington's disease. *Royal Society*, *Open Biology*, 8(12), 180165.
137. Halliwell, B. (2007). Oxidative stress and cancer: have we moved forward? *Biochemical Journal*, *401*(1), 1–11
138. Schramm, A., Ebrahimi, S. N., Raith, M., Zaugg, J., Rueda, D. C., Hering, S., & Hamburger, M. (2013). Phytochemical profiling of Curcuma kwangsiensis rhizome extract, and identification of labdane diterpenoids as positive GABAA receptor modulators. *Phytochemistry*, 96, 318–329.
139. Moniruzzaman, M., Mannan, M. A., Hossen Khan, M. F., Abir, A. B., & Afroze, M. (2018). The leaves of Crataeva nurvala Buch-Ham. modulate locomotor and anxiety behaviors possibly through GABAergic system. *BMC Complementary and Alternative Medicine*, *18*(1), 1–12.
140. Dey, P., Kundu, A., Kumar, A., Gupta, M., Lee, B. M., Bhakta, T., & Kim, H. S. (2020). Analysis of alkaloids (indole alkaloids, isoquinoline alkaloids, tropane alkaloids). In *Recent Advances in Natural Products Analysis* (pp. 505–567). Elsevier.
141. Crunkhorn, S. (2012). Nanoparticle opens door to cerebral palsy treatment. *Nature Reviews Drug Discovery*, *11*(6), 440–440.
142. Shal, B., Ding, W., Ali, H., Kim, Y. S., & Khan, S. (2018). Anti-neuroinflammatory potential of natural products in attenuation of Alzheimer's disease. *Frontiers in Pharmacology*, *9*, 548.
143. Cho, S. K., Jeong, M., Jang, D. S., & Choi, J. H. (2018). Anti-inflammatory Effects of Canthin-6-one Alkaloids from Ailanthus altissima. *Planta Medica*, *50*(08), 527–535.
144. Gao, P., Wang, L., Zhao, L., Zhang, Q. Y., Zeng, K. W., Zhao, M. B., & Guo, X. Y. (2020). Anti-inflammatory quinoline alkaloids from the root bark of Dictamnus dasycarpus. *Phytochemistry*, *172*, 112260.
145. Xu, L., Botchway, B. O., Zhang, S., Zhou, J., & Liu, X. (2018). Inhibition of NF-κB signaling pathway by resveratrol improves spinal cord injury. *Frontiers in Neuroscience*, *12*, 690.
146. Lai, W. W., Hsu, S. C., Chueh, F. S., Chen, Y. Y., Yang, J. S., Lin, J. P., & Chung, J. G. (2013). Quercetin inhibits migration and invasion of SAS human oral cancer cells through inhibition of NF-κB and matrix metalloproteinase-2/-9 signaling pathways. *Anticancer Research*, *33*(5), 1941–1950.
147. Chang, X., He, H., Zhu, L., Gao, J., Wei, T., Ma, Z., & Yan, T. (2015). Protective effect of apigenin on Freund's complete adjuvant-induced arthritis in rats via inhibiting P2X7/NF-κB pathway. *Chemical-biological Interactions*, *236*, 41–46.
148. Wang, W., Zhang, Y., Jin, W., Xing, Y., & Yang, A. (2018). Catechin weakens diabetic retinopathy by inhibiting the expression of NF-κB signaling pathway-mediated inflammatory factors. *Annals of Clinical & Laboratory Science*, 48(5), 594–600.
149. Kambli, L., Bhatt, L. K., Oza, M., & Prabhavalkar, K. (2017). Novel therapeutic targets for epilepsy intervention. *Seizure*, *51*, 27–34.
150. Aliyu, M. M., Musa, A. I. I., Kamal, M. J. A., & Mohammed, M. G. (2014). Phytochemical screening and anticonvulsant studies of ethyl acetate fraction of Globimetula braunii on laboratory animals. *Asian Pacific Journal of Tropical Biomedicine*, 4(4), 285–289.
151. Aghdash, S. N. (2021). Herbal medicine in the treatment of epilepsy. *Current Drug Targets*, *22*(3), 356–367.
152. Zhang, H., & Tsao, R. (2016). Dietary polyphenols, oxidative stress and antioxidant and anti-inflammatory effects. *Current Opinion in Food Science*, *8*, 33–42.
153. Ansari, M. Y., Ahmad, N., & Haqqi, T. M. (2020). Oxidative stress and inflammation in osteoarthritis pathogenesis: Role of polyphenols. *Biomedicine & Pharmacotherapy*, *129*, 110452.

154. Zhang, T., Qiu, F., Chen, L., Liu, R., Chang, M., & Wang, X. (2021). Identification and in vitro anti-inflammatory activity of different forms of phenolic compounds in Camellia oleifera oil. *Food Chemistry*, *344*, 128660.
155. Gabay, O., Sanchez, C., Salvat, C., Chevy, F., Breton, M., Nourissat, G., & Berenbaum, F. (2010). Stigmasterol: a phytosterol with potential anti-osteoarthritic properties. *Osteoarthritis and Cartilage*, 18(1), 106–116.
156. Jung, H. Y., Nam, K. N., Woo, B. C., Kim, K. P., Kim, S. O., & Lee, E. H. (2013). Hirsutine, an indole alkaloid of Uncaria rhynchophylla, inhibits inflammation-mediated neurotoxicity and microglial activation. *Molecular Medicine Reports*, *7*(1), 154–158.
157. Ma, J., Chan, C. C., Huang, W. C., & Kuo, M. L. (2020). Berberine inhibits pro-inflammatory cytokine-induced IL-6 and CCL11 production via modulation of STAT6 pathway in human bronchial epithelial cells. *International Journal of Medical Sciences*, *17*(10), 1464.

13 Phytoconstituents
The Natural Tonic for a Healthy Brain

Bapi Ray Sarkar, Simran Giri, Ganesh Dey, and Prosanta Pal

13.1 INTRODUCTION

In the 18th century, William James recognized that one of the most important characteristics of humans is the ability to change their behaviour in response to their surroundings. The concept of plasticity has since been expanded to include changes underlying the structure of the brain at various levels, ranging from cellular to systematic, with the intermingling of these changes aiding the phenomenon of behavioural plasticity [1]. Since then, it has been clear that the brain is the vital organ for the coordination and multifunctioning of the body. In the year 1948, according to WHO, health can be defined as 'a state of complete physical, mental, and social wellbeing and not merely the absence of disease or infirmity', whereas the key component of health is a healthy brain because it is the brain that is responsible for all actions [2]. Although the definition of a healthy brain is still controversial in many aspects, different health authorities, like the US Centres for Disease Control and Prevention, define a healthy brain as the capacity to perform all of the psychological processes, including perception, as well as counting the capacity to learn and pass judgement, use language, and recall it [3]. According to the American Heart Association/American Stroke Association (AHA/ASA), a healthy brain is defined as 'average performance levels among all people at that age who are free of known brain or other organ system diseases in terms of decline from function levels, or as adequacy to perform all activities that the individual wishes to undertake' [4]. The online survey was conducted for the very first time from June 2019 to August 2020 by the European Union-funded Life Brain Project, led in collaboration with Sweden, the National University of Ostroh Academy, national brain councils in Norway, Germany, and Belgium, the Women's Brain Project, and Brain Foundations in the Netherlands, where the condition of brain health is analysed in 14 different languages in the overall population above 18 years. An internet-based survey is proficient at contacting huge groups of people, at least at the expense and, from the result of a survey, a bit of knowledge regarding the perspectives on the brain's health. The degree of reception of positive practices can be empowered. The outcomes add to the advancement of strategy suggestions for supporting population brain health, including measures custom-made to individual requirements, information, inspirations, and life circumstances [5].

For many decades, natural products, including plant and animal products, were used to make drugs. Largely from the plant source, drugs were made because it was found that the plant produces plenty of different structured organic compounds (phytoconstituents), which are often called secondary metabolites. Many scientific approaches have been made to develop the drugs in a more effective form; with the development of scientific techniques, the isolation of active phytoconstituents has

DOI: 10.1201/9781003300502-13

become simpler. Hence, the structure of phytoconstituents was easily characterized and many of them are synthetically made in laboratories nowadays. With the advancement in technology, the modification of chemical groups in the structure can develop highly active drugs [6]. Regardless of these benefits, many big companies with pharmacy backgrounds diminish the utilization of natural sources for screening drug discovery. This has been a result of the drawbacks of natural products like trouble in supply, complex chemical constituents, and intrinsically slow working patterns with natural sources as well as worries about intellectual property rights. There is quite a large number of promising medication applicants in the current advancement pipeline that are of natural source. Specialized disadvantages related to natural item research have been reduced, and there are more chances to investigate the pharmacological activity of unexplored natural sources. By understanding and accepting the complexity of chemical compounds from natural sources, it is found that natural products are appropriate to give the centre platforms for future medications and advancements in the utilization of novel natural sources [7]. It is assumed that the risk connected with herbal products is minimal. Even though the number of adverse reactions to herbal medication is quite low, somehow, the reports of reactions point to the need for marker systems that are excellent for isolating and identifying particular components. For the identification and authentication of herbal medicinal components, marker-based standards are becoming increasingly common. Herbal medicines, on the other hand, are made from the extract of plants and require a multi-marker system to be implemented. The impact of storage conditions is a crucial consideration. A pharmacopeial comparison of standards, in general, demonstrates that there are a lot of variances in plant types, which include factors like quality and quantity assurance, specified limitation of pesticides, heavy metals, and microbial contamination in many nations. Hence, standards that are peculiar to each country, as well as regional guidelines, have been developed, and only a few of them are accepted at this time [8]. In this chapter, things will be clearer about growing natural product use, and its effect on brain health.

13.2 TRADITIONAL PREPARATIONS AND FORMULATIONS

Medicinal herbs are potential sources of therapeutic aids. Natural products from natural origin (plant, animal, and minerals) have not been used only on the basis of the treatment of disease but also play an important role in the development and function of humans and animals. Tribal healers in most of the countries are using various health tonics along with various formulations to treat diseases [9].

A healthy diet is the most important part of a mother during her pregnancy for the function and development of her child. It is important that all necessary energy and nutrients are absorbed from the diet during the development of brain structures [10]. At the fetal brain development stage, a healthy diet influences the formation of myelin sheaths, which play an important role in signal transduction. At the age of one or two years, neurons and neuronal connections are produced at peak levels [11]. During gestation and early childhood, malnutrition induces the development of the brain. Dietary components directly influence the action of neurotransmitters. The amino acids tryptophan and tyrosine are the precursors of serotonin, dopamine, and norepinephrine, respectively [12]. Tyrosine supplementation also enhances the cognitive performance of the brain [13]. Brain foods also help to optimize mental performance [14]. Micronutrients, vitamins, and trace elements are essential during early brain development [15]. Several micronutrients, vitamins, and related trace elements, including folate, vitamin B12, and choline, are related to homocysteine concentrations, providing a link to 1-carbon metabolism, biosynthesis of phospholipids, and DNA methylation, which are involved in brain development [16].

A lot of preparations or formulations have been used since ancient times for the purpose of a healthy brain. Traditional healers played a significant role in preparing brain tonics and selecting food supplements at that time. The commonly used preparation and formulations are as follows:

13.2.1 Chyawanprash

It is an Ayurvedic health tonic, which is a polyherbal formulation containing almost 50 herbal extracts. Chyawanprash is very popular in Indian households. The memory impairment protection effect of Chyawanprash is confirmed by performing animal studies. The most significant ingredients of Chyawanprash are Amla (*Phyllanthus emblica),* which is a rich natural source of vitamin C, clove bud (*Eugenia caryophyllus*), which is composed of Eugenol and used as carminative as well as flavouring agents, ginseng (*Panax ginseng*), which is composed of gensenosides and act as antioxidants [17]. Furthermore, it decreases thiobarbituric acid levels in the brain and increases the levels of glutathione. Thus, it is protected from free radicals [18].

13.2.2 Abana

It is an Indian Ayurvedic polyherbal formulation that contains 16 herbal extracts. It significantly reduces the activity of cholinesterase, which results in to increase in acetylcholine levels in the mice brain. The ingredients of Abana reflect the antioxidant properties. Abana helps to reduce brain impairment and improves functions and the development of the brain. The influence of orally administered Abana on cognitive functions was monitored in groups of aged and young mice. Testing of memory was performed by using exteroceptive behavioural models (passive avoidance apparatus and elevated plus-maze test) and interoceptive behavioural models (amnesia induced by diazepam, scopolamine, and ageing) [19].

13.2.3 Bramhi Ghrita

It contains four Ayurvedic herbs, including Brahmi (*Bacopa monnieri*), ghee, minerals, and metals. One of the most popular medicines is Brahmi Ghrita, a well-known memory booster. In the folk of Indian medicine, it has been used traditionally as brain or nerve tonics (Sahoo et al., 2014). The memory and learning effect of Brahmi Ghrita was performed in rodents in different concentrations. Evaluation of memory acquisition and interpreting spatial memory were tested using the elevated plus-maze test in rats and the Morris water maze test in mice, respectively [20].

13.2.4 Memorin

It is a polyherbal capsule that contains 21 kinds of herbal extracts. The major herbs of the formulation are *Mandookparni, Shankhpushpi, Yashtimadhu,* and *Jatamansi*. Memorin was performed using a passive avoidance learning paradigm in a shuttle box on rats. Memorin was evaluated by electroconvulsive shocks (ECS) induced amnesia [21].

13.2.5 Brain-o-brain

It is a polyherbal capsule that contains a mixture of herbs, namely Shankhpushpi, Brahmi, and Sweet Flag. to improve learning ability, and memory power and naturally helps to avoid stress and depression. This herbal preparation is made specially to decrease forgetting. Brain-o-brain is widely used to increase the power and natural ability of the brain [22].

13.2.6 Shimotsu-to

It is Japanese–Sino medicine consisting of four crude extracts in the same weight ratio prepared by decoction process. Japanese angelica root, peony root, Rehmannia root and cnidium rhizome are used in the formulation, which describes traditional Chinese medicine. Spatial memory improvement was studied in rat models by using an extract of peony root and an extract from cnidium rhizome. Thus, Shimotsu-to enhancers of cognitive functions [23].

13.3 ROLE AND IMPLEMENTATION OF VARIOUS ETHNOMEDICINAL PLANTS AS A BRAIN TONIC

Over the century, numerous ethnomedicinal plants have been used to treat neurodegenerative diseases and cognitive disorders. The list of plants, such as *Bacopa monnieri, Centella asiatica, Cannabis sativa, Glycyrrhiza glabra, Ginkgo biloba,* and *Tinospora cordifolia* enhance mental functions such as motivation, memory, intelligence, attention, and concentration. A list of ethnomedicinal plants that improve brain function as well as brain health is mentioned in Table 13.1.

13.4 CURRENT SCENARIO OF BRAIN HEALTH AND AVAILABLE MARKETED PRODUCTS

Disruptions in normal brain growth and/or functioning describe brain health disorders that develop in a person's life. Brain tumours, intellectual developmental disorders, epilepsy, dementia, autism spectrum disorders, cerebral palsy, Parkinson's disease, cerebrovascular disease, multiple sclerosis, neuro infections, traumatic injury, headache and neurological disorders caused by malnutrition are some of the conditions that they can manifest as. Active engagement of people with these neurological conditions and their families and carers, as appropriate, is required for health and social care for these conditions. On July 22, World Brain Day (WBD) is observed to honour the founding of the World Federation of Neurology (WFN) in 1957. Its goal is to raise neurological disease awareness and advocacy [48]. The worldwide burden of neurological diseases is significant, with nations having low and high income bearing over 70% of the burden. Neurological diseases are the major cause of disability-adjusted life years (DALYs) and the second biggest cause of death worldwide, with approximately 9 million deaths annually. In 2016, meningitis (7.9%), dementia (10.4%), migraine (16.3%) and Stroke (42.2%) were the four leading causes of DALYs, followed by epilepsy (5%). Parkinson's disease is the fastest-growing neurological condition, fuelled by an ageing population. Premature delivery, neuro infections and neonatal encephalopathy are especially prevalent in Southeast Asia and Africa, where the burden of neurological diseases in children is particularly high.

Even with the significant burden of brain diseases, economically low nations (28%) have a lower percentage of specialized policies for neurological disease as compared to high-income nations (64%). In most nations, funding for these disorders is insufficient, and many neurological disease conditions have unacceptable treatment gaps. Recent available marketed products for the improvement of a healthy brain are mentioned in Table 13.2.

13.5 THE INCLINATION FOR NATURAL PRODUCTS OVER SYNTHETICS

There are many reasons for the inclination of natural products over synthetic ones, such as:

Chemists might be just as innovative as a natural ecosystem, but ecological processes have been doing so for millennia. The introduction of new biological molecular skeletons and functionalities not previously thought up by humans seems to be the most crucial and noticeable price of natural product chemistry [24]. It is impossible to quantify the extent of this advantage, but according to a recent assessment, natural products have a 100-fold higher hit rate than synthetic products [61].

Secondary metabolites are organic chemicals produced by plants that are not involved directly in the organism's regular growth, development, or reproduction. Instead, they usually regulate ecological functions, which might give the organism a selection advantage by enhancing its survivorship. Regardless of their true function, it is becoming obvious that several natural bioactive

TABLE 13.1
List of ethnomedicinal plants improve brain health

Sl No	Name of Plant	Synonym	Family	Implementation	Reference
1	*Bacopa monnieri* (Linn.) Wettst	Indian pennywort water hyssop and Jalneem	*Plantaginaceae*	It is very useful in the recovery of memory power and improves learning memory and mental concentration. It is taken with 'kuth' to cure epilepsy and hysteria.	[24,25]
2	*Achyranthes aspera* Linn.	Prickly chaff-flower, Apamaga, Chirchita, and Ovotasodh	*Amaranthaceae*	The powder of the seed gives relief from stiffness and headache of migraine. It is used as an immuno-modulator.	[26]
3	*Allium cepa* Linn.	Onion, Piyaz, Kanda	*Liliaceae*	It reduces neurological deficit scores and increases grasping ability, motor performance, and grip strength.	[27]
4	*Brassica nigra* Linn.	Black mustard, Raee	*Brassicaceae*	It is used in memory, thinking, and cognitive skills. *Brassica nigra* is used for eliminating memory loss in traditional medicine caused by β-amyloid.	[28]
5	*Calotropis procera* (Aiton) Dryand	Madar, Aak, Akwan, Swallow wort, Madar and Sodom apple	*Apocynaceae*	latex proteins of the plant have central nervous system depressant activity and anticonvulsant action in the treatment of seizures.	[29]
6	*Cassia occidentalis* Linn	Negro coffee, Kasaundi	*Caesalpiniaceae*	The plant is very useful in the treatment of hysteria and epilepsy. The seeds are used in the treatment of severe brain dysfunction on outbreaks of acute childhood illnesses.	[30]
7	*Centella asiatica* (L.) Urb	Thankuni, Gotu Kola	*Apiaceae*	It has a significant effect on the learning and memory process. It also decreases lipid peroxidation and augmentation of the antioxidant enzymes of the brain.	[31]
8	*Cannabis sativa* L.	Bhang	*Cannabinaceae*	It improves neuro-degeneration, loss of learning and memory. It is also used for sleeplessness.	[32]
9	*Cassia tora* L.	Panvad, Chakravada	*Caesalpiniaceae*	The plant has cognition-enhancing properties.	[33]
10	*Convolvulus pluricaulis* Choisy	Shankhapushpi, Shankahuli	*Convolvulaceae*	It is generally used as a brain tonic and brain stimulator. Daily intake also hinders memory loss.	[34]
11	*Brassica oleracea* L.	Hari gobhi,	*Brassicaceae*	It has potential to the management of cognitive dysfunction and the ability to reduce elevated acetylcholine (Ach) levels in the brain which improved memory.	[35]

TABLE 13.1 (Continued)
List of ethnomedicinal plants improve brain health

Sl No	Name of Plant	Synonym	Family	Implementation	Reference
12	*Aegle marmelos* (L.) Corrêa	Bengal quince, Shree Fal, holy fruit and golden apple	*Rutaceae*	It reduces acetylcholinesterase (AChE) inhibitory activity in the brain, which improves the cognitive deficit.	[36]
13	*Acorus calamus* L.	Sweet flag and sweet calomel	*Acoraceae*	Sweet flag increases the overall memory of the person.	[23]
14	*Eclipta prostrata* L.	Bhangara, Bhringraj	*Asteraceae*	According to the traditional system of medicine, it has beneficial effects on learning and memory.	[37]
15	*Nardostachys Jatamansi* (D.Don) DC	Spikenard, Jatamansi	*Caprifoliaceae*	It improves learning and memory power.	[38]
16	*Panax ginseng* C.A.Mey.	Asian ginseng, Chinese ginseng	*Araliaceae*	Traditionally it is used as a brain tonic.	[39]
17	*Glycyrrhiza glabra* L.	*Licorice*	*Leguminosae*	The roots and rhizomes of the plant are effective brain tonic, which enhances the CNS system circulation.	[40]
18	*Ginkgo biloba* L.	maidenhair tree, kew tree	*Ginkgoaceae*	It improves the circulation of blood into the central nervous system and increases the distribution of nutrients and oxygen into the brain. It also eliminates free radicals from the body and improves memory.	[41]
19	*Tinospora cordifolia* (Willd.) Miers	Guduchi and Giloy)	*Menispermaceae*	It has memory-enhancing properties.	[42]
20	*Myristica fragrans* Houtt.	Nutmeg	*Myristicaceae*	The seeds of the plant have been recommended for memory enhancement.	[43]
21	*Phyllanthus emblica* L.	Indian gooseberry and amla	*Phyllanthaceae*	Fruits have significant effects on improving learning memory with antioxidant potential. This plant extract is useful in the treatment of various cognitive disorders.	[44]
22	*Curcuma longa* L.	Turmeric and Haldi	*Zingiberaceae*	Fermented *Curcuma longa* regulates memory dysfunction. The test was performed using brain cell-line on scopolamine-treated mice.	[45]
23	*Zizyphus mauritiana* Lam.	Ber	*Rhamnaceae*	The extract of seeds has anxiolytic and sedative action. It modulates choline acetyltransferase enzymes.	[46]
24	*Punica granatum* L.	Pomgranate	*Lythraceae*	It improves learning and memory impaired by diabetes.	[46]
25	*Asparagus racemosus* Willd.	Shatavari, Shatamul	*Asparagaceae*	The extract of root improves cognitive performance.	[47]

TABLE 13.2
Natural marketed products for a healthy brain

Sl.no.	Brand name	Components	Function	Reference
1	Ustukhuddus Alavi	*Lavandula stoechas* L., *Anacyclus pyrethrum* (L.) *Lag, Polypodium vulgare* L., *Cuscuta epithymum* (L.), *Paeonia officinalis* L., and *Vitis vinifera* L.	In female high school students, it helps with cognitive function and cortisol levels.	[49]
2	BacoMind	*Bacopa monnieri*	Aids in the development of children's cognitive abilities.	[50]
3	Noocube	Alphaglycerylphosphorylcholine, Huperzine-A, *Bacopa monnieri*, Oat Straw, L-theanine, and L-tyrosine.	a. Memory improvement. b. Concentration increase. c. Communication abilities that have improved. d. Increased mental stamina. e. Multitasking abilities have improved.	
4	Hunter Focus	Phosphatidylserine, Ashwagandha, Lion's Mane Mushroom, Maritime Pine Bark, Bacopa, Citicoline, L-Tyrosine and L-Theanine.	a. Improved memory and concentration. b. Creativity and learning are enriched. c. Increased energy and positive attitude.	
5	Performance Lab Mind	Cognizin, Sharp-PS Green, Ajipure L-tyrosine and Maritime Pine Bark Extract.	a. Improves cognitive functions by supporting brain cells. b. Focus, concentration, vitality, and short-term memory are just a few of the mental advantages. c. Taking 1–2 nootropic pills on an empty stomach twice a day encourages blood flow to the brain, which supports healthy brain cells. d. Promotes optimal mental performance in high-pressure situations and comes with a 30-day money-back guarantee.	
6	Brain Pill	Citicoline, Tyrosine, Phosphatidylserine, Vitamin B12, L-theanine, Vitamin B6, *Bacopa monnieri*, Vitamin B9, *Gingko biloba*, Docosahexaenoic acid (DHA),	a. Defeat brain fog and mental exhaustion. b. Improve your memory and recall. c. Enhance your learning abilities. d. Reduces forgetfulness and boosts energy levels.	
7	BR-16A (Mentat)	Brahmi, Mandookaparni, Ashwagandha, Shankapushpi, Jatamansi, Vach, Tagar, Badam, Salap, Lavang, Pearl, Malkangni, and Sonth.	Rehabilitate memory and cognitive deficiencies caused by chronic disease and ageing.	[51]
8	Trasina	*Withania somnifera, Ocimum sanctum, Eclipta alba, Tinospora cordifolia, Picrorrhiza kurroa,* and shilajit.	Enhanced memory as well as cholinergic indicators such as choline acetyl transferase activity, acetylcholine concentration and muscarinic cholinergic receptor expression.	[52]

TABLE 13.2 (Continued)
Natural marketed products for a healthy brain

Sl.no.	Brand name	Components	Function	Reference
9	Memorin	Mandookparni, Shankhpushpi, Jatamansi, Yashtimadhu, and Smruti sagar.	Positive effects in older people suffering from age-related memory loss.	[53]
10	Bramhi Ghrita	*Bacopa monneri, Evolvulus alsinoids, Acorus calamus, Saussurea lappa,* and cow's ghee.	Improved memory and learning.	[20]
11	Ben-Cha-LoKa-Wi-Chian	Unknown	Anti-pyretic and antinociceptive.	[54]
12	Wanderer plus	*Paeonia lactilora, Mentha haplocalyx, poria cocos fungus, Anglica sinensis, Atractylodes macroephala, Zingiber oficinale, Glycyrrhiza Paeonia suffruticosa, Gardenia jasminoides, and Bupleurum chinense.*	Depressive disorder.	[55]
13	Monavvem	saffron, lettuce, opium poppy, frankincense, agarwood and sugar.	Treat insomnia.	[56]
14	Itrifal Muqawwi-e-Dimagh (IMD)	*Emblica officinalis Gaertn., Terminalia chebula Retz., Terminalia bellirica Roxb., Amygdalus communis Linn., Papaver somniferum Linn, Coriandrum sativum Linn., Rosa damascene Mill,* Cane sugar, Clarified butter, and silver.	Treat Parkinson's Disease (PD).	[57]
15	POL-6	*Bacopa monnieri, Hypericum perforatum, Centella asiatica, Withania somnifera, Camellia sinesis,* and *Ocimum sanctum.*	Antianxiety effect in alcohol withdrawal.	[58]
16	Bo-yang-Hwan-O Tang (mBHT)	*Astragali Radix, Angelicae Gigantis Radix, Paeoniae Radix Rubra, Cnidii Rhizoma, Lumbricus, Percicae Semen* and *Carthami Flos.*	Neuroprotective property.	[59]
17	Xtend®	Natural royal jelly, Panax ginseng root, Clove powder, Clove oil, Cinnamon oil, Ginger oil, Nigella sativa oil and Capsicum powder and sertraline.	Antistressor and antianxiety.	[60]
18	Brain-o-brain	Sweet flag, Shankhpushpi, and Brahmi.	Enhances learning capacity, brain capacity, and helps to naturally prevent psychological distress.	[23]
19	Chyawanprash	Out of the approximately 50 plants and herbal extracts utilized in the recipe, amla is the most essential ingredient.	Protective effect against memory loss.	[23]
20	Optimized-Sopung Sunkiwon (OSS)	SopungSunkiwon has six therapeutic plants.	By inhibiting AChE activity, it enhances memory performance.	[59]
22	Shimotsu-to	Japanese angelica root, cnidium rhizome, Rehmannia root, and. peony root.	Enhancement of memory function.	[23]

compounds, like drugs, may reach receptor-active sites. The huge number of purely natural products discovered to interact with distinct mammalian receptors stands as a testament to natural products that its inherent bioactivity. For example, natural plant product ligands for the GABA receptor comprise bicuculline [62], picrotoxin [63], etc.

Natural products can be used as prototypes for combinatorial chemistry, allowing for the creation of libraries of natural substance analogues with improved drug-like characteristics (e.g., solubility, pharmacokinetics, bioavailability). This is a dramatic development because combinatorial chemistry was formerly thought to be likely to substitute the need to find new natural compounds [64].

The regulation of natural product biosynthesis can be optimized as our knowledge regarding genetics and the biosynthetic pathway of natural products improves natural product biosynthesis and can also be modified to produce novel derivatives with potentially better quality and quantity [65]. Genome mining would have a significant influence on the development and modification of natural compound production to identify the novel natural compound.

Synthetic medications are often the result of several structural transformations throughout a lengthy drug development cycle, whereas a natural compound might go from 'hit' into a drug in a relatively short time. Microbial natural compounds are noteworthy not only for their therapeutic potential capabilities but also for having the appropriate pharmacokinetic qualities essential for clinical testing.

13.6 RECENT STATUS OF RESEARCH AND DEVELOPMENT OF NATURAL BRAIN TONIC

The insight into biological complexes at the molecular level has advanced dramatically as a result of advances in computational areas, particularly in the subject of bioinformatics. Various computational tools, such as silico docking methodologies, could be used to examine enzyme interactions with various substrates, protein synthesis, ubiquitination, breakdown of multiple physiological macromolecules, and many other activities. Under the guidance of the ANS, regular homeostasis encompasses metabolic balance, which is related to many complicated biochemical pathways, and also the anticipation of disease states and therapeutic potential solutions.

In silico approaches were utilized to investigate the molecular behaviour of various traditional medicines for the treatment of Parkinson's disease. Chemical constituents such as stigmasterol, wogonin, emodin, curcumin, eriodyctiol, and baicalin are been found to have a binding affinity of roughly 7 kcal/mol, which was equivalent to tolcapone (a synthetic psychotropic substance used to improve levodopa medical intervention), noting that amentoflavone (obtain from Ginkgo biloba) and ginsenoside (obtain from Panax ginseng) are considered to be very good COMT inhibitors and also a desirable substance needed for L-dopa. Furthermore, [66] used a comparative homology model and ligand docking to examine human COMT to create anti-PD drugs.

The development and advancement of nanotechnology offer promising signs of being able to overcome the problem of BBB. Because of their modified Physico-chemical properties, nanoparticles have a size range from 1 to 100 nm, which enhances activity like specific transport, efficacy, cellular uptake, etc. The component which is formed from the conjugation of plant constituent and the nanoparticle is called phytomedicine, which has recently gained importance in the development of new neuro-therapeutics because of their low adverse effect, wide availability and efficiency of targeted drug delivery [67].

The spirocyclic alkaloid rhynchophylline (RIN) is produced from *Uncaria* species and has been reported to have several pharmacological actions, including neuroprotective properties. In Alzheimer's disease, hyperexcitability induced by soluble A in hippocampus neurons is reduced by RIN [68]. Despite advancements in natural products, there are a few drawbacks, e.g., there is a chance that NP will be absorbed in the blood or other parts of the body rather than the targeted site. Another limitation noted would be that the therapy could not be tracked efficiently and effectively, posing a significant problem for practitioners who would be unable to follow the treatment's progress and effects. Furthermore, not only is the therapy and technique ineffective, but it is also exceedingly expensive.

13.7 CONCLUSION

Plants have been used in the form of spices or foods for many years, and it was later found that they play a critical role in improving bad memory. Various medicinal plants are utilized all over the world (including TCM of China, Ayurveda of India, African medicines, etc.), and they have demonstrated significant memory and learning abilities on animal models. There is a range of phytoconstituents present in the plants, which, through enhancing cognition, intelligence, focus, and concentration, ensures that the neurotransmitter acetylcholine is at its optimal level inside the brain by allowing for controlled brain activity. Various polyherbal mixtures show effective action on CNS; even in preclinical trials, these herbs seem to have an efficient therapeutic effect. The plants' formulation shows an effect on different animal models in terms of diseases like dementia, PD, amnesia, etc. It can be assumed that in the near future, phytoconstituents can be used as an effective formulated drug for the treatment of various brain illnesses.

REFERENCES

1. Jeyam, Muthusamy, GR Raj Karthika, Vasudevan. Poornima, and Manoharan. Sharanya. *Asian Journal of Pharmaceutical and Clinical Research* 5, no. 4 (2012): 125–128.
2. Rizig, Mie, Njideka Okubadejo, Mohamed Salama, Omotola Thomas, Albert Akpalu, and Riadh Gouider. *The Lancet Neurology* 20, no. 5 (2021): 335.
3. Wang, Yongjun, Yuesong Pan, and Hao Li. *BMJ,* 371 (2020).
4. Gorelick, Philip B., Karen L. Furie, Costantino Iadecola, Eric E. Smith, Salina P. Waddy, Donald M. Lloyd-Jones, Hee-Joon Bae et al. *Stroke* 48, no. 10 (2017): e284–e303.
5. Budin-Ljøsne, Isabelle, Barbara Bodorkos Friedman, Sana Suri, Cristina Solé-Padullés, Sandra Düzel, Christian A. Drevon, William FC Baaré et al. *Frontiers in Public Health* 8 (2020): 387.
6. Dewick, Paul M. *Medicinal Natural Products: A Biosynthetic Approach.* John Wiley & Sons, 2002.
7. Harvey, Alan L. *Drug Discovery Today* 13, no. 19–20 (2008): 894–901.
8. Sahoo, Niharika, Padmavati Manchikanti, and Satyahari Dey. *Fitoterapia* 81, no. 6 (2010): 462–471.
9. Verma, Sheetal, and S. P. Singh. *Veterinary World* 1, no. 11 (2008): 347.
10. Benton, David. *European Journal of Nutrition* 47, no. 3 (2008): 25–37.
11. Keunen, Kristin, Ruurd M. Van Elburg, Frank Van Bel, and Manon JNL Benders. *Pediatric Research* 77, no. 1 (2015): 148–155.
12. Young, Simon N. *Journal of Psychiatry & Neuroscience* 32, no. 6 (2007): 394.
13. Jongkees, Bryant J., Bernhard Hommel, Simone Kühn, and Lorenza S. Colzato. *Journal of Psychiatric Research* 70 (2015): 50–57.
14. Lanni, Cristina, Silvia C. Lenzken, Alessia Pascale, Igor Del Vecchio, Marco Racchi, Francesca Pistoia, and Stefano Govoni. *Pharmacological Research* 57, no. 3 (2008): 196–213.
15. Mattei, Davide, and Angelo Pietrobelli. *Current Nutrition Reports* 8, no. 2 (2019): 99–107.
16. Moretti, Rita, and Paola Caruso. *International Journal of Molecular Sciences* 20, no. 1 (2019): 231.
17. Narayana, DB Anantha, Sharanbasappa Durg, P. Ram Manohar, Anita Mahapatra, and A. R. Aramya. *Journal of Ethnopharmacology* 197 (2017): 52–60.
18. Vasudevan, Mani, and Milind Parle. *Physiology & Behavior* 91, no. 1 (2007): 46–54.
19. Parle, Milind, and Mani Vasudevan. *Journal of Health Science* 53, no. 1 (2007): 43–52.
20. Yadav, Kapil Deo, K. R. C. Reddy, and Vikas Kumar. *Ayu* 35, no. 3 (2014): 325.
21. Sankpal, Shital D., Ashish K. Mullani, Shrinivas K. Mohite, and John I. D'Souza. *International Journal of Bioassays* 3, no. 9 (2014): 3265–3269.
22. Halder S, Anand U, Nandy S, Oleksak P, Qusti S, Alshammari EM, Batiha GE, Koshy EP, Dey A. *Saudi Pharmaceutical Journal.* 2021 Aug 1;29(8):879–907.

23. Halder, Swati, Uttpal Anand, Samapika Nandy, Patrik Oleksak, Safaa Qusti, Eida M. Alshammari, Gaber El-Saber Batiha, Eapen P. Koshy, and Abhijit Dey. *Saudi Pharmaceutical Journal* 29, no. 8 (2021): 879–907.
24. Dubey, Tushar, and Subashchandrabose Chinnathambi. “Brahmi (Bacopa monnieri): An ayurvedic herb against the Alzheimer’s disease.” *Archives of Biochemistry and Biophysics* 676 (2019): 108153.
25. Uabundit, Nongnut, Jintanaporn Wattanathorn, Supaporn Mucimapura, and Kornkanok Ingkaninan. *Journal of Ethnopharmacology* 127, no. 1 (2010): 26–31.
26. Khajuria, Arun Kumar, R. K. Manhas, Harish Kumar, and N. S. Bisht. *Journal of Ethnopharmacology* 276 (2021): 114204.
27. Saher, Sana, Syed Ayaz Ali, Nikhil S. Sakle, and Subur W. Khan. *International Journal of Pharmaceutical Research and Allied Sciences* 4 (2015): 59–68.
28. Nazari, Elahe, Mahnaz Khanavi, Leili Amani, Mohammad Sharifzadeh, Mahdi Vazirian, Mina Saeedi, Mehdi Sanati, and Seyede Nargess Sadati Lamardi. *Pharmaceutical Sciences* 26, no. 3 (2020): 261–269.
29. Paul, Amandeep, and Antul Kumar. *International Journal of Economic Plants* 5, no. 3 (2018): 157–162.
30. Yadav, J. P., Vedpriya Arya, Sanjay Yadav, Manju Panghal, Sandeep Kumar, and Seema Dhankhar. *Fitoterapia* 81, no. 4 (2010): 223–230.
31. Belwal, Tarun, Harish C. Andola, Maria S. Atanassova, Bhasker Joshi, Renu Suyal, Shinny Thakur, Arti Bisht, Arvind Jantwal, Indra D. Bhatt, and Ranbeer S. Rawal. pp. 265–275. Academic Press, 2019.
32. Zhou, Yuefang, Shanshan Wang, Jianbo Ji, Hongxiang Lou, and Peihong Fan. *ACS omega* 3, no. 11 (2018): 15988–15995.
33. Malabade, Rohit, and Taranalli Ashok. *Journal of Young Pharmacists* 7, no. 4s (2015): 455.
34. Bhowmik, Debjit, KP Sampath Kumar, Shravan Paswan, Shweta Srivatava, A. Yadav, and Amitsankar Dutta. *J Pharmacogn Phytochem* 1, no. 1 (2012): 44–51.
35. Agrawal, Aashruti, Ankur Joshi, Sapna Malviya, and Anil Kharia. *Journal of Drug Delivery and Therapeutics* 9, no. 2-A (2019): 81–84.
36. Lomate, Krushna A., Krishna Murthy, Vishal S. Adak, and Rajkumar VV Shete. *Journal of Drug Delivery and Therapeutics* 11, no. 2-S (2021): 162–166.
37. Vaidya, A. B. *Indian Journal of Pharmacology* 29, no. 5 (1997): 340.
38. Li, Ran, Zhi-Mei Wang, Yan Wang, Xueqi Dong, Li-Hua Zhang, Tao Wang, Yan Zhu, Xiu-Mei Gao, Hong-Hua Wu, and Yan-Tong Xu. *Journal of Ethnopharmacology* 268 (2021): 113601.
39. Ranade, Anagha, Mayur Surana, Shivani V. Dhokne, Sudesh Gaidhani, and Sharad D. Pawar. In *Medicinal Herbs and Fungi*, pp. 445–467. Springer, Singapore, 2021.
40. Rathee, Permender, Hema Chaudhary, Sushila Rathee, and Dharmender Rathee. *Pharmacognosy Reviews* 2, no. 4 (2008): 249.
41. Kumar Singh, Sandeep, George E Barreto, Gjumrakch Aliev, and Valentina Echeverria. *Current Drug Metabolism* 18, no. 2 (2017): 112–119.
42. Agarwal, Ashutosh, Shaya Malini, K. L. Bairy, and Muddanna S. Rao. *Indian Journal of Pharmacology* 34, no. 5 (2002): 339–349.
43. Omidpanah, Sahar, Yasaman Vahedi-Mazdabadi, Azadeh Manayi, Arezoo Rastegari, Roshanak Hariri, Elahe Mortazavi-Ardestani, Mahdieh Eftekhari, Mahnaz Khanavi, Tahmineh Akbarzadeh, and Mina Saeedi. *Natural Product Research* 36, no. 2 (2022): 610–616.
44. Uddin, Md Sahab, Abdullah Al Mamun, Md Sarwar Hossain, Farjana Akter, Mohammed Ashraful Iqbal, and Md Asaduzzaman. *Annals of Neurosciences* 23, no. 4 (2016): 218–229.
45. Eun, Cheong-Su, Jong-Soon Lim, Jihye Lee, Sam-Pin Lee, and Seun-Ah Yang. *BMC Complementary and Alternative Medicine* 17, no. 1 (2017): 1–12.
46. Mishra, Tulika, Alistair G. Paice, and Aruna Bhatia. In *Nuts and Seeds in Health and Disease Prevention*, pp. 733–739. Academic Press, 2011.
47. Majumdar, Shreyasi, Smriti Gupta, Santosh Kumar Prajapati, and Sairam Krishnamurthy. *Neurochemistry International* 145 (2021): 105013.
48. Wijeratne, Tissa, Wolfgang Grisold, David Dodick, and William Carroll. *The Lancet Neurology* 18, no. 10 (2019): 914.

49. Tefagh, Samane, Roshanak Mokaberinejad, Mehrdad Shakiba, Mahdi Jafari, Maryam Salehi, Maryam Khayatkashani, and Nezhat Shakeri. *Journal of Ethnopharmacology* 288 (2022): 114971.
50. Grebow, Jennifer. 2022. "Natural Remedies' Bacopa Brain-Health Ingredients Granted Natural Product Numbers from Health Canada". *Nutritional Outlook.* www.nutritionaloutlook.com/view/natural-remedies-bacopa-brain-health-ingredients-granted-natural-product-numbers-from-health-canada.
51. Bhattacharya, S. K., A. Kumar, and A. K. Jaiswal. *Fitoterapia (Milano)* 66, no. 3 (1995): 216–222.
52. Bhattacharya, S. K., and A. Kumar. *The Journal of Alternative and Complementary Medicine* 3, no. 4 (1997): 327–336.
53. Andrade, Chittaranjan, Srikante Gowda, and S. K. Chaturvedi. *Indian Journal of Psychiatry* 40, no. 3 (1998): 240.
54. Jongchanapong, Anusara, Chatubhong Singharachai, Chanida Palanuvej, Nijsiri Ruangrungsi, and Pasarapa Towiwat. *Journal of Health Research* 24, no. 1 (2010): 15–22.
55. Qin, Feng, Xiao-Ai Wu, Yao Tang, Qiang Huang, Zhang-Jin Zhang, and Jiu-Hong Yuan. *Journal of Psychiatric Research* 45, no. 11 (2011): 1518–1524.
56. Poursaleh, Zohreh, Ensieh Vahedi, Mina Movahhed, Mohammad Mahdi Ahmadian-Attari, Mahdi Jafari, Leila Mohammad Taghizadeh Kashani, Mahmood Khodadoost, and Amirhossein Sahebkar. *European Journal of Integrative Medicine* 49 (2022): 101608.
57. Siddique, Yasir Hasan, Falaq Naz, Mohammad Rashid, and Shariq Mian. *Phytomedicine Plus* 1, no. 4 (2021): 100131.
58. Sharma, Lalit, Aditi Sharma, Ashutosh Kumar Dash, Gopal Singh Bisht, and Girdhari Lal Gupta. *BMC Complementary Medicine and Therapies* 21, no. 1 (2021): 1–15.
59. Choi, Yookeum, Seul-Ki Kim, In-Young Choi, Chung Ju, Kung-Woo Nam, Sunyoung Hwang, Byung-Woo Kim et al. *Journal of Pharmacy and Pharmacology* 63, no. 5 (2011): 695–706.
60. El-Ashmawy, I. M., A. E. Bayad, and A. M. E. Omar. "Antidepressant activity of a polyherbal mixture in mice." *International Food Research Journal* 21, no. 6 (2014).
61. Berdy, Janos. *Journal of Antibiotics* 58, no. 1 (2005): 1–26.
62. Johnston, GAR, Beart PM, Curtis DR, Game CJA, McCulloch RM, Maclachlan RM. *Nature New Biology* 240, no. 102 (1972): 219–220.
63. Akaike, N., K. Hattori, Y. Oomura, and D. O. Carpenter. *Experientia* 41, no. 1 (1985): 70–71.
64. Lam, Kin S. *Trends in Microbiology* 15, no. 6 (2007): 279–289.
65. Bibb, Mervyn J. "Regulation of secondary metabolism in streptomycetes." *Current Opinion in Microbiology* 8, no. 2 (2005): 208–215.
66. Lee, Jee-Young, and Yang-Mee Kim. *Bulletin of the Korean Chemical Society* 26, no. 11 (2005): 1695–1700.
67. Bhattacharya, Tanima, Giselle Amanda Borges E. Soares, Hitesh Chopra, Md Mominur Rahman, Ziaul Hasan, Shasank S. Swain, and Simona Cavalu. *Materials* 15, no. 3 (2022): 804.
68. Xu, Ruiling, Junying Wang, Juanjuan Xu, Xiangrong Song, Hai Huang, Yue Feng, and Chunmei Fu. *International Journal of Nanomedicine* 15 (2020): 1149.
69. "Brain Health". 2022. *Who.Int.* www.who.int/health-topics/brain-health.
70. Azmi, Abdul Azim, Shahnawaz Jamali, Rafat Murad, and Aley Hasan Zaidi. *Pakistan Journal of Pharmacology* 27 (2010): 25–28.

Abbreviations

AChE Acetylcholine
AD Alzheimer's disease
AHA American Heart Association
ANS Autonomic nervous system
ASA American Stroke Association
BBB Blood-brain barrier
CNS Central nervous system

COMT Catecholamine-O-methyltransferase
DALYs Disability-adjusted life years
DNA Deoxyribonucleic acids
ECS Electroconvulsive shocks
GABA Gamma-aminobutyric acid
IMD Itrifal Muqawwi-e-Dimagh
L-dopa Levodopa
NUOA National University of Ostroh Academy
PD Parkinson's disease
RHB RUZU herbal bitters
RIN Rhynchophylline
Sp. Species
TCM Traditional Chinese Medicine
WBD World Brain Day
WFN World Federation of Neurology
WHO World Health Organisation

14 Effect of Edible Alkaloids on Memory and Behaviour

Kunal Bhattacharya, Dibyajyoti Das, Debabrata Nath, Nongmaithem Randhoni Chanu, and Ramen Kalita

14.1 INTRODUCTION

Alkaloids are naturally occurring organic chemicals that are often found in plants. They can also be found in marine algae and, in rare cases, animals. For example, it may be found in the deadly secretions of fire ants, ladybirds, and toads **[1]**, as well as over 150 plant species. Alkaloids account for almost 12,000 of the more than 50,000 natural chemicals known today **[2]**. In 1819, the term 'alkaloid' was coined by Meissner, a German pharmacist. Humans have been using alkaloids for over 4,000 years for a range of purposes such as pharmaceuticals, meals, cosmetics, poisons and so on, suggesting that alkaloids have an evolutionary relevance in physiological processes. They are chemically a varied group of compounds that all include at least one nitrogen atom. These are basic in nature and are referred to as alkaloids (alkali-like). In contrast to other phytochemicals, alkaloids have exceptional physiological effects on humans and other animals; they are active components in many medicinal plants and plant-derived medications. Their structural diversity and various physiological functions set them apart from any other category of natural products. Because of their physiological effects, alkaloids are important compounds in medicine. They are used to treat pain, stimulants, muscle relaxants, tranquillizers, anaesthetics, antimalarials, antibacterials, antidiabetics, anticancer, anti-HIV, antioxidants and other disorders.

The effects of alkaloids are as varied as the plants that make them, and many are palatable, healthy and essential components of the food we consume daily. Others, however, when consumed, can be intoxicating or even hazardous. However, the alkaloids present in food are generally not harmful, but in reality, a lot of the foods high in alkaloids are a crucial source of nutrients for us. Some foods that are rich in alkaloids are peppers (produce an alkaloid known as capsaicin), black pepper (piperine alkaloid), coffee and tea leaves (caffeine alkaloid), cinchona bark (quinine) and many more.

Alkaloids are well known to produce numerous physiological and psychological effects on the human brain. The first alkaloid that was identified and isolated by the French chemist Dersone in 1803 was narcotine from opium. In the same year, Serturner isolated morphine from opium. These two alkaloids that were discovered bind to the opioid receptors present in the brain, producing sleep, analgesic and hallucinogenic properties. The pharmacological or psychological effects of alkaloids are due to the similarity of structures of neurotransmitters in the human body and the alkaloids. Neurotransmitters released from the nerve endings bind to the receptors on the surface

DOI: 10.1201/9781003300502-14

of another neuron to produce physiological effects. Alkaloids can mimic these neurotransmitters to either stimulate or block the function to produce different psychological effects.

Memory is one of the most important factors in the human brain. It is necessary for the survival of an individual. It is the retention power of the brain to store information that the brain grasps. 'Memory is the ability of an individual to record sensory stimuli, events, pieces of information, etc., and retain them over short or long periods and recall the same whenever needed at a later date' **[3]**. The Indian traditional system of medicine, Ayurveda, is based on treating numerous diseases with a natural approach. Rasayana therapy, an integral part of Ayurveda, describes the uses of natural herbs and herbal medicine as a brain tonic to promote the brain's health and improve memory impairment behaviour.

The main factors that are responsible for good memory and cognitive function are the brain's hippocampus region and the level of acetylcholine. Poor memory, low retention power, difficulty recalling, lack of concentration, and weak analyzing ability are very common problems in the modern world. Besides, some of the conditions, such as stress, ageing, and emotions, may lead to memory impairment, amnesia, dementia, etc., and sometimes to some serious threats like Alzheimer's disease (AD), schizophrenia, etc. **[4]**.

The neuropharmacology of learning and memory is also significantly influenced by nicotinic acetylcholine receptors (nAChRs). Reports also claim that three known nicotinic alkaloids, such as nicotine, cotinine, and anatabine could suppress the scopolamine-induced memory deficit in rodents **[5]**. Huperzine A and B are alkaloids isolated from *Huperzia serrata* (Thunb.) that are reported to possess strong anticholinesterase activity to treat Alzheimer's dementia, schizophrenia and markedly increase efficiency in learning in animals **[6]**. Beta-carboline alkaloids were also found to stop or slow down the progressive effects of AD and Down syndrome **[7]**. In this chapter, the authors have discussed the psychological effects of edible alkaloids and their bioactive components that affect memory and behaviour.

14.2 LIST OF EDIBLE ALKALOIDS

Alkaloids are naturally occurring organic compounds found mostly in plants, especially marine algae, and only rarely in mammals. Alkaloids have extraordinary physiological effects on humans and other animals. These are the active ingredients in a variety of therapeutic plants. Many medications used by humans for both medicinal and non-medical purposes are created in nature as alkaloids, such as strychnine and caffeine. Many naturally occurring alkaloid receptors in humans and other animals have been found. The following is the list of edible alkaloids (Table 14.1).

143 BIOSYNTHESIS OF TRYPTOPHAN

Tryptophan is an indole alkaloid. Its biosynthesis in the plant starts with chorismate as the starting material (Figure 14.1).

14.4 BIOSYNTHESIS OF GALANTAMINE

Galantamine is a reversible acetylcholinesterase inhibitor and allosteric nicotinic receptor modulator. Its starting material is phenylalanine (Figure 14.2).

TABLE 14.1
List of edible alkaloids

Name	Chemical formula	Molecular weight	Structure	Reference
			INDOLE ALKALOIDS	
5-hydroxyindolin-2-one	$C_8H_7NO_2$	149.15	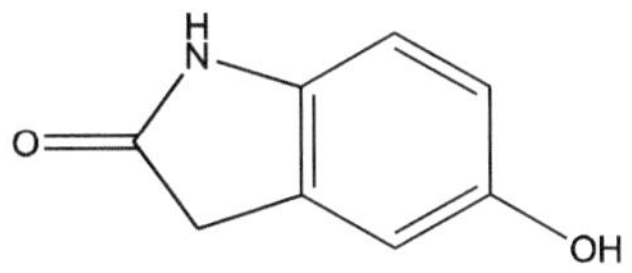	[7]
(1R,3S)-1-Methyl-1,2,3,4-tetrahydro-β-carboline-3-carboxylic acid	$C_{13}H_{14}N_2O_2$	230.26	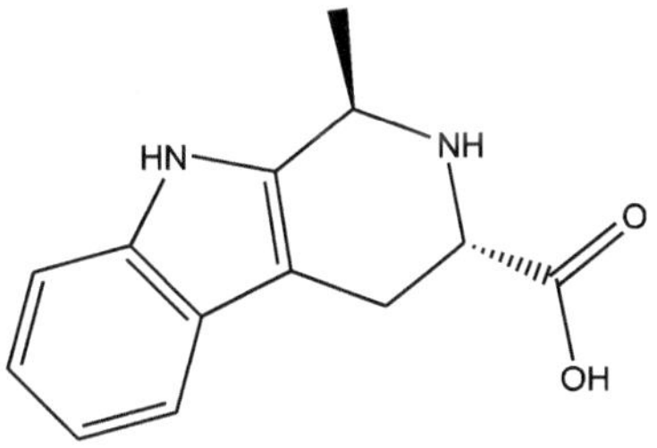	[7]
(3S)-1,2,3,4-tetrahydro-β-carboline-3-carboxylic acid	$C_{12}H_{12}N_2O_2$	216.24	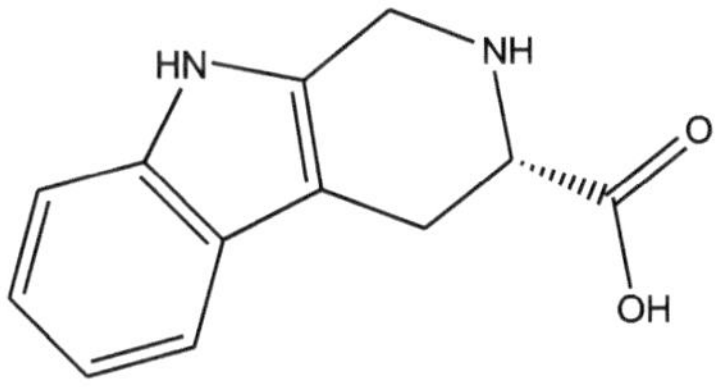	[7]
L-tryptophan	$C_{11}H_{12}N_2O_2$	204.23	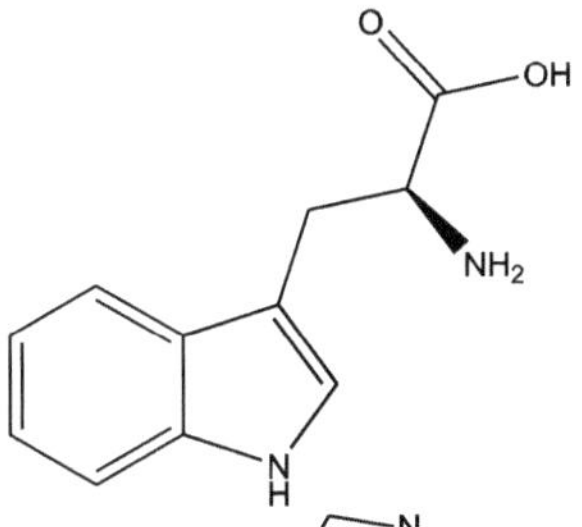	[7]
Strychnine	$C_{21}H_{22}N_2O_2$	334.4	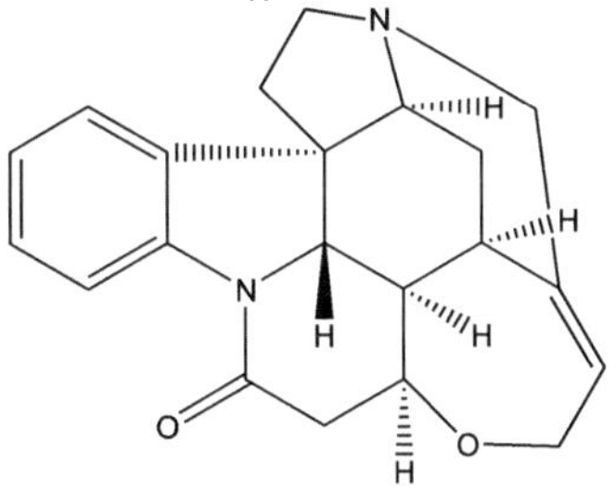	[7]
			PYRROLE ALKALOIDS	
5-(hydroxymethyl)-1-isobutyl-1*H*-pyrrole-2-carbaldehyde	$C_{10}H_{15}NO_2$	181.23	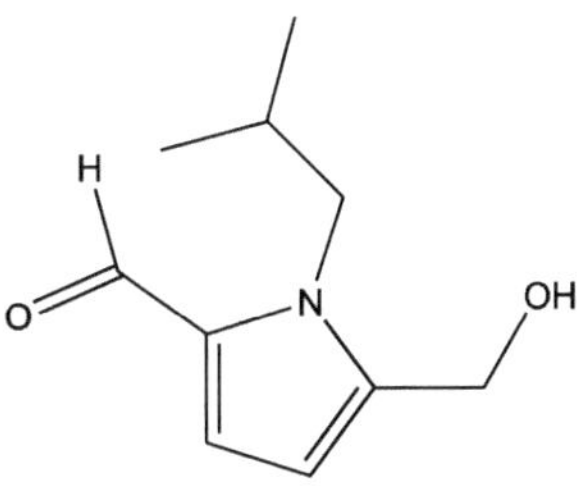	[8]

(*continued*)

TABLE 14.1 (Continued)
List of edible alkaloids

Name	Chemical formula	Molecular weight	Structure	Reference
5-(hydroxymethyl)-1-(2-methylbutyl)-1*H*-pyrrole-2-carbaldehyde	$C_{11}H_{17}NO_2$	195.26	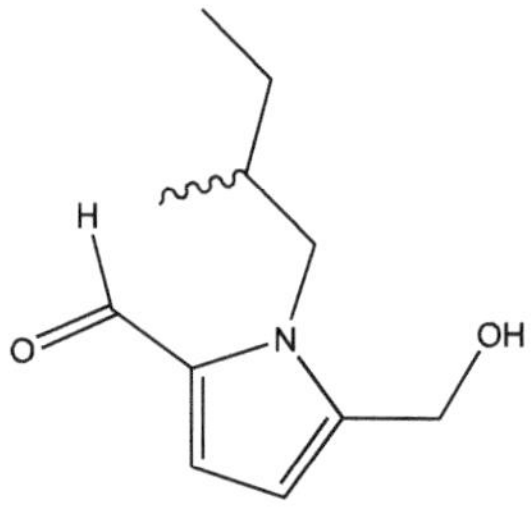	[8]
(R)-ethyl-2-(2-formyl-5-(methoxymethyl)-1*H*-pyrrol-1-yl) propanoate	$C_{12}H_{17}NO_4$	239.27	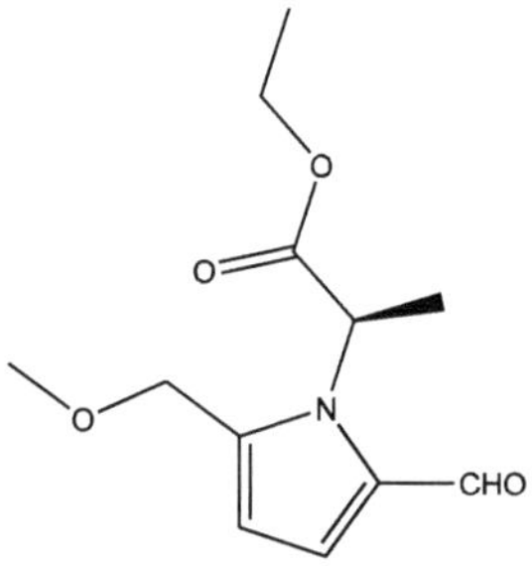	[8]
(R)-methyl-2-(2-formyl-5-(methoxymethyl)-1*H*-pyrrol-1-yl) propanoate	$C_{11}H_{15}NO_4$	225.24	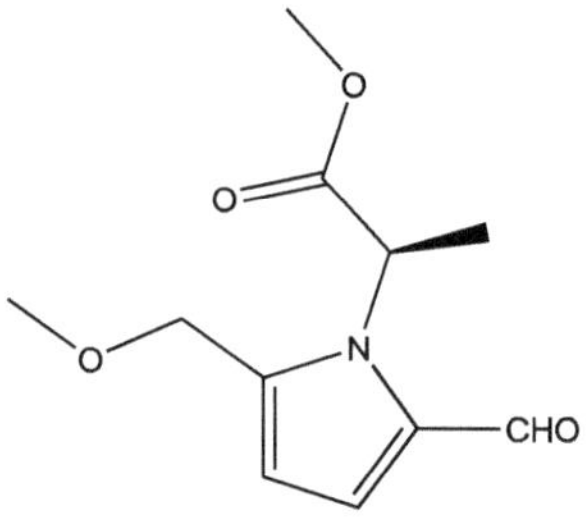	[8]
5-(hydroxymethyl)-1-phenethyl-1*H*-2-carbaldehyde	$C_{14}H_{15}NO_2$	229.27	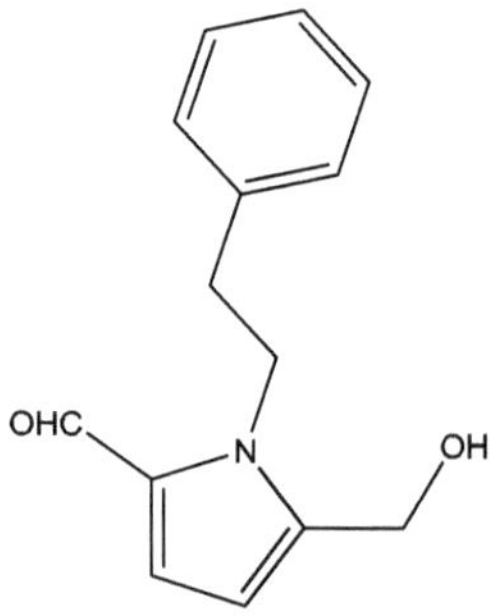	[8]
5-(hydroxymethyl)-1-isopentyl-1*H*-pyrrole-2-carbaldehyde	$C_{11}H_{17}NO_2$	195.26	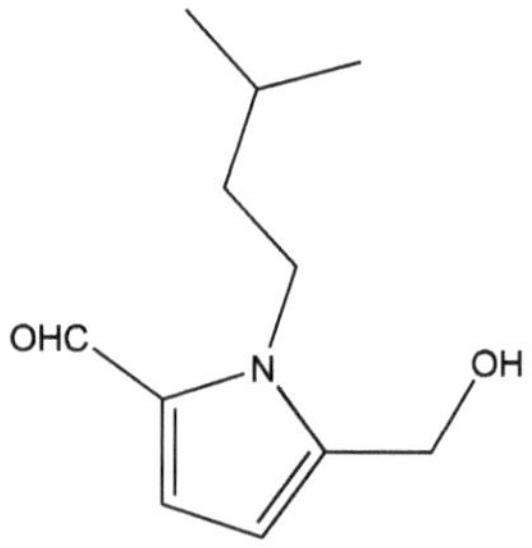	[8]

TABLE 14.1 (Continued)
List of edible alkaloids

Name	Chemical formula	Molecular weight	Structure	Reference
1-(2-(1*H*-indol-3-yl) ethyl)-5-(hydroxymethyl)-1H-pyrrole-2-carbaldehyde	$C_{16}H_{16}N_2O_2$	268.31	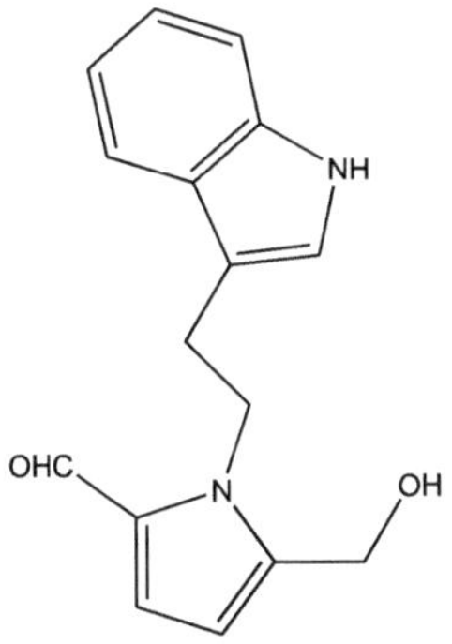	[8]
4-[2-formyl-5-(hydroxymethyl)-1*H*-pyrrol-1-yl] butanoic acid	$C_{10}H_{13}NO_4$	211.21	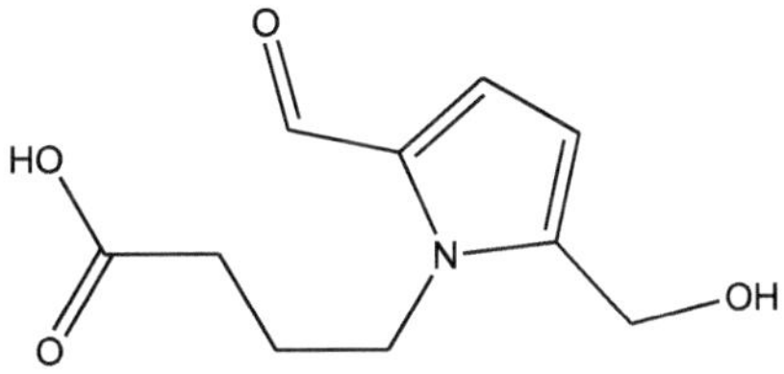	[9]
4-[2-formyl-5-(hydroxymethyl)-1*H*-pyrrol-1-yl] butanamide	$C_{10}H_{14}N_2O_3$	210.23	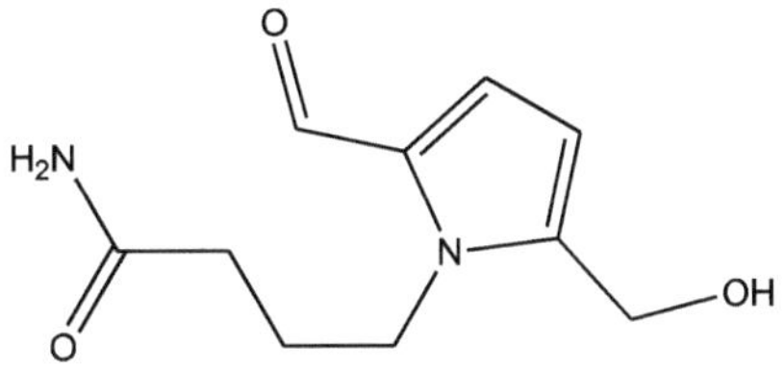	[9]
5-(hydroxymethyl)-1*H*-pyrrole-2-carboxaldehyde	$C_6H_7NO_2$	125.13	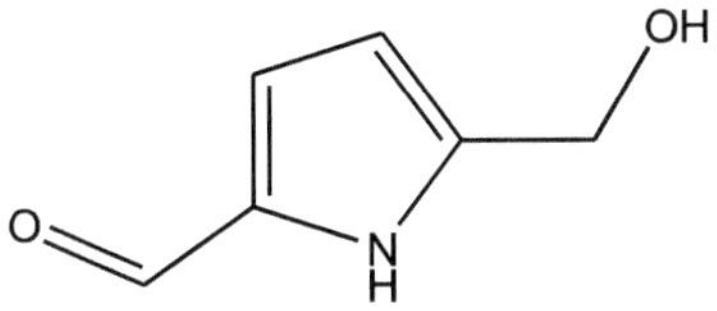	[9]
			CARBAZOLE ALKALOID	
(+)-Mahanine	$C_{23}H_{25}NO_2$	347.45	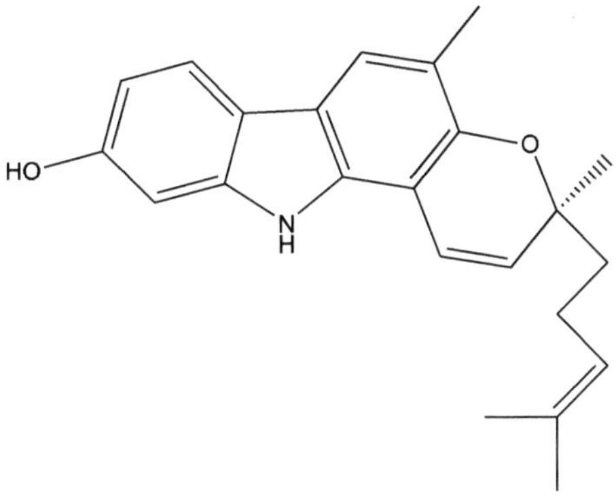	[10]

(*continued*)

TABLE 14.1 (Continued)
List of edible alkaloids

Name	Chemical formula	Molecular weight	Structure	Reference
Murrayanine	$C_{14}H_{11}NO_2$	225.24	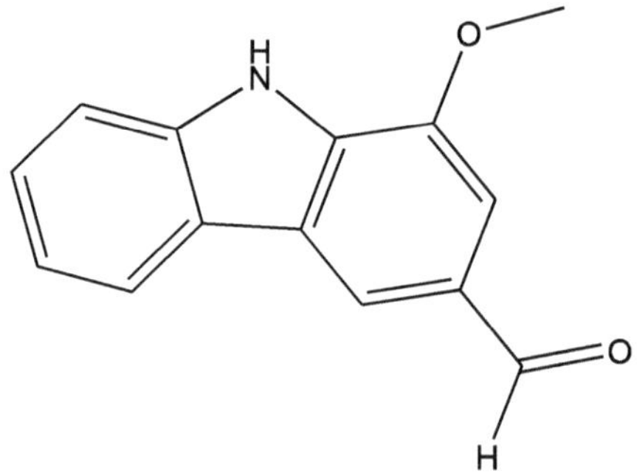	[11]
8,8'-Biskoenigine	$C_{38}H_{36}N_2O_6$	616.7	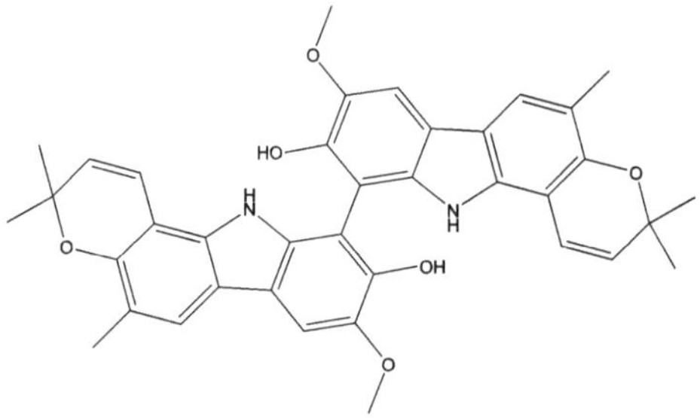	[11]
			β-CARBOLINE ALKALOIDS	
Harmol	$C_{12}H_{10}N_2O$	198.22	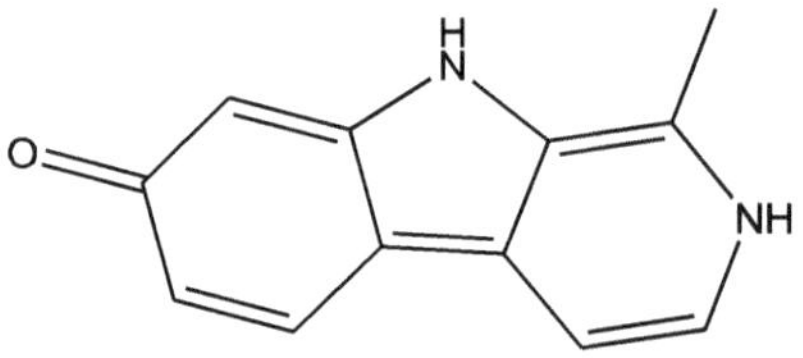	[12]
Norharman	$C_{11}H_8N_2$	168.19	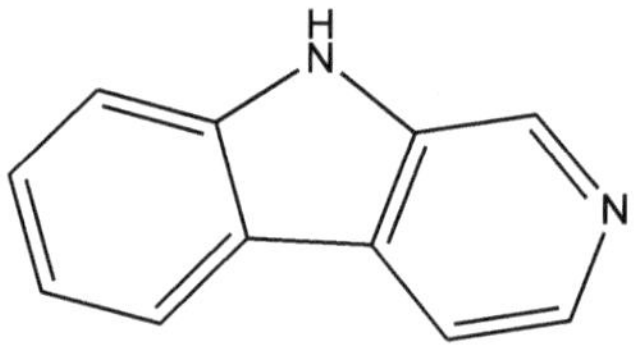	[12]
Harman	$C_{12}H_{10}N_2$	182.22	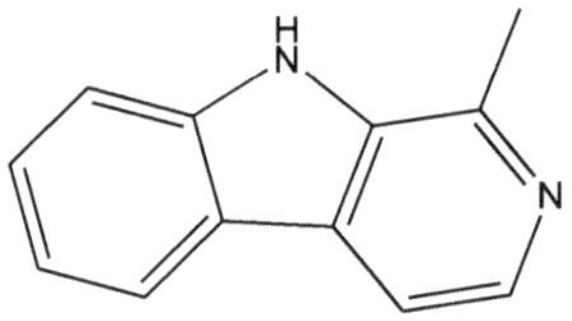	[12]
Harmine	$C_{13}H_{12}N_2O$	212.25	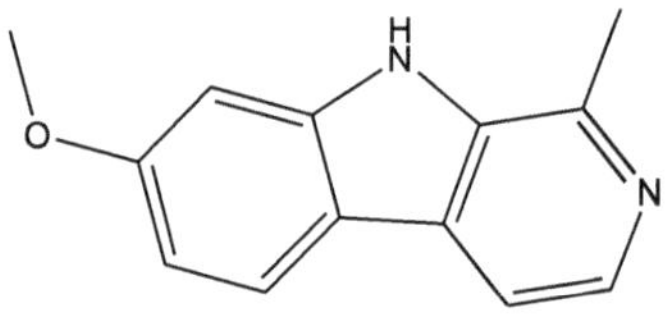	[12]

TABLE 14.1 (Continued)
List of edible alkaloids

Name	Chemical formula	Molecular weight	Structure	Reference
Harmalol	$C_{12}H_{10}N_2O$	198.22	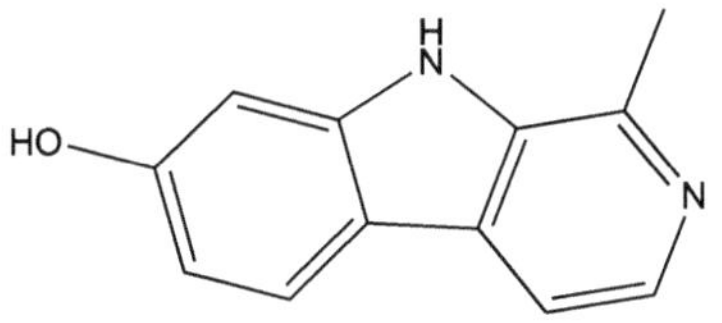	[12]
Harmaline	$C_{13}H_{14}N_2O$	214.26	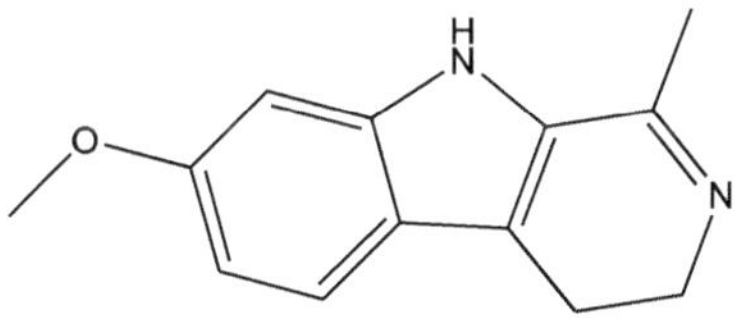	[12]
Tetrahydronorharman	$C_{26}H_{22}N_2O_2$	394.47	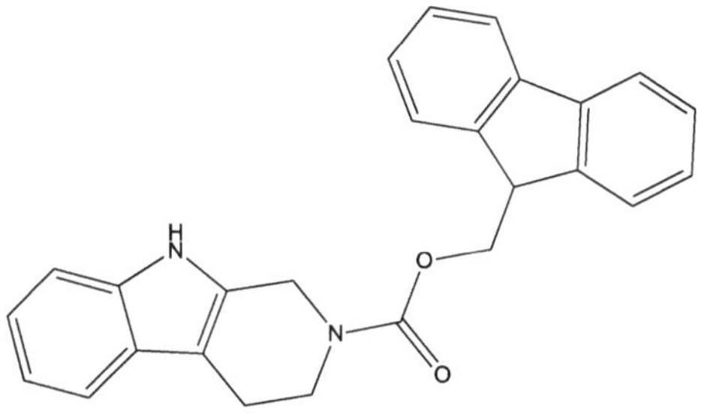	[12]
			TERTIARY ALKALOIDS	
Galantamine	$C_{17}H_{21}NO_3$	287.35	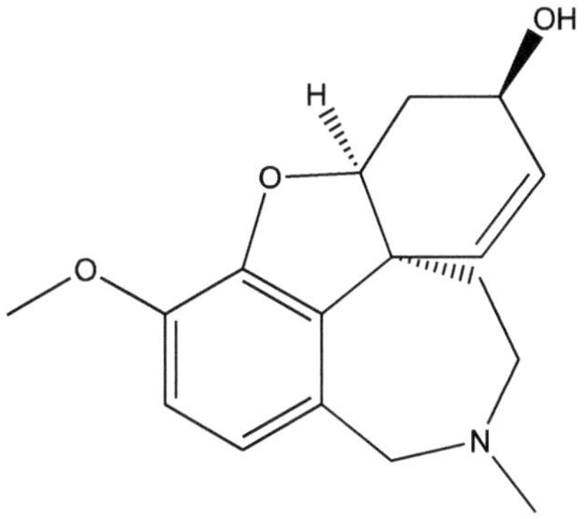	[13]
			PURINE ALKALOIDS	
Caffeine	$C_8H_{10}N_4O_2$	194.19	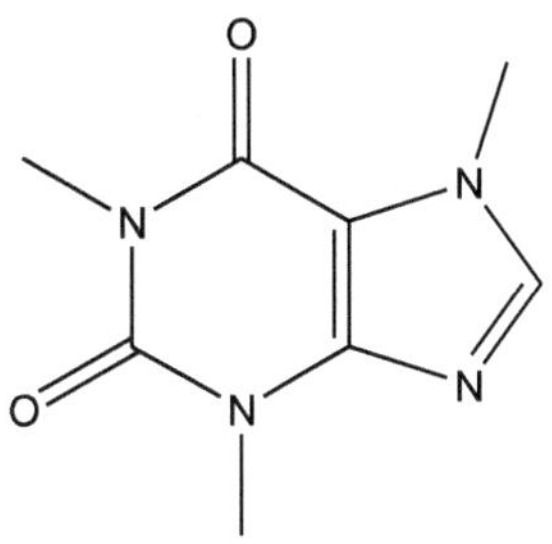	[13]

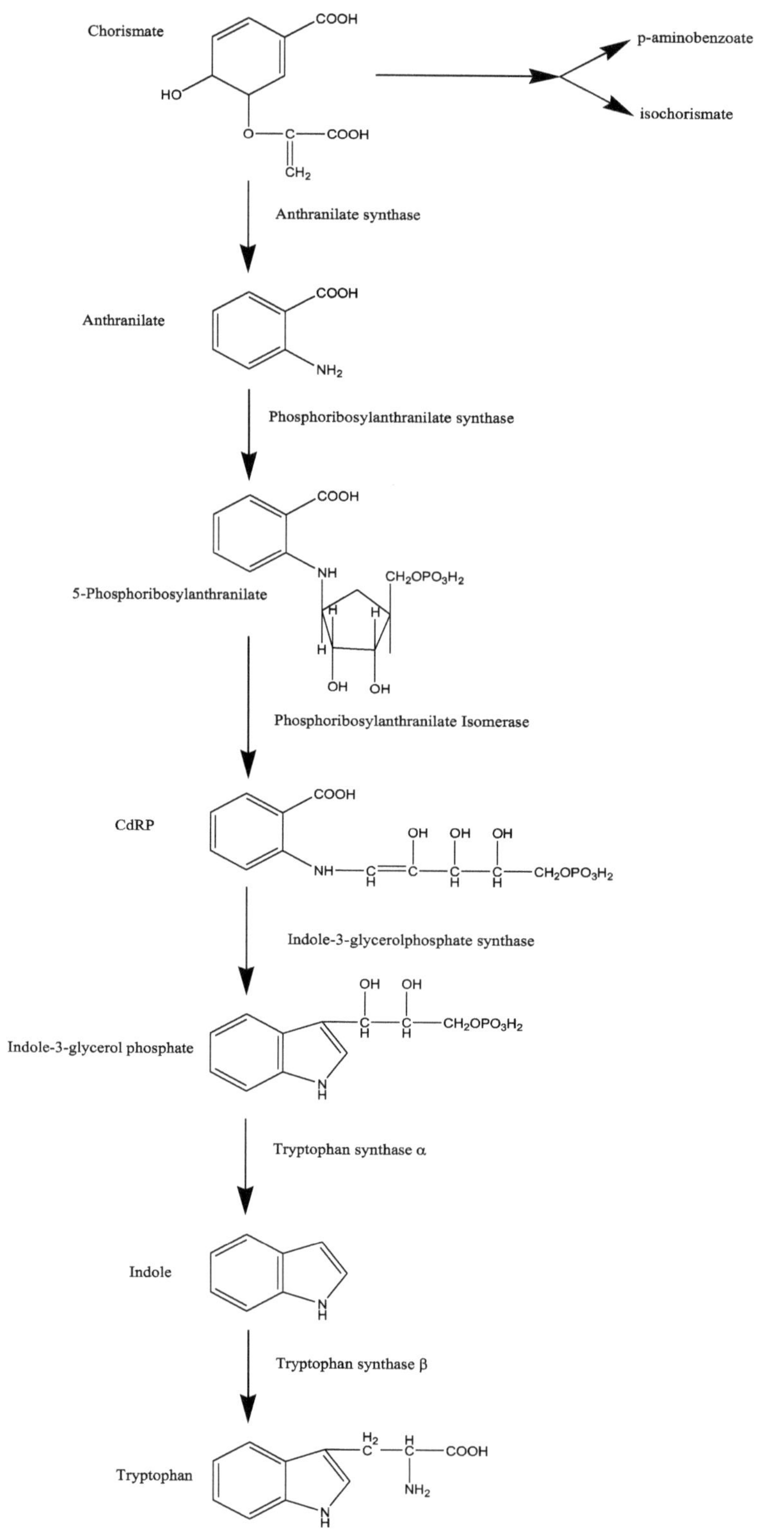

FIGURE 14.1 Biosynthesis of Tryptophan.

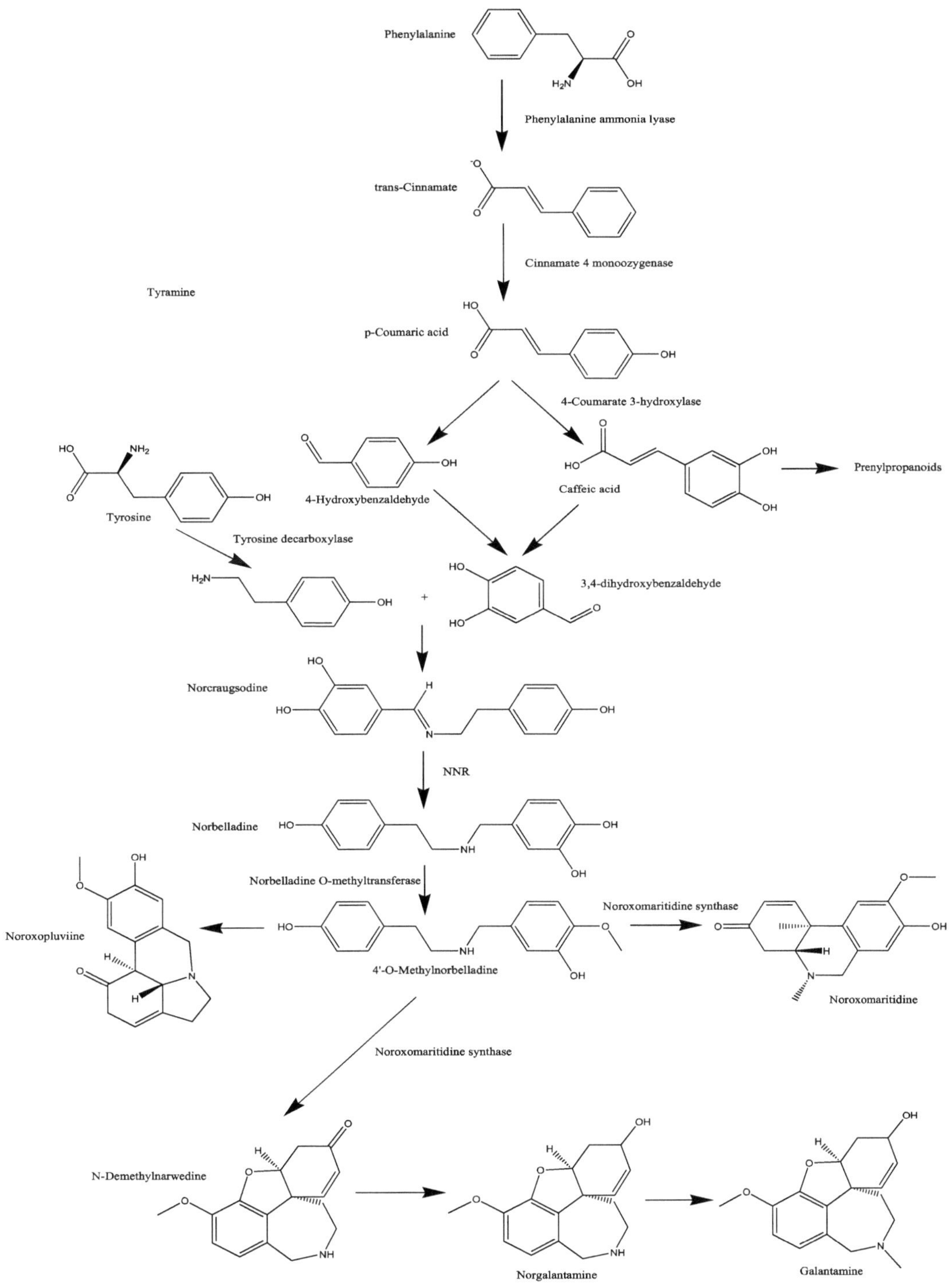

FIGURE 14.2 Biosynthesis of Galantamine.

14.5 GENERAL BIOSYNTHESIS OF INDOLE ALKALOIDS (FIGURE 14.3)

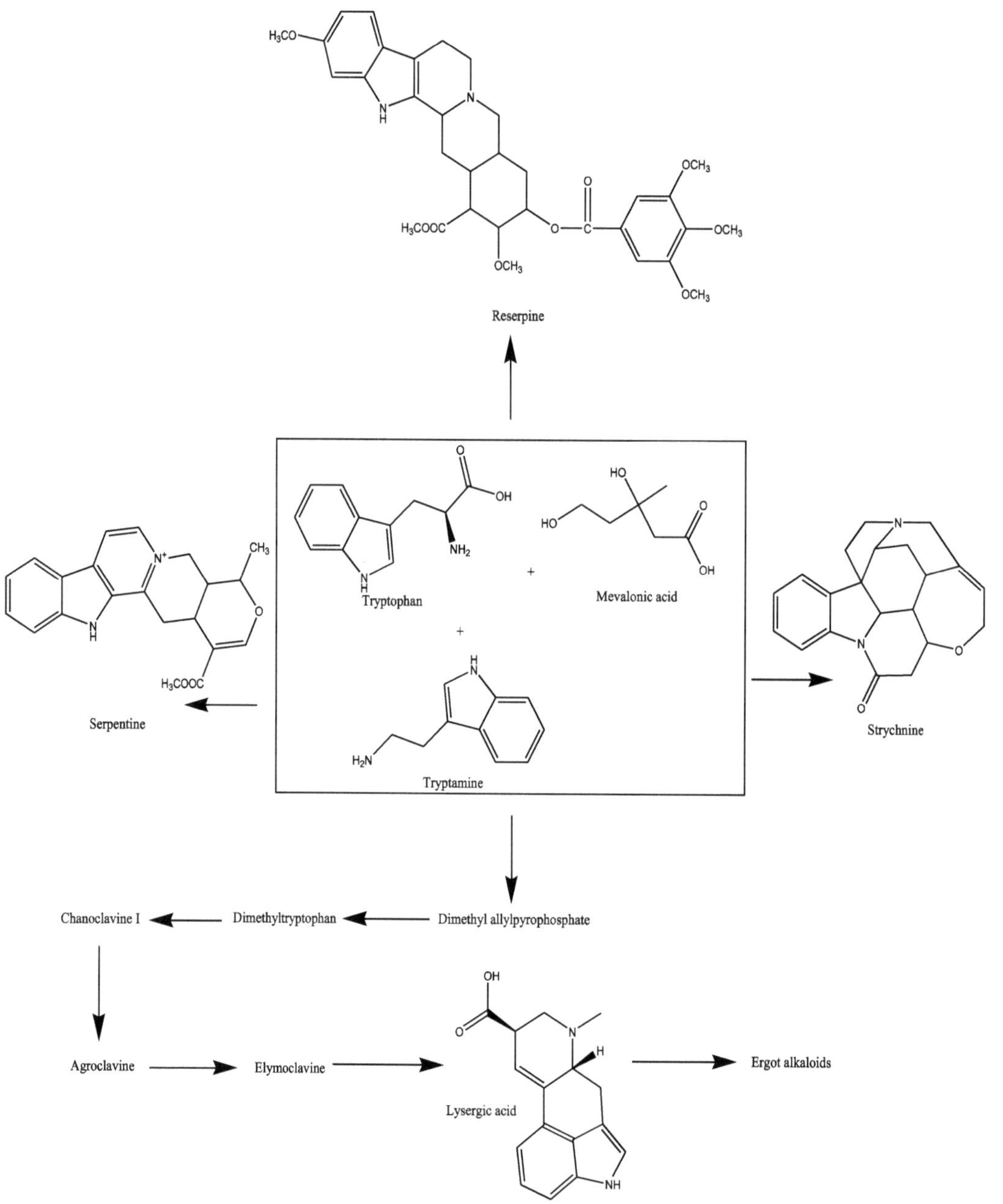

FIGURE 14.3 General Biosynthesis of Indole Alkaloids.

14.6 TOXICITY OF EDIBLE ALKALOIDS

Since the earliest time, mankind has been using different kinds of plants in various aspects of life. In today's era, it is known that plants possess numerous potent medicinal values because of the presence of different types of secondary metabolites like alkaloids, phenols, glycosides, terpenoids, etc. Among these, alkaloids is a large group of secondary metabolites, which produce different physiological changes in humans and animals when consumed. This is due to the presence of a nitrogen atom within the heterocyclic ring of alkaloids. However, some alkaloids are also found to produce toxic effects in humans and animals. Toxic Alkaloids may come in contact with animals and humans via inhalation, swallowing, or direct contact, which may produce hepatotoxicity, neurotoxicity, genotoxicity, and musculoskeletal deformities. They also produce various side effects like hypersensitivity, nausea, vomiting, psychosis, paralysis, gastrointestinal problems, arrhythmias, and sometimes death. Some alkaloids that have been reported to produce toxic effects are listed below.

1. *Pyrrolizidine Alkaloids (PA):*
   Pyrrolizidine alkaloids (PA) are ester alkaloids found in the plant families *Asteraceae, Boraginaceae and Fabaceae*. These alkaloids are toxic because they include necine and necic acid moieties as well as a 1, 2 double bond in the necine moiety. Two fused five-membered rings form a heterocyclic nucleus with a single nitrogen atom at position 4. A 1–2 double bond in the necine base, esterification of the hydroxyl group in one or more spots and a branched carbon chain in one of the ester side chains all contribute to hepatotoxicity. PAs occurs spontaneously in plants in the form of free base and n-oxide. Both of these forms are non-toxic; however, hepatic cytochrome P-450 quickly dehydrogenates the free base forms of PAs to ester pyrrole intermediates and dehydro-alkaloids (DHA), which are powerful electrophiles **[16]**. These electrophiles can generate detoxified products and alkylate liver macromolecules when they combine with water or glutathione (GSH). PAs are hepatotoxins that can act chronically or acutely depending on the dose and time of exposure **[17]**. PAs have been associated with liver centrilobular haemorrhage and necrosis, also known as veno-occlusive liver disease (VOD) **[18]**. VOD is the cause of cirrhosis and death. Several cases of PA poisoning have been reported in the United States **[19]**, Afghanistan **[20]**, Jamaica **[21]**, the United Kingdom **[22]**, Egypt **[23]**, India **[24]**, Israel **[25]**, and Scotland **[26]**. There have also been some documented cases of VODs induced by PAs: as stated by **[27]**, a newborn baby got VOD from nursing from a lady who had consumed herbal tea for almost the entire pregnancy. VOD outbreaks have occurred in places like Central India **[24]** and Afghanistan **[20]**. There is currently no antidote for natural compounds that cause liver damage, such as veno-occlusive disease. It can be treated by minimizing and managing its use.
2. *Aconitine:*
   Aconitine is a diterpene alkaloid produced by various plant species belonging to the family *Aconitum*. It is mainly found in plants like *Aconitum napellus* and *Aconitum vulparia.* It is a highly toxic alkaloid, and in our history, it is well known for its usage in arrow poisons for assassinations or common poisons in homicides. Greek mythology refers to Aconitum napellus as the first poisonous plant. In reality, the most poisonous plant in central Europe is considered to be aconite. **[28]**. Aconitine is highly lipophilic and can be readily reabsorbed through the skin. Aconitine induces quick paralysis and anaesthesia by inhibiting the presynaptic release of acetylcholine and activating Na+ channels. All the species aconitine genus are highly toxic and poisonous to humans and animals. A study was conducted **[29]** examining the neurotoxic effects of aconitine on cerebral cortex neuron cells generated from newborn SD rats and the underlying mechanisms responsible for those effects. Aconitine had an impact on the morphology of neuron cells, causing synapse fracture, membrane fragmentation, mitochondrial swelling, cytoplasmic vacuoles, and nuclear chromatin condensation and accumulation. The study reported that aconitine was neurotoxic to neuron cells of SD rats, which was caused due to the inhibition of Na+ -K+ -ATP. This influenced the concentrations of [K+], [Ca2+], [Mg2+] and neurotransmitters in the cells and resulted in the injury of these cells. Another study conducted by **[30]** reported brain tissue damage caused by aconitine. This study investigated the neurotoxicity of aconitine and the underlying mechanism. The rats were administered intragastrically with different concentrations of aconitine solution. After an interval of 30 mins, it was found that at lower concentrations of 0.5mg/kg and 1.5 mg/kg showed mild neurological symptoms like increased

oral secretion, shortness of breath and unsteady gait. At higher concentrations, like 2.5 mg/kg, the previous symptoms got worse, including weakness in the hind limb. Blood capillaries had enlarged, neurons were sparse and had a disordered arrangement, the neuronal nucleus had shrunken, and the matrix structure had become critically swollen. Evans blue (EB) extravasation and evaluation of tight junction protein expression were carried out to determine the blood-brain barrier's permeability. When compared to the control rats, the rats in the treated groups had considerably more EB in their cerebral tissue. Using the TUNEL assay, the toxic effects of aconitine on rat neurons were determined. Aconitine was found to cause rat brain tissue damage, with the severity of the damage increasing with dosage. The colour of the neurons' nucleus became even more intense in the group that received drug treatment, indicating severe apoptosis. The MTT assay was used to assess the cell damage caused by aconitine. According to the results, raising the aconitine concentration made the drug's effect on PC12 cell viability much more restrictive. The results concluded that aconitine induces brain tissue damage by increasing the penetrability of the blood-brain barrier in the cerebral region and over-activating the endoplasmic reticulum stress.

3. *Tropane alkaloids:*

   Tropane alkaloids (TA) are important secondary plant metabolites that are primarily concentrated in the Solanaceae and Erythroxylaceae groups of plants. The TAs can be categorized into hyoscyamine and scopolamine, cocaine, and calystegines. Their unique bicyclic tropane ring system distinguishes them. Although all TAs have a similar basic structure, their biological, chemical, and pharmacological characteristics vary greatly. Due to its effectiveness in treating nausea, vomiting, motion sickness, and smooth muscle spasms, scopolamine, also known as hyoscine, has the largest legitimate pharmaceutical market, whereas cocaine is the second most often used illicit substance worldwide. TA from the *Solanaceae* family are common in poisonous plants (*Hyoscyamus niger, Datura, Atropa belladonna, Scopolia lurida, Mandragora officinarum*), as well as the *Erythroxylaceae* family (*Erythroxylum coca*). These substances have anticholinergic effects because they can bind to either muscarinic or nicotinic receptors, inhibiting the function of the neurotransmitter acetylcholine in the central and peripheral nervous systems. They are utilized as bronchodilators, anaesthetics, antispasmodics, antiemetics, and mydriatics and are hence of major therapeutic value. TAs also serve as lead components to develop new drug molecules with great therapeutic value. The hallucinogenic properties of Atropa belladonna and Datura stramonium have been well-known for centuries. They are freely accessible everywhere in the globe and have been utilized by individuals for religious and medicinal causes. Accidental consumption can result in intoxication, particularly in youngsters. A study **[31]** reported the effects of Jimson weed (*Datura stramonium* L.) alkaloid extracts on cognitive performance, antioxidant state, and activity of enzymes of rats involved in the monoaminergic and cholinergic systems of neurotransmission. For 30 days, rats received extracts at doses of 100 and 200 mg/kg body weight. The rats underwent behavioural tests using a hole board and spontaneous alternation before the experiment was over. After that, rats were sacrificed, and the activities of the antioxidant enzymes catalase, glutathione S-transferase (GST), superoxide dismutase (SOD), and acetylcholinesterase (AChE), as well as malondialdehyde (MDA) and total thiol (TSH) contents of their brain homogenates were assessed. According to the results, rats that were given the extracts displayed a lower memory index and an increase in anxiety-like behaviour. Hyoscyamine (atropine), scopolamine, amphetamine, cathine, phenlyephirine, and Tropine trifluoroacetate were found in the extracts, according to the results of the GC-MS characterization. As a result, one of the main biochemical mechanisms facilitating the ability of Jimson weed fruit and leaf alkaloid extracts to induce cognitive function impairment that could be significantly linked to its constituent alkaloids is the alteration of cholinergic, monoaminergic, and antioxidant systems in the brain of rats that were administered with the alkaloid extract from leaf and fruit of Jimson weed (*Datura stramonium* L).

14.7 EFFECT OF β-CARBOLINE ON MEMORY & BEHAVIOUR

β-carboline is an indole alkaloid compound with a pyridine ring attached to an indole skeleton. It has a similar structure to tryptamine. Harmane (HA; 1-methyl-carboline), 9H-pyrido [3, 4-] indole (norharmane), and Harmine are all tremorogenic β-carboline alkaloids found in the food chain (7-methoxy-1-methyl-9H-pyrido [3, 4-b] indole). They have been found in plant-derived beverages, common plant-derived meals (rice, wheat, corn, mushrooms, barley, soybeans, vinegar and grapes) and plant-derived inhaled medications like tobacco. The first β-carbolines found in *Peganum harmala* were harmane, norharmane, and harmine **[32]**.

Peganum harmala L. (*P. harmala*) is a plant endemic to North Africa and China. Because of its anti-inflammatory properties, *P. harmala* is used in traditional medicine to treat a variety of ailments. *P. harmala* yields seven alkaloids: harmaline, harmine, harmalol, harmole, vasicine, vasicinone and deoxyvasicine. β-carbolines are widely employed these days due to their numerous pharmacological actions, including sedative, hypnotic, anxiolytic, anticonvulsant, anticancer, antithrombotic, antiseptic, antibacterial and antiviral properties. Harmine is a naturally occurring β-carboline alkaloid found in a variety of plant species. Two examples include the seeds of Syrian rue (*Peganum harmala*) and the Banisteriopsis caapi vine, both of which are often used to make ayahuasca (Hallucinogen beverage). Preclinical research suggests that harmine may have neuroprotective and cognitive-enhancing effects and retrospective/observational studies of the mental health of long-term ayahuasca users suggest that longer use of this harmine-rich hallucinogen is associated with improved neuropsychological functioning. Acute systemic treatment of aromatic β-carboline alkaloids, harmine and harmol boosted novel recognition assessed 1.5 h (STM) after training in rats, whereas dihydro-β-carboline harmaline enhanced the same to 24 h memory (LTM).

It has been demonstrated that the dihydro-β-carboline harmaline affects nonspatial and non-aversive memory tasks, enhancing principally LTM memory, and electrophysiological evidence reveals that harmaline induced tremor by activating the inferior olivary nucleus and cerebellum **[33; 34]**. It has also been studied that harmaline blocks both associative and motor learning, although it is unknown whether the inferior olive or another area mediates this effect on learning and memory. The behavioural effects of harmine on rats were investigated. The effects of harmine on animal behaviour were studied using forced swimming and open field tests, and the results indicated that harmine reduced immobility while increasing climbing and swimming duration compared to the saline group. Harmine demonstrated an anti-naloxone-induced morphine withdrawal syndrome effect in morphine-dependent rats. Harmine activates the CNS by inhibiting the metabolism of amine neurotransmitters and enhancing dopamine release, suggesting its use in treating brain illnesses **[35]**. The effects of β-carbolines on depression and memory are given in (Table 14.2)

TABLE 14.2
Effects of *β*-carbolines on depression and memory [36]

Depression	5-hydroxytryptamine (5-HT) and dopamine receptors are two more neurotransmitter receptors that β-carbolines bind to in the brain. We also know that these alkaloids increase extracellular norepinephrine, dopamine and 5-HT levels while blocking monoamine reuptake processes in specific brain locations. These data suggest that β-carbolines may have a future role in the treatment of depression.
Memory	Several β-carbolines have been demonstrated to change learning capacity in mice (promnesic effects), but they can cause memory impairment in certain investigations. Benzodiazepines and beta-CCM seem to target the acquisition (learning) phase of memory processing.

14.8 EFFECT OF *MURRAYA KOENIGII* ON MEMORY & BEHAVIOUR

Murraya koenigii Linn. In Hindi, belonging to the family *Rutaceae* is commonly known as 'curry patta'. It is popularly used as a spice in India and as a flavouring agent in various parts of Asia. Fresh or dried leaves are added to the vegetables, other curry food items, and sometimes tea to produce characteristic flavour and aroma. However, it has traditionally been used to cure various illnesses like stomach aches, traumatic injury, influenza, rheumatism, dysentery and astringent **[37]**. The leaves of this plant are reported to have hypoglycaemic and anti-diabetic **[38]**, antibacterial **[39]**, antioxidant **[40; 41]**, hepatoprotective **[42]**, chemo modulatory **[43]**, immunomodulatory **[44]**, wound healing **[45]**, and anti-inflammatory activities **[46]**. All aspects of perception, learning, reasoning, and memory are part of cognition. Behavioural issues, dementia, and delirium are examples of cognitive dysfunctions. As documented by **[47]** the alkaloidal extract of *M. koenigii* leaves (MKA) was used to determine the cognitive functions and brain cholinesterase activity in Swiss Albino mice. In this study, about (2 kg) of fresh leaves of *M. koenigii* were collected and dried in the shade. The dried leaves were crushed into coarse powder form and subjected to extraction. The extraction was carried out in a Soxhlet apparatus using methanol as the solvent. The methanol extract was then concentrated using a rotary evaporator and acidified with 0.5 M H_2SO_4, and then it was washed with chloroform. The aqueous acidic fraction was collected and again extracted with chloroform. Preliminary phytochemical evaluation and NMR analysis confirmed the presence of alkaloids. In the animal study, the mice were divided into different groups according to various interoceptive and exteroceptive memory models. MKA (10, 20, and 30 mg/kg) was given orally for 15 days. The mice were exposed to a training session utilizing an elevated plus maze and a passive avoidance apparatus on the 15th day following the last MKA treatment, and retention memory was recorded after 24 hours on the 16th day. On the 15th day, the animals were given scopolamine (0.4 mg/kg, i.p.). This well-known centrally acting cholinergic probe interferes with memory and cognitive function in human and experimental animals or diazepam (1 mg/kg, i.p.) to produce amnesia in separate groups of young mice (Interoceptive models). After 45 minutes of scopolamine and diazepam injection, the animals were subjected to a training session and the retention memory was assessed after 24 hours on the 16th day. Piracetam (400 mg/kg, i.p.) was delivered into the positive control groups in order to test its efficacy as a nootropic drug. MKA (10, 20, and 30 mg/kg) was administered orally to separate groups of young and elderly mice for 15 days as part of a biochemical study. Metrifonate (50 mg/kg, i.p., and 60 min before dissecting the brain) was used as the positive control for comparison of brain cholinesterase activity. Memory was tested using exteroceptive behavioural models such as the elevated plus maze and the passive avoidance apparatus. The interoceptive behavioural models were diazepam, scopolamine, and age-induced forgetfulness. In elevated plus maze and passive avoidance paradigm maze models, mice given MKA extract (20 and 30 mg/kg, p.o.) had inhibitory effects against scopolamine-induced memory impairment. Young and old mice both significantly improved their memory scores when given MKA (20 and 30 mg/kg, p.o.), which was confirmed that it elevated acetylcholine levels in the brain.

14.9 EFFECT OF ANONAINE ON MEMORY & BEHAVIOUR

Anonaine is a bioactive benzylisoquinoline alkaloid, largely found in members of plant families like *Magnoliaceae* and *Annonaceae*. *Annona squamosal* Linn, is one of the plants from the family *Annonaceae*. It is a native plant of the West Indies, and it is also cultivated throughout India. It is commonly known as sitaphal and custard apple or sugar apple. It has been reported to have a wide range of biological activities like vasorelaxant, antibacterial, antifungal, antioxidant and antidepressant effects **[48]** and epilepsy **[49]**. The plant leaves are reported to possess several alkaloids like annonaine, flavonoids and acetogenins **[50]**. A study was conducted on Annona squamosa's neuroprotective effects on epileptic rats' spatial learning and memory as well as GABAergic receptor binding in the cerebral cortex. The leaves of Annona squamosal were collected and extracted using a hydro-alcoholic solvent in the Soxhlet apparatus for 24 hours. Adult male Wistar rats were used to screen the neuroprotective effect of the plant's extract and its isolated content Anonaine. The rats were injected with pilocarpine to produce status epilepticus, which was allowed to continue for 1 hour, and then all the animals were treated with diazepam. Rats that showed recurrent seizures were used for further experiments. The objective of the research was to reveal that Annona squamosa, which contains the active alkaloid (-) Anonaine, protects against epilepsy by acting on GABA receptors. Confocal microscopy was used to examine the prospective uses of Annona squamosa and its isolated chemical component (Anonaine), as well as variations of GABA, GABAA, and GABA receptors in the cortical region of the brain of epileptic rats. In the radial arm and y-maze models, the extract of Annona squamosa leaves and its isolated constituent, Anonaine, demonstrated memory-improving and memory-recovering properties. The data from the results concluded that Anonaine, an alkaloidal component of Annona squamosa leaves, improves the behaviour of epileptic rats and their diminished GABA receptors **[51]**.

14.10 CONCLUSION

Alkaloids are considered one of the major plant sources that can produce marked physiological effects in humans and animals and are also major contributors to different types of drugs and pharmaceuticals. Throughout history, alkaloids have been found to produce several toxicity effects. The authors have enlisted some of the major alkaloids known to produce toxic effects and some edible alkaloids with their structures. The authors finally concluded that despite some toxic effects, alkaloids had been found to promote brain health and behaviour. The information provided in this chapter mainly focused on the edible alkaloids that produce marked effects on memory and behaviour, which will support and strengthen research in exploration of potent memory enhancers and the treatment of major neurodegenerative diseases.

REFERENCES

1. Güven, K. C., Percot, A., & Sezik, E. (2010). Alkaloids in marine algae, *Mar Drugs*, *8*(2), 269–284.
2. Wink, M. (1998). A short history of alkaloids. In *Alkaloids* (pp. 11–44). Springer, Boston, MA.
3. Zlotnik, G., & Vansintjan, A. (2019). *Frontiers in Psychology*, *10*, 2523.
4. Gur, R. C., & Gur, R. E. (2022). *Dialogues in Clinical Neuroscience*.
5. Callahan, P. M., Terry, A. V., Peitsch, M. C., Hoeng, J., & Koshibu, K. (2021). *Scientific Reports*, *11*(1), 1–9.
6. Liu, J. S., Zhu, Y. L., Yu, C. M., Zhou, Y. Z., Han, Y. Y., Wu, F. W., & Qi, B. F. (1986). *Canadian Journal of Chemistry*, *64*(4), 837–839.
7. Lee J., Lee W., Kim M. A., Hwang J. S., Na M., & Bae J. S. (2017). *Journal of Cellular and Molecular Medicine*. Jun;*21*(6), 1217–1227.
8. Sun Z., Hu M., Sun Z., Zhu N., Yang J., Ma G., & Xu X. (2018). *Molecules*. May 17;*23*(5), 1198.
9. Sakamoto T., Nishida A., Wada N., Nakamura Y., Sato S., Konishi T., & Matsugo S. (2020). *Molecules*. Oct 22;*25*(21), 4879.
10. Nakahara K., Trakoontivakorn G., Alzoreky N. S., Ono H., Onishi-Kameyama M., & Yoshida M. (2002). *Journal of Agricultural and Food Chemistry*. Aug 14;*50*(17), 4796–4802.
11. Mani V., Ramasamy K., Ahmad A., Parle M., Shah S. A., & Majeed A. B. (2012). *Food and Chemical Toxicology*. Mar 1;*50*(3–4), 1036–1044.
12. Tsuchiya H., Hayashi H., Sato M., Shimizu H., & Iinuma M. (1999). *Phytochemical Analysis*. Sep;*10*(5), 247–253.
13. Howes M. J., & Houghton P. J. (2003). *Pharmacology Biochemistry and Behavior*. Jun 1;*75*(3), 513–527.
14. Radwanski E. R., & Last R. L. (1995). *The Plant Cell*. Jul;*7*(7), 921.
15. Park C. H., Yeo H. J., Park Y. E., Baek S. A., Kim J. K., & Park S. U. (2019). *Biology*. Aug 29;8(3), 63.

16. Culvenor, C. C. J., Edgar, J. A., Smith, L. W., & Tweeddale, H. J. (1970). *Australian Journal of Chemistry*, *23*(9), 1853–1867.
17. Wainwright, J. & Schonland, M. M. (1977). *South African Medical Journal*, *51*(17), 571–573.
18. Willmot, F., & Robertson, G. (1920). *The Lancet*, *196*(5069), 848–849.
19. Huxtable, R. J. (1980). *Perspectives in Biology and Medicine*, *24*(1), 1–14.
20. Tandon, H. D., Tandon, B. N., & Mattocks, A. R. (1978). *American Journal of Gastroenterology (Springer Nature)*, *70*(6), 607–613.
21. Hill, H. R. (1952). *West Indian Medical Journal*, *1*(3), 243–264.
22. Weston, C. F., Cooper, B. T., Davies, J. D., & Levine, D. F. (1987). *British Medical Journal (Clinical Research ed.)*, *295*(6591), 183.
23. Safouh M., Shehata A. H., & Elwi A. (1965). *Archives of Pathology*, 79, 505–511.
24. Tandon, B. N., Tandon, H. D., Tandon, R. K., Narndranathan, M., & Joshi, Y. K. (1976). *The Lancet*, *308*(7980), 271–272.
25. McLEAN, E. K. (1974). *Isr. J. Med. Sci*, *10*, 436–440.
26. Bateman, J., Chapman, R. D., & Simpson, D. (1998). *Scottish Medical Journal*, *43*(1), 7–15.
27. Roulet, M., Laurini, R., Rivier, L., & Calame, A. (1988). *The Journal of Pediatrics*, *112*(3), 433–436.
28. Knight, A. P. (2006). *Native Plants Journal*, *7*(1), 102–103.
29. Beyer, J., Drummer, O. H., & Maurer, H. H. (2009). *Forensic Science International*, *185*(1–3), 1–9.
30. Zhang, Y., Chen, S., Fan, F., Xu, N., Meng, X. L., Zhang, Y., & Lin, J. M. (2022). *Journal of Pharmaceutical Analysis,* 13(1), 88–98.
31. Ogunsuyi, O. B., Ademiluyi, A. O., Oboh, G., & Agbebi, O. J. (2020). *African Journal of Biomedical Research*, *23*(1), 87–96.
32. Oakley, N. R., & Jones, B. J. (1982). *Neuropharmacology*, *21*(6), 587–589.
33. Lamarre Y., Montigny C., de Dumont M., & Weiss M. (1971). Brain Res, *32*, 246–250.
34. De Montigny C., & Lamarre Y. (1975). *Can J Physiol Pharmacol*, *53*, 845–849.
35. Olena V Bogdanova, Shami Kanekar, Kristen E D'Anci, Perry F Renshaw (2013). *Physiol Behav*, *13*(118), 227–239.
36. Pooneh Heshmati, Mohammad Nasehi, Mohammad Reza Zarrindast (2014). *Journal of Paramedical Sciences*, *5*(1), 99–104
37. Joseph, S., & Peter, K. V., (1985). *Econ. Bot. 39*, 68–73.
38. Saritha, K., & Usha, P. T. A. (2009). *Indian Journal of Animal Sciences*, *79*(10), 986–987.
39. Ningappa, M. B., Dhananjaya, B. L., Dinesha, R., Harsha, R., Srinivas, L., (2010). *Food Chem*. 118, 747–750.
40. Tachibana, Y., Kikuzaki, H., Lajis, N. H., Nakatani, N., (2001). *J. Agric. Food Chem*. 49, 5589–5594.
41. Tachibana, Y., Kikuzaki, H., Lajis, N. H., Nakatani, N., (2003). *J. Agric. Food Chem.* 51, 6461–6467.
42. Sathaye, S., Bagul, Y., Gupta, S., Kaur, H., Redkar, R., (2010). *Exp. Toxicol. Pathol*. 63, 587–591.
43. Dasgupta, T., Rao, A. R., Yadava, P. K., (2003). *Nutr. Res*. 23, 1427–1446.
44. Shah, A. S., Wakade, A. S., Juvekar, A. R., (2008). *Indian J. Exp. Biol*. *46*, 505–509.
45. Gupta, S., George, M., Singhal, M., Garg, V., (2009). *Pharmacologyonline 3*, 915–923.
46. Mani, V., Ramasamy, K., Ahmad, A., Parle, M., Shah, S. A. A., & Majeed, A. B. A. (2012). *Food and Chemical Toxicology*, *50*(3–4), 1036–1044.
47. Manoj, N. G., Archana, R. J., Sachin, S. S., (2009). *Pharmacol. Online 1*, 1072–1094.
48. Porwal, M., Sharma, K., & Sharma, P. K. (2013). *Asian Journal of Pharmaceutical Research and Development*, 92–95.
49. Kurioka, S., Kimura, Y., & Matsuda, M. (1981). *Journal of Neurochemistry*, *37*(2), 418–421.
50. Li H. T., Wu H. M., Chen H. L., Liu C. M., & Chen C. Y. (2013). *Molecules*, Jul 12, *18*(7), 8257–8263.
51. Porwal, M., & Kumar, A. (2015). *Journal of Applied Pharmaceutical Science*, *5*(1), 018–023.

15 Caffeine and Motor Skills

Robert Lalonde and Catherine Strazielle

15.1 INTRODUCTION

15.1.1 Pharmacological Profile and Pharmacokinetics

Over 80% of people in the world drink caffeine products every day [1], all the more because of its blood half-life of 4 hours among regular drinkers [2–4]. It is mainly catabolized in the liver by cytochrome P450 and excreted in urine [5]. Also named 1,3,7-methylxanthine or methyltheobromine, caffeine is an alkaloid from the methylxanthine class [6] known for its bronchodilator properties due to antagonism of adenosine receptors and inhibition of phosphodiesterase [5, 7].

15.1.2 Stimulant Properties

Caffeine is known as a stimulant by virtue of its antagonism of adenosine receptors, the neuromodulator acting as an inhibitor of neuronal activity [8–11]. More precisely, caffeine is an antagonist at A_1, $A_{2A,}$ $A_{2B,}$ and A_3 adenosine receptors [12, 13] widely distributed in the brain (Figure 15.1), with high levels found for A_1 receptors and lower levels for A_3 ones in cerebral cortex, striatum, cerebellum, and thalamus, while A_{2A} receptors are abundant in basal ganglia and blood vessels and A_{2B} in hippocampus, nucleus accumbens, and hypothalamus [14-16], all of these furnishing explanations as to why caffeine affects sensorimotor functions, arousal, selective attention, emotions, and cognition. In humans, moderate plasma concentrations (10 – 50 μM) of caffeine are sufficient to stimulate A_1 and A_{2A} receptors, but higher concentrations are necessary to activate A_{2B} and A_3 ones [17]. Its appealing stimulant properties by day are hampered as a disrupter of circadian rhythms due to the delayed release of melatonin at night [18].

Open-field activity increased after administration of caffeine dissolved in drinking water (≈ 10 mg/kg) every other day over a 2-week period in rats [19] or caffeine injected acutely at 10 mg/kg in mice [20] or rats [21], but this dose was without effect in the latter after one week of injections [22]. Caffeine-induced hyperactivity was also evident when rats explored a Y-maze [23]. Hyperactivity was correlated with A_1 receptor binding in mouse whole brain membranes when a series of methylxanthines were examined [24]. Caffeine-induced hypermobility in mice was prevented by removing A_{2A} receptors from the shell region of the nucleus accumbens via the Cre/loxP gene technique [25]. At higher doses, caffeine sometimes increased [26,27] and sometimes decreased [28] open-field activity, the latter presumably as a result of increased anxiety, as shown in social interaction [28, 29], hole-board [29], elevated plus-maze [28, 30], and light-dark box [30] tests. Because most students regularly consume the beverage, analyses of caffeine's action in the open field constitute an appropriate initiation into the field of behavioural neuroscience [31].

DOI: 10.1201/9781003300502-15

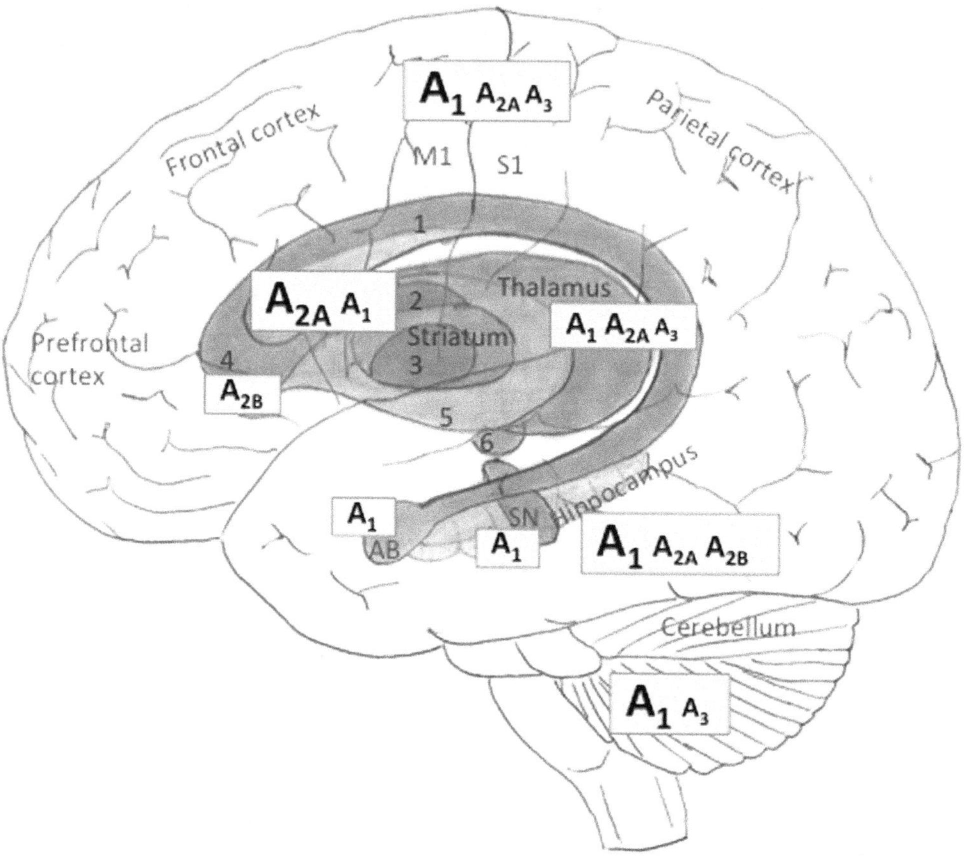

FIGURE 15.1 Cartography of the adenosine subtype receptors in the sensorimotor control regions of the brain, i.e. the fronto-parietal (M1 as primary motor and S1 primary sensory areas) and prefrontal cortices, the cerebellum, and the striatum with the caudate nucleus (1), the lenticular nucleus (2), the pallidal regions with the globus pallidus (3), the nucleus accumbens (4), the putamen (5) as well as the subthalamic nucleus (6) and substantia nigra (SN). Two limbic structures, the amygdaloid body (AB) and a part of the hippocampal formation presented as a complement. The densities of different receptor-subtypes (A1, A2A, A2B, and A3) appear as a function of the size of the letters used, the highest density presented with the larger letters.

Since adenosine receptors are heavily concentrated in sensorimotor brain regions, the effects of caffeine on motor coordination are a worthwhile point of investigation, all the more so when considering that 10 mg/kg caffeine increased glucose utilization in rat brains at the level of substantia nigra, dorsal striatum, globus pallidus, red nucleus, pontine grey, cerebellar cortex, and vestibular nucleus [32]. Moreover, coffee drinkers displayed different levels of connectivity from non-coffee drinkers in motor-related regions based on functional magnetic resonance imaging [33]. The main purpose of the present review is to answer the following interrogations: 1) caffeine is known to improve selective attention and vigilance tasks [34–36], but does it help to coordinate movements in normal subjects? 2) Is caffeine helpful in coordinating the movements of patients with neural disorders? 3) Under which conditions does caffeine worsen motor coordination? The inclusion of animal studies targeting motor coordination should contribute more precisely to our understanding of caffeine's role in human motor skills.

15.2 MOTOR SKILLS IN NORMAL SUBJECTS

15.2.1 Human Subjects

Regular caffeine drinkers are less likely to have been in traffic accidents than non-consumers [37], and caffeine may improve performances during sporting competitions, results perhaps due to better motor coordination but also sharper vigilance, endurance, and memory [38–41]. In particular, 6 mg/kg caffeine enhanced ball-passing accuracy and jump performance in simulated soccer [42], accelerated reaction times to the non-dominant side in a reactive agility test by soccer players [43], and quickened reactive agility, decision, and movement times in simulated team-sport exercise [44] in men. Likewise, 250 mg of caffeine increased letter search though not navigational chart accuracy in experienced seamen [45], and 4.5 mg/kg of caffeine improved handwriting movements in skilled writers [46]. On the contrary, 3 mg/kg caffeine had no effect on the accuracy of épée fencers despite a lesser degree of fatigue [47].

Among standard laboratory tests, various results have been obtained in regard to reaction or movement times, digit symbol substitution, and Stroop colour-word conflict tasks that require selective attention and visuomotor speed, functions that underlie most motor skills. On one hand, 300 though not 600 mg caffeine speeded up reaction times and movement times in normal subjects [48], and 250 or 500 mg [49] as well as 250 or 500 mg/70 kg [50] caffeine improved performances on the digit symbol substitution task, tapping speed [49], and the repeated acquisition and performance procedure [50] in healthy adults. Moreover, 100 mg of caffeine speeded up reaction times in the Stroop word-colour conflict and rapid information processing tasks in men [51]. A dose of 300 – 350 mg of caffeine improved balance scores in healthy adults [52]. On the other hand, 120 mg caffeine had no effect on the digit symbol substitution task in women [53] and neither did 3 mg/kg caffeine affect the Stroop test in fencer athletes [47]. Moreover, in healthy young adults, 60 mg caffeine had no effect on visuomotor tracking and pegboard tasks [54], nor did 200 or 400 mg caffeine affect bimanual tapping, time discrimination, or grip and lift tasks despite speedier simple reaction times [55], nor did 3, 6, 9 mg/kg caffeine affect postural stability [56], nor did 5 mg/kg caffeine affect three motor skill tasks comprising standing balance, manual dexterity, and running time [57], nor did 6 mg/kg caffeine affect standing balance during perceptual-cognitive tasks [58]. A dose of 30 mg caffeine even deteriorated hand steadiness [59] in healthy women, as did 6 mg/kg caffeine in adult subjects [60, 61], in all three cases without disturbing tapping rate, in line with reports of hand tremor felt by 2% of coffee drinkers [62]. Moreover, 5 mg/kg caffeine decreased hand steadiness as well as retarding hand tracing and pin-placement dexterity tests in caffeine-naïve though not caffeine-familiarized women [63, 64]. Caffeine has been given after training sessions to test the retention of sensorimotor training. Post-trial 200 mg caffeine increased memory as measured by reaction times of a keyboard sequence [65] without affecting a visuomotor skill task [66] in adult subjects.

In contrast to minimal effects on the dexterity of younger subjects, 100 mg caffeine accelerated the gait speed of middle-aged women in single and dual tasks [67], and 3 mg/kg caffeine improved the Minnesota manual dexterity turning test score as well as upper and lower limb performance in a senior fitness test among aged subjects [68]. Yet when two age groups in the range of 20–25 or 50–65 years were compared, 250 mg caffeine improved digit symbol substitution and speeded up reaction times to target letters in both groups but improved rapid information processing and tapping rate only in young subjects [69]. Thus, at a dose as low as 250 mg/kg, caffeine improves the speed of responding underlying all motor skills in young subjects, but results are more limited on tests of manual dexterity or postural stability when it may have a more favourable impact in older people, a field worth continuing to be explored in the future.

15.2.2 Mice and Rats

Motor coordination in mice and rats is evaluated with inclined screen, pole-climbing, tube (a.k.a. chimney), stationary beam, bar-holding, and rotorod tests [70–74]. As found in human subjects, variable outcomes have been obtained concerning the effects of caffeine on motor skills in normal animals. On one hand, an acute 2 mg/kg dose of caffeine [75] as well as 20 (though not 200) mg/kg per day given for one week [76] or 150 mg/kg per day given for 6 weeks [77] increased latencies before falling from the rotorod in mice. On the other hand, latencies decreased in mice after higher acute doses of 50 or 100 mg/kg caffeine [78] or caffeine dissolved in drinking water (≈ 20 or 200 mg/kg per day) for one month [79]. Further deterioration in mouse performance was found after acute 30, 60, or 120 mg/kg caffeine on stumbles in a hole-board [80]. Contrariwise, a single dose of 5, 10, or 20 mg/kg caffeine had no effect on the rotorod in mice [81]. An 8-week regimen of 60 mg/kg per day caffeine reduced latencies before mice descended from a vertical pole, but neither this dose nor 20 mg/mg per day caffeine affected latencies before falling from an inverted grid [82]. The higher dose subgroup increased cerebellar Purkinje cell density, and both caffeine subgroups increased dendritic length in pyramidal neurons of the sensorimotor cortex and dendritic arborization in striatal medium spiny neurons. Thus, in young animals, caffeine effects appear to depend on narrow dose ranges, perhaps with an optimal acute dose of 2 mg/kg and 2 to 20 mg/kg per day as a weekly regimen, a result that could explain the lack of a caffeine effect on standing balance in young human subjects.

In ageing research, various levels of coffee extract were included in the diet for two months in 19-month-old rats [83]. The group with 0.55% (but not 0.165%, 0.275%, 0.825%) coffee had higher latencies before falling from the rotorod than placebo. A similar 0.553% group (but also a 0.387% group) had higher latencies before falling from an inclined screen. As found in human subjects, caffeine may have a more favourable impact on aged animals.

15.2.3 Human Development

Caffeine is used to treat apnoea in pre-term infants [84, 85], a group vulnerable to neurodevelopmental deficits [86]. A dose of 80 but not 20 mg/kg caffeine increased deviant neurologic signs (hand posture, tremors, and startles) in pre-term infants at the neonatal stage without affecting Bayley-III scores of motor development at two years of age [87]. In contrast, improvements were found when caffeine at various doses starting from postnatal day ten to week six reduced the rate of disorders assessed in the Movement Assessment Battery for Children (MABC) at five years of age [88]. Moreover, caffeine started at a loading dose of 20 mg/kg and maintained at a median dose of 5 mg/kg from postnatal day three to the term equivalent age of 35 weeks improved manual dexterity in the MABC and motor coordination scores on the Beery-Buktenica test in five [89] and 11 [90, 91] year old children.

Regarding the mother's caffeine intake during early pregnancy, the heavier the intake, the more their offspring committed errors in a motor steadiness test evaluated at four years of age, but yet the offspring were faster to correct these errors and successfully dropped more pegs on a grooved pegboard [92]. On the contrary, the mother's caffeine intake during pregnancy had no effect on their offspring's Bayley scores of motor development evaluated at eight months of age [93].

15.2.4 Rodent Development

Caffeine administration during development in mice and rats has mostly been given at high doses. When mouse dams were given caffeine mixed with drinking water (≈ 60 mg/kg per day) from day eight of pregnancy until weaning, their six-week-old male offspring fell sooner from the rotorod than controls [94]. After rat dams received 50 or 100 mg/kg caffeine per day throughout pregnancy and during three weeks of lactation, the offspring of the lower dose subgroup had faster latencies than controls before turning up the slope of the inclined screen on P9 and P14 while the higher dose subgroup had weaker forelimb grip strength [95]. While the external cerebellar granule cell layer of the control group declined in size, both caffeine groups presented with thicker layers from P19 to P21, indicating a delay in cell migration. In contrast to the faster latencies reported in the previous study, rat pups had slower latencies before turning on the inclined screen following a lower estimated dose (0.3 g/L) of caffeine mixed in drinking water throughout gestation and lactation until the pups were 14 days old [96].

Several sensorimotor tests were given to two groups of rats injected with 20 mg/kg caffeine on either postnatal day 7 to 11 (P7–P11) or P13 to P17 [97]. For the P7–P11 group, caffeine impaired motor skills by increasing latencies before turning upward on an inclined surface (negative geotaxis) and reaching the top of an inclined screen on the 4th post-injection day (P15). However, for the P13–P17 group, caffeine improved motor skills by decreasing latencies before reaching the top of the inclined screen on the 4th and 8th post-injection day (P21 and P25). Once again, caffeine deteriorated motor performances in the P7–P11 group by decreasing latencies before falling from a thin bar on P15 and P18 and from the rotorod on P21, while the P13–P17 group was unaffected in both tasks. Thus, opposite effects were obtained with caffeine during development, with younger animals being adversely affected and more mature animals being unaffected or even improved. As found in human subjects, the effects of high-dose caffeine on motor development in animals vary depending on dose and postnatal age.

15.3 DRUG INTERACTIONS

15.3.1 Ethanol

Coffee is popularly sought to ward off the sedative effects of alcohol [98], but does it help to correct movement disorders associated with ethanol? On the one hand, 120 mg caffeine [99] and 250 or 500 mg/70 kg caffeine [49] countered the decrease in performance caused by ethanol on the digit symbol substitution task and, at the two latter doses, the repeated acquisition and performance procedure in young adults. Likewise, 400 mg of caffeine attenuated the ethanol-induced delay in simple reaction times [100] and 3.3 mg/kg caffeine the ethanol-induced delay in reaction times of a subject-paced rapid information processing task [101] in men. Though no statistical analysis was presented, 200 mg caffeine was reported to mitigate the ethanol-induced decrement in simulated automobile-driving performance [102]. On the other hand, 300 mg/70 kg caffeine had no effect on the ethanol-induced decrease in standing steadiness and manual dexterity in university students despite mitigating ethanol-induced slowing of simple reaction times [103]. Moreover, 250 and 500 mg caffeine had no effect on the ethanol-induced decline in equilibrium, visuomotor coordination, hand steadiness, or choice reaction times in male medical students [104], nor did 300 mg caffeine affect equilibrium or hand steadiness in adult volunteers [105]. More paradoxically, 150 mg of caffeine even potentiated the ethanol-induced increase in reaction times of the Sternberg memory set task in adult volunteers [106]. Fudin and Nicastro [107] concluded that 'legally intoxicated individuals cannot antagonize alcohol-induced, driving-related decrements with caffeine prior to driving an automobile'. At the present time, 120 to 500 mg caffeine appears to mitigate ethanol-induced delays in responding, while there is little or no evidence of it affecting the loss in movement accuracy or standing steadiness.

Nevertheless, on the one hand, 20 mg/kg of caffeine countered the ethanol-induced decrease in latencies before falling from the rotorod in rats [108], as did 2.5 or 5 mg/kg of intracerebroventricular caffeine in mice [109]. On the other hand, still in mice, higher doses of 40 or 62.5 mg/kg caffeine potentiated the ethanol-induced decrease in latencies before falling from a thin bar [110], and 62.5 mg/kg potentiated the ethanol-induced decrease in latencies before falling from the rotorod [109], as did caffeine mixed in drinking water ($\approx$ 45 or 90 mg/kg per day) for ten days [111]. Thus, 2.5 – 20 mg/kg caffeine appears to minimize ethanol-induced incoordination in animals, while higher doses may potentiate them, which should prompt more detailed studies on standing balance in human subjects.

15.3.2 Other Sedatives and Benzodiazepines

No information is available on other sedatives except for an $\alpha2$ noradrenergic agonist when 100 (but not 25 or 50) mg/kg caffeine countered the increase in landing foot splay as a measure of ataxia in medetomidine-treated rats [112]. In an opposite manner to adenosine receptor agonists, 10 or 20 mg/kg caffeine potentiated diazepam-induced tolerance on rotorod and tube tests [113].

15.3.3 Glutamate

Interactions between caffeine and glutamate receptor agonists and antagonists have only been examined in animals. Administration of 10, 20, and 40 mg/kg caffeine for 21 days attenuated the decrease in latencies before falling from the rotorod in rats injected in the striatum with the n-methyl-d-aspartate (NMDA) receptor agonist, quinolate [114]. These results stand in contrast with the potentiation caused by 20 mg/kg caffeine on the decrease in latencies before falling from the rotorod in mice injected with the NMDA receptor antagonist, ketamine [115], whereas 186 mg/kg caffeine was without effect on another NMDA receptor antagonist, dizocilpine (MK-801), on an observed measure of ataxia [116]. Caffeine has also been examined in a diabetes mellitus model regarding its effect on glutamate. In particular, caffeine dissolved in drinking water (0.5 g/L) for three days but not three months countered the streptozotocin-induced decrease in latencies before falling from the rotorod in mice as well as minimizing the reduction in the NMDA-component of excitatory postsynaptic potentials recorded in the hippocampus [117], a study that should be pursued by measuring NMDA responses in basal ganglia and cerebellum.

15.3.4 Pesticides

On testing the effects of an organophosphate insecticide/acaricide/miticide, it was found that 20 and 40 mg/kg caffeine potentiated the chlorpyrifos-induced decrease in latencies before falling from the rotorod in mice [118]. In contrast, 20 mg/kg caffeine attenuated the loss of tyrosine hydroxylase positive neurons in the substantia nigra of mice after combined treatment of the herbicide, paraquat, and the fungicide, maneb [119]. Moreover, 10 mg/kg caffeine + l-dopa countered better than l-dopa alone the decrease in latencies before falling from the rotorod and brain dopamine levels caused by paraquat in mice [120]. Further studies of this type should be forthcoming, as these have an environmental impact.

15.4 HYPOXIA/ISCHAEMIA AND HYPOGLYCAEMIA

15.4.1 Hypoxia/ischaemia

Adenosine is released after hypoxia/ischaemia and hypoglycaemia to protect against nerve damage, justifying the use of A_1 receptor agonists under these conditions [9, 10, 121–123]. However, A_{2A} receptors are implicated in glutamate release and excitotoxicity, so that A_{2A} receptor antagonists such as caffeine are a second viable option. Caffeine has also been considered for treating hypoxia/ischaemia based on its bronchodilator and antioxidant properties [7, 124]. In mice exposed to mild hypoxia/ischaemia given at P10, 5 mg/kg caffeine countered the decrease in latencies before falling from the rotorod tested on P24 [125], but not severe hypoxia/ischaemia with the rotorod tested on P27 [126]. In addition, 20 mg/kg caffeine countered the P6 hypoxia/ischaemia-induced rotorod defect in rats tested on P35 [127]. Because caffeine mixed in drinking water (0.2%) for one month attenuated hippocampal damage after carotid artery occlusion in adult gerbils together with upregulating A_1 receptor sites [128], further experimentation on sensorimotor functions seems worthwhile. Under some conditions, chronic caffeine administration upregulated

TABLE 15.1
Caffeine effect on motor performances after ischaemic and traumatic lesions in the central nervous system

	Caffeine dose	Rotorod	Other tests	References
		Hypoxia-Ischaemia		
-mild form	5 mg/kg (P10)	↺(P24)	untested	125
-severe form	5 mg/kg (P10)	= (P27)	untested	126
-severe form	20 mg/kg (P6)	↺(P35)	untested	127
-large infact volume	10 mg+0.65 g/kg caffeinol	untested	Stat. beam ↺ (adult)	134
-small infact volume	10 mg+0.65 g/kg caffeinol	untested	Stat. beam ↺ (adult)	135
		Traumatic injury		
-TBI	0.25 g/L (21d pre-injury)	untested	SM score↑	146
-moderate TBI	0.25 g/L (30d pre-injury)	↑	untested	147
-severe TBI	0.25 g/L (30d pre-injury)	=	untested	147
-TBI	5–15 mg/kg (acute)	untested	SM score =	146
-TBI	25 mg/kg (acute)	=	untested	147
-TBI	100–150 mg/kg (pre-injury)	untested	SM score↓	148
-TBI	0.3 g/L (7d post-injury)	↓	untested	149
-TBI	25 mg/kg (4w post-injury)	untested	SM score =	150
-TSI	25 mg/kg (1w post-injury)	untested	LM score↓	151

LM score: locomotor score; P: postnatal day; SM score: sensorimotor score; Stat. beam: stationnary beam; d: day; w: week; TBI: traumatic brain injury; TSI: traumatic spinal injury.

↓: decreased latencies, ↑: increased latencies or ↺: counteracting decreased latencies before falling from rotorod; =: no effect.

A_1 but not A_{2A} receptors [129–131], whereas under other conditions, caffeine upregulated both receptors [132].

A dose of 10 mg/kg caffeine combined with 0.65 g/kg ethanol (designated as caffeinol) reduced infarct volume in adult rats after common carotid artery (CCA) and middle cerebral artery (MCA) occlusion [133]. This caffeinol solution countered the increase in missteps on the stationary beam in rats after CCA and MCA occlusion [134]. On the contrary, the same caffeinol solution had no effect on missteps or latencies before falling on the stationary beam in the same rat model with a smaller infarct volume [135], perhaps because of a floor effect on the task measure. All these data are summarized in Table 15.1.

The role of more selective adenosine receptor agonists or antagonists has come under scrutiny. The success of adenosine receptor agonists or antagonists on ischaemia depends on the receptor subtype and whether administration is acute or chronic. Under some conditions, acute administration of A_1 receptor agonists protects against brain damage, while chronic administration potentiates it, whereas A_1 receptor antagonists have the reverse pattern [15]. A selective A_{2A} receptor antagonist, SCH 58261, improved a neurological score after unilateral MCA occlusion in rats [136].

15.4.2 Hypoglycaemia

The caffeine reversal of the hypoglycaemia-induced decrement in hippocampal CA1 population spikes [137] is a further indication of its potential use in neuroprotection and remains to be examined in regard to motor functions.

15.5 TRAUMATIC INJURY

15.5.1 Brain Trauma

Because traumatic brain injury (TBI) may result in apnoea or other breathing difficulties [138], adenosine receptor antagonists have been considered for pharmacotherapy of this condition [139]. In patients with TBI, a favourable outcome on the Glasgow coma scale was more often seen in patients with elevated cerebrospinal caffeine concentrations than those with lower concentrations [140], a result which should prompt interest in pharmacological studies of the methylxanthine in this field.

Favourable outcomes have been obtained with caffeine [141, 142] or nanocoffee particles [143] on cognition and anxiety in animals exposed to TBI, but sometimes the effects have been unfavourable [144, 145]. Only a few reports concern neurological scores or specific motor coordination tests (Table 15.1). On one hand, pre-treatment with caffeine dissolved in drinking water (0.25 g/L) for 21 days improved the sensorimotor score of mice exposed to TBI, whereas acute 5 or 15 mg/kg caffeine had no effect [146]. The same pre-treatment for 30 days increased latencies before falling from the rotorod in rats under moderate though not severe TBI, whereas acute 25 mg/kg caffeine had no effect [147]. On the other hand, still in rats, 100 or 150 mg/kg caffeine prior to TBI increased righting latencies and worsened sensorimotor scores [148], caffeine dissolved in drinking water (0.3 g/L) for seven days after TBI decreased latencies before falling from the rotorod [149], and 25 mg/kg caffeine for four weeks after TBI had no effect on a sensorimotor score [150]. Thus, the outcome appears favourable with long-term caffeine pre-treatment rather than short- or long-term post-treatment. During development, caffeine dissolved in drinking water (1 g/L) from P21 to P69 had no effect on the increase in hindlimb slips on the stationary beam in rats exposed to repetitive mild TBI [144, 145].

15.5.2 Spinal Trauma

One week of 30 mg/kg caffeine worsened the Bresnahan locomotor rating scale score in rats exposed to traumatic spinal injury (TSI) [151]. Conversely, an A_{2A} receptor agonist, ATL-313, reduced motor deficits in mice exposed to TSI [152]. Likewise, another A_{2A} receptor agonist, ATL-146e, improved the neurologic outcome after TSI in rabbits [153-155]. Yet the A_{2A} receptor antagonist, SCH 58261, improved hindlimb movement in mice after TSI [156]. As found in other neurological conditions, diverse effects of adenosine receptor agonists or antagonists depend on the dose, time of assessment, and age of the animal [157].

15.6 NON-CEREBELLAR NEURODEGENERATIVE DISEASES

15.6.1 Amyotrophic Lateral Sclerosis

Patients with amyotrophic lateral sclerosis (ALS) had lower lifetime coffee intake than neurologic and non-neurologic controls [158], but there was no relation between caffeine intake and ALS risk [159] or between coffee, tea, or caffeine intake and ALS-linked mortality [160]. Nevertheless, methylxanthine pharmacotherapy is possibly indicated by the finding that cerebrospinal levels of adenosine were higher in ALS patients than controls [161].

Caffeine has been examined in several animal models of neurodegenerative disease, including ALS [121, 122, 162]. Caffeine dissolved in drinking water (0.3 g/L) from P70 until death had no effect on latencies before falling from the rotorod but decreased its variability toward a more severe phenotype in *SOD1*^{G93} transgenic mice overexpressing an ALS-linked mutation of superoxide dismutase-1 [163]. A_{2A} but not A_1 adenosine receptor expression was higher than wild-type controls in the spinal cord of these *SOD1*^{G93} transgenic mice, the former receptor subtype being more abundant in normal mice [164]. Evoked endplate potentials in the neuromuscular junction of

SOD1^{G93} transgenics were higher than those of wild-type after administration of the A_{2A} receptor agonist, CGS 21680, an effect prevented by the A_{2A} receptor antagonist, SCH 58261 [165], data justifying further tests of such agents. Chronic but not acute administration of divers A_{2A} adenosine receptor antagonists protected motoneurons from excitotoxic damage in rat spinal cord cultures [166]. At 10 μM, caffeine increased neurite length in a motoneuron-enriched dissociated cell culture from the ventral horn of the spontaneous *wobbler* mouse mutation of *Vps54*, encoding vacuolar protein sorting 54 and characterized by motoneuron degeneration [167], so that behavioural testing of this mutant appears like a worthwhile endeavour.

15.6.2 Parkinson's Disease

High caffeine intake is associated with a lower risk of Parkinson's disease [168, 169], so pharmacological studies are of interest in this condition [170, 171]. In an open-label escalation study, up to 400 mg per day of caffeine improved scores on the Unified Parkinson's Disease Rating Scale (UPDRS) and the Parkinson's Disease Questionnaire-39 items (PDQ-39) [172]. In a randomized double-blind trial, caffeine doses of 200 mg per day for three weeks and 400 mg per day for three more weeks also improved the UPDRS [173]. But in another randomized double-blind trial, caffeine doses gradually increased to 400 mg per day in week nine and lasting a total of 6 months had no effect on the revised UPDRS [174]. In two more double-blind studies, acute 1 g/kg caffeine had no effect on the antiparkinsonian action of bromocriptine [175], nor did caffeine gradually increase to a maximal level of 1.4 g/kg per day for seven to ten days on the antiparkinsonian action of l-dopa or piribedil [176].

Yet several experiments show favourable effects of caffeine in dopamine-depleted animal models (Table 15.2). Deficits in the vertical pole test in mice occurred after peripherally injected 1-methyl-4-phenyl-1,2,3,6-tetrahydropyridine (MPTP) combined with 3-nitropropionic acid, an inhibitor of succinate dehydrogenase [177] or MPTP combined with probenecid, a substance that delays MPTP metabolism [178] as well as unilateral 6-OHDA injections in the medial forebrain bundle [179]. Still in mice, 20 mg/kg caffeine countered the increase in pole-climbing time [180] and the decrease in latencies before falling from the rotorod, stride length [181], and striatal dopamine levels [180,

TABLE 15.2
Behavioural and structural effects of caffeine in animal models of Parkinson's disease

Animal model	Caffeine dose	Behaviour	DA neurons	References
MPTP mice	20 mg/kg	Vertical pole test↺	DA level ↺ (striatum)	180
MPTP mice	20 mg/kg	Rotorod lat.↺ Stride length↺	DA level ↺ (striatum)	181
MPTP mice	30 mg/kg	Grip strength↺	untested	170
MPTP rats	1 g/L (28d)	untested	TH-neurons ↺ (SN)	182
MPTP mice	10 mg/kg	untested	DA level and reuptake sites ↺ (striatum)	183
6-OHDA rat	15 mg/kg	Contralateral stepping ↺	untested	184
6-OHDA rat	1 – 3 mg/kg (3 – 14d)	Contralateral stepping↺	untested	185 – 187
MDMA mice	10 mg/kg	untested	TH-neurons↓ (SN)	188
SNCA transgenic mice	50 mg/kg	Rotorod lat. = Bar hanging =	untested	189

DA: dopamine; SN: substantia nigra; TH: tyrosine hydroxylase; lat.: latency; ↓: decreased number of cells; ↺: counteracting the decrease of performance, cell numbers, or DA level; =: no effect.

181] caused by MPTP, while 30 mg/kg caffeine countered the decrease in grip strength caused by MPTP [170]. Caffeine mixed in drinking water (1 g/L) for 28 days mitigated the decrease in tyrosine hydroxylase positive cells in the substantia nigra of MPTP-treated rats [182] and 10 mg/kg caffeine attenuated the decrease in striatal dopamine levels and the density of dopamine reuptake sites of MPTP-treated mice, the former effect being reproduced after administration of SCH 58261 as well as two other selective A_{2A} receptor antagonists: 3,7-dimethyl-1-propargylxanthine (DMPX) and KW-6002 [183].

In the rat 6-OHDA model of Parkinson's disease injected unilaterally in the medial forebrain bundle, 15 mg/kg caffeine increased the number of contralateral steps, and caffeine + l-dopa increased the same behaviour better than l-dopa alone [184]. Likewise, 1 or 3 mg/kg caffeine for three or 14 days alleviated the decrease in contralateral forepaw steps caused by a unilateral 6-OHDA injection in the substantia nigra [185-187].

In contrast to MPTP and 6-OHDA, 10 mg/kg caffeine potentiated the loss of tyrosine-hydroxylase positive cells in the substantia nigra of mice injected with 3,4-methylenedioxymethamphetamine (MDMA) [188]. An estimated dose of 50 mg/kg caffeine per day in drinking water had no effect on rotorod and bar-hanging tests unless combined with another coffee component, eicosanoyl-5-hydroxytryptamide, in transgenic mice overexpressing *SNCA*, encoding α-synuclein, implicated in synapse trafficking and Parkinson's disease [189].

A dose of 25 mg/kg caffeine mitigated the loss in striatal dopamine content in wild-type but not A_{2A} adenosine receptor null mutant mice caused by MPTP [190], data in line with the reversal of cell loss in the substantia nigra pars compacta in MPTP-treated mice after administration of selective A_{2A} receptor antagonists, SCH 58261 [191] and ST1535 [192], the SCH 58261-induced reversal of muscle rigidity caused by haloperidol or reserpine in rats [193], and the faster reaction times in 6-OHDA-treated rats with another A_{2A} receptor antagonist, 1,3,7-trimethyl-8-(3-chlorostyryl) caffeine [194], all of these effects underlining the importance of the A_{2A} receptor in parkinsonian pharmacotherapy [195, 196].

15.6.3 Alzheimer's Disease

In view of caffeine-induced neuroprotection in Alzheimer's disease [122, 197–200] and its action as an amyloid-β inhibitor [201], the methylxanthine has been examined in murine models of the disorder, with improvements often found in cognitive functions [202–210], though sometimes with decrements [211]. Few have examined motor coordination, except when caffeine dissolved in drinking water (300 mg/L) for four months rescued the decrease in latencies before falling from the stationary beam in Tg4-42. Transgenic mice overexpressing the Aβ4-42 peptide sequence [210]. Chronic administration of a coffee component other than caffeine, eicosanoyl-5-hydroxytryptamide, mitigated the spatial deficit in the Morris water maze caused by viral expression of the inhibitor protein (I2-PP2A) of protein phosphatase 2 catalytic subunit alpha without affecting stationary beam, bar-holding, and foot-print analyses [212].

15.7 CEREBELLAR DISORDERS

The effects of caffeine on cerebellar-related motor functions have been examined in episodic ataxia types 1 and 2, spinocerebellar atrophy (SCA), and ataxia telangiectasia.

15.7.1 Episodic Ataxia Types 1 and 2

Episodic ataxia type 1 is an autosomal dominant disorder characterized by motor coordination deficits, vertigo, epilepsy, and spastic muscular contractions caused by variants of the *KCNA1* gene, encoding potassium voltage-gated channel subfamily A member 1 [213, 214]. The disease is triggered by caffeine, stress, abrupt movements, and loud noises. Episodic ataxia type 2 is an autosomal dominant disorder characterized by episodic cerebellar symptoms caused by loss-of-function mutations of the *CACNA1A* gene, encoding the α1A subunit of the CaV2.1 voltage-gated (P/Q-type) Ca^{2+} channel and also precipitated by caffeine and stress [213, 215–218]. More recently, case studies were presented of a disease similar to these two and characterized by late-onset episodic gait ataxia, downbeat nystagmus, and vertical oscillopsia also precipitated by caffeine and stress [219].

In a similar fashion to episodic ataxia type 2, the *tottering* (*tg*) mouse is a spontaneous mutation of the autosomal recessive type involving a missense mutation of *Cacna1a* [220–222], characterized by ataxia and absence seizures along with dystonia and dyskinesia triggered by caffeine and stress [223–231], the dystonia being the result of interactions between the cerebellum and the basal ganglia [232]. Low-frequency oscillations in the *Cacna1a*^{tg} cerebellar cortex were dampened by extracellular Ca^{2+} or by blocking L-type Ca^{2+} channels [233]. A dose of 15 mg/kg caffeine not only triggered motor attacks but accentuated such oscillations. In particular, *Cacna1a*^{tg} exhibited high-frequency burst firing in Purkinje cells [231]. It is proposed that caffeine initiates oscillations by mobilizing Ca^{2+} from Purkinje cell stores, which in turn depolarizes the cell membrane and opens up L-type Ca^{2+} channels. The L-type calcium channel antagonist, nimodipine, and the NMDA receptor antagonist, dizocilpine, prevented motor attacks caused by 15 mg/kg caffeine [227] and 5 or 15 mg/kg caffeine-induced motor attacks were also inhibited by K^+ channel blockers [234]. In line with several animal models of cerebellar atrophy sensitive to the rotorod and other motor coordination tasks [235], *Cacna1a* knock-in mice with a missense mutation causing human episodic ataxia type 2 were generated with defects in rotorod and vertical pole tests, though without affecting the vertical grid test [236].

15.7.2 Spinocerebellar Atrophy

In contrast to episodic ataxia types 1 and 2, in which caffeine triggers attacks, the methylxanthine has been assessed to improve symptoms of spinocerebellar atrophy (SCA), consisting of several subtypes of ataxin genes. Relative to wild-type controls, *ATX3* transgenic mice overexpressing ataxin-3, a model of SCA3 Machado-Joseph disease, caffeine dissolved in drinking water (1 g/L) for periods from eight to 20 weeks attenuated the decrease in latencies before falling from the rotorod and the stationary beam, the increase in latencies before descending from a vertical pole, and the weakening in grip strength [237]. Caffeine also attenuated the loss in Purkinje cells and the thinning of the cerebellar molecular layer. Likewise, the same caffeine solution for three weeks attenuated synaptotoxicity, astrogliosis, and microgliosis in another SCA3 murine model caused by the unilateral striatal injection of 72Q-mutated *ATXN3* with lentiviral vectors [238].

15.7.3 Ataxia Telangiectasia

Ataxia telangiectasia is an autosomal recessive disorder characterized by cerebellar degeneration, the appearance of small, widened blood vessels on the skin (telangiectasia), immunodeficiency, and increased risks of cancer and radiation sensitivity associated with DNA damage [239]. The disease is caused by mutations of the *ATM* gene, encoding the ataxia telangiectasia mutated (ATM) protein. Caffeine serves as an inhibitor in the activity of the ATM enzyme [240, 241]. Caffeine increased cell viability in the MPP+ model of apoptosis on primary cultures of cerebellar granule cells by attenuating the activation of the ATM/p53 pathway [242]. The caffeine-mediated neuroprotection occurred independently of its blockade of adenosine receptors because two other antagonists of these receptors failed to attenuate apoptosis produced by MPP+ treatment. Conversely, when acetaminophen caused oxidative stress, mitochondrial membrane potential impairments, and DNA strand breaks requiring ATM-dependent repair in human immortalized hepatocytes and liver organoids, caffeine exacerbated toxicity by its ATM-suppressing action [243]. Thus, caffeine can either mitigate or potentiate ATM-related anomalies.

15.7.4 Joubert Syndrome

Joubert syndrome is an autosomal recessive or chromosome X-linked disorder characterized by cerebellar ataxia, oculomotor incoordination, hypotonia, developmental delay, and breathing difficulties caused by one of 21 genes encoding proteins of the primary cilium or its apparatus [244, 245]. Caffeine is used to help breathing patterns in Joubert syndrome [246], but, to our knowledge, it remains to be evaluated in regard to motor coordination.

15.8 HUNTINGTON'S DISEASE

The favourable effect of caffeine on Parkinson's disease may be extended to other basal ganglia-related diseases since subjects with Huntington's disease manifest motor symptoms, but also their controls had better scores on the Unified Huntington's Disease Rating Scale (UHDRS) [247].

15.9 CONCLUDING REMARKS

The main purpose of the present review was to answer three interrogations. First of all, does caffeine help to coordinate movements in normal subjects? The answer appears to be that favourable effects on response accuracy appear confined to narrow dose ranges. Second of all, is caffeine helpful in coordinating the movements of patients with neural disorders? This appears true in pre-term infants and animal models of hypoxia/ischaemia, TBI, Parkinson's disease, and spinocerebellar atrophy. Third of all, under which conditions does caffeine worsen motor coordination? Worsening occurred at high doses in normal subjects as well as those under the influence of ethanol and various neurological conditions, including hypoxia/ischaemia, TBI, episodic ataxia types 1 and 2, and ataxia telangiectasia. Experiments regarding ALS and Alzheimer's disease are in the preliminary stages.

ACKNOWLEGMENTS

The authors thank laboratory members for the discussions.

REFERENCES

[1] Fredholm BB, Bättig K, Holmén J, Nehlig A, Zvartau EE. Actions of caffeine in the brain with special reference to factors that contribute to its widespread use. *Pharmacol Rev* 1999;51:83–133. PMID: 10049999

[2] Axelrod J, Reichenthal J. The fate of caffeine in man and a method for its estimation in biological material. *J Pharmacol Exp Ther* 1953;107(4):519–23. PMID: 13053413

[3] Beach CA, Bianchine JR, Gerber N. The excretion of caffeine in the semen of men: comparison of the concentration in blood and semen. *Proc West Pharmacol Soc* 1982;25:377–80. PMID: 7122524

[4] Cook CE, Tallent CR, Amerson EW, Myers MW, Kepler JA, Taylor GF, Christensen HD. Caffeine in plasma and saliva by a radioimmunoassay procedure. *J Pharmacol Exp Ther* 1976;199(3):679–86. PMID: 1033273

[5] Burdan F. Pharmacology of caffeine: the main active compound of coffee. In: *Preedy V (ed) Coffee in Health and Disease Prevention*. New York: Academic Press, 2015, 823–9.

[6] dePaula J, Farah A. Caffeine consumption through coffee: content in the beverage, metabolism, health benefits and risks. *Beverages* 2019;5:37. doi: 10.3390/beverages5020037.

[7] Gong H Jr, Simmons MS, Tashkin DP, Hui KK, Lee EY. Bronchodilator effects of caffeine in coffee. A dose-response study of asthmatic subjects. *Chest* 1986;89(3):335–42. doi: 10.1378/chest.89.3.335. PMID: 3948545

[8] Borea PA, Gessi S, Merighi S, Vincenzi F, Varani K. Pharmacology of adenosine receptors: the state of the art. *Physiol Rev* 2018;98(3):1591–625. doi: 10.1152/physrev.00049.2017. PMID: 29848236

[9] Dunwiddie TV. The physiological role of adenosine in the central nervous system. *Int Rev Neurobiol* 1985;27:63–139. doi: 10.1016/s0074-7742(08)60556-5. PMID: 2867982

[10] Dunwiddie TV, Masino SA. The role and regulation of adenosine in the central nervous system. *Annu Rev Neurosci* 2001;24:31–55. doi: 10.1146/annurev.neuro.24.1.31. PMID: 11283304

[11] Fisone G, Borgkvist A, Usiello A. Caffeine as a psychomotor stimulant: mechanism of action. *Cell Mol Life Sci* 2004;61(7-8):857–72. doi: 10.1007/s00018-003-3269-3. PMID: 15095008

[12] Daly JW, Shi D, Nikodijevic O, Jacobson KA. The role of adenosine receptors in the central action of caffeine. *Pharmacopsychoecologia (Berl)* 1994;7(2):201–13. PMID: 25821357

[13] Ribeiro JA, Sebastião AM, de Mendonça A. Adenosine receptors in the nervous system: pathophysiological implications. *Prog Neurobiol.* 2002;68(6):377–92. doi: 10.1016/s0301-0082(02)00155-7. PMID: 12576292

[14] Dixon AK, Gubitz AK, Sirinathsinghji DJ, Richardson PJ, Freeman TC. Tissue distribution of adenosine receptor mRNAs in the rat. *Br J Pharmacol* 1996;118(6):1461–8. doi: 10.1111/j.1476-5381.1996.tb15561.x. PMID: 8832073

[15] von Lubitz DK, Carter MF, Beenhakker M, Lin RC, Jacobson KA. Adenosine: a prototherapeutic concept in neurodegeneration. *Ann NY Acad Sci* 1995;765:163–78; discussion 196–7. doi: 10.1111/j.1749-6632.1995.tb16573.x. PMID: 7486604

[16] Wardas J. Neuroprotective role of adenosine in the CNS. *Pol J Pharmacol* 2002;54(4):313–26. PMID: 12523485

[17] Watson J. Caffeines. In: Caballero B, Finglas PM, Toldra F (eds) *Encyclopedia of Food Sciences and Nutrition.* Cambridge: Academic Press, 2003, 745–50.

[18] Landolt HP. Caffeine, the circadian clock, and sleep. *Science* 2015;349(6254):1289. doi: 10.1126/science.aad2958. PMID: 26383940

[19] Ball KT, Poplawsky A. Low-dose oral caffeine induces a specific form of behavioral sensitization in rats. *Pharmacol Rep* 2011;63(6):1560–3. doi: 10.1016/s1734-1140(11)70721-6. PMID: 22358105

[20] Kuribara H, Asami T, Ida I, Tadokoro S. Characteristics of the ambulation-increasing effect of the noncompetitive NMDA antagonist MK-801 in mice: assessment by the coadministration with central-acting drugs. *Jpn J Pharmacol* 1992;58(1):11–8. doi: 10.1254/jjp.58.11. PMID: 1640658

[21] Imaki T, Shibasaki T, Masuda A, Demura H, Shizume K, Ling N. Effects of adrenergic blockers on corticotropin-releasing factor-induced behavioral changes in rats. *Regul Pept* 1987;19(3–4):243–51. doi: 10.1016/0167-0115(87)90280-1. PMID: 3501600

[22] Assis MS, Soares AC, Sousa DN, Eudes-Filho J, Faro LRF, Carneiro FP, Silva MV, Motoyama AB, Souza GMR, Marchiori S, Lima NT, Boëchat-Barros R, Ferreira VM. Effects of caffeine on behavioural and cognitive deficits in rats. *Basic Clin Pharmacol Toxicol* 2018;123(4):435–42. doi: 10.1111/bcpt.13036. PMID: 29736913

[23] Ettarh RR, Okoosi SA, Eteng MU. The influence of kolanut (*Cola nitida*) on exploratory behaviour in rats. *Pharm Biol* 2000;38:281–3. https://doi.org/10.1076/1388-0209(200009)3841-AFT281. PMID: 21214476s

[24] Snyder SH, Katims JJ, Annau Z, Bruns RF, Daly JW. Adenosine receptors and behavioral actions of methylxanthines. *Proc Natl Acad Sci USA* 1981;78(5):3260–4. doi: 10.1073/pnas.78.5.3260. PMID: 6265942

[25] Lazarus M, Shen HY, Cherasse Y, Qu WM, Huang ZL, Bass CE, Winsky-Sommerer R, Semba K, Fredholm BB, Boison D, Hayaishi O, Urade Y, Chen JF. Arousal effect of caffeine depends on adenosine A2A receptors in the shell of the nucleus accumbens. *J Neurosci* 2011;31(27):10067–75. doi: 10.1523/JNEUROSCI.6730-10.2011. PMID: 21734299

[26] Molinengo L, Orsetti M, Pastorello B, Scordo I, Ghi P. Habituation of exploratory activity in rats: action of N6phenylisopropyladenosine, caffeine and their combination. *Prog Neuropsychopharmacol Biol Psychiatry* 1995;19(7):1189–200. doi: 10.1016/0278-5846(95)00236-7.PMID: 8787042

[27] Schulz D. Acute food deprivation separates motor-activating from anxiolytic effects of caffeine in a rat open field test model. *Behav Pharmacol* 2018;29(6):543–6. doi: 10.1097/FBP.0000000000000396. PMID: 29543609

[28] Bhattacharya SK, Satyan KS, Chakrabarti A. Anxiogenic action of caffeine: an experimental study in rats. *J Psychopharmacol* 1997;11(3):219–24. doi: 10.1177/026988119701100304. PMID: 9305413
[29] File SE, Baldwin HA, A L Johnston AL, Wilks LJ. Behavioral effects of acute and chronic administration of caffeine in the rat. *Pharmacol Biochem Behav* 1988;30(4):809–15. doi: 10.1016/0091-3057(88)90104-9. PMID: 3227034
[30] El Yacoubi M, Ledent C, Parmentier M, Costentin J, Vaugeois JM. The anxiogenic-like effect of caffeine in two experimental procedures measuring anxiety in the mouse is not shared by selective A(2A) adenosine receptor antagonists. *Psychopharmacology (Berl)* 2000;148(2):153–63. doi: 10.1007/s002130050037. PMID: 10663430
[31] Pham K, Romeo RD. An undergraduate laboratory exercise examining the psychomotor stimulant effects of caffeine in laboratory rats. *J Undergrad Neurosci Educ* 2012;10(2):A132–9. PMID: 23504612
[32] Nehlig A, Lucignani G, Kadekaro M, Porrino LJ, Sokoloff L. Effects of acute administration of caffeine on local cerebral glucose utilization in the rat. *Eur J Pharmacol* 1984;101(1–2):91–100. doi: 10.1016/0014-2999(84)90034-7. PMID: 6086367
[33] Magalhaes R, Picó-Pérez M, Esteves M, Vieira R, Castanho TC, Amorim L, Sousa M, Coelho A, Fernandes HM, Cabral J, Moreira PS, Sousa N. Habitual coffee drinkers display a distinct pattern of brain functional connectivity. *Mol Psychiatry* 2021;26(11):6589–98. doi: 10.1038/s41380-021-01075-4. PMID: 33875801
[34] Einöther SJ, Giesbrecht T. Caffeine as an attention enhancer: reviewing existing assumptions. *Psychopharmacology (Berl)* 2013;225(2):251–74. doi: 10.1007/s00213-012-2917-4. PMID: 23241646
[35] McLellan TM, Caldwell JA, Lieberman HR. A review of caffeine's effects on cognitive, physical and occupational performance. *Neurosci Biobehav Rev* 2016;71:294–312. doi: 10.1016/j.neubiorev.2016.09.001. PMID: 27612937
[36] Nehlig A, Daval JL, Debry G. Caffeine and the central nervous system: mechanisms of action, biochemical, metabolic and psychostimulant effects. *Brain Res Rev* 1992;17(2):139–70. doi: 10.1016/0165-0173(92)90012-b. PMID: 1356551
[37] Smith AP. Caffeine and road traffic accidents. *World J Pharmaceut Med Res* 2021;7(12):1–4.
[38] Baker LB, Nuccio RP, Jeukendrup AE. Acute effects of dietary constituents on motor skill and cognitive performance in athletes. *Nutr Rev* 2014;72(12):790–802. doi: 10.1111/nure.12157. PMID: 25400063
[39] Burke LM. Caffeine and sports performance. *Appl Physiol Nutr Metab* 2008;33(6):1319–34. doi: 10.1139/H08-130. PMID: 19088794
[40] Contreras-Barraza N, Madrid-Casaca H, Salazar-Sepúlveda G, Garcia-Gordillo MÁ, Adsuar JC, Vega-Muñoz A. Bibliometric analysis of studies on coffee/caffeine and sport. *Nutrients* 2021;13(9):3234. doi: 10.3390/nu13093234. PMID: 34579111
[41] Soos R, Gyebrovszki A, Tóth A, Jeges S, Wilhelm M. Effects of caffeine and caffeinated beverages in children, adolescents and young adults: short review. *Int J Environ Res Public Health* 2021;18(23):12389. doi: 10.3390/ijerph182312389. PMID: 34886115
[42] Foskett A, Ali A, Gant N. Caffeine enhances cognitive function and skill performance during simulated soccer activity. *Int J Sport Nutr Exerc Meta* 2009;19:410–23. doi: 10.1123/ijsnem.19.4.410; PMID: 19827465
[43] Jordan JB, Korgaokar A, Farley RS, Coons JM, Caputo JL. Caffeine supplementation and reactive agility in elite youth soccer players. *Pediatr Exerc Sci* 2014;26(2):168–76. doi: 10.1123/pes.2013-0134. PMID: 24277927
[44] Duvnjak-Zaknich DM, Dawson BT, Wallman KE, Henry G. Effect of caffeine on reactive agility time when fresh and fatigued. *Med Sci Sports Exerc* 2011;43(8):1523–30. doi: 10.1249/MSS.0b013e31821048ab. PMID: 21266929
[45] Marsden G, Leach J. Effects of alcohol and caffeine on maritime navigational skills. *Ergonomics* 2000;43(1):17–26. doi: 10.1080/001401300184639. PMID: 10661690
[46] Tucha O, Walitza S, Mecklinger L, Stasik D, Sontag TA, Lange KW. The effect of caffeine on handwriting movements in skilled writers. *Hum Mov Sci* 2006;25(4–5):523–35. doi: 10.1016/j.humov.2006.06.001. PMID: 17010462

[47] Bottoms L, Greenhalgh A, Gregory K. The effect of caffeine ingestion on skill maintenance and fatigue in epée fencers. *J Sports Sci* 2013;31(10):1091–9. doi: 10.1080/02640414.2013.764466. PMID: 23383959

[48] Jacobson BH, Edgley BM. Effects of caffeine on simple reaction time and movement time. *Aviat Space Environ Med* 1987;58(12):1153–6. PMID: 3426488

[49] Rush CR, Higgins ST, Hughes JR, Bickel WK, Wiegner MS. Acute behavioral and cardiac effects of alcohol and caffeine, alone and in combination, in humans. *Behav Pharmacol* 1993;4:562–72. PMID: 11224225

[50] Kaplan GB, Greenblatt DJ, Ehrenberg BL, Goddard JE, Cotreau MM, Harmatz JS, Shader RI. Dose-dependent pharmacokinetics and psychomotor effects of caffeine in humans. *J Clin Pharmacol* 1997;37(8):693–703. doi: 10.1002/j.1552-4604.1997.tb04356.x. PMID: 9378841

[51] Hogervorst E, Bandelow S, Schmitt J, Jentjens R, Oliveira M, Allgrove J, Carter T, Gleeson M. Caffeine improves physical and cognitive performance during exhaustive exercise. *Med Sci Sports Exerc* 2008;40(10):1841–51. doi: 10.1249/MSS.0b013e31817bb8b7. PMID: 18799996

[52] Çildir B, Altin B, Aksoy S. Caffeine enhances the balance system and postural balance in short time in healthy individuals. *Turk Arch Otorhinolaryngol* 2021;59(4):253–260. doi: 10.4274/tao.2021.2021-4-17. PMID: 35262042

[53] Mitchell ES, Slettenaar M, vd Meer N, Transler C, Jans L, Quadt F, Berry M. Differential contributions of theobromine and caffeine on mood, psychomotor performance and blood pressure. *Physiol Behav* 2011;104(5):816–22. doi: 10.1016/j.physbeh.2011.07.027. PMID: 21839757

[54] Smith AP. Caffeine, chocolate, performance and mood. *World J Pharmaceut Res* 2021b;10(14)180–8. doi: 10.20959/wjpr202114-22277

[55] Terry P, Doumas M, Desai RI, Wing AM. Dissociations between motor timing, motor coordination, and time perception after the administration of alcohol or caffeine. *Psychopharmacology (Berl)* 2008;202(4):719–29. doi: 10.1007/s00213-008-1352-z. PMID: 18846366

[56] Glowka N, Malik J, Podgórski T, Stemplewski R, Maciaszek J, Ciążyńska J, Zawieja EE, Chmurzynska A, Nowaczyk PM, Durkalec-Michalski K. The dose-dependent effect of caffeine supplementation on performance, reaction time and postural stability in CrossFit- a randomized placebo-controlled crossover trial. *J Int Soc Sports Nutr* 2024;21(1):2301384. doi: 10.1080/15502783.2023.2301384. PMID: 38226646

[57] Gomes AC, Tosini L, Oliveira D, Lage G, Franchini E, Meira C Jr. Caffeine produces neutral effects on extraverts' and introverts' performance of fundamental motor skills. *J Hum Kinet* 2021;78:229–37. doi: 10.2478/hukin-2021-0038

[58] Kara M, Patlar S, Stofferen TA, Erkmen N. Effect of caffeine on standing balance during perceptual-cognitive tasks. *Malaysian J Mov Health Exercise* 2018;7(2):167–75. doi: 10.15282/MOHE.V7I2.181

[59] Bovim G, Naess P, Helle J, Sand T. Caffeine influence on the motor steadiness battery in neuro-psychological tests. *J Clin Exp Neuropsychol* 1995;17(3):472–6. doi: 10.1080/01688639508405138. PMID: 7650108

[60] Hughes JR, Higgins ST, Bickel WK, Hunt WK, Fenwick JW, Gulliver SB, Mireault GC. Caffeine self-administration, withdrawal, and adverse effects among coffee drinkers. *Arch Gen Psychiatry* 1991;48(7):611–7. doi: 10.1001/archpsyc.1991.01810310029006. PMID: 2069491

[61] Loke WH, Hinrichs JV, Ghoneim MM. Caffeine and diazepam: separate and combined effects on mood, memory, and psychomotor performance. *Psychopharmacology (Berlin)* 1985;87(3):344–50. doi: 10.1007/BF00432719. PMID: 3936091

[62] Koller W, Cone S, Herbster G. Caffeine and tremor. *Neurology* 1987;37(1):169–72. doi: 10.1212/wnl.37.1.169. PMID: 3796831

[63] Jacobson BH, Thurman-Lacey SR. Effect of caffeine on motor performance by caffeine-naive and -familiar subjects. *Percept Mot Skills* 1992;74(1):151–7. doi: 10.2466/pms.1992.74.1.151. PMID: 1561022

[64] Jacobson BH, Winter-Roberts K, Gemmell HA. Influence of caffeine on selected manual manipulation skills. *Percept Mot Skills* 1991;72(3 Pt 2):1175–81. doi: 10.2466/pms.1991.72.3c.1175. PMID: 1961665

[65] Mednick SC, Cai DJ, Kanady J, Drummond SP. Comparing the benefits of caffeine, naps and placebo on verbal, motor and perceptual memory. *Behav Brain Res* 2008;193(1):79–86. doi: 10.1016/j.bbr.2008.04.028. PMID: 18554731

[66] Hussain SJ, Cole KJ. No enhancement of 24-hour visuomotor skill retention by post-practice caffeine administration. *PLoS One* 2015;10(6):e0129543. doi: 10.1371/journal.pone.0129543. PMID: 26053288

[67] Laatar R, Ben Waer F, Rebai H, Sahli S. Caffeine consumption improves motor and cognitive performances during dual tasking in middle-aged women. *Behav Brain Res* 2021;412:113437. doi: 10.1016/j.bbr.2021.113437. PMID: 34182008

[68] Duncan MJ, Clarke ND, Tallis J, Guimarães-Ferreira L, Leddington Wright S. The effect of caffeine ingestion on functional performance in older adults. J *Nutr Health Aging* 2014;18(10):883–7. doi: 10.1007/s12603-014-0474-8. PMID: 25470803

[69] Rees K, Allen D, Lader M. The influences of age and caffeine on psychomotor and cognitive function. *Psychopharmacology (Berl)* 1999;145(2):181–8. doi: 10.1007/s002130051047. PMID: 10463319

[70] Boissier J-R, Tardy J, Diverres J-C. Une nouvelle méthode simple pour explorer l'action "tranquillisante": le test de la cheminée. *Med Exp* 1960;3(1):81–4. doi: 10.1159/000134913.

[71] Brooks SP, Trueman RC, Dunnett SB. Assessment of motor coordination and balance in mice using the rotarod, elevated bridge, and footprint tests. *Curr Protoc Mouse Biol* 2012;2(1):37–53. doi: 10.1002/9780470942390.mo110165. PMID: 26069004

[72] Deacon RM. Measuring motor coordination in mice. *J Vis Exp* 2013;(75):e2609. doi: 10.3791/2609. PMID: 23748408

[73] Huston JP, Bures J. Innate and motivated behavior. In: Bures J, Buresova O, Huston JP (eds) *Techniques and Basic Experiments for the Study of Brain and Behavior*. Amsterdam: Elsevier, 1976, 37–89.

[74] Jones BJ, Roberts DJ. A rotarod suitable for quantitative measurements of motor incoordination in naive mice. *Naunyn Schmiedebergs Arch Exp Pathol Pharmakol* 1968;259(2):211. doi: 10.1007/BF00537801. PMID: 4232694

[75] Adeniyi PA, Omatsuli EP, Akinyemi AJ, Ishola AO. Caffeine plus nicotine improves motor function, spatial and non-spatial working memory and functional indices in BALB/c male mice. *Pathophysiology* 2016;23(4):251–8. doi: 10.1016/j.pathophys.2016.08.002. PMID: 27596362

[76] Almosawi S, Baksh H, Qareeballa A, Falamarzi F, Alsaleh B, Alrabaani M, Alkalbani A, Mahdi S, Kamal A. Acute administration of caffeine: the effect on motor coordination, higher brain cognitive functions, and the social behavior of BLC57 mice. *Behav Sci (Basel)* 2018;8(8):65. doi: 10.3390/bs8080065. PMID: 30044406

[77] Estler CJ, Ammon HP, Herzog C. Swimming capacity of mice after prolonged treatment with psychostimulants. I. Effects of caffeine on swimming performance and cold stress. *Psychopharmacology (Berl)* 1978;58(2):161–6. doi: 10.1007/BF00426900. PMID: 98784

[78] Lopez-Cruz L, Pardo M, Salamone JD, Correa M. Differences between the nonselective adenosine receptor antagonists caffeine and theophylline in motor and mood effects: studies using medium to high doses in animal models. *Behav Brain Res* 2014;270:213–22. doi: 10.1016/j.bbr.2014.05.020. PMID: 24859174

[79] Mahdi S, Almosawi S, Baksh H, Qareeballa A, Alsaleh B, Falamarzi F, Alrabaani M, Alkalbani A, Kamal A. Effect of chronic administration and withdrawal of caffeine on motor function, cognitive functions, anxiety, and the social behavior of BLC57 mice. *Int J Health Sci (Qassim)* 2019;13(2):10–6. PMID: 30983940

[80] Meyer L, Caston J. Repeated stress alters caffeine action on motor coordination in C57Bl6/J male mice. *Brain Res* 2005;1039(1–2):171–6. doi: 10.1016/j.brainres.2005.01.053. PMID: 15781059

[81] Plotnikoff N, Reinke D, Fitzloff J. Effects of stimulants on rotarod performance of mice. *J Pharm Sci* 1962;51:1007–8. doi: 10.1002/jps.2600511025. PMID: 13944087

[82] Olopade FE, Shokunbi MT. Chronic caffeine ingestion improves motor function and increases dendritic length and arborization in the motor cortex, striatum, and cerebellum. *J Caff Aden Res* 2021;11(1):27–36. doi: 10.1089/caff.2020.0017

[83] Shukitt-Hale B, Miller MG, Chu YF, Lyle BJ, Joseph JA. Coffee, but not caffeine, has positive effects on cognition and psychomotor behavior in aging. *Age (Dordr)* 2013;35(6):2183–92. doi: 10.1007/s11357-012-9509-4. PMID: 23344884

[84] Doyle J, Davidson D, Katz S, Varela M, Demeglio D, DeCristofaro J. Apnea of prematurity and caffeine pharmacokinetics: potential impact on hospital discharge. *J Perinatol* 2016;36(2):141–4. doi: 10.1038/jp.2015.167. PMID: 26562367

[85] Kumar VHS, Lipshultz SE. Caffeine and clinical outcomes in premature neonates. *Children (Basel)* 2019;6(11):118. doi: 10.3390/children6110118. PMID: 31653108

[86] Maitre NL, Stark AR. Neuroprotection for premature infants? another perspective on caffeine. *JAMA* 2012;307(3):304–5. doi: 10.1001/jama.2011.2026. PMID: 22253397

[87] McPherson C, Neil JJ, Tjoeng TH, Pineda R, Inder TE. A pilot randomized trial of high-dose caffeine therapy in preterm infants. *Pediatr Res* 2015;78(2):198–204. doi: 10.1038/pr.2015.72. PMID: 25856169

[88] Doyle LW, Schmidt B, Anderson PJ, Davis PG, Moddemann D, Grunau RE, O'Brien K, Sankaran K, Herlenius E, Roberts R. Caffeine for Apnea of Prematurity Trial investigators. Reduction in developmental coordination disorder with neonatal caffeine therapy. *J Pediatr* 2014;165(2):356–9.e2. doi: 10.1016/j.jpeds.2014.04.016. PMID: 24840756

[89] Schmidt B, Anderson PJ, Doyle LW, Dewey D, Grunau RE, Asztalos EV, Davis PG, Tin W, Moddemann D, Solimano A, Ohlsson A, Barrington KJ, Roberts RS, Caffeine for Apnea of Prematurity (CAP) Trial Investigators. Survival without disability to age 5 years after neonatal caffeine therapy for apnea of prematurity. *JAMA* 2012;307(3):275–82. doi: 10.1001/jama.2011.2024. PMID: 22253394

[90] Mürner-Lavanchy IM, Doyle LW, Schmidt B, Roberts RS, Asztalos EV, Costantini L, Davis PG, Dewey D, D'Ilario J, Grunau RE, Moddemann D, Nelson H, Ohlsson A, Solimano A, Tin W, Anderson PJ, Caffeine for Apnea of Prematurity (CAP) Trial Group. Neurobehavioral outcomes 11 years after neonatal caffeine therapy for apnea of prematurity. *Pediatrics* 2018;141(5):e20174047. doi: 10.1542/peds.2017-4047. PMID: 29643070

[91] Schmidt B, Roberts RS, Anderson PJ, Asztalos EV, Costantini L, Davis PG, Dewey D, D'Ilario J, Doyle LW, Grunau RE, Moddemann D, Nelson H, Ohlsson A, Solimano A, Tin W, Caffeine for Apnea of Prematurity (CAP) Trial Group. Academic performance, motor function, and behavior 11 years after neonatal caffeine citrate therapy for apnea of prematurity: an 11-year follow-up of the CAP randomized clinical trial. *JAMA Pediatr* 2017;171(6):564–72. doi: 10.1001/jamapediatrics.2017.0238. PMID: 28437520

[92] Barr HM, Streissguth AP, Darby BL, Sampson PD. Prenatal exposure to alcohol, caffeine, tobacco, and aspirin: effects on fine and gross motor performance in 4-year-old children. *Dev Psychol* 1990;26(3):339–48. doi: 10.1037/0012-1649.26.3.339.

[93] Streissguth AP, Barr HM, Martin DC, Herman CS. Effects of maternal alcohol, nicotine, and caffeine use during pregnancy on infant mental and motor development at eight months. *Alcohol Clin Exp Res* 1980;4(2):152–64. doi: 10.1111/j.1530-0277.1980.tb05630.x. PMID: 6990818

[94] Alsayed A, Jamal M, Al Hoty N, Eqab S, Ahmadi N, Sameer M, Alathem A, Al Saqer M, Alsulami B, Kamal A. Perinatal administration of caffeine in mice affects the spatial memory and motor coordination later in adult life. *J Brain Disord* 2017;1(1):4–3. doi: 10.36959/524/324

[95] Olopade FE, Shokunbi MT. The development of the external granular layer of the cerebellum and neurobehavioral correlates in neonatal rats following intrauterine and postnatal exposure to caffeine. *J Caff Aden Res* 2018;8(1):27–36. doi: 10.1089/caff.2017.0010

[96] De Souza AC, Souza A, Medeiros LF, De Oliveira C, Scarabelot VL, Da Silva RS, Bogo MR, Capiotti KM, Kist LW, Bonan CD, Caumo W, Torres IL. Maternal caffeine exposure alters neuromotor development and hippocampus acetylcholinesterase activity in rat offspring. *Brain Res* 2015;1595:10–8. doi: 10.1016/j.brainres.2014.10.039. PMID: 25451122

[97] Tchekalarova J, Kubova H, Mares P. Postnatal caffeine exposure: effects on motor skills and locomotor activity during ontogenesis. *Behav Brain Res* 2005;160(1):99–106. doi: 10.1016/j.bbr.2004.11.018. PMID: 15836904

[98] Lopez-Cruz L, Salamone JD, Correa M. The impact of caffeine on the behavioral effects of ethanol related to abuse and addiction: a review of animal studies. *J Caffeine Res* 2013;3(1):9–21. doi: 10.1089/jcr.2013.0003. PMID: 24761272

[99] Mackay M, Tiplady B, Scholey AB. Interactions between alcohol and caffeine in relation to psychomotor speed and accuracy. *Hum Psychopharmacol* 2002;17:151–6. doi: 10.1002/hup.371. PMID: 12404692

[100] Azcona O, Barbanoj MJ, Torrent J, Jane F. Evaluation of the central effects of alcohol and caffeine interaction. *Br J Clin Pharmacol* 1995;40:393–400. doi: 10.1111/j.1365-2125.1995.tb04562.x. PMID: 8554942

[101] Hasenfratz M, Bunge G, Dal Pra G, Bättig K. Antagonistic effects of caffeine and alcohol on mental performance parameters. *Pharmacol Biochem Behav* 1993;46:463–5. doi: 10.1016/0091-3057(93)90380-c. PMID: 8265702

[102] Rutenfranz J, Jansen G. On compensation of the alcohol effect by caffeine and pervitin in a psychomotor performance. *Int Z Angew Physiol* 1959;18:62–81. PMID: 14440356

[103] Franks HM, Hagedorn H, Hensley VR, Hensley WJ, Starmer GA. The effect of caffeine on human performance alone and in combination with ethanol. *Psychopharmacology (Berl)* 1975;45:171–81. doi: 10.1007/BF00429058. PMID: 1215448

[104] Nuotto E, Mattila MJ, Seppala T, Konno K. Coffee and caffeine and alcohol effects on psychomotor function. *Clin Pharmacol Ther* 1982;31:68–76. doi: 10.1038/clpt.1982.11. PMID: 7053307

[105] Newman HW, Newman EJ. Failure of dexedrine and caffeine as practical antagonists of the depressive effects of ethyl alcohol in man. *Q J Stud Alc* 1956;17:406–11. doi: https://doi.org/10.15288/qjsa.1956.17.406. PMID: 13359661

[106] Oborne DJ, Rogers Y. Interactions of alcohol and caffeine on human reaction time. *Aviat Space Env Med* 1983;54:528–34. PMID: 6882312

[107] Fudin R, Nicastro R. Can caffeine antagonize alcohol-induced performance decrements in humans? *Percept Mot Skills* 1988;67:375–91. doi: 10.2466/pms.1988.67.2.375. PMID: 3064045

[108] Connole L, Harkin A, Maginn M. Adenosine A_1 receptor blockade mimics caffeine's attenuation of ethanol-induced motor incoordination. *Basic Clin Pharmacol Toxicol* 2004;95(6):299–304. doi: 10.1111/j.1742-7843.2004.pto950509.x. PMID: 15569276

[109] Dar MS. The biphasic effects of centrally and peripherally administered caffeine on ethanol-induced motor incoordination in mice. *J Pharm Pharmacol* 1988;40(7):482–7. doi: 10.1111/j.2042-7158.1988.tb05282.x. PMID: 2904988

[110] Dar MS, Jones M, Close G, Mustafa SJ, Wooles WR. Behavioral interactions of ethanol and methylxanthines. *Psychopharmacology (Berl)* 1987;91(1):1–4. doi: 10.1007/BF00690916. PMID: 2434967

[111] Dar MS, Wooles WR. Effect of chronically administered methylxanthines on ethanol-induced motor incoordination in mice. *Life Sci* 1986;39(16):1429–37. doi: 10.1016/0024-3205(86)90547-3. PMID: 2430156

[112] Mohammad FK, Dawood KA, al-Dabbagh SG. Antagonism of medetomidine-induced ataxia in rats by yohimbine, aminophylline and caffeine. *Pharmacol Res* 1993;27(1):81–8. doi: 10.1006/phrs.1993.1008. PMID: 8096078

[113] Listos J, Talarek S, Fidecka S. Adenosinergic system is involved in development of diazepam tolerance in mice. *Pharmacol Biochem Behav* 2010;94(4):510–5. doi: 10.1016/j.pbb.2009.11.005. PMID: 19941886

[114] Mishra J, Kumar A. Improvement of mitochondrial NAD(+)/FAD(+)-linked state-3 respiration by caffeine attenuates quinolinic acid induced motor impairment in rats: implications in Huntington's disease. *Pharmacol Rep* 2014;66(6):1148–55. doi: 10.1016/j.pharep.2014.07.006. PMID: 25443748

[115] Hsu HR, Mei YY, Wu CY, Chiu PH, Chen HH. Behavioural and toxic interaction profile of ketamine in combination with caffeine. *Basic Clin Pharmacol Toxicol* 2009;104(5):379–83. doi: 10.1111/j.1742-7843.2009.00382.x. PMID: 19413657

[116] de Oliveira RV, Dall'Igna OP, Tort AB, Schuh JF, Neto PF, Santos Gomes MW, Souza DO, Lara DR. Effect of subchronic caffeine treatment on MK-801-induced changes in locomotion, cognition and ataxia in mice. *Behav Pharmacol* 2005;16(2):79–84. doi: 10.1097/00008877-200503000-00002. PMID: 15767842

[117] Marshad RA, Khatib RA, Amer H, Shammari MA, Otaibi AA, Otaibi FA, Behbehani N, Sayed AA, Hoty NA, Hassan Z, Kamal A. Streptozotocin-induced diabetes mellitus affects the NMDA receptors: role of caffeine administration in enhancing learning, memory and locomotor deficits. *Int J Health Sci (Qassim)* 2018;12(3):10–7. PMID: 29896066

[118] Lukawski K, Raszewski G, Kruszyński K, Czuczwar SJ. Neurological effects of acute exposure to caffeine and organophosphates in mice. *Pol J Vet Sci* 2021;24(2):211–6. doi: 10.24425/pjvs.2021.137655. PMID: 34250774

[119] Kachroo A, Irizarry MC, Schwarzschild MA. Caffeine protects against combined paraquat and maneb-induced dopaminergic neuron degeneration. *Exp Neurol* 2010;223(2):657–61. doi: 10.1016/j.expneurol.2010.02.007. PMID: 20188092

[120] Aboutabl ME, Salman AM, Gamal el Din AA, Maklad YA. Simultaneous administration of coffee and rasagiline/l-dopa protects against paraquat-induced neurochemical and motor behavior impairments *in vivo*. *Bull Natl Res Centre* 2021;45:219. doi: 10.1186/s42269-021-00678-9

[121] Gomes CV, Kaster MP, Tomé RA, Agostinho PM, Rodrigo A, Cunha RA. Adenosine receptors and brain diseases: neuroprotection and neurodegeneration. *Biochim Biophys Acta* 2011;1808(5):1380–99. doi: 10.1016/j.bbamem.2010.12.001. PMID: 21145878

[122] Kolahdouzan M, Hamadeh MJ. The neuroprotective effects of caffeine in neurodegenerative diseases. *CNS Neurosci Ther* 2017;23(4):272–90. doi: 10.1111/cns.12684. PMID: 28317317

[123] Pedata F, Gianfriddo M, Turchi D, Melani A. The protective effect of adenosine A_{2A} receptor antagonism in cerebral ischemia. *Neurol Res* 2005;27(2):169–74. doi: 10.1179/016164105X21913. PMID: 15829180

[124] Shi X, Dalal NS, Jain AC. Antioxidant behaviour of caffeine: efficient scavenging of hydroxyl radicals. *Food Chem Toxicol* 1991;29(1):1–6. doi: 10.1016/0278-6915(91)90056-d. PMID: 1847890

[125] Winerdal M, Urmaliya V, Winerdal ME, Fredholm BB, Winqvist O, Aden U. Single dose caffeine protects the neonatal mouse brain against hypoxia ischemia. *PLoS One* 2017;12(1):e0170545. doi: 10.1371/journal.pone.0170545. PMID: 28129361

[126] Di Martino E, Bocchetta E, Tsuji S, Mukai T, Harris RA, Blomgren K, Ådén U. Defining a time window for neuroprotection and glia modulation by caffeine after neonatal hypoxia-ischaemia. *Mol Neurobiol* 2020;57(5):2194–205. doi: 10.1007/s12035-020-01867-9. PMID: 31974940

[127] Potter M, Rosenkrantz T, Fitch RH. Behavioral and neuroanatomical outcomes in a rat model of preterm hypoxic-ischemic brain injury: effects of caffeine and hypothermia. *Int J Dev Neurosci* 2018;70:46–55 https://doi.org/10.1016/j.ijdevneu.2018.02.001. PMID: 29476789

[128] Rudolphi KA, Keil M, Fastbom J, Fredholm BB. Ischaemic damage in gerbil hippocampus is reduced following upregulation of adenosine (A1) receptors by caffeine treatment. *Neurosci Lett* 1989;103(3):275–80. doi: 10.1016/0304-3940(89)90112-2. PMID: 2812514

[129] Jacobson KA, von Lubitz DK, Daly JW, Fredholm BB. Adenosine receptor ligands: differences with acute versus chronic treatment. *Trends Pharmacol Sci* 1996;17(3):108–13. doi: 10.1016/0165-6147(96)10002-x. PMID: 8936347

[130] Shi D, Nikodijević O, Jacobson KA, Daly JW. Effects of chronic caffeine on adenosine, dopamine and acetylcholine systems in mice. *Arch Int Pharmacodyn Ther* 1994;328(3):261–87. PMID: 7625882

[131] Shi D, Nikodijevic O, Jacobson KA, Daly JW. Chronic caffeine alters the density of adenosine, adrenergic, cholinergic, GABA, and serotonin receptors and calcium channels in mouse brain. *Cell Mol Neurobiol* 1993;13(3):247–61. doi: 10.1007/BF00733753. PMID: 8242688

[132] Hawkins M, Dugich MM, Porter NM, Urbancic M, Radulovacki M. Effects of chronic administration of caffeine on adenosine A_1 and A_2 receptors in rat brain. *Brain Res Bull* 1988;21(3):479–82. doi: 10.1016/0361-9230(88)90162-1. PMID: 3214753

[133] Strong R, Grotta JC, Aronowski J. Combination of low dose ethanol and caffeine protects brain from damage produced by focal ischemia in rats. *Neuropharmacology* 2000;39(3):515–22. doi: 10.1016/s0028-3908(99)00156-2. PMID: 10698017

[134] Aronowski J, Strong R, Shirzadi A, Grotta JC. Ethanol plus caffeine (caffeinol) for treatment of ischemic stroke: preclinical experience. *Stroke* 2003;34(5):1246–51. doi: 10.1161/01.STR.0000068170.80517.B3. PMID: 12690223

[135] Dash PK, Moore AN, Moody MR, Treadwell R, Felix JL, Clifton GL. Post-trauma administration of caffeine plus ethanol reduces contusion volume and improves working memory in rats. *J Neurotrauma* 2004;21(11):1573–83. doi: 10.1089/neu.2004.21.1573. PMID: 15684650

[136] Melani A, Gianfriddo M, Vannucchi MG, Cipriani S, Baraldi PG, Giovannini MG, Pedata F. The selective A_{2A} receptor antagonist SCH 58261 protects from neurological deficit, brain damage and activation of p38 MAPK in rat focal cerebral ischemia. *Brain Res* 2006;1073–4:470–80. doi: 10.1016/j.brainres.2005.12.010. PMID: 16443200

[137] Zhu PJ, Krnjević K. Adenosine release is a major cause of failure of synaptic transmission during hypoglycaemia in rat hippocampal slices. *Neurosci Lett* 1993;155(2):128–31. doi: 10.1016/0304-3940(93)90689-i. PMID: 8397349

[138] Atkinson JL. The neglected prehospital phase of head injury: apnea and catecholamine surge. *Mayo Clin Proc* 2000;75(1):37–47. doi: 10.4065/75.1.37. PMID: 10630756
[139] Lusardi TA. Adenosine neuromodulation and traumatic brain injury. *Curr Neuropharmacol* 2009;7(3):228–37. doi: 10.2174/157015909789152137. PMID: 20190964
[140] Sachse KT, Jackson EK, Wisniewski SR, Gillespie DG, Puccio AM, Clark RS, Dixon CE, Kochanek PM. Increases in cerebrospinal fluid caffeine concentration are associated with favorable outcome after severe traumatic brain injury in humans. *J Cereb Blood Flow Metab* 2008;28(2):395–401. doi: 10.1038/sj.jcbfm.9600539. PMID: 17684518
[141] Ning YL, Yang N, Chen X, Zhao ZA, Zhang XZ, Chen XY, Li P, Zhao Y, Zhou YG. Chronic caffeine exposure attenuates blast-induced memory deficit in mice. *Chin J Traumatol* 2015;18(4):204–11. doi: 10.1016/j.cjtee.2015.10.003. PMID: 26764541
[142] Zhao ZA, Zhao Y, Ning YL, Yang N, Peng Y, Li P, Chen XY, Liu D, Wang H, Chen X, Bai W, Chen JF, Zhou YG. Adenosine A(2A) receptor inactivation alleviates early-onset cognitive dysfunction after traumatic brain injury involving an inhibition of tau hyperphosphorylation. *Transl Psychiatry* 2017;7(5):e1123. doi: 10.1038/tp.2017.98. PMID: 28485728
[143] Ratliff WA, Saykally JN, Mervis RF, Lin X, Cao C, Citron BA. Behavior, protein, and dendritic changes after model traumatic brain injury and treatment with nanocoffee particles. *BMC Neurosci* 2019;20(1):44. doi: 10.1186/s12868-019-0525-5. PMID: 31438853
[144] Christensen J, Yamakawa GR, Salberg S, Wang M, Kolb B, Mychasiuk R. Caffeine consumption during development alters spine density and recovery from repetitive mild traumatic brain injury in young adult rats. *Synapse* 2020;74(4):e22142. doi: 10.1002/syn.22142. PMID: 31705690
[145] Yamakawa GR, Lengkeek C, Salberg S, Spanswick SC, Mychasiuk R. Behavioral and pathophysiological outcomes associated with caffeine consumption and repetitive mild traumatic brain injury (RmTBI) in adolescent rats. *PLoS One* 2017;12(11):e0187218. doi: 10.1371/journal.pone.0187218. PMID: 29108016
[146] Li W, Dai S, An J, Li P, Chen X, Xiong R, Liu P, Wang H, Zhao Y, Zhu M, Liu X, Zhu P, Chen JF, Zhou Y. Chronic but not acute treatment with caffeine attenuates traumatic brain injury in the mouse cortical impact model. *Neuroscience* 2008;151(4):1198–207. doi: 10.1016/j.neuroscience.2007.11.020. PMID: 18207647
[147] Sanjakdar SS, Flerlage WJ, Kang HS, Napier DA, Dougherty JR, Mountney A, Gilsdorf JS, Shear DA. Differential effects of caffeine on motor and cognitive outcomes of penetrating ballistic-like brain injury. *Mil Med* 2019;184(Suppl 1):291–300. doi: 10.1093/milmed/usy367. PMID: 30901408
[148] Al Moutaery K, Al Deeb S, Ahmad Khan H, Tariq M. Caffeine impairs short-term neurological outcome after concussive head injury in rats. *Neurosurgery* 2003;53(3):704–11; discussion 711-2. doi: 10.1227/01.neu.0000079487.66013.6f. PMID: 12943586
[149] Lusardi TA, Lytle NK, Gebril HM, Boison D. Effects of preinjury and postinjury exposure to caffeine in a rat model of traumatic brain injury. *J Caffeine Adenosine Res* 2020;10(1):12–24. doi: 10.1089/caff.2019.0012. PMID: 32181443
[150] Lusardi TA, Lytle NK, Szybala C, Boison D. Caffeine prevents acute mortality after TBI in rats without increased morbidity. *Exp Neurol* 2012;234(1):161–8. doi: 10.1016/j.expneurol.2011.12.026. PMID: 22226594
[151] Yang CC, Jou IM. Caffeine treatment aggravates secondary degeneration after spinal cord injury. *Brain Res* 2016;1634:75–82. doi: 10.1016/j.brainres.2015.12.053. PMID: 26746340
[152] Genovese T, Melani A, Esposito E, Paterniti I, Mazzon E, Di Paola R, Bramanti P, Linden J, Pedata F, Cuzzocrea S. Selective adenosine A(2a) receptor agonists reduce the apoptosis in an experimental model of spinal cord trauma. *J Biol Regul Homeost Agents* 2010;24(1):73–86. PMID: 20385073
[153] Cassada DC, Tribble CG, Young JS, Gangemi JJ, Gohari AR, Butler PD, Rieger JM, Kron IL, Linden J, Kern JA. Adenosine A_{2A} analogue improves neurologic outcome after spinal cord trauma in the rabbit. *J Trauma* 2002;53(2):225–9. doi: 10.1097/00005373-200208000-00005. PMID: 12169926
[154] Okonkwo DO, Reece TB, Laurent JJ, Hawkins AS, Ellman PI, Linden J, Kron IL, Tribble CG, Stone JR, Kern JA. A comparison of adenosine A_{2A} agonism and methylprednisolone in attenuating neuronal damage and improving functional outcome after experimental traumatic spinal cord injury in rabbits. *J Neurosurg Spine* 2006;4(1):64–70. doi: 10.3171/spi.2006.4.1.64. PMID: 16506468

[155] Reece TB, Davis JD, Okonkwo DO, Maxey TS, Ellman PI, Li X, Linden J, Tribble CG, Kron IL, Kern JA. Adenosine A_{2A} analogue reduces long-term neurologic injury after blunt spinal trauma. *J Surg Res* 2004;121:130–4. doi: 10.1016/j.jss.2004.04.006. PMID: 15313386

[156] Paterniti I, Melani A, Cipriani S, Corti F, Mello T, Mazzon E, Esposito E, Bramanti P, Cuzzocrea S, Pedata F. Selective adenosine A_{2A} receptor agonists and antagonists protect against spinal cord injury through peripheral and central effects. *J Neuroinflammation* 2011;8:31. doi: 10.1186/1742-2094-8-31. PMID: 21486435

[157] Rivera-Oliver M, Díaz-Ríos M. Using caffeine and other adenosine receptor antagonists and agonists as therapeutic tools against neurodegenerative diseases: a review. *Life Sci* 2014;101(1–2):1–9. doi: 10.1016/j.lfs.2014.01.083. PMID: 24530739

[158] Beghi E, Pupillo E, Messina P, Giussani G, Chiò A, Zoccolella S, Moglia C, Corbo M, Logroscino G, EURALS Group. Coffee and amyotrophic lateral sclerosis: a possible preventive role. *Am J Epidemiol* 2011;174(9):1002–8. doi: 10.1093/aje/kwr229. PMID: 21946385

[159] Fondell E, O'Reilly ÉI, Fitzgerald KC, Falcone GJ, Kolonel LN, Park Y, Gapstur SM, Ascherio A. Intakes of caffeine, coffee and tea and risk of amyotrophic lateral sclerosis: Results from five cohort studies. *Amyotroph Lateral Scler Frontotemporal Degener* 2015;16(5–6):366–71. doi: 10.3109/21678421.2015.1020813. PMID: 25822002

[160] Petimar J, O'Reilly É, Adami HO, van den Brandt PA, Buring J, English DR, Freedman DM, Giles GG, Håkansson N, Kurth T, Larsson SC, Robien K, Schouten LJ, Weiderpass E, Wolk A, Smith-Warner SA. Coffee, tea, and caffeine intake and amyotrophic lateral sclerosis mortality in a pooled analysis of eight prospective cohort studies. *Eur J Neurol* 2019;26(3):468–75. doi: 10.1111/ene.13840. PMID: 30326172

[161] Yoshida Y, Une F, Utatsu Y, Nomoto M, Furukawa Y, Maruyama Y, Machigashira N, Matsuzaki T, Osame M. Adenosine and neopterin levels in cerebrospinal fluid of patients with neurological disorders. *Intern Med* 1999;38(2):133–9. doi: 10.2169/internalmedicine.38.133. PMID: 10225668

[162] Sebastião AM, Rei N, Ribeiro JA. Amyotrophic lateral sclerosis (ALS) and adenosine receptors. *Front Pharmacol* 2018;9:267. doi: 10.3389/fphar.2018.00267. PMID: 29713276

[163] Potenza RL, Armida M, Ferrante A, Pèzzola A, Matteucci A, Puopolo M, Popoli P. Effects of chronic caffeine intake in a mouse model of amyotrophic lateral sclerosis. *J Neurosci Res* 2013;91(4):585–92. doi: 10.1002/jnr.23185. PMID: 23361938

[164] Ng SK, Higashimori H, Tolman M, Yang Y. Suppression of adenosine 2a receptor ($A_{2a}R$)-mediated adenosine signaling improves disease phenotypes in a mouse model of amyotrophic lateral sclerosis. *Exp Neurol* 2015;267:115–22. doi: 10.1016/j.expneurol.2015.03.004. PMID: 25779930

[165] Nascimento F, Pousinha PA, Correia AM, Gomes R, Sebastião AM, Ribeiro JA. Adenosine A_{2A} receptors activation facilitates neuromuscular transmission in the pre-symptomatic phase of the SOD1(G93A) ALS mice, but not in the symptomatic phase. *PLoS One* 2014;9(8):e104081. doi: 10.1371/journal.pone.0104081. PMID: 25093813

[166] Mojsilovic-Petrovic J, Jeong GB, Crocker A, Arneja A, David S, Russell DS, Kalb RG. Protecting motor neurons from toxic insult by antagonism of adenosine A2a and Trk receptors. *J Neurosci* 2006;26(36):9250–63. doi: 10.1523/JNEUROSCI.1856-06.2006. PMID: 16957081

[167] Zwilling M, Theiss C, Matschke V. Caffeine and NAD(+) improve motor neural integrity of dissociated wobbler cells *in vitro*. *Antioxidants (Basel)* 2020;9(6):460. doi: 10.3390/antiox9060460. PMID: 32471290a

[168] Hong CT, Chan L, Bai CH. The effect of caffeine on the risk and progression of Parkinson's disease: a meta-analysis. *Nutrients* 2020;12(6):1860. doi: 10.3390/nu12061860. PMID: 32580456

[169] Ross GW, Abbott RD, Petrovitch H, Morens DM, Grandinetti A, Tung KH, Tanner CM, Masaki KH, Blanchette PL, Curb JD, Popper JS, White LR. Association of coffee and caffeine intake with the risk of Parkinson disease. *JAMA* 2000;283:2674–9. doi: 10.1001/jama.283.20.2674. PMID: 10819950

[170] Bagga P, Patel AB. Pretreatment of caffeine leads to partial neuroprotection in MPTP model of Parkinson's disease. *Neural Regen Res* 2016;11(11):1750–1. doi: 10.4103/1673-5374.194716. PMID: 28123409

[171] Prediger RD. Effects of caffeine in Parkinson's disease: from neuroprotection to the management of motor and non-motor symptoms. *J Alzheimers Dis* 2010;20 Suppl 1:S205–20. doi: 10.3233/JAD-2010-091459. PMID: 20182024

[172] Altman RD, Lang AE, Postuma RB. Caffeine in Parkinson's disease: a pilot open-label, dose-escalation study. *Mov Disord* 2011;26(13):2427–31. doi: 10.1002/mds.23873. PMID: 21953603

[173] Postuma RB, Lang AE, Munhoz RP, Charland K, Pelletier A, Moscovich M, Filla L, Zanatta D, Rios Romenets S, Altman R, Chuang R, Shah B. Caffeine for treatment of Parkinson disease: a randomized controlled trial. *Neurology* 2012;79(7):651–8. doi: 10.1212/WNL.0b013e318263570d. PMID: 22855866

[174] Postuma RB, Anang J, Pelletier A, Joseph L, Moscovich M, Grimes D, Furtado S, Munhoz RP, Appel-Cresswell S, Moro A, Borys A, Hobson D, Lang AE. Caffeine as symptomatic treatment for Parkinson disease (Café-PD): a randomized trial. *Neurology* 2017;89(17):1795–1803. doi: 10.1212/WNL.0000000000004568. PMID: 28954882

[175] Kartzinel R, Shoulson I, Calne DB. Studies with bromocriptine: III. Concomitant administration of caffeine to patients with idiopathic parkinsonism. *Neurology* 1976;26(8):741–3. doi: 10.1212/wnl.26.8.741. PMID: 181692

[176] Shoulson I, Chase T. Caffeine and the antiparkinsonian response to levodopa or piribedil. *Neurology* 1975;25(8):722–4. doi: 10.1212/wnl.25.8.722. PMID: 125389

[177] Diguet E, Fernagut PO, Scherfler C, Wenning G, Tison F. Effects of riluzole on combined MPTP + 3-nitropropionic acid-induced mild to moderate striatonigral degeneration in mice. *J Neural Transm* 2005;112:613–31. doi: 10.1007/s00702-004-0206-z. PMID: 15503195

[178] Luchtman DW, Meng Q, Song C. Ethyl-eicosapentaenoate (E-EPA) attenuates motor impairments and inflammation in the MPTP-probenecid mouse model of Parkinson's disease. *Behav Brain Res* 2012;226(2):386–96. doi: 10.1016/j.bbr.2011.09.033. PMID: 21971013

[179] Glajch KE, Fleming SM, Surmeier DJ, Osten P. Sensorimotor assessment of the unilateral 6-hydroxydopamine mouse model of Parkinson's disease. *Behav Brain Res* 2012;230:309–16. doi: 10.1016/j.bbr.2011.12.007. PMID: 22178078

[180] Duan WJ, Liang L, Pan MH, Lu DH, Wang TM, Li SB, Zhong HB, Yang XJ, Cheng Y, Liu B, Li WX, Kurihara H, Bi W, Li YF, He RR. Theacrine, a purine alkaloid from kucha, protects against Parkinson's disease through SIRT3 activation. *Phytomedicine* 2020;77:153281. doi: 10.1016/j.phymed.2020.153281. PMID: 32707370

[181] Karuppagounder SS, Uthaythas S, Govindarajulu M, Ramesh S, Parameshwaran K, Dhanasekaran M. Caffeine, a natural methylxanthine nutraceutical, exerts dopaminergic neuroprotection. *Neurochem Int* 2021;148:105066. doi: 10.1016/j.neuint.2021.105066. PMID: 34004240

[182] Sonsalla PK, Wong LY, Harris SL, Richardson JR, Khobahy I, Li W, Gadad BS, German DC. Delayed caffeine treatment prevents nigral dopamine neuron loss in a progressive rat model of Parkinson's disease. *Exp Neurol* 2012;234(2):482–7. doi: 10.1016/j.expneurol.2012.01.022. PMID: 22309831

[183] Chen JF, Xu K, Petzer JP, Staal R, Xu YH, Beilstein M, Sonsalla PK, Castagnoli K, Castagnoli N Jr, Schwarzschild MA. Neuroprotection by caffeine and A2A adenosine receptor inactivation in a model of Parkinson's disease. *J Neurosci* 2001;21(10):RC143. doi : 10.1523/JNEUROSCI.21-10-j0001.2001. PMID: 1131924

[184] Kelsey JE, Langelier NA, Oriel BS, Reedy C. The effects of systemic, intrastriatal, and intrapallidal injections of caffeine and systemic injections of A_{2A} and A_1 antagonists on forepaw stepping in the unilateral 6-OHDA-lesioned rat. *Psychopharmacology (Berl)* 2009;201(4):529–39. doi: 10.1007/s00213-008-1319-0. PMID: 18791705

[185] Acuna-Lizama MM, Bata-García JL, Alvarez-Cervera FJ, Góngora-Alfaro JL. Caffeine has greater potency and efficacy than theophylline to reverse the motor impairment caused by chronic but not acute interruption of striatal dopaminergic transmission in rats. *Neuropharmacology* 2013;70:51–62. doi: 10.1016/j.neuropharm.2013.01.002. PMID: 23321687

[186] Bata-Garcia JL, Tun-Cobá L, Alvarez-Cervera FJ, Villanueva-Toledo JR, Heredia-López FJ, Góngora-Alfaro JL. Improvement of postural adjustment steps in hemiparkinsonian rats chronically treated with caffeine is mediated by concurrent blockade of A_1 and A_{2A} adenosine receptors. *Neuroscience* 2010;166(2):590–603. doi: 10.1016/j.neuroscience.2009.12.072. PMID: 20056138

[187] Bata-Garcia JL, Villanueva-Toledo J, Gutiérrez-Ospina G, Alvarez-Cervera FJ, Heredia-López FJ, Góngora-Alfaro JL. Sustained improvement of motor function in hemiparkinsonian rats chronically treated with low doses of caffeine or trihexyphenidyl. *Pharmacol Biochem Behav*. 2007;86(1):68–78. doi: 10.1016/j.pbb.2006.12.009. PMID: 17250882

[188] Frau L, Costa G, Porceddu PF, Khairnar A, Castelli MP, Ennas MG, Madeddu C, Wardas J, Morelli M. Influence of caffeine on 3,4-methylenedioxymethamphetamine-induced dopaminergic neuron degeneration and neuroinflammation is age-dependent. *J Neurochem* 2016;136(1):148–62. doi: 10.1111/jnc.13377. PMID: 26442661
[189] Yan R, Zhang J, Park HJ, Park ES, Oh S, Zheng H, Junn E, Voronkov M, Stock JB, Mouradian MM. Synergistic neuroprotection by coffee components eicosanoyl-5-hydroxytryptamide and caffeine in models of Parkinson's disease and DLB. *Proc Natl Acad Sci USA* 2018;115(51):E12053–62. doi: 10.1073/pnas.1813365115. PMID: 30509990
[190] Xu K, Di Luca DG, Orrú M, Xu Y, Chen JF, Schwarzschild MA. Neuroprotection by caffeine in the MPTP model of Parkinson's disease and its dependence on adenosine A_{2A} receptors. *Neuroscience* 2016;322:129–37. doi: 10.1016/j.neuroscience.2016.02.035. PMID: 26905951
[191] Carta AR, Kachroo A, Schintu N, Xu K, Schwarzschild MA, Wardas J, Morelli M. Inactivation of neuronal forebrain A receptors protects dopaminergic neurons in a mouse model of Parkinson's disease. *J Neurochem* 2009;111(6):1478–89. doi: 10.1111/j.1471-4159.2009.06425.x. PMID: 19817968
[192] Frau L, Borsini F, Wardas J, Khairnar AS, Schintu N, Morelli M. Neuroprotective and anti-inflammatory effects of the adenosine A_{2A} receptor antagonist ST1535 in a MPTP mouse model of Parkinson's disease. *Synapse* 2011;65(3):181–8. doi: 10.1002/syn.20833. PMID: 20665698
[193] Wardas J, Konieczny J, Lorenc-Koci E. SCH 58261, an A_{2A} adenosine receptor antagonist, counteracts parkinsonian-like muscle rigidity in rats. *Synapse* 2001;41(2):160–71. doi: 10.1002/syn.1070. PMID: 11400182
[194] Coccurello R, Breysse N, Amalric M. Simultaneous blockade of adenosine A_{2A} and metabotropic glutamate mGlu5 receptors increase their efficacy in reversing Parkinsonian deficits in rats. *Neuropsychopharmacology* 2004;29(8):1451–61. doi: 10.1038/sj.npp.1300444. PMID: 15039773
[195] Morelli M, Wardas J. Adenosine A_{2A} receptor antagonists: potential therapeutic and neuroprotective effects in Parkinson's disease. *Neurotox Res* 2001;3(6):545–56. doi: 10.1007/BF03033210. PMID: 15111244
[196] Morelli M, Di Paolo T, Wardas J, Calon F, Xiao D, Schwarzschild MA. Role of adenosine A_{2A} receptors in parkinsonian motor impairment and l-DOPA-induced motor complications. *Prog Neurobiol* 2007;83(5):293–309. doi: 10.1016/j.pneurobio.2007.07.001. PMID: 17826884
[197] Arendash GW, Cao C. Caffeine and coffee as therapeutics against Alzheimer's disease. *J Alzheimers Dis* 2010;20(Suppl 1):S117–26. doi: 10.3233/JAD-2010-091249. PMID: 20182037
[198] Maia L, de Mendonça A. Does caffeine intake protect from Alzheimer's disease? *Eur J Neurol* 2002;9(4):377–82. doi: 10.1046/j.1468-1331.2002.00421.x. PMID: 12099922
[199] Rahman A. The role of adenosine in Alzheimer's disease. *Curr Neuropharmacol* 2009;7(3):207–16. doi: 10.2174/157015909789152119. PMID: 20190962
[200] Rosso A, Mossey J, Lippa CF. Caffeine: neuroprotective functions in cognition and Alzheimer's disease. *Am J Alzheimers Dis Other Dementia* 2008;23:417–22. doi: 10.1177/1533317508320083. PMID: 19230121
[201] Sharma B, Paul S. Action of caffeine as an amyloid inhibitor in the aggregation of Abeta16-22 peptides. *J Phys Chem B* 2016;120(34):9019–33. doi: 10.1021/acs.jpcb.6b03892. PMID: 27487451
[202] Arendash GW, Schleif W, Rezai-Zadeh K, Jackson EK, Zacharia LC, Cracchiolo JR, Shippy D, Tan J. Caffeine protects Alzheimer's mice against cognitive impairment and reduces brain beta-amyloid production. *Neuroscience*. 2006;142(4):941–52. doi: 10.1016/j.neuroscience.2006.07.021. PMID: 16938404
[203] Arendash GW, Mori T, Cao C, Mamcarz M, Runfeldt M, Dickson A, Rezai-Zadeh K, Tane J, Citron BA, Lin X, Echeverria V, Potter H. Caffeine reverses cognitive impairment and decreases brain amyloid-beta levels in aged Alzheimer's disease mice. *J Alzheimers Dis* 2009;17(3):661–80. doi: 10.3233/JAD-2009-1087. PMID: 19581722
[204] Cao C, Wang L, Lin X, Mamcarz M, Zhang C, Bai G, Nong J, Sussman S, Arendash G. Caffeine synergizes with another coffee component to increase plasma GCSF: linkage to cognitive benefits in Alzheimer's mice. *J Alzheimers Dis* 2011;25(2):323–35. doi: 10.3233/JAD-2011-110110. PMID: 21422521
[205] Chu YF, Chang WH, Black RM, Liu JR, Sompol P, Chen Y, Wei H, Zhao Q, Cheng IH. Crude caffeine reduces memory impairment and amyloid beta(1–42) levels in an Alzheimer's mouse model. *Food Chem* 2012;135(3):2095–102. doi: 10.1016/j.foodchem.2012.04.148. PMID: 22953961

[206] Faivre E, Coelho JE, Zornbach K, Malik E, Baqi Y, Schneider M, Cellai L, Carvalho K, Sebda S, Figeac M, Eddarkaoui S, Caillierez R, Chern Y, Heneka M, Sergeant N, Müller CE, Halle A, Buée L, Lopes LV, Blum D. Beneficial effect of a selective adenosine A(2A) receptor antagonist in the APPswe/PS1dE9 mouse model of Alzheimer's disease. *Front Mol Neurosci* 2018;11:235. doi: 10.3389/fnmol.2018.00235. PMID: 30050407

[207] Han K, Jia N, Li J, Yang L, Min LQ. Chronic caffeine treatment reverses memory impairment and the expression of brain BNDF and TrkB in the PS1/APP double transgenic mouse model of Alzheimer's disease. *Mol Med Rep* 2013;8(3):737–40. doi: 10.3892/mmr.2013.1601. PMID: 23900282

[208] Laurent C, Eddarkaoui S, Derisbourg M, Leboucher A, Demeyer D, Carrier S, Schneider M, Hamdane M, Müller CE, Buée L, Blum D. Beneficial effects of caffeine in a transgenic model of Alzheimer's disease-like tau pathology. *Neurobiol Aging* 2014;35(9):2079–90. doi: 10.1016/j.neurobiolaging.2014.03.027. PMID: 24780254

[209] Londzin P, Zamora M, Kąkol B, Taborek A, Folwarczna J. Potential of caffeine in Alzheimer's disease: a review of experimental studies. *Nutrients* 2021;13(2):537. doi: 10.3390/nu13020537. PMID: 33562156

[210] Stazi M, Lehmann S, Sakib MS, Pena-Centeno T, Büschgens L, Fischer A, Weggen S, Wirths O. Long-term caffeine treatment of Alzheimer mouse models ameliorates behavioural deficits and neuron loss and promotes cellular and molecular markers of neurogenesis. *Cell Mol Life Sci* 2021;79(1):55. doi: 10.1007/s00018-021-04062-8. PMID: 34913091

[211] Baeta-Corral R, Johansson B, Giménez-Llort L. Long-term treatment with low-dose caffeine worsens BPSD-like profile in 3xTg-AD mice model of Alzheimer's disease and affects mice with normal aging. *Front Pharmacol* 2018;9:79. doi: 10.3389/fphar.2018.00079. PMID: 29497377

[212] Basurto-Islas G, Blanchard J, Tung YC, Fernandez JR, Voronkov M, Stock M, Zhang S, Stock JB, Iqbal K. Therapeutic benefits of a component of coffee in a rat model of Alzheimer's disease. *Neurobiol Aging* 2014;35(12):2701–12. doi: 10.1016/j.neurobiolaging.2014.06.012. PMID: 25034344

[213] Baloh RW. Episodic ataxias 1 and 2. *Handb Clin Neurol* 2012;103:595–602. doi: 10.1016/B978-0-444-51892-7.00042-5. PMID: 21827920

[214] Hasan SM, D'Adamo MC. Episodic ataxia type 1. In: Adam MP, Ardinger HH, Pagon RA, Wallace SE, Bean LJH, Mirzaa G, Amemiya A (eds) GeneReviews® [Internet]. Seattle: University of Washington, 2010 Feb 9 [updated 2018 Nov 1]. PMID: 20301785 https://pubmed.ncbi.nlm.nih.gov/20301785/

[215] Baloh RW, Yue Q, Furman JM, Nelson SF. Familial episodic ataxia: clinical heterogeneity in four families linked to chromosome 19p. *Ann Neurol* 1997;41:8–16. PMID: 9005860. doi: 10.1002/ana.410410105

[216] Jen J, Kim GW, Baloh RW. Clinical spectrum of episodic ataxia type 2. *Neurology* 2004;62:17–22. doi: 10.1212/01.wnl.0000101675.61074.50. PMID: 14718690

[217] Ophoff RA, Terwindt GM, Vergouwe MN, van Eijk R, Oefner PJ, Hoffman SM, Lamerdin JE, Mohrenweiser HW, Bulman DE, Ferrari M, Haan J, Lindhout D, van Ommen GJ, Hofker MH, Ferrari MD, Frants RR. Familial hemiplegic migraine and episodic ataxia type-2 are caused by mutations in the Ca2+ channel gene *CACNL1A4*. *Cell* 1996;87:543–52. doi: 10.1016/s0092-8674(00)81373-2. PMID: 8898206

[218] Rajakulendran S, Kaski D, Hanna MG. Neuronal P/Q-type calcium channel dysfunction in inherited disorders of the CNS. *Nat Rev Neurol* 2012;8:86–96. doi: 10.1038/nrneurol.2011.228. PMID: 22249839

[219] Cha YH, Lee H, Jen JC, Kattah JC, Nelson SF, Baloh RW. Episodic vertical oscillopsia with progressive gait ataxia: clinical description of a new episodic syndrome and evidence of linkage to chromosome 13q. *J Neurol Neurosurg Psychiatry* 2007;78(11):1273–5. doi: 10.1136/jnnp.2006.111138. PMID: 17522101

[220] Doyle J, Ren X, Lennon G, Stubbs L. Mutations in the *Cacnl1a4* calcium channel gene are associated with seizures, cerebellar degeneration, and ataxia in *tottering* and *leaner* mutant mice. *Mamm Genome* 1997;8(2):113–20. doi: 10.1007/s003359900369. PMID: 9060410

[221] Li W, Zhou Y, Tian X, Kim TY, Ito N, Watanabe K, Tsuji A, Niimi K, Aoyama Y, Arai T, Takahashi E. New ataxic *tottering*-6j mouse allele containing a *Cacna1a* gene mutation. *PLoS One* 2012;7(8):e44230. doi: 10.1371/journal.pone.0044230. PMID: 22952933

[222] Miki T, Zwingman TA, Wakamori M, Lutz CM, Cook SA, Hosford DA, Herrup K, Fletcher CF, Mori Y, Frankel WN, Letts VA. Two novel alleles of *tottering* with distinct Ca(v)2.1 calcium channel neuropathologies. *Neuroscience* 2008;155(1):31–44. doi: 10.1016/j.neuroscience.2008.05.028. PMID: 18597946

[223] Campbell DB, Hess EJ. Cerebellar circuitry is activated during convulsive episodes in the *tottering* (*tg/tg*) mutant mouse. *Neuroscience* 1998;85:773–83. PMID: 9639271. doi: 10.1016/s0306-4522(97)00672-6

[224] Campbell DB, Hess EJ. L-type calcium channels contribute to the *tottering* mouse dystonic episodes. *Mol Pharmacol* 1999;55:23–31. PMID: 9882694 DOI: 10.1124/mol.55.1.23

[225] Fletcher CF, Lutz CM, O'Sullivan TN, Shaughnessy JD Jr, Hawkes R, Frankel WN, Copeland NG, Jenkins NA. Absence epilepsy in *tottering* mutant mice is associated with calcium channel defects. *Cell* 1996;87:607-17. doi: 10.1016/s0092-8674(00)81381-1. PMID: 8929530

[226] Fletcher CF, Tottene A, Lennon VA, Wilson SM, Dubel SJ, Paylor R, Hosford DA, Tessarollo L, McEnery MW, Pietrobon D, Copeland NG, Jenkins NA. Dystonia and cerebellar atrophy in *Cacna1a* null mice lacking P/Q calcium channel activity. *FASEB J* 2001;15:1288–90. doi: 10.1016/s0092-8674(00)81381-1. PMID: 8929530

[227] Fureman BE, Jinnah HA, Hess EJ. Triggers of paroxysmal dyskinesia in the calcium channel mouse mutant *tottering*. *Pharmacol Biochem Behav* 2002;73:631–7. doi: 10.1016/s0091-3057(02)00854-7. PMID: 12151038

[228] Noebels JL. A single gene error of noradrenergic axon growth synchronizes central neurones. *Nature* 1984;310:409–11. doi: 10.1038/310409a0. PMID: 6462226

[229] Noebels JL, Sidman RL. Inherited epilepsy: spike-wave and focal motor seizures in the mutant mouse *tottering*. *Science* 1979;204:1334–6. doi: 10.1126/science.572084. PMID: 572084

[230] Raike RS, Weisz C, Hoebeek FE, Terzi MC, De Zeeuw CI, van den Maagdenberg AM, Jinnah HA, Hess EJ. Stress, caffeine and ethanol trigger transient neurological dysfunction through shared mechanisms in a mouse calcium channelopathy. *Neurobiol Dis* 2013;50:151–9. doi: 10.1016/j.nbd.2012.09.005. PMID: 23009754

[231] Tara E, Vitenzon A, Hess E, Khodakhah K. Aberrant cerebellar Purkinje cell activity as the cause of motor attacks in a mouse model of episodic ataxia type 2. *Dis Model Mech* 2018;11(9):dmm034181. doi: 10.1242/dmm.034181. PMID: 30279196

[232] Neychev VK, Fan X, Mitev VI, Hess EJ, Jinnah HA. The basal ganglia and cerebellum interact in the expression of dystonic movement. *Brain.* 2008;131(Pt 9):2499–509. doi: 10.1093/brain/awn168. PMID: 18669484

[233] Chen G, Popa LS, Wang X, Gao W, Barnes J, Hendrix CM, Hess EJ, Ebner TJ. Low-frequency oscillations in the cerebellar cortex of the *tottering* mouse. *J Neurophysiol* 2009;101(1):234–45. doi: 10.1152/jn.90829.2008. PMID: 18987121

[234] Weisz CJ, Raike RS, Soria-Jasso LE, Hess EJ. Potassium channel blockers inhibit the triggers of attacks in the calcium channel mouse mutant *tottering*. *J Neurosci* 2005;25(16):4141–5. doi: 10.1523/JNEUROSCI.0098-05.2005. PMID: 1584361

[235] Lalonde R, Strazielle C. Motor performances of spontaneous and genetically modified mutants with cerebellar atrophy. *The Cerebellum* 2019;18:615–34. doi: 10.1007/s12311-019-01017-5. PMID: 30820866

[236] Rose SJ, Kriener LH, Heinzer AK, Fan X, Raike RS, van den Maagdenberg AM, Hess EJ. The first knockin mouse model of episodic ataxia type 2. *Exp Neurol* 2014;261:553–62. doi: 10.1016/j.expneurol.2014.08.001. PMID: 25109669

[237] Goncalves N, Simões AT, Prediger RD, Hirai H, Cunha RA, Pereira de Almeida L. Caffeine alleviates progressive motor deficits in a transgenic mouse model of spinocerebellar ataxia. *Ann Neurol* 2017;81(3):407–18. doi: 10.1002/ana.24867. PMID: 28032667

[238] Goncalves N, Simões AT, Cunha RA, de Almeida LP. Caffeine and adenosine A(2A) receptor inactivation decrease striatal neuropathology in a lentiviral-based model of Machado-Joseph disease. *Ann Neurol* 2013;73(5):655–66. doi: 10.1002/ana.23866. PMID: 23625556

[239] Rothblum-Oviatt C, Wright J, Lefton-Greif MA, McGrath-Morrow SA, Crawford TO, Lederman HM. Ataxia telangiectasia: a review. *Orphanet J Rare Dis* 2016;11(1):159. doi: 10.1186/s13023-016-0543-7. PMID: 27884168

[240] Blasina A, Price BD, Turenne GA, McGowan CH. Caffeine inhibits the checkpoint kinase ATM. *Curr Biol* 1999;9(19):1135–8. doi: 10.1016/s0960-9822(99)80486-2. PMID: 10531013

[241] Kozlov S, Gueven N, Keating K, Ramsay J, Lavin MF. ATP activates ataxia-telangiectasia mutated (ATM) *in vitro*. Importance of autophosphorylation. *J Biol Chem* 2003;278(11):9309–17. doi: 10.1074/jbc.m300003200. PMID: 12645530

[242] Alvira D, Yeste-Velasco M, Folch J, Casadesús G, Smith MA, Pallàs M, Camins A. Neuroprotective effects of caffeine against complex I inhibition-induced apoptosis are mediated by inhibition of the Atm/p53/E2F-1 path in cerebellar granule neurons. *J Neurosci Res* 2007;85(14):3079–88. doi: 10.1002/jnr.21427. PMID: 17638302

[243] Viswanathan P, Gupta P, Sharma Y, Maisuradze L, Bandi S, Gupta S. Caffeine disrupts ataxia telangiectasia mutated gene-related pathways and exacerbates acetaminophen toxicity in human fetal immortalized hepatocytes. *Toxicology* 2021;457:152811. doi: 10.1016/j.tox.2021.152811. PMID: 33971260

[244] Cantani A, Lucenti P, Ronzani GA, Santoro C. Joubert syndrome: review of the fifty-three cases so far published. *Ann Genet* 1990;33(2):96–8. PMID: 224109

[245] Romani M, Micalizzi A, Valente EM. Joubert syndrome: congenital cerebellar ataxia with the molar tooth. *Lancet Neurol* 2013;12(9):894–905. doi: 10.1016/S1474-4422(13)70136-4. PMID: 23870701

[246] Parisi M, Glass I. Joubert syndrome. In: Adam MP, Ardinger HH, Pagon RA, Wallace SE, Bean LJH, Mirzaa G, Amemiya A (eds) GeneReviews® [Internet]. Seattle: University of Washington, 2003. www.ncbi.nlm.nih.gov/books/NBK1325/. PMID: 20301500

[247] Achenbach J, Matusch A, Elmenhorst D, Bauer A, Saft C. Divergent effects of the nonselective adenosine receptor antagonist caffeine in pre-manifest and motor-manifest Huntington's disease. *Biomedicines* 2022;10(6):1258. doi: 10.3390/biomedicines10061258. PMID: 35740281

16 Melatonin and the Brain

Shabana Bibi, Sana Javed, Farah Alam Gill, Shopnil Akash, Hitesh Chopra, Ammara Saleem, and Rashid Hussain

16.1 INTRODUCTION

Melatonin (MT) is a molecule observed in plants and animals that grows naturally. It is well-studied in living things as a hormone naturally produced at night by the pineal gland located in the brain, and it has been connected to the regulation of the sleep-wake feedback loop [1]. MT is a chemical that has been connected to the body's intrinsic immune system. MT has already been proven to be beneficial to health in a broad range of abnormalities [2].

MT was first separated from a cow's pineal gland. MT's physiological functions have been studied widely, so it is recognized as having a new therapeutic impact, and its medical benefits might very well span a broad spectrum [3]. MT has the ability to restrict human physiological rhythms, lessen related conditions such as lethargy and sleepiness, clean free radical living things, improve the immune system, and produce cytotoxic activity [4].

MT-containing food products are famous and recognized as favourable health supplements. Biological rhythms have evolved in order to modify natural roles to particular times of daytime or nighttime [5]. The said time helps to regulate the diurnal sleep and wake cycles, in addition to regulating body temperature and secretion of different hormones as per their requirement. Neural signals are sent to nucleus cells from the membrane of the retina, which transmits messages regarding the photoperiodic condition to the pineal gland, which in turn creates and emits hormones, transferring the clock signal all through the body [6].

MT synthesis takes place 24 hours a day. At night, even so, more is generated and released into the bloodstream. The living human reconstructs about 30 g of per day, with the highest concentration in the blood happening during the mid-dark period of time. This hormone is not crystalized in the pineal gland; it is absorbable into the bloodline and well-decomposed in the liver [7]. MT can be connected with haemoglobin and albumin protein inside the systemic spreading, but it is primarily translocated off albumin. MT's amphiphilic nature enables it to easily travel through normal cells and morpho-physiological, including the blood-brain barrier. MT has been discovered to rearrange lipid membranes utilizing two-dimensional X-ray diffraction readings, with the last impacts relying on its accumulation [8].

Thereby, at small concentrations, MT-enriched spots were discovered in the cell tissue layer; nevertheless, at high concentrations, an extremely involved homogeneous MT structure was detectable all across the tissue layer [9]. Several research findings indicate that consuming foods that contain MT could help to raise the serum levels of this compound [5]. MT is produced by organ systems, as well as the endocrine gland, retina, bone marrow, skin, gastrointestinal tract, and lymphocytes, and it is digested through diets rich in this substance [9].

DOI: 10.1201/9781003300502-16

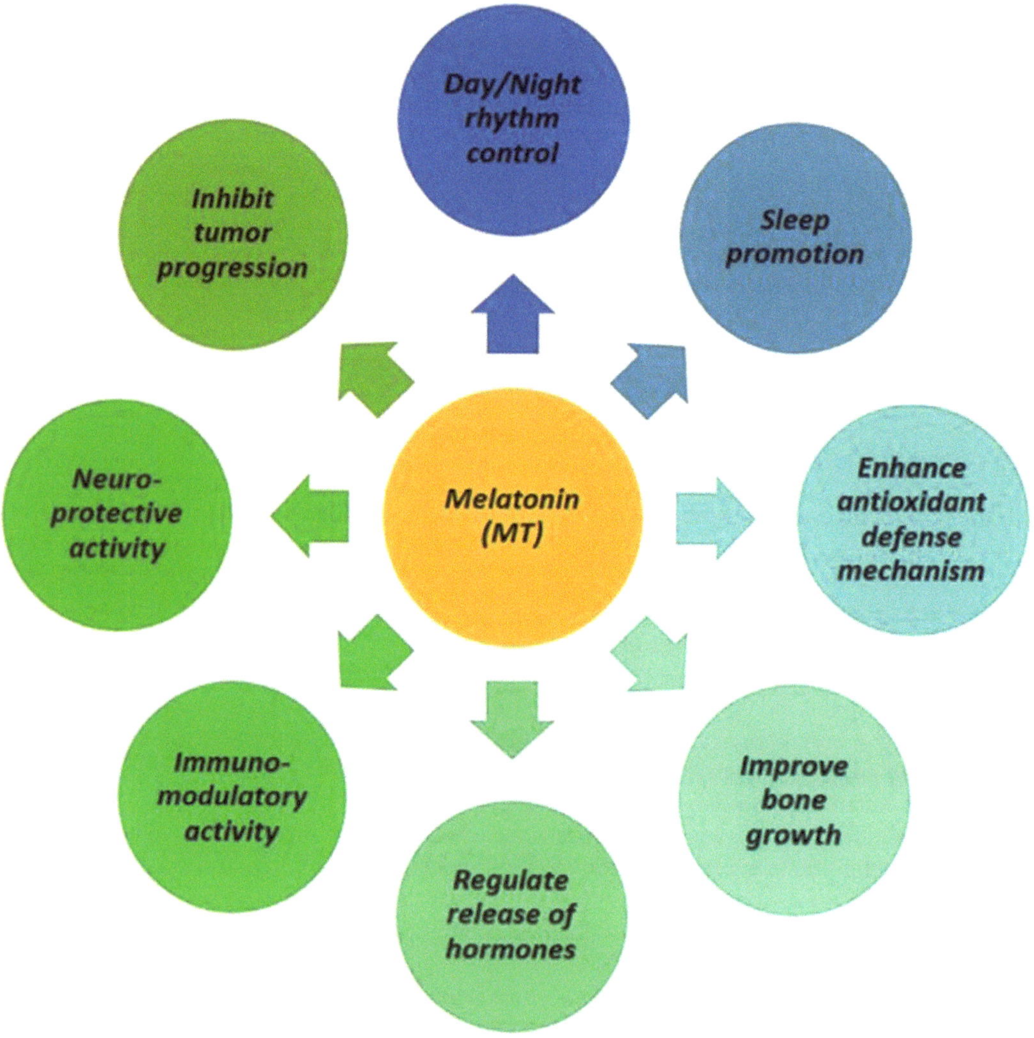

FIGURE 16.1 The primary functions of melatonins in human beings.

Figure 16.1 depicts the MT's primary roles and mechanisms in the human body; it promotes health by controlling the regulation of different pathways in the body. It may also be recognized as a curative choice for handling diseases. MT is normally non-toxic and secure; however, at larger concentrations, only slight adverse effects such as lightheadedness, migraine, vomiting, and tiredness have been observed [8].

16.2 DIETARY SOURCES OF MELATONIN

16.2.1 Animal Foods

The material of MT in food products, on the other hand, varies considerably among organisms. MT levels in nuts and medical herbal products have been discovered to be significantly greater. Furthermore, due to the scientific effort features in organelles, MT may allocate irregularly in one individual living thing in both animal and plant foods [5]. Plant foods with the minimum MT material are fruits, while seedlings and leaves possess the greatest [10].

Moreover, MT was discovered in colostrum at equivalent concentrations to that discovered in plasma, which might advantage newborn babies who possess the found repetitive absorption of MT in the first few weeks of their life, despite the fact that neither synthetic formulas nor dairy-free liquors enclosed detectable MT [11].

16.2.2 Plant Foods

MT has been observed in a diverse array of widely consumed fruits. Strawberries, cherries, and grapes were among the most well-known MT item study results, which also confirmed variety distinctions. MT levels in grape skin, tart cherry skin, and strawberry skin varied from 8.9 to 158.9 ng/g DW. MT levels have been significantly small in other fruits and vegetables. MT has been discovered in a broad scope of day-to-day utilized vegetables, but it is unperceivable in potatoes and cost-effective in beetroots [5]. Tomatoes and peppers are among the most researched, and they contain the top levels of MT accumulation in every vegetable group [12].

MT has been observed throughout many dicotyledonous and vegetative growth plants, with levels ranging from 189 ng/g DW in white mustard and 129 ng/g DW in black mustard [13].

16.2.3 Juices and Beverages

MT levels were measured in famous hard liquor such as drinks and champagne [14]. MT was assumed to be produced synthetically sometimes during yeast due to yeast fermentation, in addition to becoming obtained from essential factors [15].

16.3 BIOAVAILABILITY OF MELATONIN IN HUMANS

As it relates to MT bioactivity, a variety of studies have been conducted, with outcomes varying from verbal to intravenous injection. Moreover, and possibly most relevantly, significant variations in MT items relating to absorption, metabolism, and eradication have also been reported for diverse persons [16]. These outcomes obviously show the need for serious research into the pharmacokinetics and pharmacodynamics of MT [17].

16.4 MELATONIN SYNTHESIS AND METABOLISM

MT is generated in pinealocytes from tryptophan. This synthesis is accomplished in four steps. Tryptophan is transformed into serotonin first, and thereafter MT is synthesized in a two-step process. These enzymes are N-acetyltransferase (NAT), an antagonist of MT, and hydroxyindole-O-methyltransferase (HIOMT). The regulation of that kind of synthesis differs according to the day/night rhythm, with the formulation reaching its highest point at night [18] (Figure 16.2).

The synthesis of MT takes place in the pinealocytes, the cellular units of the pineal gland. The biosynthesis of MT starts with the active absorption of the amino acid tryptophan from the blood. First, 5-hydroxy-tryptophan is formed through tryptophanehydroxylase (TH). Subsequently, the transformation into serotonin (5-hydroxy-tryptamine) takes place by the enzyme aromatic-amino-acid-decarboxylase (AAMD). The next step is the transformation of serotonin into N-acetylserotonin through N-acetyltransferase (NAT), which is the limiting step for MT synthesis. Finally, the synthesis of MT (N-acetyl-5-methoxy-tryiptamine) through the enzyme hydroxyindole-O-methyltransferase (HIOMT) takes place. A variety of variables affect the frequency of this procedure. The existence of b1 adrenoceptors, for example, or the participation of noradrenaline, both play significant functions in these biosynthetic pathways [19]. MT levels increase in such circumstances. Nevertheless, the nutrients folate and B6 are needed for this procedure. Folate is necessary even throughout the

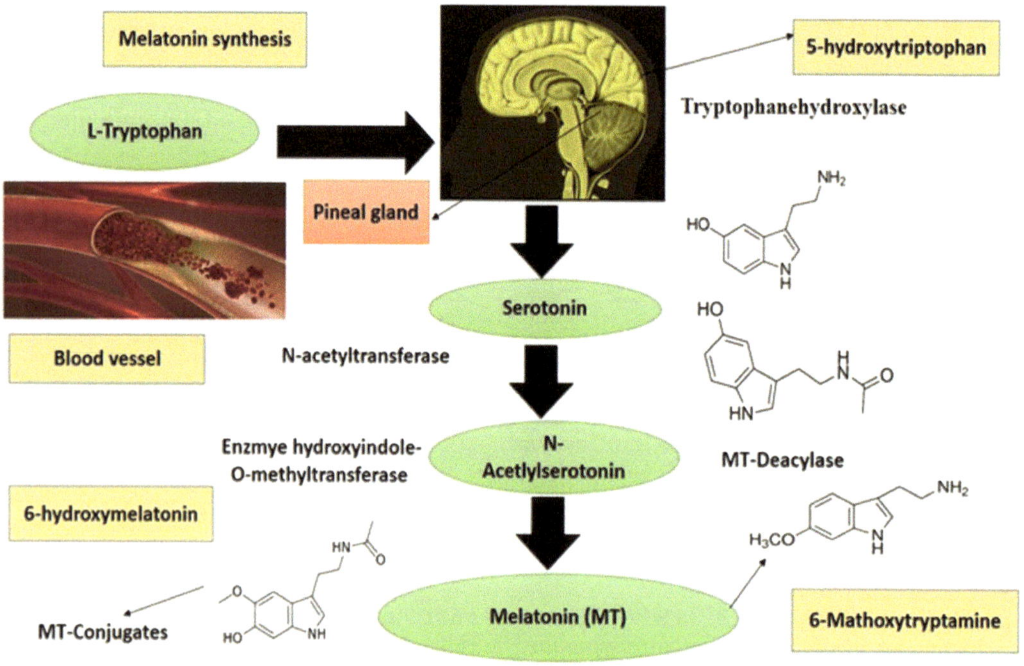

FIGURE 16.2 Process of synthesis of melatonin in the human body.

methylation procedure. B6 vitamins are frequently linked to tryptophan decarboxylation. MT formulation will not occur if they are not present in the environment [20].

16.5 PHARMACOKINETIC PROPERTIES OF MELATONIN

MT metabolic activity is less widely acknowledged than fabrication. 6-OH MT is one of the more important MT metabolites [21]. MT has been suggested as the ultimate biosynthesis of MT for so many centuries, but now it is comprehended that there is no separate breakdown product [22]. MT metabolic activity is, in reality, a very process ensures. MT is hydrolyzed in three ways: enzymatic, pseudo enzymatic, and through reactive oxygen species (ROS) interactions [23]. AFMK may be the next MT biological substance produced through one pseudo enzymatic procedure. It has been proved that haemoglobin and some other iron-containing hemoproteins communicate with MT to establish AFMK as well as other metabolic products. In this performing action, the oxophoryl hemoprotein damages MT and attempts to break down to form AFMK [24].

The main protein involved in the conversion of MT in animals is CP450 [16]. This enzyme is noticed in a wide range of tissues, most noticeably the liver. Indoleamine 2, 3- dioxygenase is the main protein necessary for MT metabolism in the brain (IDO). AFMK is generated as a side effect of enzyme production. MT could also be digested nonenzymatically, which is shocking. As a consequence of the interaction with ROS, so many substances are constituted. It is hard to evaluate which route is more present in the body. Metabolism is the consequence of connections with ROS under serious oxidative stress. Sixty-seven percent of MT throughout plasma adheres to albumin. MT has a half-life of 3–45 minutes. As it is water soluble and has fatty acids, it can directly enter connective tissue. Furthermore, because hepatocellular first throw takes place with intravenous infusion, the impact is significantly larger [25].

16.6 MELATONIN RECEPTORS

MT is readily soluble in both water and lipids. MT this property allows it to easily cross the membranes. Recent studies showed that the cell nucleus is the site where MT is abundantly found because of the presence of specific binding sites for MT [26]. Researchers observed during their in vitro quantitative studies on MT agonist 2- iodomelatonin that MT receptors are present in the brain and peripheral tissues [27].

MT is also found in many other tissues of the body, like veins, ovaries, small and large intestines, liver, eyes, heart, and brain. MT1, MT2, and MT3 are the MT receptors identified to date [28].

16.6.1 MT1 Receptor

The MT1 receptor, which is classified into two subsets called MT1A and MT1B, imposes its action by blocking the adenylate cyclase enzyme in the target organ [26]. It is discovered in the cardiovascular vessels, and it governs circadian rhythms and regulates vessel compression. The MT1 receptor is recognized to be manufactured in sectors beyond the peripheral and brain tissue [29].

16.6.2 MT2 Receptor

The enhancement of cardiovascular vessels is caused by a binding site involved in retinal human biology. But unlike the MT1 receptor, the MT2 receptor is confined to a specific region of the tissue. They are almost always discovered in the hypothalamus, retina, kidney, and heart tissue [29].

16.6.3 MT3 Receptor

In definitions of high-affinity characteristics and detoxification from the Syrian hamster kidney, the MT3 receptor seems to be similar to the MT2 receptor. Its structure is 95% exactly equivalent to that of the living thing quinone reductase 2 enzyme [30]. This enzyme is also involved in purification. The MT3 receptor's action diminishes intraocular pressure by the ability to inhibit leukotriene-induced leukocyte adhesion [31].

16.6.4 Retinoic Acid Receptor-related Orphan Receptor (ROR)

Nuclear hormone receptors (NRs) are a group of signal transduction factors that regulate metabolic activity, inflammatory cells, and stressors and are presently the biochemical objectives of an estimated 13% of Approved drugs. RORa, RORb, and ROR are NR subgroups [32]. The most significant of all these types of receptors is RORa. RZR/RORa is discovered within both the central and peripheral nerves. MT unites with RORs, uncovering its bioactivities. RORs also have immune and metabolic regulatory oversight features [36]. RORs (particularly the RORa subtype) are essential for controlling everyday rhythmic patterns. MT has been shown to possess anti-inflammatory and antioxidant properties by affecting transcription factors via RORs discovered in the cell nucleus. MT is believed to have anti-neurodegenerative characteristics via ROR receptor sites due to these characteristics [33].

16.7 MECHANISM OF ACTION OF MT

Light and dark exposure affects the secretion of MT, which in turn regulates the circadian rhythm or the sleep-wake cycle. Apart from regulating other functions of the body like immune activity, dreaming, fertility and reproduction, learning and memory, and mood control, the main function of

MT is regulating the sleep-wake cycle. MT performs this activity by lowering the body temperature and chemically causing drowsiness [34].

Although there are different pathways involved in MT functioning, the most popular one is the membrane-specific receptor activation pathway [3]. It includes melatonin receptor 1 (MT1) and melatonin receptor 2 (MT2). MT1 inhibits the cAMP signal transduction pathway by acting on decreased or no activity of adenylate cyclase and activating phospholipase C, thus releasing the arachidonate. MT receptors are expressed in multiple tissues of the body and are basically G protein-coupled receptors. In the central nervous system (CNS), expression sites for MT 1 receptors include the suprachiasmatic nucleus (SCN), hippocampus, substantia nigra, cerebellum, and central dopaminergic pathway. Apart from the aforementioned areas, MT1 is also expressed in the retina, coronary circulation, aorta, immune system, skin, kidney, liver, gall bladder, ovary, testis, and memory glands [35]. Similarly, the MT2 receptor expression area is mainly CNS. But it is also found in cardiac, aortic, and coronary tissue, myometrium and granulose cells, lungs, adipocytes, and duodenum. MT2 receptor binding with MT increases protein kinase C (PKC) activity by affecting phospholipase C (PLC). It also hinders guanylyl cyclase activity by forming cyclic guanosine monophosphate (cGMP) and decreasing the formation of cAMP by inhibiting adenylyl cyclase. Receptor activation heads in ion flux channel activation inside the cell [36]. Along with these pathway activities, melatonin also participates in some other intracellular actions. It binds to cytosolic calmodulin on one hand and with two receptors of the Z retinoid nuclear receptors family on the other hand and thus plays a role in immune modulation, different enzyme regulations, and detoxification. MT receptor's expression regions and the role they are playing are summarized in Table 16.1 [37].

Nuclear hormone receptors i.e. ROR, belong to the transcription factor family [38]. It is located in the brain and peripheral nervous system and helps in generating stress responses [39], regulating metabolism and immune reaction after interacting with MT [40]. It also has a crucial role in circadian rhythm regulation. Anti-inflammatory and antioxidant effects of MT are also conferred because of these RORs present in the cell nucleus [41].

MT also exerts some receptor-independent functions besides its receptor-dependent activities [38]. These independent functions include reduction in free radical formation in mitochondria, inhibition of the activities of oxidative enzymes as well as detoxification of radicals and radical's products. All these beneficial effects result in the anti-inflammatory and antioxidant qualities of MT [1] (Table 16.1).

TABLE 16.1
Location, types, and properties of melatonin receptors [37]

Receptor Independent		**Receptor Dependent**	
Membrane (MT1, MT2)	ROS and RNS like free radical scavenging	Cytosol (Calmodulin)	Nucleus (RZR, ROR)
Circadian modulation	Ischaemia and repurfusion	Enzyme regulation	Antioxidant enzyme regulation
Sleep promotion	Protection against UV radiation	Detoxification	Immune modulation
Retinal physiology	Ionizing radiation		
Blood pressure modulation	Drug toxicity		
Seasonal reproduction	Alcohol toxicity		
Bone growth	Heavy metal toxicity		

*RZR: retinoid Z receptor, *ROS: reactive oxygen species, *ROR: retinoic acid orphan receptor, *RNS: reactive nitrogen species, *MT1 & 2: melatonin receptor 1 & 2.

16.8 MELATONIN IN THE CENTRAL NERVOUS SYSTEM

MT regulation and synthesis and its synchronization in the human body are dependent upon and influenced by environmental factors, i.e., light and dark [42]. The dark period of the day allows the release of MT, while the light period disseminates information through the central pacemaker through the retinohypothelamic tract to inhibit MT synthesis by the retinohypothelamic pathway [43]. Artificial light at night does not allow and support MT synthesis. MT controls circadian rhythm by producing its secretion in peaks throughout the 24 hr. These secretions are maximum at night, particularly from 3 to 4 AM. This peak is at its lowest level at 6 to 7 AM in the morning. MT secretion rises on a daily basis with a subsequent increase in sleep predisposition to occur. Once it starts, the 'sleep gate' opens, and it forces the person to sleep by producing some chemical changes in the body and brain, like lowering the mind's activity. Such sleep inclination takes place about two hours with regard to the individual's regular bedtime [44]. This rhythmic secretion of MT by the brain is regulated by the suprachiasmatic nucleus (SCN) of the anterior hypothalamus. MT is endogenously secreted by the pineal gland but in some quantity, it is also synthesized by the eyes, skin, lymphocytes, bone marrow, thymus, and gastrointestinal tract. The SCN is affected by light through retinal ganglion cells. The information from the eyes to the SCN not only allows MT secretion but also delivers information to other areas of the brain regulating pupil constriction and sleep induction [33]. Once information from melanopsins starts, it then travels from the SCN to the pineal gland to the upper parts of cervical spinal cords, which form the preganglionic part of the superior cervical ganglion (SCG) and possess sympathetic chains. For the regulation of the pineal gland, norepinephrine is the most important neurotransmitter [43]. SCN stimulated its release at night, while b1 adrenergic system blockers inhibit MT secretion at night. Similarly, reduction in MT secretion level is also done by clonidine and methylparatyrosine which inhibit presynaptic catecholamine. MT levels are elevated by using catecholamine and neuropeptide Y. However, gamma-aminobutyric acid (GABA), glutamate, and dopamine inhibit MT synthesis [45].

16.8.1 Neuroprotective Capacity

MT acts as a neuroprotective agent in multiple neurodegenerative diseases like amyotrophic lateral sclerosis, Huntington's disease, ischaemic brain injury, Parkinson's disease and Alzheimer's disease [45]. MT is also involved in maintaining proper balance and controlling body posture. It also helps in brain development and prevents any adverse effects as a result of accidental damage to the brain in the adult and prenatal stages since it is capable of crossing the blood-brain barrier and placenta alike [34]. By crossing the blood-brain barrier, it inhibits certain inflammatory and excitotoxic pathways and also protects against the pathogenesis of oxidative stress because of the specific interaction of its indole moiety with its receptors. These receptors, when activated, play a role in neuronal cell maintenance by either down-regulating the levels of brain-derived neurotrophic factor (BDNF) [14]. In post-traumatic conditions, administration of MT individually, as well as with uridine, results in decreasing brain oedema and lowering malondialdehyde levels. It also stimulates the release of the blood-brain barrier and intracranial stresses, also the actions of superoxide dismutase (SOD) and glutathione peroxidase (GPx).

The severity of 6-OHDA-induced Hemi-disease Parkinson's can be reduced by increasing MT bioavailability [16]. MT wants to treat these maladies by affecting cells by initiating several signal pathways and blocking intrinsic apoptotic pathways [3]. MT has an impressive neuroprotective part in decreasing cognitive impairment caused by sleep deprivation (SD), as well as restoring the number of oxidative stress markers such as SOD and melonaldehyde (MDA) as well as NO activity [46]. Figure 16.3 shows the action of melatonin in anxiety, restricting the neuro-inflammatory pipeline.

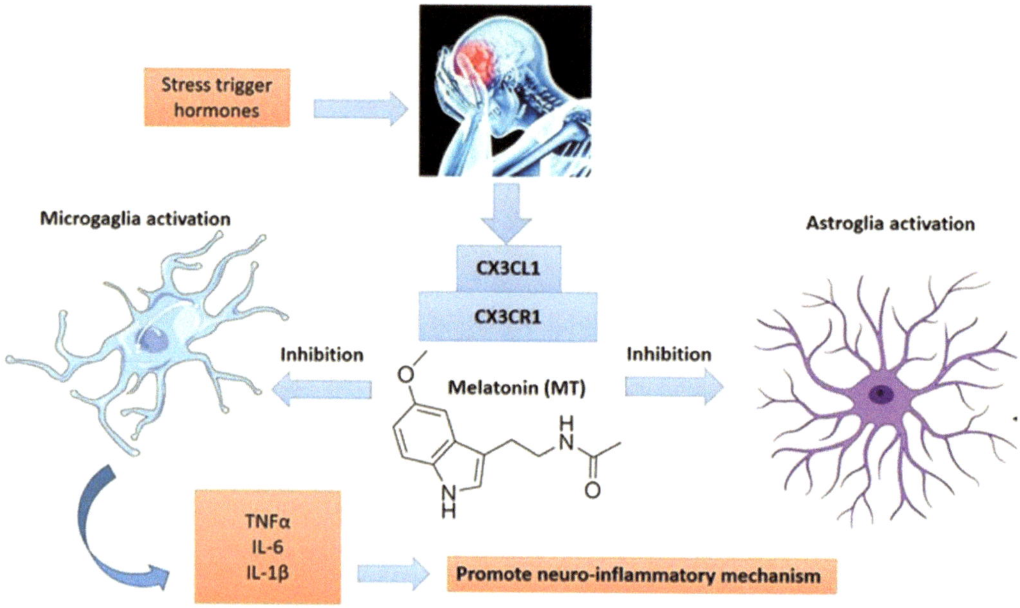

FIGURE 16.3 Action of melatonin in anxiety restricting the neuro-inflammatory pipeline.

16.8.2 Anti-ischaemic and Anti-oxidative Effects of Melatonin

The most prevalent type of stroke is ischaemic stroke, which happens as a result of thrombotic and embolic clots, which then obstruct the blood supply to the brain from that particular cerebellar artery [28]. Ischaemic stroke not only causes neuronal necrosis but also initiates a series of pathophysiological events, which include oxidative stress, excitotoxicity, and inflammation [47]. To treat this damage, MT has been vastly taken into consideration because of the well-known anti-oxidative effects of this molecule. As MT is a potential free radicals scavenger, its prophylactic and post-disease effects for ischaemic stroke are hotspots of discussion [47]. MT has the capacity to forge various free radicals such as O_2 radicals, singlet oxygen (O_{12}), hydroxyl (OH) radicals, peroxyl radicals, hydrogen peroxide (H_2O_2), nitric oxide (NO), and peroxynitrite (ONOO_) in physiological conditions [48]. By targeting such a wide variety of free radicals, MT protects the brain from inflammation, oxidative stress, and neuronal death and alleviates brain oedema [48]. Post-stroke neurological defects are also ameliorated. For all these beneficial effects, MT and its various metabolites, such as N_1-acetyl-N_2-formyl-5-methoxykynuramine (AFMK) and N1-acetyl-5-methoxykynuramine (AMK), are widely studied and considered as most potent antioxidants, playing an important protective role in ischaemic injury. MT not only scavenges the free radicals but also improves motor behavioural outcomes [28].

16.8.3 Anti-oxidative Effects of Melatonin in Alzheimer's Disease

Alzheimer's is a neurodegenerative disorder caused by the deposition of a toxic protein, b-amyloid, in the brain. This abnormal structural protein produces neurofibrillary tangles, which in turn cause hurdles for synaptic functioning leading to decreased cognitive performance. The cascade arises from disrupted b-amyloid accumulation in the brain resulting in neuronal loss to increase oxidative stress. AD is an age-dependent problem [48]. So, the effect of oxidative stress is more severe in older age AD patients. The antioxidant, anti-inflammatory, as well as anti-amyloidogenic properties of MT make it a ray of hope for AD patients [25]. Although the exact cause of AD has not been elucidated yet, it has been observed that AD is a multifactorial disease. These factors include gender, nutrition, age, and genetics. The studies show the enhanced production of presenilins 1 and 2, apolipoprotein E (ApoE), and amyloid precursor proteins (APP) in AD patients as compared to normal people. MT inhibits the synthesis and activation of APP and b-amyloid along with the protection against fibril production in the hippocampus and cortex. Thus, MT plays a dominant and beneficial role in mitigating neurodegenerative changes [25].

16.9 ANTI-DEPRESSANT-LIKE ACTION OF MELATONIN

Many presymptomatic studies have identified an association between stress and MT, about which circadian rhythms of Mongolian Gerbils were subjected to blue light for long periods of time [48]. This action lessens corticosterone levels, lowers corticotrophin-releasing hormone (CRH) in the hypothalamus, and induces the protein melanopsin to be dopamine-depleted in those living creatures. MT has been shown in research to have therapeutic value against insomnia, sleep problems, and anxiety. Neurogenesis was stimulated by administrating a low dose of ketamine and MT around each other. This also overturns neuronal muscle wastage, which enhances symptoms of depression [49].

MT, an intrinsic hormone, is essential for circadian, immune regulation, emotion control, synapse formation, and neurohormonal efflux [50]. Depression caused by MT secretion abnormalities. To cure this anxiety, an external supply of MT is administered orally [49][50][51]. MT supplements have presented the association with neuro-inflammation, cholinergic disturbance, and structural abnormalities in the hippocampus (Figures 16.3 and 16.4).

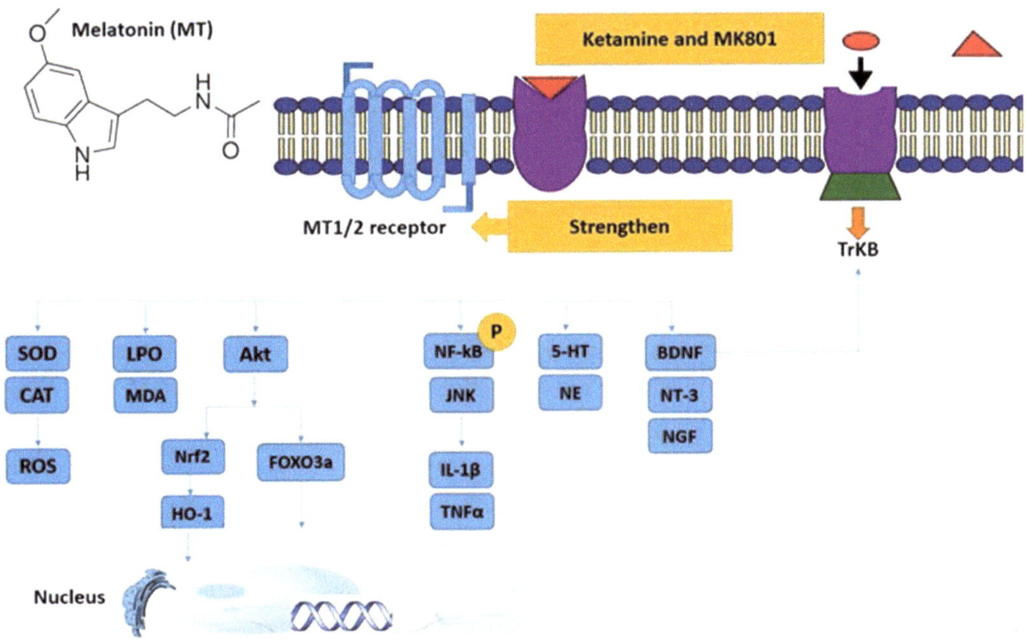

FIGURE 16.4 Features of melatonin presenting anti-depressant activates in the human body.

*TrkB: tropomyosin receptor kinase B; *AKT: protein kinase B, *Nrf2: NF-E2-related factor 2, *FOXO3a: forkhead box transcription factor 3a, *SOD: superoxide dismutase, *HO-1: haem oxygenase-1, *CAT: catalase, *ROS: reactive oxygen species, *LPO: lipid peroxide; *MDA: malondialdehyde, *NT-3: neurotrophin-3, and *NGF: nerve growth factor.

16.10 CONCLUSION

To successfully survive and live harmoniously in the environment by maintaining a proper balance with outer and inner environmental conditions, MT plays an important role. MT not only helps in homeostasis of the body but also synchronizes the central and peripheral nervous system functioning. Studies prove that MT might be the most potent biomarker for human circadian rhythms and all other activities linked to the biological clock. Its beneficial effects as an antioxidant, anti-ageing, anti-inflammatory, and anti-depressant agent cannot be denied. Rather, it can be used as a significant prophylactic factor against many neurodegenerative diseases like ischaemia-reperfusion, Alzheimer's, Parkinson's, depression, oxidative stress and cancer. Thus, melatonin's role as a potential therapeutic agent against many other disorders must be researched in the future.

REFERENCES

1. Reiter, R.J.; Tan, D.X.; Galano, A. *Physiology* **2014**, *29*, 325–333.
2. De Bodinat, C.; Guardiola-Lemaitre, B.; Mocaër, E.; Renard, P.; Muñoz, C.; Millan, M.J. *Nat. Rev. Drug Discov.* **2010**, *9*, 628–642.
3. Majidinia, M.; Reiter, R.J.; Shakouri, S.K.; Yousefi, B. *Ageing Res. Rev.* **2018**, *47*, 198–213.
4. Neubauer, D.N. *Neuropsychiatr. Dis. Treat.* **2008**, *4*, 69.
5. Meng, X.; Li, Y.; Li, S.; Zhou, Y.; Gan, R.-Y.; Xu, D.-P.; Li, H.-B. *Nutrients* **2017**, *9*, 367.
6. Penev, P.D.; Zee, P.C. *Ann. Neurol. Off. J. Am. Neurol. Assoc. Child Neurol. Soc.* **1997**, *42*, 545–553.
7. Lockley, S.W.; Dressman, M.A.; Licamele, L.; Xiao, C.; Fisher, D.M.; Flynn-Evans, E.E.; Hull, J.T.; Torres, R.; Lavedan, C.; Polymeropoulos, M.H. *Lancet* **2015**, *386*, 1754–1764.
8. Kato, K.; Hirai, K.; Nishiyama, K.; Uchikawa, O.; Fukatsu, K.; Ohkawa, S.; Kawamata, Y.; Hinuma, S.; Miyamoto, M. *Neuropharmacology* **2005**, *48*, 301–310.
9. Miyamoto, M.; Nishikawa, H.; Doken, Y.; Hirai, K.; Uchikawa, O.; Ohkawa, S. *Sleep* **2004**, *27*, 1319–1325.
10. Lavedan, C.; Forsberg, M.; Gentile, A.J. Tasimelteon: *Neuropharmacology* **2015**, *91*, 142–147.
11. Ebisawa, T.; Karne, S.; Lerner, M.R.; Reppert, S.M. *Proc. Natl. Acad. Sci.* **1994**, *91*, 6133–6137.
12. Liu, C.; Weaver, D.R.; Jin, X.; Shearman, L.P.; Pieschl, R.L.; Gribkoff, V.K.; *Neuron* 1997, *19*, 91–102.
13. Clemens, J.W.; Jarzynka, M.J.; Witt-Enderby, P.A. *Life Sci.* **2001**, *69*, 27–35.
14. Ram, P.T.; Dai, J.; Yuan, L.; Dong, C.; Kiefer, T.L.; Lai, L.; Hill, S.M. *Cancer Lett.* **2002**, *179*, 141–150.
15. Brzezinski, A. *From Mol. to Ther. SR Pandi-Perumal, DP Cardinali Melatonin. NY Nov. Sci. Publ.* 2007, 219–225.
16. Mannino, G.; Pernici, C.; Serio, G.; Gentile, C.; Bertea, C.M. *Int. J. Mol. Sci.* **2021**, *22*, 9996.
17. Doolen, S.; Krause, D.N.; Dubocovich, M.L.; Duckles, S.P. *Eur. J. Pharmacol.* **1998**, *345*, 67–69.
18. Dubocovich, M.L.; Rivera-Bermudez, M.A.; Gerdin, M.J.; Masana, M.I. *Front. Biosci.* **2003**, *8*, 1093–1108.
19. Hernández-Flores, T. de J.; Pedraza-Brindis, E.J.; Cárdenas-Bedoya, J.; Ru\'\iz-Carrillo, J.D.; Méndez-Clemente, A.S.; Mart\'\inez-Guzmán, M.A.; Iñiguez-Gutiérrez, L. *Int. J. Mol. Sci.* 2022, *23*, 12324.
20. Marciano, D.P.; Chang, M.R.; Corzo, C.A.; Goswami, D.; Lam, V.Q.; Pascal, B.D.; Griffin, P.R. *Cell Metab.* 2014, *19*, 193–208.
21. Koss-Mikołajczyk, I.; Todorovic, V.; Sobajic, S.; Mahajna, J.; Gerić, M.; Tur, J.A.; Bartoszek, A. *Int. J. Mol. Sci.* **2021**, *22*, 10037.
22. Malika, A.S.B.; Hajar, K.; Loubna, B.; El Hiba, O.; Rachida, M.; Leila, B.; Souad, S.; Hanane, R.S. In *Research Anthology on Advancements in Women's Health and Reproductive Rights*; IGI Global, **2022**; pp. 357–397.
23. Overington, J.P.; Al-Lazikani, B.; Hopkins, A.L. *Nat. Rev. Drug Discov.* **2006**, *5*, 993–996.
24. Tan, D.-X.; Manchester, L.C.; Terron, M.P.; Flores, L.J.; Reiter, R.J. *Journal Pineal Res.* **2007**, *42*, 28–42.
25. Parlakpinar, H.; Orum, M.H. *Med. Sci.* **2014**, *3*, 1766–1780.

26. Dubocovich, M.L.; Markowska, M. *Endocrine* **2005**, *27*, 101–110.
27. Tan, D.-X.; Hardeland, R.; Manchester, L.C.; Galano, A.; Reiter, R.J. *Curr. Med. Chem.* **2014**, *21*, 1557–1565.
28. Hardeland, R.; Tan, D.-X.; Reiter, R.J. *J. Pineal Res.* **2009**, *47*, 109–126.
29. Gunata, M.; Parlakpinar, H.; Acet, H.A. *Rev. Neurol. (Paris).* **2020**, *176*, 148–165.
30. Acuña-Castroviejo, D.; Escames, G.; Venegas, C.; Díaz-Casado, M. E.; Lima-Cabello, E.; López, L. C.; Rosales-Corral, S.; Tan, D. X.; Reiter, R. J. .*Cell Mol. Life Sci.* 2014, *71*, 2997–3025.
31. Sack, R.L.; Lewy, A.J.; Hughes, R.J. *Ann. Med.* **1998**, *30*, 115–121.
32. Ladurner, A.; Schwarz, P.F.; Dirsch, V.M. *Nat. Prod. Rep.* **2021**, *38*, 757–781.
33. Di, W.-L.; Kadva, A.; Johnston, A.; Silman, R. *N. Engl. J. Med.* **1997**, *336*, 1028–1029.
34. Cardinali, D.P.; Golombek, D.A.; Rosenstein, R.E.; Cutrera, R.A.; Esquifino, A.I. *J. Pineal Res.* **1997**, *23*, 32–39.
35. Anadón, A.; Ares, I.; Mart\'\inez-Larrañaga, M.-R.; Mart\'\inez, M.-A. In *Nutraceuticals*; Elsevier, **2021**; pp. 537–553.
36. Serpa, A.S.R.P. *Implications on Neurotransmission and Synaptic Plasticity*, Universidade da Beira Interior (Portugal), **2015**.
37. Morgan, P.J.; Williams, L.M. *Experientia* **1989**, *45*, 955–965.
38. Luchetti, F.; Canonico, B.; Betti, M.; Arcangeletti, M.; Pilolli, F.; Piroddi, M.; Canesi, L.; Papa, S.; Galli, F. *FASEB J.* **2010**, *24*, 3603–3624.
39. Emet, M.; Ozcan, H.; Ozel, L.; Yayla, M.; Halici, Z.; Hacimuftuoglu, A. *Eurasian J. Med.* **2016**, *48*, 135.
40. Gege, C. *Expert Opin. Drug Discov.* **2021**, *16*, 1517–1535.
41. Huang, H.; Liu, X.; Chen, D.; Lu, Y.; Li, J.; Du, F.; Zhang, C.; Lu, L. *Int. Immunopharmacol.* **2020**, *83*, 106365.
42. Minich, D.M.; Henning, M.; Darley, C.; Fahoum, M.; Schuler, C.B.; Frame, J. *Nutrients* **2022**, *14*, 3934.
43. Reddy, V.D.K.; Dalai, M.; Khan, M.S.; Jagota. *Biogerontology* **2022**, 1–18.
44. Skene, D.J.; Arendt, J. *Ann. Clin. Biochem.* **2006**, *43*, 344–353.
45. Trudeau, V.L. *Rev. Reprod.* **1997**, *2*, 55–68.
46. Yang, Y.; Islam, M.S.; Wang, J.; Li, Y.; Chen, X. *Int. J. Biol. Sci.* **2020**, *16*, 1708–1717.
47. Andrabi, S.S.; Parvez, S.; Tabassum, H. *Adv. Pharmacol. Sci.* **2015**, 384750. https://doi.org/10.1155/2015/384750
48. Reiter, R.J.; Tan, D.; Osuna, C.; Gitto, E. *J. Biomed. Sci.* **2000**, *7*, 444–458.
49. Estrada-Reyes, R.; Quero-Chávez, D.B.; Alarcón-Elizalde, S.; Cercós, M.G.; Trueta, C.; Constantino-Jonapa, L.A.; Oikawa-Sala, J.; Argueta, J.; Cruz-Garduño, R.; Dubocovich, M.L.; et al. *Molecules* **2022**, *27*, 5650.
50. Kholghi, G.; Eskandari, M.; Saadlou, M.-S.S.Q.; Zarrindast, M.-R.; Vaseghi, S. *Physiol. & Behav.* **2022**, 113835.
51. Wang, Y.; Jiang, Y.; Zou, M.; Liu, J.; Zhao, H.; Wang, Y. *Behav. Brain Res.* **2022**, *420*, 113724.

17 Narcotics and Brain Health

Rashid Hussain, Shabana Bibi, Muhammad Asad Ali, Shopnil Akash, Nobendu Mukerjee, Ammara Saleem, and Athanasios Alexiou

17.1 INTRODUCTION

Americans spend more than seven hundred billion dollars annually on higher healthcare expenditures, violence, and lost productivity as a consequence of the use and abuse of illegal and prescribed drugs [1–3]. Alcohol is a factor in the fatalities of more than 90,000 Americans each year, while illegitimate and narcotic high intake cause tens of thousands of mortalities each year (almost 70,000 in 2018). Tobacco usage is thought to be responsible for 480,000 deaths annually [4, 5].

The negative effects of drug usage and addiction affect individuals of all ages;

- Teenagers who use drugs may disobey, poor performance in school, or drop out [6].
- Mature drug addicts may have difficulties paying proper attention, remembering things, and thinking effectively.
- Drug use by parents can result in stressful families and abuse and neglect of children [7]. These things are harmful to the development and well-being of children at home, and they may pave the way for descendants to use drugs [8].
- Fetuses exposed to drugs have a higher chance of being underdeveloped and underweight when they are born [9].

Researchers study the impact of drugs on the brain and behaviour. They make use of this information to plan campaigns that support those fighting drug addiction and deter drug use. Putting these ideas into practice in the community will benefit from more research. Addiction is a disease, just like other disorders like cardiovascular disease. Both have significant detrimental effects on an organ's normal, healthy functioning and are usually preventable and treatable. If not treated, they may persist for a lifetime and possibly cause death [10]. The following are common reasons why people use narcotics: drugs can cause extremely intense pleasure. Depending on the medication taken, there can be further negative effects after this first pleasure. However, heroin and other opioids provide pleasure along with calmness and a sense of fulfilment. Some individuals who are stressed, depressed, or anxious begin taking medications in an effort to feel less anxious [11]. Some people feel the need to focus more intently on their work, studies, or sporting endeavours. This might affect someone's decision to try or continue taking drugs like amphetamine or prescription stimulants. If risk factors, such as the presence of companions who use drugs, are present, substance use may begin at any point during the adolescent growth period [12].

When a person takes medicine for the first time, they might notice what seem to be positive effects. They can also believe that they are in charge of their consumption. However, abusing drugs may fundamentally alter a person's life. Other fun activities soon become less enjoyable with continued drug use, and the user has to use the narcotic just to feel 'normal' [13]. They strive

DOI: 10.1201/9781003300502-17

TABLE 17.1
Risk factors and defending elements of individual related to drugs

Risk Factors	Defending Elements
Childhood aggression [18, 19]	Self-efficacy (confidence in one's own ability) [20]
Inadequate parental oversight [19, 21]	Parental supervision and assistance [21 – 23]
Low peer rejection abilities [18, 22, 23]	Connections that are fruitful [22, 24]
Testing for drugs [19, 25, 26]	Excellent academic performance [23, 27]
Narcotics usage in schools [25, 28]	Drug-free schools policies [22]
Inflation in society [29, 30]	Community resources [31]

to control their addiction despite the harm that drug use poses to them and their loved ones. Most users may start to feel the need to use drugs more frequently or in larger doses, even in the early stages of their addiction. Addiction is indicated by these symptoms. Narcotic use is risky, even in very moderate amounts. Think about how a social drinker may get drunk, get in a car, and rapidly transform a fun pastime into a catastrophe that destroys many people. Even casual drug use can still have devastating consequences, including driving while intoxicated, such as when an opioid is misused to get high [14].

Initially, deciding to take drugs is often a free choice. Regular use, however, can significantly reduce a person's ability to exercise self-control. Self-control is lost in addiction, which is its defining feature [15]. Research on brain imaging in addicts reveals physical changes in brain areas crucial for judgement, decision-making, memories, cognition, and behavioural control [16]. The obsessive character of addiction is explained by these modifications. The likelihood of addiction to narcotics varies from individual to individual, just like it does with other diseases and disorders, and no one characteristic indicates whether a person will become hooked to narcotics. The more risk factors an individual has, the more likely they are to use drugs and become addicted [17] (Table 17.1).

17.2 BIOLOGICAL ELEMENTS MAKE ADDICTION TO DRUGS MORE LIKELY

An individual's propensity to become addicted might depend on a variety of factors, including their genetic makeup, developmental stage, colour, and gender. It is believed that genes have a role in between 40–60 % of a person's likelihood of becoming addicted, including how environmental factors affect a person's gene expression [32]. In addition, drug use and addiction are more prevalent in young individuals and those with mental problems than in the general population [33].

17.2.1 Environmental Risk Factors

Home, work, and community factors all fall under the category of the environment. The following elements may have an impact on one's risk. The home environment is very important, especially in early development. When parents or other older family members abuse drugs, alcohol, or the law, children are more likely to experience drug problems in the future [34]. During the adolescent years, colleagues and other companions may have an ever-greater influence. If a youngster struggles academically or socially, their propensity to use drugs or become addicted to them may rise [35].

17.2.2 Additional Elements Boost the Chances of Drug Addiction

Although consuming narcotics at any time can lead to addiction, data suggests that the sooner a person begins using narcotics, the more likely they will encounter serious troubles [36]. This may be due to the detrimental effects that medications may have on the growth of the brain [37]. Early genetic and social risk factors, such as an unstable home or family, having experienced physical or sexual abuse, possessing certain genes, or suffering from a mental disease, may all play a role in how it develops. A drug's potential for addiction is increased when it is smoked or injected into a vein [38, 39]. Drugs taken orally or intravenously both quickly enter the brain and cause a strong surge of pleasure. This powerful high may, however, pass quickly. According to scientists, this stark difference causes some people to frequently take drugs in an attempt to recreate the transitory sense of pleasure.

The brain experiences significant changes during puberty and continues to grow until maturity. Teens are more likely to experiment with drugs or continue using them since their prefrontal cortex is still maturing in the brain. Drug use at this stage of brain development has the potential to alter the brain significantly and permanently [40]. In spite of the negative effects that follow from taking a substance or drugs, addiction is defined by a need to continue doing so. Previously believed to be a moral issue and a result of making poor decisions, addiction is now increasingly acknowledged as a chronic, relapsing condition that entails potentially long-lasting alterations to the brain [41, 42].

Getting clean is considerably harder than just saying 'no' to drugs because of the brain alterations brought on by prolonged substance use. Drug use is initially a decision made by the user, but when brain chemistry changes and addiction set in, the user's capacity to manage their usage decreases [43]. According to NIDA, both addiction and cardiovascular disease cause an important body organ to stop working efficiently, and both conditions can be fatal if left untreated [41]. To achieve long-lasting recovery, an addict frequently needs therapy, sometimes numerous courses of treatment [44]. People in therapy can learn how to control their addiction and acquire the tools and abilities they need to deal with impulses, prompts, and other difficulties that come with rehabilitation.

17.3 DRUGS HARM THE BRAIN CELLS PERMANENTLY

The long-term outcomes of various drugs on the brain might differ. While most people experience the alterations that lead to addiction, several drug classes are linked to additional, distinctive effects on the brain. The alterations to the brain that are covered below are not a complete list of all possible changes brought on by using these medicines [45]. Narcotics affect the brain in the following ways [46]: variations in the white matter tracts of the brain will be affected, such as antisocial conduct, violence and aggressiveness, which may be related to abnormalities of the white matter pathways in the brain [47]. Alterations in the functional connectivity of certain brain areas take place, and issues with cognitive processing may result from disrupted interconnection. It could also be a sign of alterations in how the brain is structured. Both a reduction in amygdala volume and a problem with how well the amygdala processes information [48].

17.3.1 Opioid Overdose and Brain Injury

The prevalence of very strong synthetic opioids like fentanyl on the black market puts those who misuse opioids in particular danger of overdosing. Opioid overdoses cause significant respiratory depression that, if not addressed right away, can lead to damage from hypoxia. Hypoxia describes a lack of oxygen in the tissues [49]. Insufficient oxygen over an extended period can harm the brain. Hypoxia-related brain damage may lead to [50]:

- Ambiguity.
- Memory issues, such as loss of recent memories.
- A shift in behaviour.
- Deteriorated cognitive abilities.
- Slower movement and reaction times.

- Issues with walking.
- Damage to the nerves.
- Chronic stress.

17.3.2 Alcoholic Effects on the Brain

Alcohol affects the brain in many different ways during a short period of time, including slowing reaction times and memory loss, but it can also cause serious long-term damage. For instance, prolonged alcohol consumption may cause brain shrinkage, with women possibly being more susceptible [51].

Alcoholism can indirectly impact the brain as well. Alcoholism frequently causes inadequate nutrition, which can result in a vitamin B1 deficit (thiamine). In fact, according to the National Institute on Alcohol Abuse and Alcoholism, up to 80% of people with alcohol use disorders have a thiamine deficiency [52]. Unfortunately, the brain may suffer greatly as a result of this: Wernicke-Korsakoff syndrome, a condition that combines Wernicke's encephalopathy with Korsakoff's psychosis, can be brought on by vitamin B1 deficiency [51]. Its symptoms include uncertainty, neurological eye paralysis, a decline in motor coordination and serious issues with learning and memory. Alcohol is also well recognized to be particularly bad for the brain's development. Drinking alcohol while pregnant can lead to a variety of challenges, from behavioural problems to learning difficulties. A kid may have fetal alcohol syndrome in the most extreme situations. This disorder may result in distinctive facial characteristics, a smaller brain, and fewer healthy brain cells [51].

17.3.3 Stimulant Effects on the Brain

Substances like Adderall, which are frequently misused as 'study medications', actually reduce brain plasticity, impair executive function, and reduce cognitive and behavioural flexibility. As behavioural flexibility is essential to recovery from drug use disorders, this might be especially worrisome for someone who has been addicted to stimulants. Young people with growing brains may be particularly vulnerable to these alterations [45].

Abuse of illegal psychostimulants like cocaine and methamphetamine is associated with depressed symptoms. When stimulants are not used, symptoms like depression or irritability frequently go away. Even after extended periods of sobriety, additional symptoms like anhedonia (the inability to experience pleasure) and lack of desire might yet endure [53].

Paranoia, delusions, and other psychotic symptoms are common in long-term methamphetamine users and might last even after they stop taking the substance. In previous meth users, stressful circumstances may have resulted in spontaneous psychosis [54]. Persistent methamphetamine use can change the shape and working of the brain in ways that affect memory and emotional processing. Meth usage may also result in flexibility issues that are comparable to those brought on by prescription use [55]. These people could find it extremely difficult to cease engaging in pointless or unproductive activities, which would make long-term addiction treatment much more challenging [56].

17.4 NARCOTICS AND BRAIN HEALTH

The Greek Alcmaeon made the following significant advancements in neuroanatomy when he discovered that the brain, not the heart, was in charge of controlling the body and that the senses were reliant on it. Following Alcmaeon, several researchers, thinkers, and medical professionals from different countries proceeded to advance our knowledge of neuroanatomy. The most significant Greek cognitive scientists were probably Herophilus and Erasistratus of Alexandria,

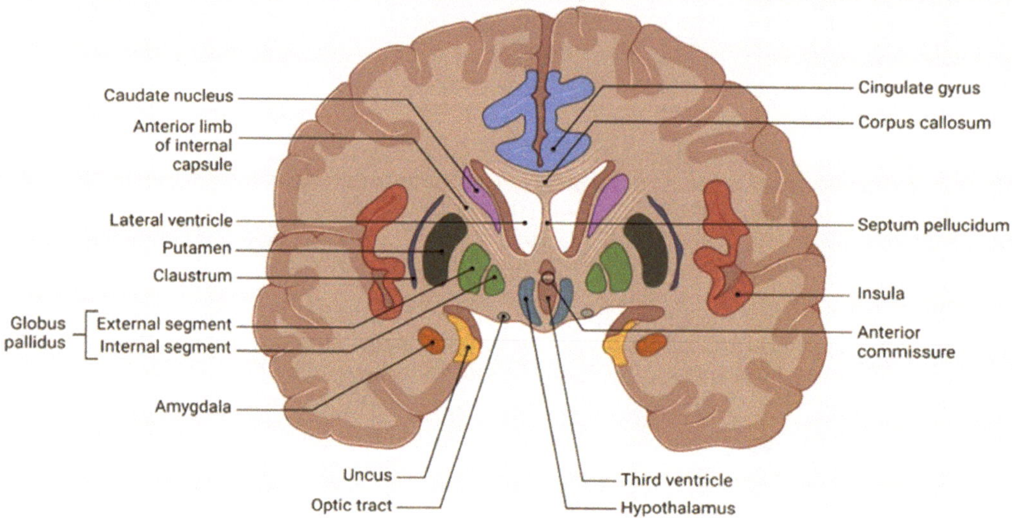

FIGURE 17.1 Graphical view of the brain presenting different dissectional compartments.

whose research involved brain dissection [57]. The traditional taboo against dissection persisted for some hundred years after that, preventing significant advancement in neuroscience. However, in the Renaissance, it was once more permitted to autopsy human beings, reviving the study of neuroscience (Figure 17.1). The brain system has been extensively studied and documented over the following centuries. But we are nowhere near the limit of our comprehension at this point [58].

The brain consists of billions of neurons, and they transmit information. If a neuron receives enough impulses from the other neurons it is connected to, it will fire and send its own message to the other neurons in the circuit. The brain is made up of several interconnected circuits that work together as a unit. Task coordination and execution are carried out by many brain circuits [59]. Networks of neurons that exchange messages (the peripheral nervous system) connect the spinal cord, several brain areas, and the rest of the body's nerves [60] (Figure 17.1).

The delivery, reception, and processing of signals by neurons via neurotransmitters are all impacted by drugs [61]. Because certain drugs' molecular structures resemble those of the body's natural neurotransmitters, such as heroin and marijuana, they can increase brain activity. This makes the drugs' ability to bind to and excite the neurons dependable [62] (Figure 17.2).

Cocaine and amphetamine are two examples of additional drugs that may either inhibit the normal recycling of these neurotransmitters in the brain by interfering with transporters or cause exceptionally high levels of natural neurotransmitter release from neurons. Additionally, this intensifies or obstructs normal brain connection [63].

Addictive drug use that is characterized by compulsive drug use can be brought on by narcotics, which have the ability to alter critical brain areas necessary for vital functioning [64]. Narcotic usage has an effect on the following areas of the brain:

- The basal ganglia are responsible for the development of routines and habits, as well as positive motivation. These areas play a key role in the 'reward circuit' of the brain, as it is sometimes referred to. When this circuit is strongly impacted by drugs, the euphoria of the narcotic high is created. But with repeated exposure, the circuit loses sensitivity to drug use and struggles more to get pleasure from sources other than drugs [65].

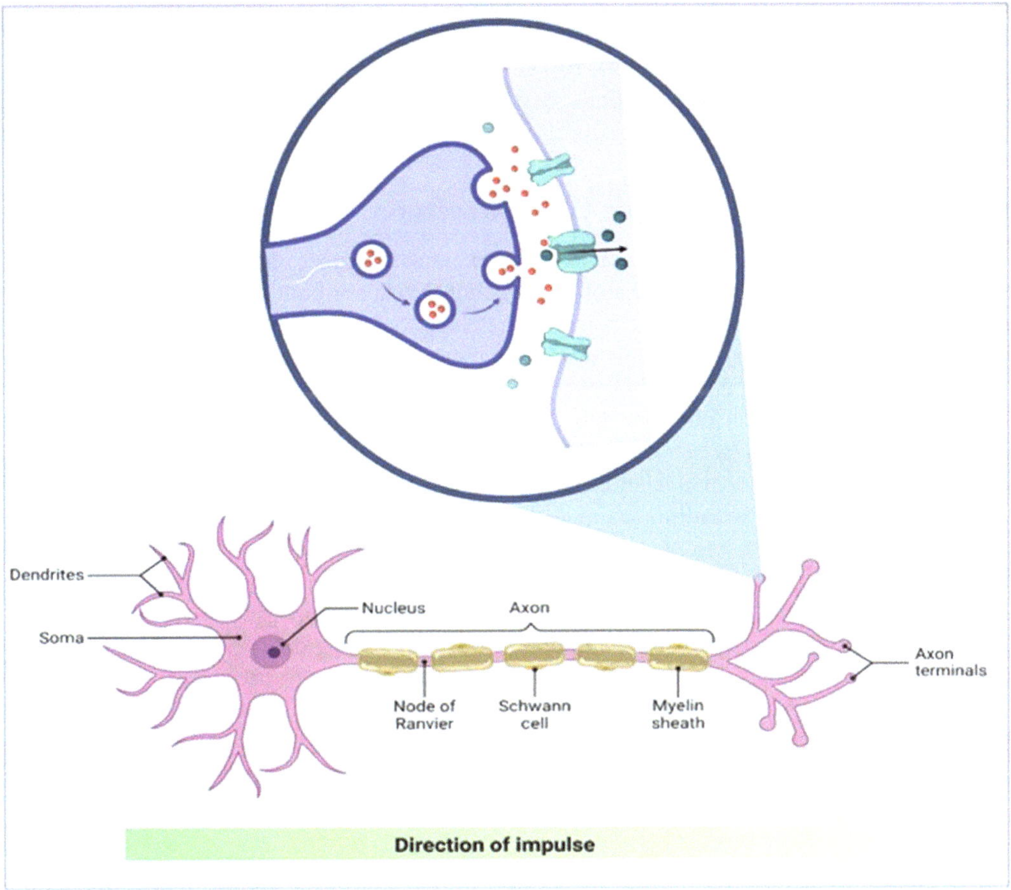

FIGURE 17.2 Process of cell signalling proceeded by neuron networks.

- Stressful emotions are influenced by the enlarged amygdala. This circuit gets more sensitive as drug use rises. Drug usage by an addict changes over time from being primarily about getting high to also serving as a means of short-term pain relief [66].
- Thinking, planning, problem-solving, making decisions, and exerting self-control over desires are all functions of the prefrontal cortex. Teenagers are especially vulnerable since the prefrontal cortex is the last part of the brain to mature. When the balance between this circuit, the basal ganglia, and an enlarged amygdala shifts, the result is obsessive drug-seeking and impaired impulsive behaviour in an individual with a substance use problem [67].

Some drugs, such as opioids, also have an impact on other regions of the brain, including the brain stem, which controls basic physiological functions like breathing, relaxation, and heart rate. Overdosing might induce delayed breathing and possibly death as a result of this interference [68].

17.5 HOW DO NARCOTICS MAKE YOU FEEL GOOD?

Although the reason for the joy or euphoria experienced after consuming drugs is yet unknown, it is believed to be brought on by increases in the body's endogenous opioids (endorphins) and other neurotransmitters in specific areas of the basal ganglia (the reward circuit). In contrast to the lesser bursts that naturally follow pleasurable events like eating, listening to or playing music, engaging in creative pursuits, or socializing with others, certain medications have the capacity to significantly boost these neurotransmitter surges [69, 70]. Dopamine, formerly thought to be directly responsible for pleasure via drug-induced spikes in the neurotransmitter, is now thought to have a more indirect role by motivating us to engage in pleasurable actions [71].

17.5.1 Dopamine's Role in Drug Use Reinforcement

A healthy brain identifies and encourages pleasurable experiences, such as eating, engaging with people, and having sex [72]. Because of the way our brains are built, we are more prone to repeat rewarding actions. A crucial neurotransmitter in this is dopamine [73]. When a rewarding, healthy experience activates the reward circuit, dopamine levels increase, notifying the brain that something important is happening and should be remembered. The dopamine signal promotes the formation of habits by altering brain circuitry and making it easier to repeat an activity without thinking [74] (Figure 17.3). Drugs cause extreme euphoria and significantly greater dopamine surges, which together strongly reinforce the relationship between using the drug, the associated pleasure, and all of the environmental stimuli. Dopamine 'teaches' the brain to pursue narcotics instead of other, healthier objectives and activities when dopamine levels are high [75].

A person may have uncontrolled cravings even in the absence of drugs by being exposed to cues in their surroundings or daily routine that have been linked to narcotic use as a result of modifications to the reward circuit. This entrenched 'reflex' may last a very long time, even in those who have not used drugs in a while. For example, individuals who have abstained from drugs for ten years may still have cravings when they go back to a former residence or an area where they used drugs. The brain recalls things like how to ride a bike [76].

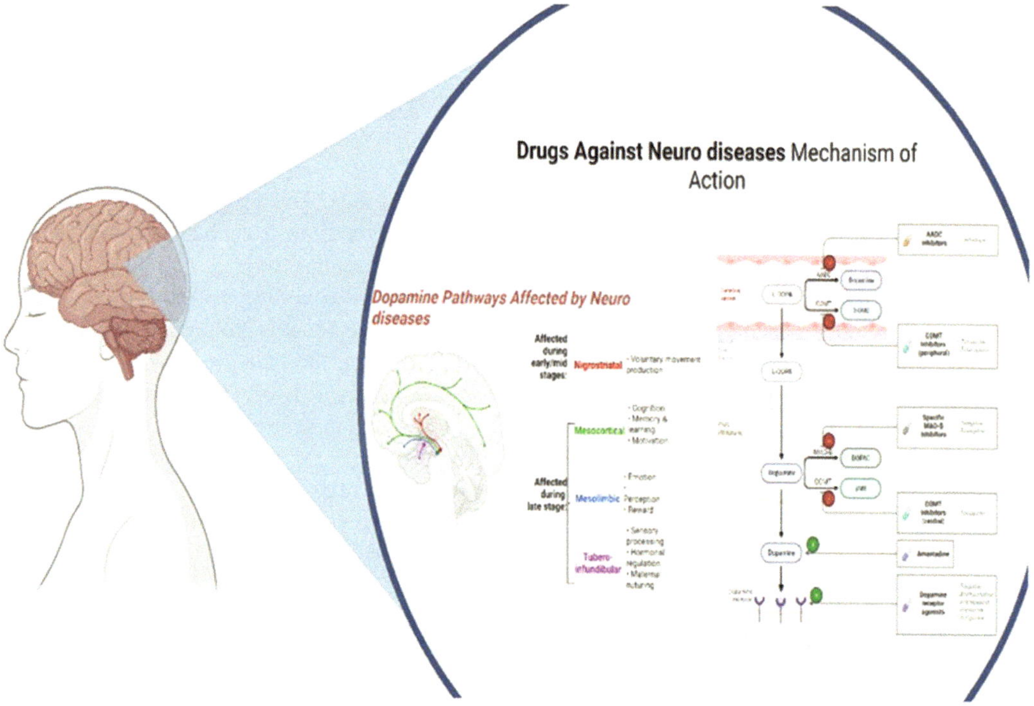

FIGURE 17.3 Graphical representation of dopamine in neurodegenerative diseases.

17.6 WHAT MAKES DRUGS MORE ADDICTING THAN REWARDS FROM NATURE?

The difference between obtaining rewards normally and receiving them as a result of drugs is comparable to that between hearing someone whisper in your ear and hearing them yell through a brain microphone [77]. The brain of a drug user regulates by creating fewer neurotransmitters in the reward circuit or by restricting the number of receptors that can accept impulses, similar to how we lower the volume on an exceptionally loud radio [78]. Consequently, the person is less able to take pleasure in naturally rewarding (i.e., reinforcing) activities. This is why drug users eventually become lifeless, lack motivation, are depressed, and/or are unable to appreciate activities that they used to find delightful [79]. The person now needs to use drugs in order to experience even a small amount of pleasure, which just makes the problem worse and starts a vicious cycle.

17.7 WHAT ARE THE ADDITIONAL NEGATIVE HEALTH EFFECTS OF DRUG ABUSE?

Addicts typically have one or more comorbid medical conditions, such as heart, lung, or blood vessel disease, stroke, tumours, or mental health problems. Imaging scans, chest X-rays, and blood tests can all show the negative effects of prolonged narcotic use throughout the body [80]. The use of narcotics may increase your risk of contracting infections. Human immunodeficiency virus (HIV) and hepatitis C (HCV), a fatal liver disease, can spread through the sharing of injectable supplies and dangerous behaviours such as sexual contact without the use of condoms [81, 82]. Addiction can have wide-reaching effects:

- Heart disease
- Stroke
- Tumour
- AIDS
- HBV/HCV
- Lung diseases
- Mental diseases

The following are some of the most serious effects of addiction: A baby may have withdrawal symptoms, which are referred to as Neonatal Abstinence Syndrome (NAS), as a consequence of a mother's drug or medication use during gestation. Tremors, issues with sleeping and eating, and even seizures are possible symptoms, but they may vary depending on the drug consumed [83]. Some kids who have been around drugs will develop behaviour, attention, and thinking issues. Current studies are examining whether these mental and behavioural consequences last throughout adolescence and contribute to ongoing developmental issues. Additionally, a mother's breast milk may include several chemicals. Researchers are still discovering the long-term effects of drug use on a kid who is breastfed [84]. Passive smoking exposes bystanders to at least 250 substances that are known to be hazardous, especially to children [85]. Accidental exposure to passive smoking raises the risk of heart disease and lung cancer in nonsmokers [86].

Drug injection causes one out of every ten cases of HIV. Injection drug usage has been associated with endocarditis, cellulitis, and a substantial contribution to the spread of hepatitis C [87]. Driving while under the influence of drugs or prescribed medicines can be dangerous, much like driving after drinking. More than 12 million individuals (16 and older) admitted to driving while high on marijuana or other illicit drugs in 2016 [88].

17.8 ADDICTION: IS IT CURABLE?

Similar to other chronic conditions like cardiovascular disease or asthma, therapy for drug misuse is often not a cure. However, addiction can be successfully controlled. People who are in treatment may regain control over their lives and fight the harmful effects that addiction has on their brains and actions [89].

17.8.1 Is a Drug Relapse a Sign that Therapy was Ineffective?

No. Relapse, or a comeback to narcotic use after such a try to prevent or stop, can be a part of the recovery process for certain people since addiction is a chronic condition, but modern therapies are made to assist in avoiding relapse. Relapse rates for drug use are in line with those for other chronic diseases. If someone stops taking their health treatment programme, they are more likely to relapse [90]. Pharmaceuticals are used to aid with drug detoxification, even though this is not the same as therapy and is insufficient to aid in recovery. In most cases, stopping narcotics use after detoxification but before receiving therapy results in resuming that use.

17.9 DEVICES AND MEDICATIONS FOR THE TREATMENT OF DRUG ABUSE

A patient may benefit from different medication types at different stages of therapy to help them quit abusing drugs, adhere to their treatment plan, and avoid relapsing [44].

17.9.1 Treatment for Withdrawal

When patients stop using drugs for the first time, they may experience a variety of physical and psychological symptoms, such as agitation or sleeplessness, as well as depression, worry, and other mental health problems. When these symptoms are lessened by particular therapy, medications, and equipment, quitting taking drugs is easier [33].

17.9.2 Remaining in Therapy

Numerous narcotic therapies and smartphone applications are employed to help the brain gradually get acclimated to the drug's absence. These treatments have a calming effect on the body's systems and gradually reduce drug cravings. Focusing on counselling and other psychotherapies can help patients get pharmaceutical treatment [30].

17.9.3 Avoiding Relapse

According to scientific research, the most frequent catalysts for relapsing are stress cues related to drug use (i.e., persons, locations, objects, and emotions). To assist patients in maintaining their recovery, researchers have been working on medicines that interfere with these factors [91].

17.9.4 Treatment with Behavioural Therapy

In drug addiction therapy, behavioural treatments assist patients in changing their attitudes and drug-related behaviours [92]. Patients are, therefore, better prepared to handle challenging situations and other triggers that can trigger another relapse. Additionally, behavioural therapy can help patients stay in treatment longer and improve the efficacy of drugs [93].

- Cognitive-behavioural therapy (CBT) [94]: Its goal is to assist patients in understanding, preventing, and coping with the situations that contribute to drug use.

- Contingency Management: It uses positive reinforcement by providing rewards or benefits for refraining from drug use, actively participating in treatment sessions, or taking prescribed medication for a disease.
- Motivational Enhancement Therapy: It makes use of strategies to profit from people's willingness to modify their actions and seek therapy.
- Twelve-Step Facilitation (TSF): To prepare individuals to take part in 12-step mutual support groups, this type of individual treatment is frequently given throughout twelve (12) weekly sessions. The 12-step concepts of tolerance, surrender and active involvement in healing are upheld by TSF.

17.10 BEST HEALTH PROGRAMMES FOR PATIENT RECOVERY FROM ADDICTION

The road to recovery is a long and difficult one, and drug abstinence is simply the first step. Addiction has frequently had major effects on people's lives when they enter treatment, perhaps affecting their healthcare and ability to participate in their families, at work, and in society [95]. To address all sides of narcotic usage and its drastic impacts, NIDA's research programme spans a wide variety of issues, from fundamental investigations into the brain and behaviour of drug addicts to therapeutic strategies and health care research. The research programmes at NIDA create preventative and treatment strategies and make sure they are effective in real-world situations. To achieve this objective, NIDA is dedicated to supporting studies that address the health disparities and vulnerabilities experienced by racial or ethnic minorities or those resulting from gender inequalities [96].

This evidence-based manual may be used by teachers, parents, and community leaders to aid in the reduction of drug use among kids and teenagers. The NIDA's evidence-based guide for reducing drug use among kids and teenagers provides recommendations based on successful drug-prevention research and includes answers to questions on risk and protective factors as well as on neighbourhood planning and implementation [97]. A guide based on research provides examples of scientifically supported and tried-and-true therapy's components, outlines the fundamentals of efficient treatment, responds to frequently asked questions, and covers various treatment options [98].

17.11 CONCLUSION

Substance abuse disorders are acute illnesses that include disturbances of important neural systems, and drug use has complicated biological and social factors. Evidence-based behavioural and medication-assisted therapies (MAT) implemented with a chronic disease management strategy have been demonstrated to enhance other outcomes, such as lowering criminal behaviour and the transmission of infectious illnesses and helping recovery from drug use disorders. The most severe drug use disorders may require treatment using a chronic disease management strategy, and former drug users who have access to recovery support programmes can attain and maintain long-term well-being. An extensive system is required to manage drug addiction and abuse from lower to higher levels in each and every country. Underdeveloped countries require more proper plans and financial assistance to manage such cases in all cities.

REFERENCES

[1] J. Rehm, C. Mathers, S. Popova, M. Thavorncharoensap, Y. Teerawattananon, *J. Lancet*. 373 (2009) 2223–2233. https://doi.org/10.1016/S0140-6736(09)60746-7.

[2] Centers for Disease Control, and Prevention (US). Best practices for comprehensive tobacco control programs. US Department of Health and Human Services, Centers for Disease Control and Prevention, National Center for Chronic Disease Prevention and Health Promotion, Office of Smoking and Health (2007).

[3] National Drug Intelligence Center, The Economic Impact of Illicit Drug Use on American Society, Washington, DC: United States Department of Justice (2011): Q0317-Q0002.
[4] D.K. Goyal , J.R. Neil, S.D. Simmons, F. Mansab, S. Benjamin et al., Insights Biomed. 4(3) (2019) 12. https://doi.org/10.36648/2572-5610.4.3.63
[5] N. Wilson, M. Kariisa, P. Seth, H. Smith, N.L. Davis, *MMWR. Morb. Mortal. Wkly. Rep.* 69 (2020) 290–297. https://doi.org/10.15585/mmwr.mm6911a4.
[6] P. Tice, R.N. Lipari, S.L. Van Horn, Substance Use among 12th Grade Aged Youths, in: Center for Behavioral Health Statistics and Quality, Substance Abuse and Mental Health Services Administration. The CBHSQ Report, Rockville, MD. (2013).
[7] L. Lander, J. Howsare, M. Byrne, *Soc. Work Public Health.* 28 (2013) 194–205. https://doi.org/10.1080/19371918.2013.759005.
[8] L.A. Simmons, J.R. Havens, J.B. Whiting, J.L. Holz, H. Bada, *Ann. Epidemiol.* 19 (2009) 187–193. https://doi.org/10.1016/j.annepidem.2008.12.007.
[9] S. Shankaran, B.M. Lester, A. Das, C.R. Bauer, H.S. Bada, L. Lagasse, R. Higgins, *Semin. Fetal Neonatal Med.* 12 (2007) 143–150. https://doi.org/10.1016/j.siny.2007.01.002.
[10] R.Z. Goldstein, N.D. Volkow, *Nat. Rev. Neurosci.* 12 (2011) 652–669. https://doi.org/10.1038/nrn3119.
[11] M.B. First, L.H. Yousif, D.E. Clarke, P.S. Wang, N. Gogtay, P.S. Appelbaum, *World Psychiatry.* 21 (2022) 218–219. https://doi.org/10.1002/wps.20989.
[12] D.D. Abelman, *Harm Reduct. J.* 14 (2017) 1–7. https://doi.org/10.1186/s12954-017-0194-6.
[13] S. Minozzi, L. Amato, C. Bellisario, M. Ferri, M. Davoli, *Cochrane Database Syst. Rev.* 11 (2020). https://doi.org/10.1002/14651858.CD006318.pub3.
[14] N. Heather, *Addict. Behav. Rep.* 6 (2017) 15–38. https://doi.org/10.1016/j.abrep.2017.05.002.
[15] G.M. Heyman, *Addict. Behav. Rep.* 5 (2017) 85–93. https://doi.org/10.1016/j.abrep.2017.02.001.
[16] J.S. Fowler, N.D. Volkow, C.A. Kassed, L. Chang, *Sci. Pract. Perspect.* 3 (2007) 4–16. https://doi.org/10.1151/spp07324.
[17] D. Eddie, L. Hoffman, C. Vilsaint, A. Abry, B. Bergman, B. Hoeppner, C. Weinstein, J.F. Kelly, *Front. Psychol.* 10 (2019) 1052. https://doi.org/10.3389/fpsyg.2019.01052.
[18] B.F. Grant, F.S. Stinson, T.C. Harford, J. Subst. *Abuse.* 13 (2001) 493–504. https://doi.org/10.1016/S0899-3289(01)00096-7.
[19] R.A. Zucker, J.E. Donovan, A.S. Masten, M.E. Mattson, H.B. Moss, *Pediatrics* 121 (2008) S252–272. https://doi.org/10.1542/peds.2007-2243B.
[20] K. Von Sternberg, C.C. DiClemente, M.M. Velasquez, *Psychol. Addict. Behav.* 32 (2018) 749–758. https://doi.org/10.1037/adb0000417.
[21] K.G. Hill, J.D. Hawkins, R.F. Catalano, R.D. Abbott, J. Guo, FJ. *Adolesc. Heal.* 37 (2005) 202–210. https://doi.org/10.1016/j.jadohealth.2004.08.014.
[22] M. Wood, *Addiction.* 102 (2007) 1019–1020. https://doi.org/10.1111/j.1360-0443.2007.01920.x.
[23] J.O. livi. Lee, K.G. Hill, K. Guttmannova, L.A. Hartigan, R.F. Catalano, J.D. *J. Stud. Alcohol Drugs.* 75 (2014) 684–694. https://doi.org/10.15288/jsad.2014.75.684.
[24] G.M. Houvèssou, I.O. Bierhals, B.D. Flesch, M.F. da Silveira, *Rev. Saude Publica.* 54 (2020) 57. https://doi.org/10.11606/S1518-8787.2020054002176.
[25] M. Lemoine, G. Gmel, S. Foster, S. Marmet, J. Studer, *PLoS One.* 15 (2020) e0220232. https://doi.org/10.1371/journal.pone.0220232.
[26] D.F. Shell, I.M. Newman, L. wa Yuen, *C J. Am. Coll. Heal.* 68 (2020) 762–773. https://doi.org/10.1080/07448481.2019.1590369.
[27] L. Bond, H. Butler, L. Thomas, J. Carlin, S. Glover, G. Bowes, G. Patton, J. *Adolesc. Heal.* 40 (2007) 357.e9–357.e18. https://doi.org/10.1016/j.jadohealth.2006.10.013.
[28] R.F. Farmer, D.B. Kosty, J.R. Seeley, S.C. Duncan, M.T. Lynskey, P. Rohde, D.N. Klein, P.M. Lewinsohn, *Psychol. Med.* 45 (2015) 63–72. https://doi.org/10.1017/S003329171400107X.
[29] K. Calhoun, T. Burns, A. Smith, E. Ehrlich, L. Gultekin, K. Asabigi, M. Hijazi, N. Buckhoy, Z. Rowe, M. Neff, D. Vaughter, T. Voepel-Lewis, T. Piechowski, D. Jazdzyk, D. Cordova, *Prog. Community Health Partnersh.* 16 (2022) 45–58. https://doi.org/10.1353/cpr.2022.0038.
[30] R.B. Cox, I. Washburn, K. Greder, S. Sahbaz, H. Lin, *Am. J. Drug Alcohol Abuse.* 48 (2022) 69–77. https://doi.org/10.1080/00952990.2021.1981357.

[31] C.S. Kramer, K.C. Wilcox, H.A. Lawson, *Prev. Sch. Fail.* 64 (2020) 301–315. https://doi.org/10.1080/1045988X.2020.1769011.
[32] L. Bevilacqua, D. Goldman, *Clin. Pharmacol. Ther.* 85 (2009) 359–361. https://doi.org/10.1038/clpt.2009.6.
[33] E.F. McCance-Katz, *Psychiatr. Serv.* 69 (2018) 1046–1048. https://doi.org/10.1176/appi.ps.201800281.
[34] B.C. Hohl, S. Wiley, D.J. Wiebe, A.J. Culyba, R. Drake, C.C. Branas, *JAMA Intern. Med.* 177 (2017) 317–324. https://doi.org/10.1001/jamainternmed.2016.8180.
[35] SAMHSA, Mental Health and Substance Use Disorders SAMHSA, *U.S. Dep. Heal. Hum. Serv.* 7 (2020) 20857.
[36] A.M. Nawi, R. Ismail, F. Ibrahim, M.R. Hassan, M.R.A. Manaf, N. Amit, ... N.S. Shafurdin, *BMC Public Health* 21, (2021) 1–15.
[37] L.R. Few, J.D. Grant, E.C. Nelson, T.J. Trull, R.A. Grucza, K.K. Bucholz, K.J.H. Verweij, N.G. Martin, D.J. Statham, P.A.F. Madden, A.C. Heath, M.T. Lynskey, A. Agrawal, *J. Stud. Alcohol Drugs.* 77 (2016) 873–880. https://doi.org/10.15288/jsad.2016.77.873.
[38] C.Y. Chen, J.C. Anthony, *Psychopharmacology (Berl).* 172 (2004) 78–86. https://doi.org/10.1007/s00213-003-1624-6.
[39] L.M. Squeglia, J. Jacobus, S.F. Tapert, *Clin. EEG Neurosci.*, NIH Public Access. 40 (2009) 31–38. https://doi.org/10.1177/155005940904000110.
[40] K.C. Winters, A. Arria, *Prev. Res.* 18 (2011) 21–24.
[41] S. Tabish, N. Syed, N. Syed, P.S. Shah, *Int J Sci Res (Ahmedabad)*, *4*(7) (2015) 4–438.
[42] C. Cunningham, M.J. Edlund, M. Fishman, A.J. Gordon, H.E. Jones, D. Langleben, J. Femino, The ASAM national practice guideline for the treatment of opioid use disorder: 2020 focused update. *J Addict. Med.*, *14*(2S Suppl 1) (2020) 1–91.
[43] S. Satel, S.O., *Front. Psychiatry.* 4 (2014) 141. https://doi.org/10.3389/FPSYT.2013.00141/BIBTEX.
[44] N.D. Volkow, *J. Drug Addict. Educ. Erad.* 7 (2018) 157–195.
[45] N.D. Volkow, *Drugs, brains, and behavior: The science of addiction* (2010). Retrieved on March, 23, 2011, 255–169.
[46] J. Upadhyay, N. Maleki, J. Potter, I. Elman, D. Rudrauf, J. Knudsen, D. Wallin, G. Pendse, L. McDonald, M. Griffin, J. Anderson, L. Nutile, P. Renshaw, R. Weiss, L. Becerra, D. Borsook, *Brain.* 133 (2010) 2098–2114. https://doi.org/10.1093/brain/awq138.
[47] R. Waller, H.L. Dotterer, L. Murray, A.M. Maxwell, L.W. Hyde, *NeuroImage Clin.* 14 (2017) 201–215. https://doi.org/10.1016/j.nicl.2017.01.014.
[48] B.S. McEwen, C. Nasca, J.D. Gray, *Neuropsychopharmacology.* 41 (2016) 3–23. https://doi.org/10.1038/npp.2015.171.
[49] W. Gossman, F. Alghoula, I. Berim, *Anoxia (Hypoxic hypoxia)*. StatPearls Publishing. (2019).
[50] J. Zibbell, J. Howard, S. Duhart Clarke, A. Ferrell, S.L. Karon, *US Dep. Health Hum. Serv.* 33 (2019) 1–23.
[51] D. Nutt, A. Hayes, L. Fonville, R. Zafar, E.O. Palmer, L. Paterson, A. Lingford-Hughes, *Nutrients*, 13(11) (2021), 3938.
[52] C. Harper, I. Matsumoto, *Cur. Opi. Pharmacol.* 5 (1) (2005) 73–78.
[53] A.M. Leventhal, C.W. Kahler, L.A. Ray, K. Stone, D. Young, I. Chelminski, *M. Am. J. Addict.* 17 (2008) 218–223. https://doi.org/10.1080/10550490802019774.
[54] T.A. Wearne, J.L. Cornish, *Front. Psychiatry.* 9 (2018) 491. https://doi.org/10.3389/fpsyt.2018.00491.
[55] A. Bernheim, R.E. See, C.M. Reichel, *Neurosci. Biobehav. Rev.* 69 (2016) 36–48. https://doi.org/10.1016/j.neubiorev.2016.07.020.
[56] B.D. Marshall, D. Werb, *Addiction*, 105(6) (2010) 991–1002.
[57] A.M. Zemelka, *J. Neurol. Neurosci.* 08 (2017) 0–0. https://doi.org/10.21767/2171-6625.1000190.
[58] W.J. Hamilton, *BMJ.* 1 (1965) 980–980. https://doi.org/10.1136/bmj.1.5440.980-b.
[59] P. Johns, *Clin. Neurosci.* (2014) 1–17.
[60] J.C.M. Brust, *Neurological Aspects of Substance Abuse* (2004). Butterworth-Heinemann.
[61] F.M. Filbey, S. Aslan, V.D. Calhoun, J.S. Spence, E. Damaraju, A. Caprihan, J. Segall, *Proc. Nat. Acad. Sci.*, 111(47) (2014), 16913–16918.
[62] B.A. Krumm, *Am. J. Nurs.* 98 (1998) 20–21. https://doi.org/10.1097/00000446-199805000-00034.

[63] V. Verma, *Clin. Psychopharmacol. Neurosci.* 13 (2015) 227–238. https://doi.org/10.9758/cpn.2015.13.3.227.
[64] G. V Rebec, *Cocaine and Amphetamines*, in: *Encyclopedia of Life Sciences*, John Wiley & Sons, Ltd, (2012). https://doi.org/10.1002/9780470015902.a0000042.pub3.
[65] P.E. Sharp, *Relig. Brain Behav.* 4 (2014) 202–229. https://doi.org/10.1080/2153599X.2013.826717.
[66] L. Jones, D. Vigo, Mental health and substance abuse. In *Global Health Essentials* (2023; pp. 197–201). Cham: Springer International Publishing.
[67] B.J. Casey, K. Caudle, *Curr. Dir. Psychol. Sci.* 22 (2013) 82–87. https://doi.org/10.1177/0963721413480170.
[68] I. Bachmutsky, X.P. Wei, E. Kish, K. Yackle, *Elife.* 9 (2020) e52694. https://doi.org/10.7554/eLife.52694.
[69] R.I.M. Dunbar, R. Baron, A. Frangou, E. Pearce, E.J.C. van Leeuwen, J. Stow, G. Partridge, I. Macdonald, V. Barra, M. van Vugt, *Proc. R. Soc. B Biol. Sci.* 279 (2012) 1161–1167. https://doi.org/10.1098/rspb.2011.1373.
[70] J.M. Van Ree, Endorphin Systems, Pain and Addiction, in: *Comparative Veterinary Pharmacology, Toxicology and Theraphy: Proceedings of the 3rd Congress of the European Association for Veterinary Pharmacology and Toxicology, August 25–29 1985, Ghent, Belgium Part II, Invited Lectures.* (1986) 455–466. Dordrecht: Springer Netherlands. https://doi.org/10.1007/978-94-009-4153-3_43.
[71] D. Sulzer, *Neuron.* 69 (2011) 628–649. https://doi.org/10.1016/j.neuron.2011.02.010.
[72] I.N. Karatsoreos, Bruce S. McEwen. In: R.J. Nelson, Z.M. Weil (eds) *Biographical History of Behavioral Neuroendocrinology* (2022). Springer, Cham. https://doi.org/10.1007/978-3-031-12970-4_21
[73] E.D. Wallert, E. van de Giessen, R.J.J. Knol, M. Beudel, R.M.A. de Bie, J. Booij, *J. Nucl. Med.* 63 (2022) 27S–32S.
[74] H. Juárez Olguín, D. Calderón Guzmán, E. Hernández García, G. Barragán Mejía, *Oxid. Med. Cell. Longev.* 2016 (2016) 9730467. https://doi.org/10.1155/2016/9730467.
[75] D. Belin, S. Jonkman, A. Dickinson, T.W. Robbins, B.J. Everitt, *Brain Res.* 199 (2009) 89–102. https://doi.org/10.1016/j.bbr.2008.09.027.
[76] P.J. Rogers, *Pharmacol. Biochem. Behav.* 153 (2017) 182–190. https://doi.org/10.1016/j.pbb.2017.01.001.
[77] I.W. Leigh, J.F. Andrews, R.L. Harris, T.G. Ávila, *Deaf Culture: Exploring Deaf Communities in the United States* (2020). San Diego, CA: Plural Publishing.
[78] R. Zheng, L. Hao, Y. Li, T. Zhang, D. Bai, L. Zhang, D. Li, W. Hao, *Front. Psychiatry.* 12 (2021) 592. https://doi.org/10.3389/fpsyt.2021.615921.
[79] R. Sinha, *Ann. N. Y. Acad. Sci.* 1141 (2008) 105–130. https://doi.org/10.1196/annals.1441.030.
[80] N. Jain, C. Avanthika, A. Singh, S. Jhaveri, I. De la Hoz, G. Hassen, G.P. Camacho L, K.G. Carrera, *Cureus.* 13 (2021) e18457. https://doi.org/10.7759/cureus.18457.
[81] N. El-Bassel, S.A. Shaw, A. Dasgupta, S.A. Strathdee, *Curr. Opin. HIV AIDS.* 9 (2014) 150–155. https://doi.org/10.1097/COH.0000000000000035.
[82] B. Jacka, É. Roy, S. Høj, N. Minoyan, A.A Artenie, G. Zang ... J. Bruneau, *Journal of Viral Hepatitis*, 26 (12) (2019) 1413–1422.
[83] S. Ross, E. Peselow, *Clin. Neuropharmacol.* 35 (2012) 235–243. https://doi.org/10.1097/WNF.0b013e318261e193.
[84] E.J. Ross, D.L. Graham, K.M. Money, G.D. Stanwood, *Neuropsychopharmacology.* 40 (2015) 61–87. https://doi.org/10.1038/npp.2014.147.
[85] J.Y. Ko, S. Wolicki, W.D. Barfield, S.W. Patrick, C.S. Broussard, K.A. Yonkers, R. Naimon, J. Iskander, *MMWR. Morb. Mortal. Wkly. Rep.* 66 (2017) 242–245. https://doi.org/10.15585/mmwr.mm6609a2.
[86] A.J. Alberg, D.R. Shopland, K.M. Cummings, *Am. J. Epidemiol.* 179 (2014) 403–412. https://doi.org/10.1093/aje/kwt335.
[87] E.J. Cone, G.E. Bigelow, E.S. Herrmann, J.M. Mitchell, C. LoDico, R. Flegel, *R. J. Anal. Toxicol.* 39 (2015) 1–12. https://doi.org/10.1093/jat/bku116.
[88] J.E. Zibbell, A.K. Asher, R.C. Patel, B. Kupronis, K. Iqbal, J.W. Ward, D. *Am. J. Public Health.* 108 (2018) 175–181. https://doi.org/10.2105/AJPH.2017.304132.
[89] C.Y. Chen, C.L. Storr, J.C. Anthony, *Addict. Behav.* 34 (2009) 319–322. https://doi.org/10.1016/j.addbeh.2008.10.021.

[90] G.J. Wang, N.D. Volkow, L. Chang, E. Miller, M. Sedler, R. Hitzemann, W. Zhu, J. Logan, Y. Ma, J.S. Fowler, *Am. J. Psychiatry*. 161 (2004) 242–248. https://doi.org/10.1176/appi.ajp.161.2.242.

[91] M.N. Potenza, M. Sofuoglu, K.M. Carroll, B.J. Rounsaville, *Neuron*. 69 (2011) 695–712. https://doi.org/10.1016/j.neuron.2011.02.009.

[92] D.F.K. Wong, X.Y. Zhuang, T.K. Ng, *Child. Youth Serv. Rev*. 113 (2020) 104970.

[93] R.K. McHugh, B.A. Hearon, M.W. Otto, *The Psych. Clin. N. Am.*, 33(3) (2010) 511–525. https://doi.org/10.1016/j.psc.2010.04.012.

[94] C. Otte, *Dialogues Clin. Neurosci*. 4 (2022) 413–421.

[95] Substance Abuse and Mental Health Services Administration, Office of the Surgeon General. Early Intervention, Treatment, and Management of Substance Use Disorders, in: *Facing Addiction in America: The Surgeon General's Report on Alcohol, Drugs, and Health [Internet]*. US Department of Health and Human Services. (2016) 1–63.

[96] W.M. Aklin, E.S. Herrmann, *Preven. Med.* 176 (2023) 107650.

[97] UNODC, *Youth Participation in Drug Prevention Work Handbook on.* (2020). www.unodc.org/documents/commissions/CND/CND_Sessions/CND_63/CRPs/ECN72020_CRP12_e_V2001489.pdf

[98] B.L. Duncan, R.J. Reese, *Handbook of Psychol*ogy. Second Ed. (2012).

18 Nutraceuticals in Brain Health

Pratap Kalita, Bedanta Bhattacharjee,
Ram Kumar Sahu, and Satyabrat Sarma

18.1 INTRODUCTION

Neurodegenerative diseases/disorders (NDDs) belong to a heterogeneous group of various chronic debilitating conditions or diseases that affect the peripheral nervous system or central nervous system [1]. Disorders or diseases occur due to the damage or declining function of neurons, which leads to neuronal death, called a neurodegenerative disorder. The incidence rates increase every year, especially in countries with an ageing population. Common NDDs include Alzheimer's disease, depression, Huntington's disease, prion disease, motor neuron disease (MND), psychotic disorders, and Parkinson's disease [2]. All the NDDs caused due to decreased levels of neuronal counts, which are also known as neural progenitor cells and due to loss of neuronal integrity (protein tangle/aggregation) as well as lack of communication (decreased neurotransmitters), which eventually results in loss of cognitive (memory impairment), sensory and, motor functions. Globally, approximately 50 million people are living with dementia, with 60 – 70 % diagnosed with Alzheimer's disease. In America, 5.8 million individuals over the age of 65 have Alzheimer's disease, and this is projected to increase by 6.7% between 2020 and 2025 [3]. Parkinson's disease is the second most widespread neurodegenerative disease after Alzheimer's disease and affects movement. The prevalence of Parkinson's disease is approximately 2% of the population older than 65 years old, involving 6.1 million people in 2016 globally. Although Alzheimer's disease and Parkinson's disease affect different areas in the brain and display distinctive clinical features, they share a strikingly similar pathophysiology mediated by oxidative stress [4].

Neurological diseases are generated with a wide array of chronic diseases comprising a highly complicated aetiology. It was reported that malnutrition or nutrient-deficient food behaviour may cause an imbalance in the central or peripheral nervous system. With increased age, brain function used to alter and deterioration used to occur through the neurodegenerative processes, hence leading to the identification of cellular and molecular targets that ultimately leverage better functioning of the brain [5]. The use of phytoconstituents abundantly available in both plant and non-plant sources leads to improved human health and the fight against NDDs. Therefore, dietary substances in regular diets are very important to add, and the available nutraceuticals support the NDDs. The American Nutraceutical Association defines nutraceuticals as a food or its product possessing health-benefiting properties. They range from dietary nutrient supplements to genetically designed foods, herbal products, beverages, soups, vegetables, fruits, and processed foods, like cereals [6]. Nutraceuticals are mainly represented by vitamins, minerals, and amino acids, and over 1000 other probiotic compounds have been identified to date. Nutraceuticals are directly involved in the scavenging of free radicals, prevent damage to cell membranes by inducing levels of enzymatic and non-enzymatic antioxidants and fight against oxidative stress. In addition, nutraceuticals can act as modulators of pro-survival, pro-apoptotic, and inflammatory signalling pathways [7].

 DOI: 10.1201/9781003300502-18

Elevation of mortality and morbidity rates are increasing and creating an enormous economic burden globally. Earlier studies reported that the NDD rate was found to be higher rate in aged/older people compared to young ones. It was also reported that in older people, the NDDs were generated or produced due to decline or altered hormone secretion, enhanced oxidative stress and neuro-inflammation [8]. NDDs are directly associated with the socioeconomic status of any community or country as they hamper cognition and physical (unhealthy ageing) or motor movement (quality of life) as well as an incurable condition. From the statistical analysis, it was found that NDDs are increasing in good numbers in Western countries as compared to Asian countries due to modified lifestyles (food habits and sedentary lifestyles) [9]. Changes in food behaviour and lifestyle modification are the main factors that cause NDDs. Several types of drugs are used to treat NDDs, but adverse effects, side effects and toxic reactions remain major issues. Therefore, it is very important to focus on the nutritional and functional moieties that are abundantly available in the daily diet. It is essential to focus on the nutraceuticals that are available in vegetables, food grains, herbs, etc. and rich with antioxidant and other bioactive molecules responsible for exerting neuroprotective activities [10]. This chapter is designed to highlight the different nutraceutical's roles in neurodegenerative disorders.

18.2 INTRODUCTION, CLASSIFICATION AND STRUCTURE OF NUTRACEUTICAL

Drug nutrition intervention is becoming one of the popular therapies for the treatment of neurodegenerative diseases. The combination therapies were first developed by Stephen L DeFelice. Earlier research works concluded that whole food without polishing is used to produce several beneficial effects, including different neurodegenerative diseases like PD, AD, HD and prion disease, as well as depression [11]. Bioactive compounds like antioxidants, vitamins, minerals, phenolics, flavonoids, anthocyanins, proanthocyanins, etc., have health benefits and fight against various neurodegenerative via regulating energy metabolism, neuro-oxidative stress, neuro-inflammation, and enhance neurogenesis (improve NPCs proliferation, growth factor, neurotrophins) through various signalling pathways [7]. Antioxidants are abundantly available in herbal or natural sources, and these bioactive components are responsible for fighting against ROS, RNS and different free radicals, which are directly involved in producing several degenerative disorders. It is known that oxidative stress develops due to the overproduction of ROS, and these ROS rapidly damage the neurons and nerve cells. Apart from this, lipid production is also a crucial factor towards the generation of ROS in the body [12]. Therefore, nutraceuticals remain the important and potential sources of antioxidants, which are obtained from green plants and natural sources. Nutraceuticals are the responsible components for neutralizing the ROS to prevent oxidative stress in the body. Therefore, in current times, herbal or traditional therapy is becoming an advancing therapy for the betterment of neurological disorders.

Natural substances such as herbs, leaves, roots, and food grains remain important dietary supplements and are regulated as foods. Plant or animal-derived nutraceuticals or antioxidants have received significant consideration because of their supposed safety and multiple nutritional and therapeutic effects [9]. Phytochemicals and non-phytochemicals remain the two main categories that are used for the treatment of neurological disorders. Alkaloids, carotenoids, flavonoid polyphenols, non-flavonoid polyphenols, vitamins, and polyphenols come under phytochemicals, and creatine and polyunsaturated fatty acids (from fish) come under non-phytochemicals [13]. Other bioactive compounds have also been reported to assert neuroprotective activities through participating in oxidative stress-mediated pathways that involve mitochondrial damage, autophagy, apoptosis, and neuroinflammation. Overall, the antioxidant rice foods or diet produced significant antioxidant properties, which are useful for controlling the generation of neurodegenerative disorders [14, 8].

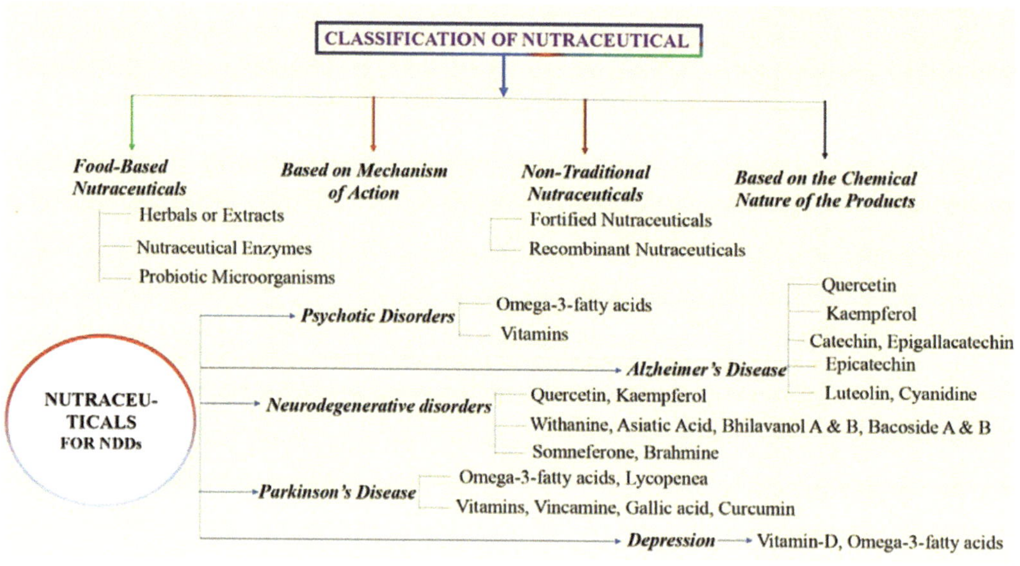

FIGURE 18.1 Classification of nutraceutical.

18.3 NUTRACEUTICALS IN THE MANAGEMENT OF VARIOUS NEUROLOGICAL BRAIN DISEASES

Natural substances have fewer unwanted risks. Hence, nutraceuticals in neurological brain diseases (NDs) have indeed proven to reveal themselves as a smarter and safer holistic strategy. In most circumstances, they must be used in conjunction with conventional medication treatment and contemporary professional supervision of NDs to enhance the effectiveness of the existence of the patient. Focusing on the evidence for their application as an adjunct therapy in NDs, various nutraceuticals' efficacy on several pathological networks (associated with NDs) is highlighted [15 – 17]. Nutraceuticals act through a multitude of pathways, which are categorized as follows: (1) modulating neuronal signalling circuits; (2) altering mitochondrial dysfunction; (3) anti-apoptotic and hindering caspase; (4) metal complexation; (5) inhibiting neuro-oxidative stress and antioxidants; (6) anti-inflammatory; (7) improve cognition; (8) anti-excitotoxic; (9) healthy ageing. The majority of nutraceuticals function via numerous molecular channels rather than clinging to a specific circuit or pathway. Resveratrol, curcumin, coenzyme Q10, α-lipoic acid, isothiocyanate, and astaxanthin are a couple of nutraceuticals that have proved effective efficacy towards a spectrum of NDs. Numerous investigations have also exhibited that nutraceuticals have therapeutic benefits for a range of NDs [18]. Figure 18.1 illustrates a graphical depiction of the pathway that is implicated in NDs and the personalized paradigm modulated by nutraceuticals. Figure 18.2 depicts the effectiveness of nutraceuticals on NDs.

18.4 NUTRACEUTICALS IN ALZHEIMER'S DISEASE

The highest prevalent kind of cognitive impairment is Alzheimer's disease (AD), sometimes defined as progressive neurodegeneration of the Alzheimer's type (PNAT) or predominant progressive neurodegeneration of the Alzheimer's type (PPNAT). Extremely pertinent antioxidants are well-known nutraceuticals that are advantageous in the treatment of AD. These antioxidants are effective in the therapy of numerous long-term ailments induced by oxidative damage, which plays a significant role in neurological conditions, especially AD [19]. Ageing and a deficiency

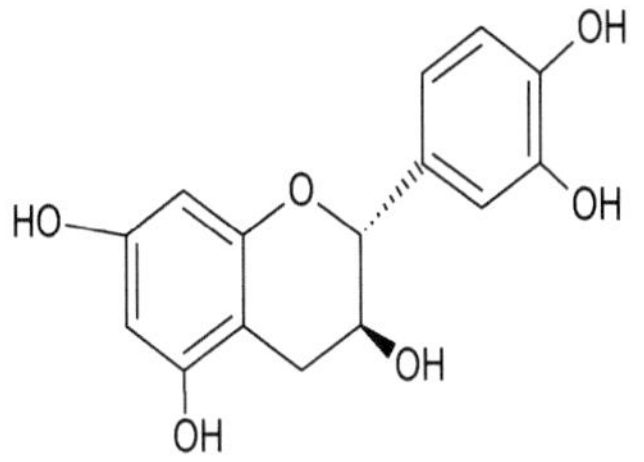

Catechin

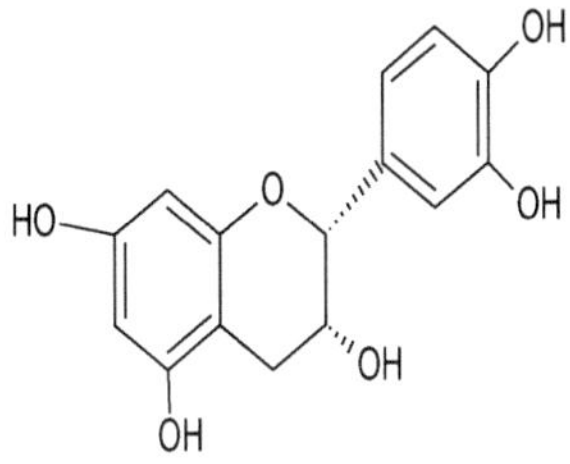

Epicatechin

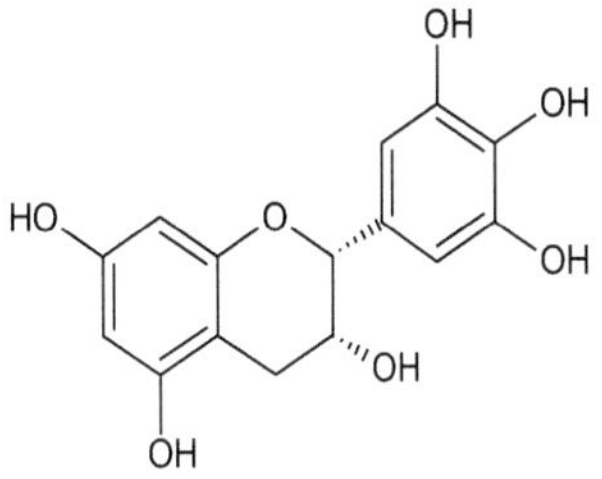

Epigallocatechin

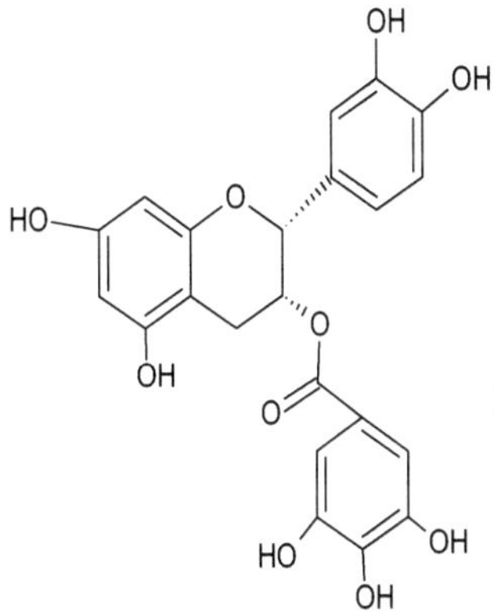

Epicatechin Gallate

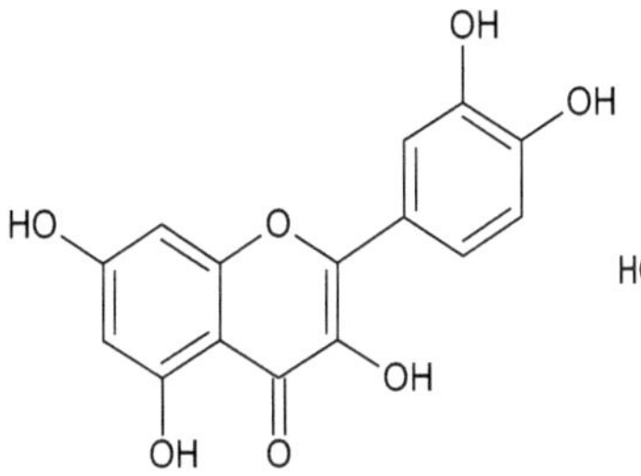

Quercetin

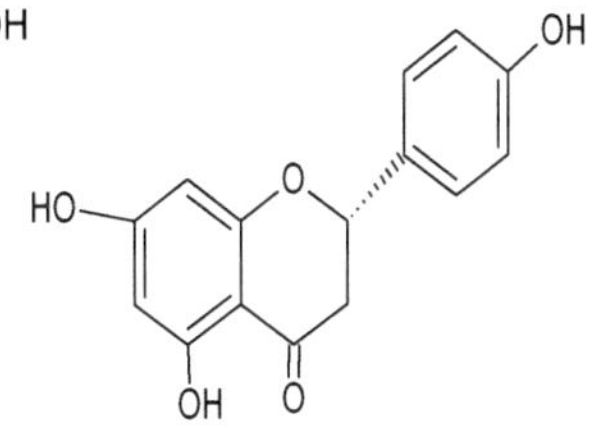

Naringenin

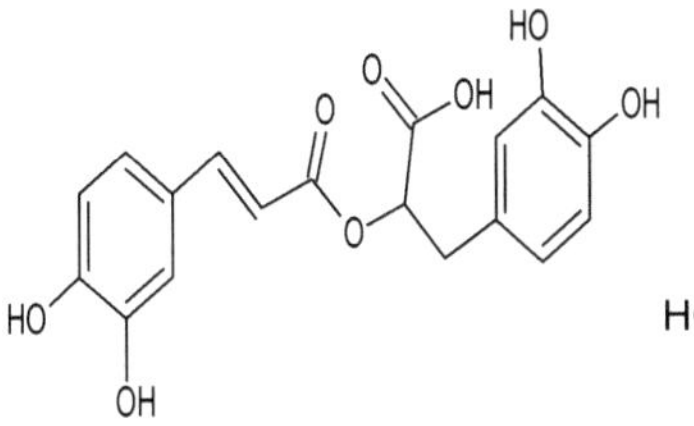

Rosmarinic Acid

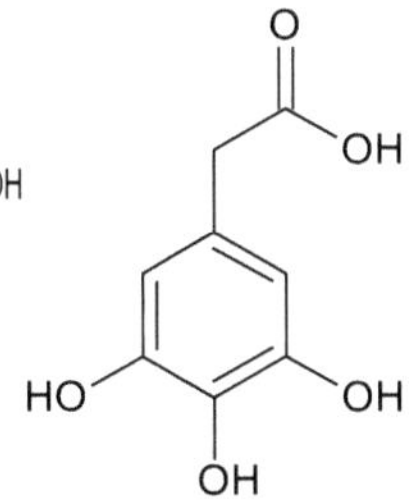

Gallic Acid

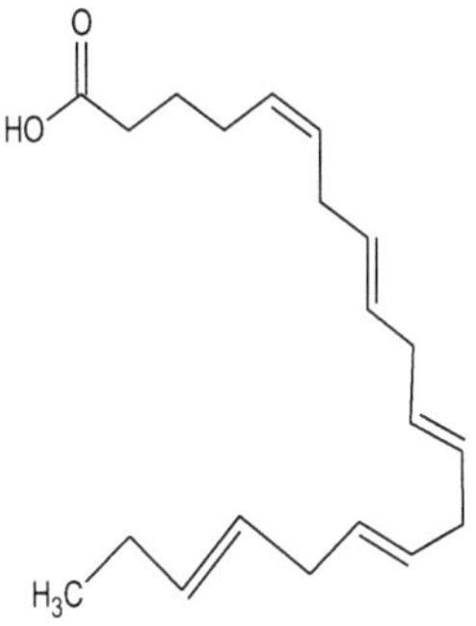

Linoleic Acid

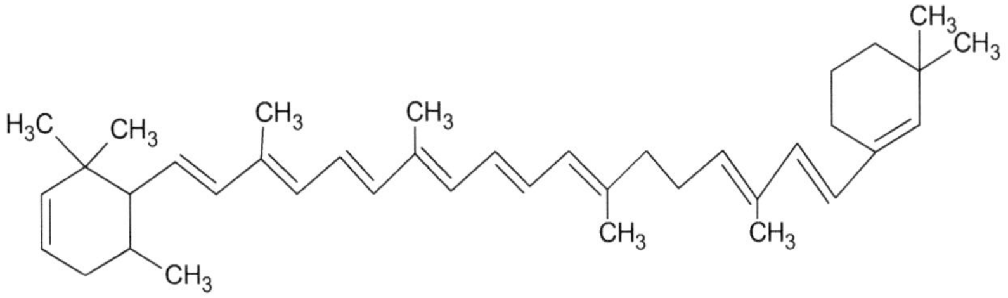

FIGURE 18.2 Structure of nutraceuticals in brain health.

β-carotene

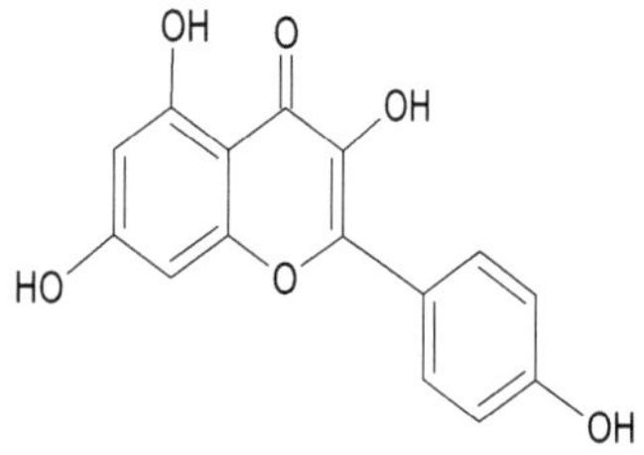

Kaempherol

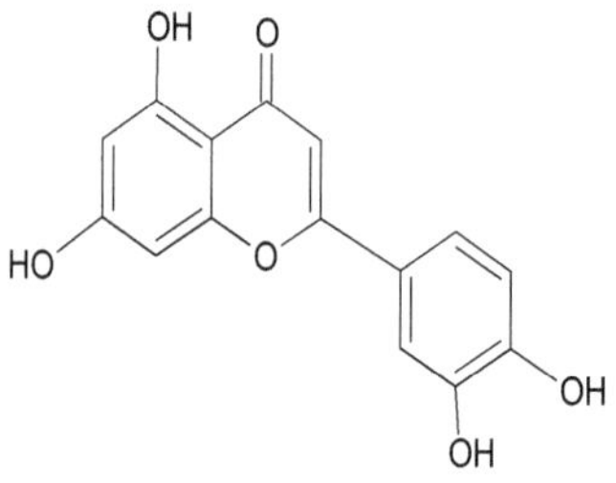

Luteolin

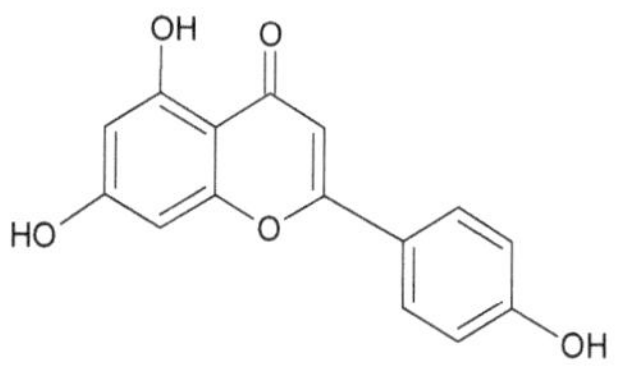

Apigenin

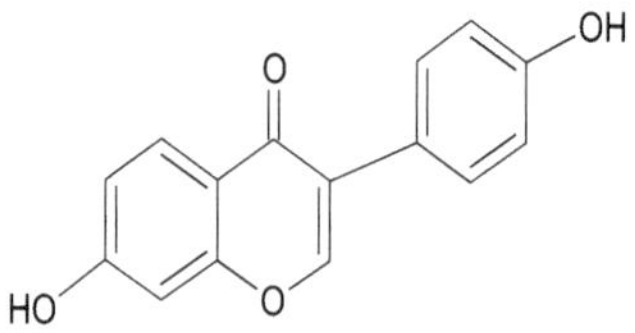

Daidzein

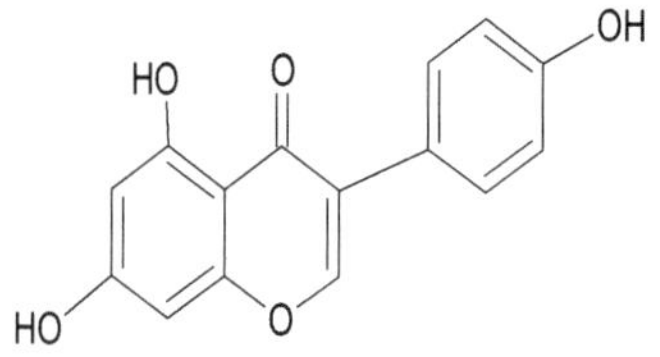

Genistein

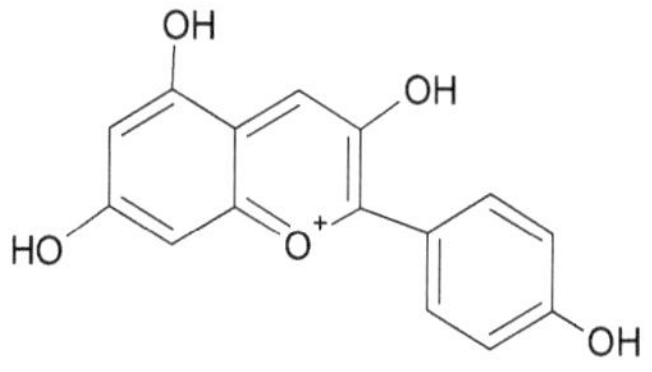

Pelargonidin

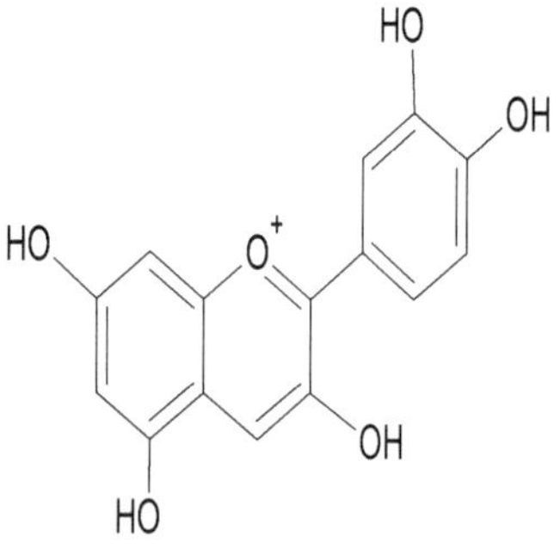

Cyanidin

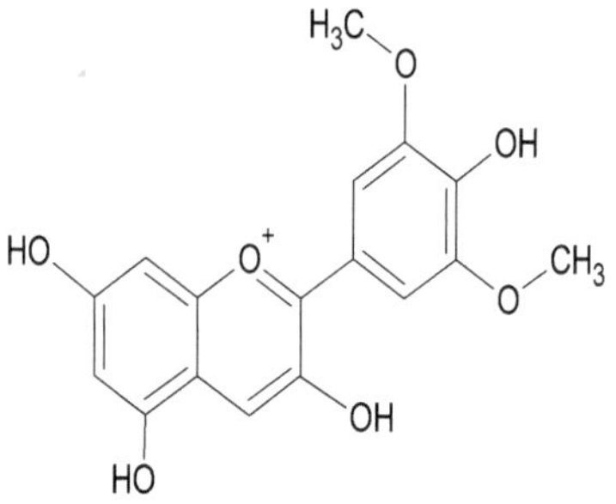

Malvidin

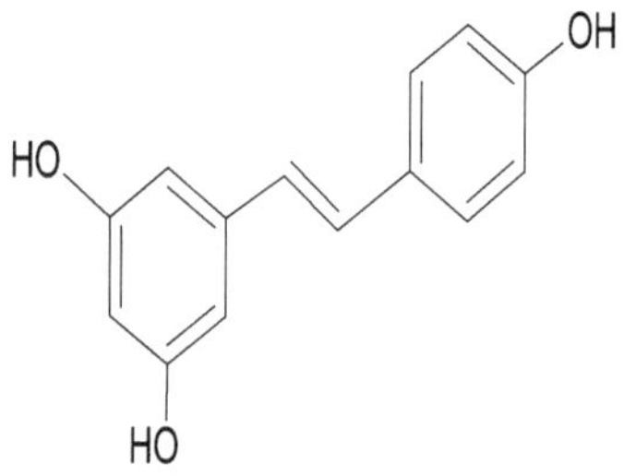

Resveratrol

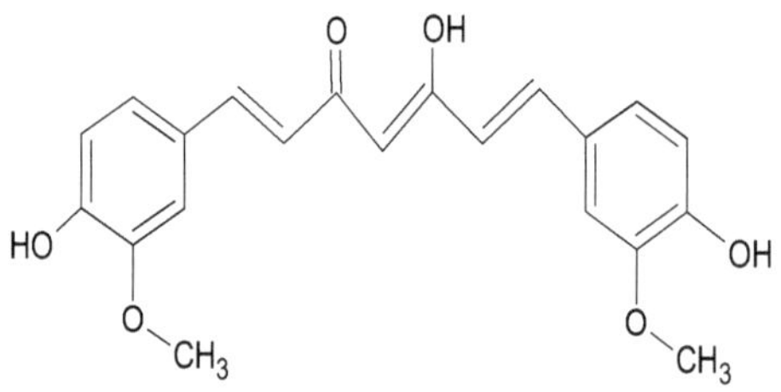

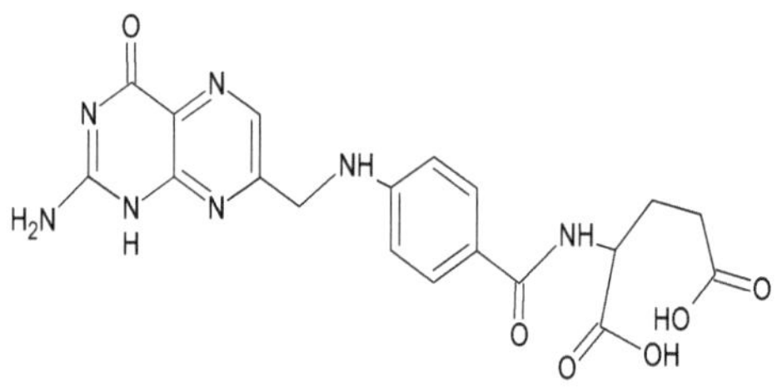

FIGURE 18.2 (Continued)

Curcumin Folate

Pyridoxine Vit. B6 Carnosic Acid

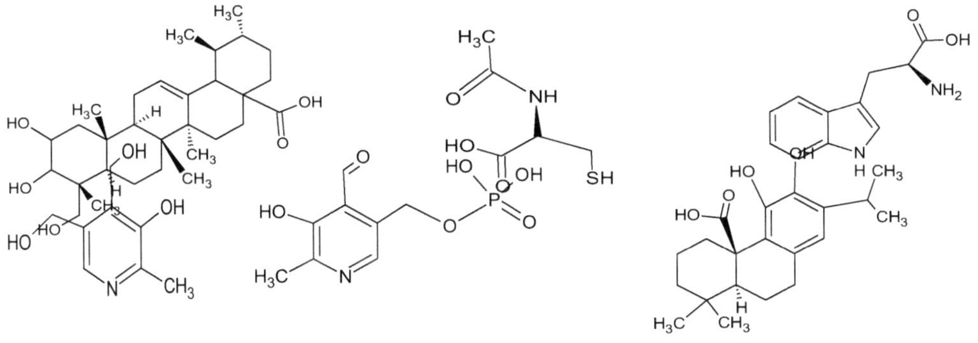

Asiatic Acid N-acetyl Cysteine L-Tryptophan

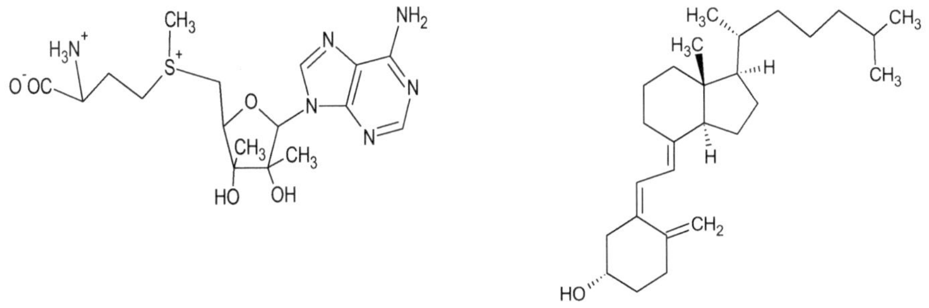

S-adenyl methionine Vit. D

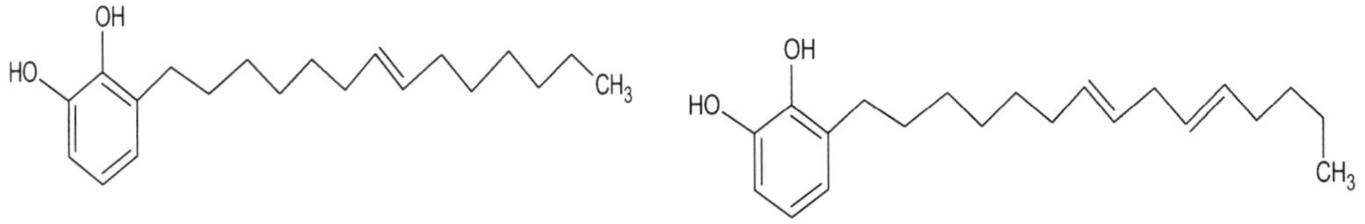

Bhilawanol A Bhilawanol B

FIGURE 18.2 (Continued)

of natural antioxidants exacerbate mitochondrial dysfunction, which promotes the incidence and advancement of the illness. Multiple investigations have established a linkage between consuming more nutritional antioxidants and a lower incidence of AD, which is critical since illness prophylaxis is far superior to disease treatment [20]. Consuming diets abundant in trans-fatty acids (TFA), saturated fatty acids (SFA), and polyunsaturated fatty acids (PUFA) has been proven to alleviate the incidence of neurodegeneration, whereas trans-fat-rich diets have been shown to augment the incidence [21]. The below-mentioned sections encompass a few of the molecules that are advantageous for AD.

18.4.1 Apigenin and Luteolin

Flavones like apigenin (trihydroxyflavone) and luteolin (tetrahydroxyflavone) have outstanding neuroprotective potential. Such flavone-containing molecules' predominant origins are celery, parsley, and rosemary. The astounding therapeutic repercussions of these bioactive molecules include their propensity to shield DNA from hydrogen peroxide-induced neurotoxicity, which further inhibits neurodegeneration and neuro-inflammation in Alzheimer's [22, 23].

18.4.2 Flavonoids

The primary flavonoids effective in treating NDs, particularly Alzheimer's, are epigallocatechin gallate, epigallocatechin, epicatechin, and catechin. This is a collection of frequently occurring polyphenolic molecules that are mostly derived from human nutrition. Vegetables, fruits, and beverages, including coffee, tea, and wine, are the major sources of flavonoids. The neuro-glia transmission cascade, which is predominantly important in the survivability and functionality of neurons, has indeed been examined to talk with flavonoids and their metabolite intermediates, which have been shown to have neurological-regulatory effects [24]. Overexpression activity of antioxidant enzymes and enzymes also alters the circulation of blood to the brain, promoting neuroplasticity and the restoration of cognitive processes by preventing the development of neurotoxicity, which is mostly linked to AD [25].

18.4.3 Crocin

The primary phytometabolite derived from saffron (Crocus sativus) is called crocin. Crocin is indeed reported to augment cognition and recall due to its extended ability being hindered via ethanol, and it is thus utilized in neurodegenerative conditions, including Alzheimer's. Several investigations have proven that crocin may drastically modify the concentrations of oxidative mediators in the hippocampal area and restore the negative impacts of long-term stress on memory and cognition [26].

18.4.4 Carotenoids

So far, approximately 700 distinct species of the carotenoid class have indeed been documented; 40 of them are identified in human cells and the bloodstream. The primary carotenoids prevalent in humans comprise β-cryptoxanthin, zeaxanthin, lutein, lycopene, and both α and β carotenes. According to their molecular framework, carotenoids can be distinguished by their antioxidant potential [27]. In-vivo and in-vitro animal models have been used to undertake a comprehensive exploration of the antioxidant and anti-inflammatory prospects of the carotenoid astaxanthin, which is extracted from seafood. Prophylactic outcomes on the cellular organelles and microcirculatory have been outlined, revealing that astaxanthin is a stimulatory neuroprotective molecule. Similar to individuals with mild AD, those with extreme or progressive AD lack key carotenoids, including β-carotene and lutein [28].

18.4.5 Anthocyanidins

Anthocyanidins are most abundant in strawberries and cranberries and have strong neuroprotective and anti-inflammatory effects by inhibiting the stimulation of pro-inflammatory cytokines mediators (TNF-α, IL-6, and IL-1β). Phospholipase A2 has significant involvement in the propagation of oxidative stress markers and pro-inflammatory cytokines mediators and whose inactivation exhibits impressive neuroprotection [19].

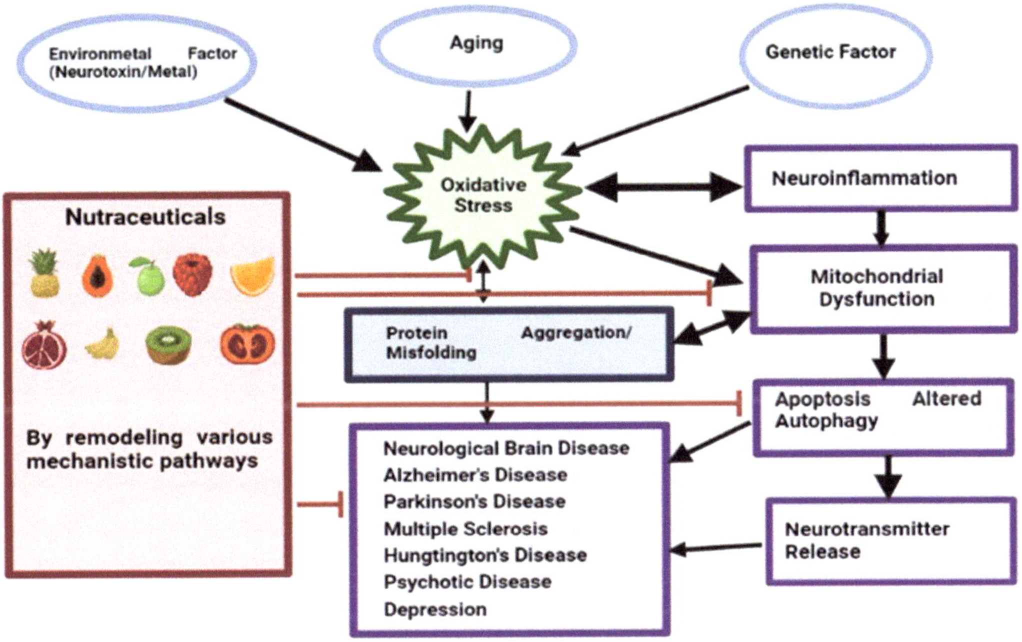

FIGURE 18.3 A graphical depiction of the pathway that is implicated in NDs and personalized paradigm modulated by nutraceuticals.

18.5 NUTRACEUTICALS IN PARKINSON'S DISEASE

Parkinson's disease (PD) is a neurodegenerative condition characterized by malfunctioning dopaminergic neurons in the substantia nigra portion of the mid-brain, which results in a substantial reduction in dopamine (DA) levels. Neurodegeneration that develops in PD is triggered by a variety of conditions, including mitochondrial dysfunction, antioxidant deprivation, and oxidative damage [29]. As an outcome, a variety of pharmacological moieties are employed, each of which acts by stimulating a different route of the same pharmacology. There is a demand for greater investigation and proven data on the benefits of vitamins on PD because several research on vitamin intake and animal and clinical trials have been undertaken, which have shown inconsistent findings in controlling PD manifestations [30]. To treat PD, a variety of vitamins, such as vitamin B3, vitamin B6, vitamin B9, vitamin B12, vitamin C, vitamin D, and vitamin E, can be administered [31]. By simply treating symptoms, the anti-Parkinson medications now in use impede the development of the condition. The greatest difficulty is identifying the optimum lead compound, which not only stimulates numerous systems and treats illness but is also minimal destructive to patients. To investigate and elucidate whether specific natural compounds may be used as a mono or combination prophylaxis in disease control, a wide variety of medicinal and traditional medicines have been researched experimentally for therapy in PD, with this as the underlying assumption [32]. Herbal medicines benefit PD by suppressing the several pathophysiological systems that underlie it, such as oxidative damage, apoptosis, neuro-inflammation, and mitochondrial dysfunction.

Polyphenolic compounds exhibit multifunctional properties that may effectively cross the blood-brain barrier (BBB) and provide positive activities by alleviating neurological and postural impairments in individuals by shielding dopaminergic neurons and restricting oxidative stress. Lycopene is a non-provitamin A carotenoid that is lipid-soluble and found in red fruits

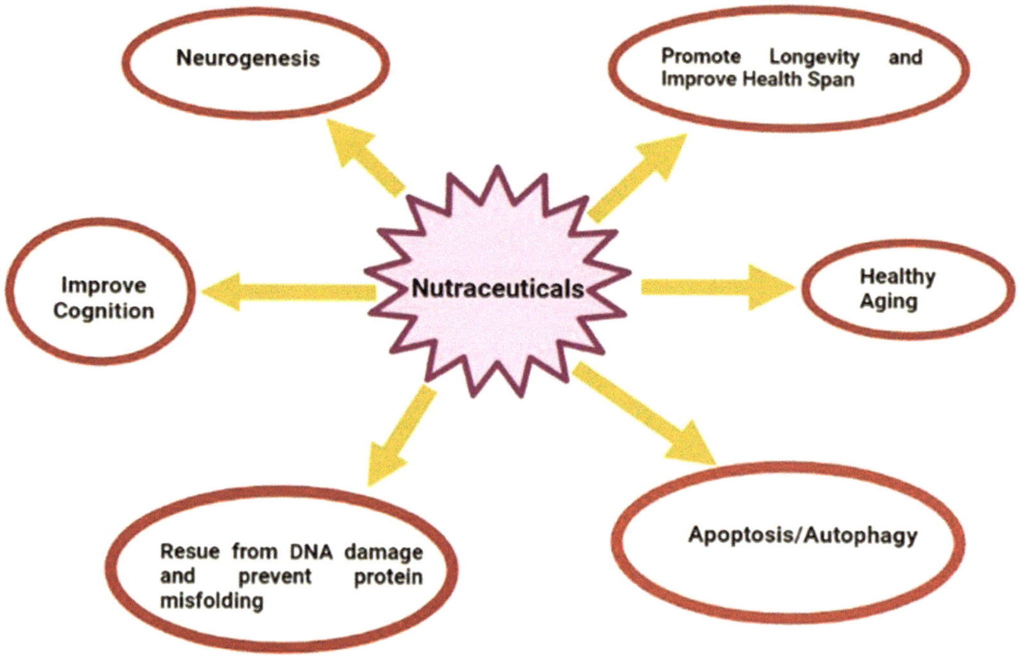

FIGURE 18.4 Depicts the effectiveness of nutraceuticals on NDs.

and vegetables, mostly tomatoes. It has been reported to possess antioxidant and neuroprotective benefits in experiments using 1-methyl-4-phenyl-1,2,3,6- tetrahydropyridine (MPTP) induced rodents, as well as to augment dopamine concentrations in the dorsal striatum [33]. The anti-oxidative properties of lycopene are predominantly responsible for its restorative effects, which also include neurocognitive shortfalls, an improvement in striatal nicotinamide adenine dinucleotide (reduced form-NADH) dehydrogenase, and superoxide dismutase (SOD) action, as well as lower malonaldehyde levels and higher glutathione.

Fish oil has a high concentration of Ω-3 fatty acids, including docosahexaenoic acid (DHA) and eicosapentaenoic acid (EPA), which have been shown to have cognitive benefits through a variety of mechanisms [34]. Among the most prevalent phenolic antioxidants found in *Camellia sinensis*, epigallocatechin-3 gallate (EGCG-ester of gallic acid and epigallocatechin) has demonstrated neurocognitive properties by effectively crossing the BBB [35]. Oxidative scavenger action and metal complexation properties of the phytometabolites are associated with the catechol ring moiety of EGCG [36]. It significantly enhanced DA concentrations in the basal striatum of the brain, which prevented neurodegeneration and markedly restored cognitive abilities in diseased individuals [35]. Ginsenoside (triterpene saponins and steroid glycosides), obtained from *Panax ginseng*, is an example of a nutraceutical supplement that has both anti-inflammatory and anti-oxidative actions [36]. Its anti-inflammatory activities derive from the modulation of various inflammatory networks, such as the oestrogen hormone circuits and Phosphatidylinositol-3-kinase (PI3K/ Protein kinase B (Akt)/ Reactive oxygen species (ROS)-Nuclear factor-κB (NF-κB), Insulin-like growth factor 1 (IGF-1), c-Jun N-terminal kinases (JNK), and Extracellular signal-regulated kinase (ERK) receptor signalling mechanisms. It carries out its anti-oxidative activities by sustaining glutathione concentrations [38]. Ginsenoside is effective at mitigating neurotoxicity and the impairment of dopaminergic neurons caused by PD chemicals like 6-hydroxydopamine (OHDA) and MPTP [37, 38]. Vincamine, a medicinal alkaloid derived from the vinca plant, has demonstrated anti-Parkinson's effectiveness through multiple molecular pathways. It has vasodilatory action and relaxes the muscles surrounding the blood vessels in nerve cells, increasing the transport of glucose and nutrients to the brain and accelerating the production of ATP via the Krebs cycle in combination [39].

18.6 NUTRACEUTICALS IN DEPRESSION

Depression is a neurological disease that is primarily distinguished by a negative or gloomy state, along with diminished excitement in most social interactions, which finally leads to a disrupting lifestyle. It exhibits a 15% frequency and a 7% yearly mortality. Folic acid and Ω-3 fatty acids have traditionally proven successful in treating atypical depression, especially when administered in combination with other treatments. There is also growing scientific proof that these treatments are useful independently [40]. A whole-grain diet enriched in minerals, including Ω-3 fatty acids, vitamin B9, zinc, and numerous other crucial micro and macronutrients, can stimulate brain activity and has proven validated outcomes in the treatment of depression [41]. Table 18.1 summarizes the pharmacological action of various dietary nutraceuticals for addressing depression.

18.7 NUTRACEUTICALS IN MULTIPLE SCLEROSIS

A progressive, inflammation and recurrent disorder of the central nervous system is known as multiple sclerosis (MS). It is characterized by myelin degeneration and successive accumulation of damaged tissue, and it has a devastating impact on quality of life (QoL) due to the catastrophic physiological and neurological impairments it presents. The most prevalent indications are stiffness, paraesthesia, double vision, numbness in the limbs, ataxia, mental abnormalities, tiredness, and bladder dysfunction [42]. Table 18.2 outlines the various dietary nutraceuticals for addressing MS.

TABLE 18.1
As a combination approach, few of the extensively exploited nutraceuticals in the treatment of depression offer therapeutic potential

Molecule	Pharmacological action	Ref.
Vitamin D	The hypothalamus, prefrontal cortex, and substantia nigra of the brain all have receptors that are substrates for the neuroactive steroid molecule known as vitamin D. It primarily modulates the genomic imprinting that results in protein tyrosine hydroxylase translation.	[40, 66]
Ω-3 fatty acid	They prevent monoamine neurotransmitters' absorption throughout neurobiological communication, which helps neurotransmitter release by enhancing neuronal barrier permeability. These compounds improve neuroplasticity, suppress the production of inflammatory markers, and protect against severe depression.	[40, 66]
S-adenosyl Methionine	As a prominent methyl donor, it primarily affects the biosynthesis and metabolism of neurotransmitters. Additionally, it enhances phosphatidylcholine synthesis while reducing prolactin release.	[40, 66, 67]
Zinc	Zinc is the highest abundant essential mineral in the brain's neocortex, amygdala, and hippocampus. It predominantly promotes neuroplasticity in hippocampus areas by augmenting brain-derived neurotrophic factor (BNDF). A change is also induced in the N-methyl-D-aspartate (NMDA) and glutamate receptors functioning.	[40, 66]
N-acetyl Cysteine	It principally consists of antioxidant and anti-inflammatory properties that augment neurogenesis and restore glutathione contents. Additionally, by moderating the glutamate cascade and preventing oxidative damage, it minimizes depression in the person.	[40, 66]

TABLE 18.2
Nutraceuticals with neuroprotective implications against the signs and symptoms of relapses in multiple sclerosis

Dietary nutraceuticals	Research plan	Patients count and MS subtype	Reported variables	Nutraceuticals administration	Outcomes	Ref.
Vitamin A	Randomized controlled experimental trial	101 Relapsing-Remitting MS (RRMS)	Recurrence probability: Multiple Sclerosis Functional Composite (MSFC) and Kurtzke Expanded Disability Status Scale (KEDSS)	10,000 IU Retinyl Palmitate for an additional 6 months completing the initial 6 months of 25,000 IU Retinyl Palmitate	Mitigation of impairment development and improvement of cognitive and upper limb functioning.	[31]
Vitamin D	Randomized controlled double-blinded experimental trial	94 RRMS	IL-17 contents	50,000 IU of vitamin D_3 per 5 days for 12 weeks	Following administering a large dosage of vitamin D_3 for 12 weeks, individuals with RRMS had a substantial shift in their IL-17 contents.	[32]
Coenzyme Q10 (CoQ10)	Randomized controlled double-blinded experimental trial	48 RRMS	Inflammatory biomarker (MMP-9, IL-6, and TNF-α)	For 12 weeks, administer 500 mg of CoQ10 daily or a placebo	Individuals with MS may experience less fatigued and depression after receiving 500 mg/day of CoQ10 supplements.	[33]
α-Calcidiol	Randomized controlled double-blinded experimental trial	600 MS	Fatigue Impact Scale (FIS) assessment	During the period of 6 months, 80 individuals were administered α-calcidiol (1 μg), whereas 78 took a placebo	Compared to those obtaining a placebo, QoL enhanced in those receiving α-calcidiol.	[34]
Vitamin B_7	Non-randomized, double-blinded experimental trial	154 RRMS	KEDSS assessment: recovery from MS-induced impairment	Orally, three times per day, either MD1003 (Biotin 100 mg) or the placebo	KEDSS advancement was reduced, and the psychological perception of improvement was superior with a placebo.	[35]
Fatty acid	Double-blinded control experimental trial	116 MS	Relapse assessment	Linoleic acid was administered to two groups, whereas oleic acid was administered to two reference groups	The length and intensity of MS relapses were very slightly altered by 20 g of linoleic acid, but total impairment was unaffected.	[36]

18.8 NUTRACEUTICALS IN HUNGTINGTON'S DISEASE

The neurodegenerative condition Huntington's disease (HD) is an uncommon genetic syndrome that impairs cognition, produces gradual neurological abnormalities, and has neuropsychiatric manifestations. Huntingtin is a protein that is expressed by the HD genome and is prevalent in innumerable bodily tissues, particularly the central nervous system (CNS). Demyelination, primarily in the striatum and cortex, is a hallmark of HD. The initial indications often start in midlife and comprise persistent motor and intellectual problems [43, 44]. There are relatively few nutraceutical supplements that have been proven to be beneficial for HD *viz* taurine, quercetin, curcumin, levocarnitine, anthocyanins, and resveratrol.

18.9 NUTRACEUTICALS IN PSYCHOTIC DISORDERS

Beyond the physiological features that have been explored, nutritional supplements, i.e., nutraceuticals, are essential for the treatment of depressive disturbances and psychotic conditions like bipolar disorder (extreme mood swings) and schizophrenia. They are frequently prescribed in individuals who are in severe demand of psychotic medication as a combination therapy and occasionally as a singly used [45]. By augmenting the suppressed re-uptake mechanism of monoamine neurotransmitters and enhancing neuroprotective effects, nutraceuticals substantially expand the pharmacological efficiency of the pharmaceuticals utilized. This enhances the effectiveness of psychiatric therapies and lessens their negative consequences [46, 47]. Vitamins and Ω-3 fatty acids are among the nutraceuticals that are frequently employed in psychosis. In the human system, polyunsaturated fatty acids are primarily classified into two groups: (1) Ω-3 category, which is acquired as an α-linolenic acid, and (2) Ω-6 category, which includes arachidonic acid (AA), which is derived from linoleic acid. The latter category comprises of docosahexaenoic acid (DHA) and eicosapentaenoic acid (EPA) [48]. They are all fundamental constituents of the phospholipid cellular membrane and necessary for the existence of the human system. Moreover, they should be acquired from meals because the system cannot synthesize them. Ω-3 DHA and EPA exhibit characteristics at the cellular level that are relevant to psychotic ailments. They enhance dopamine and serotonin neural activity. They mitigate cellular damage and brain microinflammation complaints. They regulate mitochondrial metabolism, which is the massive trigger of neuro-oxidative damage [49]. The human system cannot develop sufficient chemical substances known as vitamins; thus, they must be consumed via meals. Current research has examined the effectiveness of vitamin treatments in schizophrenia.

18.10 MECHANISM OF NUTRACEUTICALS IN AMELIORATING NEURODEGENERATION

18.10.1 Nutraceuticals Enhance Biogenesis and Maintain Mitochondrial Health

The primary source of cellular energy and a need for all biological functioning is mitochondria. AD, PD, HD, ageing, psychotic disorders, and stroke all undergo neurodegenerative carried on by the impairments, which also induce oxidative damage and trigger death-signalling cascades [50]. The neuropathological markers tau and amyloid-β (Aβ) in AD and α-synuclein in PD are deposited in mitochondria, respectively, and Aβ hinders complex IV (Cytochrome c oxidase) activity of the electron transport chain (ETC), whereas α-synuclein dysregulates complex I (NADH oxidoreductase) activity. Complex I dysfunction in the ETC has been linked to PD, and the dopaminergic neurotoxin MPTP inhibits complex I activity and triggers parkinsonism in individuals [51]. Mitochondrial biogenesis, cell death signalling, destruction through autophagy (mitophagy), ATP production, ETC functioning, and neuroprotection are all modulated by nutraceuticals [52]. The ETC efficiency was enhanced by proanthocyanidins, luteolin, EGCG, huperzine (obtained from *Huperzia serrata*), myricitrin, and quercetin [53].

18.10.2 Protein Clusters Associated with Neurodegenerative Conditions are Effectively Addressed by Nutraceuticals

The predominant hallmarks in AD are the aggregation of neurofibrillary tangle (NFT) amyloid lesions, while in PD, it is Lewy complexes. Proteins that have undergone aberrant mutations and modifications can cluster through polymerization, and oligomerization, alterations to the secondary configuration (development of β-sheets), the production of insoluble aggregation complexes, and malfunction of the proteolytic cascades. In an attempt to reduce the neurological impairment induced by Aβ, nutraceuticals have been revealed to enhance the formation of non-toxic aggregates, restrict the formation of amyloidogenic complexes, and hinder the development of fibrillar and oligomeric complexes [54, 55]. Peptides having 40–42 amino acids that are amyloidogenic synthesized by the successive fragmentation of amyloid precursor protein by β-secretase (BACE1) and γ-secretase. It has been reported that the ferulic acid (polyphenolic antioxidant) and tannic acid (polyphenol) as well as the polymethoxyflavones class of phytometabolites (sinensetin and nobiletin), and galangin, can hinder BACE1 enzyme and ameliorate behavioural impairment in mouse models of AD [56, 57]. The activity of BACE1 was attenuated by 7,8-dihydroxyflavine, icariin, EGCG, rutin, curcumin, and luteolin. Myricetin, rosmarinic acid, and curcumin reduced the cytotoxic fibrils and oligomers while producing soluble Aβ oligomers [58, 59]. The presynaptic α-synuclein deposition has been hypothesized as a deleterious factor in α-synucleinopathies, especially in PD. The occurrence of oxidative exposure, point mutations (E46K/A53T/A30P), metal cation (Cu^{2+} and Fe^{3+}), autophosphorylation, and association with lipid matrix all contribute to the induction of clustering and fibrillization. The production of α-synuclein fibrils and their integration into clusters was hindered by myricetin, curcumin, baicalein, ferulic acid, theaflavins, rosmarinic acid, and kaempferol [60, 61]. Advanced α-synuclein oligomers were transformed by EGCG into thinner, non-toxic, and unstructured clusters [62].

18.10.3 Nutraceuticals Mediated Upregulation of Neurotrophic Factors (NFTs) and Neurotrophic Receptors

In healthy human subjects, rodents, and biological prototypes, polyphenolic nutraceuticals have been proven to augment glial cell-lined derived neurotrophic factor (GDNF) and brain-derived neurotrophic factor (BDNF) to alleviate neurochemistry and psychological deficits. *Ginkgo biloba* therapy has been demonstrated to improve BDNF levels and cardiopulmonary endurance in healthy and active males [62]. Flavonoids (like phytometabolites naringenin, hesperidin, and EGCG), piperine (an alkaloid obtained from *Piper nigrum*), astaxanthin, resveratrol, and curcumin enhanced BDNF content in rodent paradigms of ageing, depression, AD, brain ischaemia, ageing, PD, and depression. Catalpol (iridoid glucoside), harpagoside (terpene glycoside), naringenin, and smilagenin (steroidal sapogenin) all improved GDNF activity in PD mouse models [63]. Tropomyosin receptor kinase (Trk) activity and neuroplasticity are enhanced by polyphenolic nutraceuticals, indicating neuroprotective effects. In experimental models of depression, olive polyphenols upregulated TrkA and TrkB in the hippocampal but not in the prefrontal lobe and had benefits comparable to those of antidepressants [64]. In a rat paradigm of depression, nobiletin (derived from citrus fruits) and deoxygedunin (derived from *Azadirachta indica*) stimulated hippocampal neuroplasticity. They also improved BDNF and TrkB activity in mice models of depression [65].

18.11 CONCLUSION

Nature has endowed mankind with phenomenal pharmacological compounds that have a great commitment to the treatment and management of lifestyle-induced and life-threatening conditions such as neurodegeneration. Multiple investigations have explicitly documented how bioactive components help to mitigate neurodegeneration and avert cognitive impairment. Nutraceutical's therapeutic benefits can be linked to their hypolipidemic, anti-inflammatory, antioxidant, neuroprotective, and restorative functionality, which targeted multiple mediators and channels to increase protein expression, resulting in neuroprotection. Protein assembly and destruction can be suppressed, resulting in a powerful neurological network. In this period of several catastrophes confronting the public healthcare sector, scientific work on plant extracts has offered novel perspectives for the effective therapy of neurological conditions. A modern lifestyle has impaired the body's defence system against reactive oxygen radicals by inhibiting antioxidants, leading to excessive oxidative damage. Age-related declines in antioxidant content in the body make people more vulnerable to long-term medical problems. As an outcome, for many years the attention has been on focusing on a range of nutraceuticals for their pharmacological effects. Vitamins and other substances with anti-oxidative activity work by arresting free radicals and promoting the body's endogenous production of antioxidants. The present investigation emphasized the effectiveness of nutraceutical treatment to stop neurological conditions from advancing. They are the greatest choices for addressing lifestyle-mediated psychological disorders like depression and neurological disorders since utilizing them in appropriate and approved doses fosters excellent brain functioning and prevents diseases.

REFERENCES

[1] T. Farkhondeh, H.S. Yazdi, S. Samarghandian. The protective effects of green tea catechins in the management of neurodegenerative diseases: A review. *Current Drug Discovery Technologies*. 2019 March 1;16(1):57–65. https://doi.org/10.2174/1570163815666180219115453.

[2] R.J. Williams, K.P. Mohanakumar, P.M. Beart. Neuro-nutraceuticals: Further insights into their promise for brain health. *Neurochemistry International*. 2016 May 1;95:1–3. https://doi.org/10.1016/j.neuint.2016.03.016.

[3] M.M. Abdel-Daim, O.S. El-Tawil, S.G. Bungau, A.G. Atanasov. Applications of antioxidants in metabolic disorders and degenerative diseases: Mechanistic approach. *Oxidative Medicine and Cellular Longevity*. 2019 Jul 29;2019. https://doi.org/10.1155/2019/4179676.

[4] K.P. Van der Burg, L. Cribb, J. Firth,. et al. EPA and DHA as markers of nutraceutical treatment response in major depressive disorder. *European Journal of Nutrition* 2020;59:2439–47. https://doi.org/10.1007/s00394-019-02090-6.

[5] A. Durazzo, M. Lucarini, E. Camilli, S. Marconi, P. Gabrielli, S. Lisciani, L. Gambelli, A. Aguzzi, E. Novellino, A. Santini, A Turrini. Dietary lignans: Definition, description and research trends in databases development. *Molecules*. 2018 Dec 8;23(12):3251. https://doi.org/10.3390/molecules23123251.

[6] A. Agnihotri, O.I. Aruoma. Alzheimer's disease and Parkinson's disease: A nutritional toxicology perspective of the impact of oxidative stress, mitochondrial dysfunction, nutrigenomics and environmental chemicals. *Journal of the American College of Nutrition*. 2020 Jan 2;39(1):16–27. https://doi.org/10.1080/07315724.2019.1683379.

[7] H.F. Chiu, Y.C. Shen, K. Venkatakrishnan, C.K. Wang. Popular functional foods and nutraceuticals with lipid lowering activity and in relation to cardiovascular disease, dyslipidemia, and related complications: An overview. *Journal of Food Bioactives*. 2018 Jun 30;2:16–27. https://doi.org/10.31665/JFB.2018.2137.

[8] P. Dadhania, P. Trivedi, A. Vikram, D. Nand Tripathi. Nutraceuticals against neurodegeneration: A mechanistic insight. *Current Neuropharmacology*. 2016 Aug 1;14(6):627–40.

[9] A. Santini, E. Novellino. Nutraceuticals-shedding light on the grey area between pharmaceuticals and food. *Expert Review of Clinical Pharmacology*. 2018 Jun 3;11(6):545–7. https://doi.org/10.1080/17512433.2018.1464911.

[10] L. Perrone, T. Squillaro, F. Napolitano, Terracciano C, Sampaolo S, Melone MA. The autophagy signaling pathway: a potential multifunctional therapeutic target of curcumin in neurological and neuromuscular diseases. *Nutrients*. 2019 Aug 13;11(8):1881. https://doi.org/10.3390/nu11081881.

[11] K. Venkatakrishnan, H.F. Chiu, C.K. Wang. Impact of functional foods and nutraceuticals on high blood pressure with a special focus on meta-analysis: Review from a public health perspective. *Food & Function*. 2020;11(4):2792–804.https://doi.org/10.1039/D0FO00357C.

[12] P. Daliu, A. Santini, E. Novellino. A decade of nutraceutical patents: where are we now in 2018?. *Expert Opinion on Therapeutic Patents*. 2018 Dec 2;28(12):875–82. https://doi.org/10.1080/13543776.2018.1552260.

[13] H.F Chiu, K. Venkatakrishnan, C.K. Wang. The role of nutraceuticals as a complementary therapy against various neurodegenerative diseases: A mini-review. *Journal of Traditional and Complementary Medicine*. 2020 Sep 1;10(5):434–9. https://doi.org/10.1016/j.jtcme.2020.03.008.

[14] M.D. Pandareesh, H.K. Kandikattu, S. Razack, N. Amruta, R. Choudhari, A. Vikram, P. Doddapattar. Nutrition and nutraceuticals in neuroinflammatory and brain metabolic stress: implications for neurodegenerative disorders. *CNS & Neurological Disorders-Drug Targets* (Formerly *Current Drug Targets-CNS & Neurological Disorders*). 2018 Nov 1;17(9):680–8. https://doi.org/10.2174/1871527317666180625104753.

[15] R. Makkar, T. Behl, S. Bungau, G. Zengin, V. Mehta, A. Kumar, M.S. Uddin, G.M. Ashraf, M.M. Abdel-Daim, S. Arora, R. Oancea. Nutraceuticals in neurological disorders. *International Journal of Molecular Sciences*. 2020 Jun 22;21(12):4424. https://doi.org/10.3390/ijms21124424.

[16] N.S. Mohd Sairazi, Sirajudeen KN. Natural products and their bioactive compounds: Neuroprotective potentials against neurodegenerative diseases. *Evidence-Based Complementary and Alternative Medicine*. 2020 Oct;2020. https://doi.org/10.1155/2020/6565396.

[17] D.J. McClements. Enhanced delivery of lipophilic bioactives using emulsions: A review of major factors affecting vitamin, nutraceutical, and lipid bioaccessibility. *Food & Function*. 2018;9(1):22–41. 10.1039/C7FO01515A.

[18] T. Behl, K. Kumar, C. Brisc, M. Rus, D.C. Nistor-Cseppento, C. Bustea, R.A. Aron, C. Pantis, G. Zengin, A. Sehgal, R. Kaur. Exploring the multifocal role of phytochemicals as immunomodulators. *Biomedicine & Pharmacotherapy*. 2021 Jan 1;133:110959. https://doi.org/10.1016/j.biopha.2020.110959.

[19] N.S. Gowthaman, P. Arul, N.A. Jailani. Scientific basis and developments in the clinical aspects of nutraceutical and dietary supplements for neurological and cognitive dysfunction. *Clinical Studies on Nutraceuticals and Dietary Supplements*. 2022 Oct 29; 99–115. CRC Press.

[20] B. Polis, A.O. Samson. Neurogenesis versus neurodegeneration: The broken balance in Alzheimer's disease. *Neural Regeneration Research*. 2021 Mar 1;16(3):496–7. https://doi.org/10.4103/1673-5374.293138.

[21] D.D. Dohrmann, P .Putnik, D.B. Kovačević, J. Simal-Gandara, J.M. Lorenzo, F.J. Barba. Japanese, Mediterranean and Argentinean diets and their potential roles in neurodegenerative diseases. *Food Research International*. 2019 Jun 1;120:464–77. https://doi.org/10.1016/j.foodres.2018.10.090.

[22] H.Y. Tsai, M.Y. Chen, C. Hsu, K.Y. Kuan, C.F. Chang, C.W. Wang, C.P. Hsu, N.W. Su. Luteolin phosphate derivatives generated by cultivating Bacillus subtilis var. Natto BCRC 80517 with luteolin. *Journal of Agricultural and Food Chemistry*. 2022 Jul 7;70(28):8738–45. https://doi.org/10.1021/acs.jafc.2c03524.

[23] S. Ahmad, M.H. Jo, M. Ikram, A .Khan, M.O. Kim. Deciphering the potential neuroprotective effects of luteolin against Aβ1–42-Induced alzheimer's disease. *International Journal of Molecular Sciences*. 2021 Sep 3;22(17):9583. https://doi.org/10.3390/ijms22179583.

[24] I. Glevitzky, G.A. Dumitrel, M. Glevitzky, B. Pasca, P. Otrisal, S. Bungau, G. Cioca, C. Pantis, M.J. Popa. Statistical analysis of the relationship between antioxidant activity and the structure of flavonoid compounds. *Revista de Chimie*. 2019 Sep 1;70(9):3103–7.

[25] P. Mecocci, C. Tinarelli. Nutraceuticals in cognitive impairment and Alzheimer's disease. *Frontiers in Pharmacology*. 2014 Jun 23;5:91372. https://doi.org/10.3389/fphar.2014.00147.

[26] V. Boccardi, C. Tinarelli, P. Mecocci. Nutraceuticals and cognitive dysfunction: Focus on Alzheimer's disease. *Neuroprotective Effects of Phytochemicals in Neurological Disorders*. 2017 Feb;28:561–79. https://doi.org/10.1002/9781119155195.ch28.

[27] S. Bungau, M.M. Abdel-Daim, D.M. Tit, Ghanem, S. Sato, M. Maruyama-Inoue, S. Yamane, K. Kadonosono. Health benefits of polyphenols and carotenoids in age-related eye diseases. *Oxidative Medicine and Cellular Longevity*. 2019 Feb 12;2019, 9783429. https://doi.org/10.1155/2019/9783429.

[28] K.W. Hu, H.F. Fan, H.C. Lin, J.W. Huang, Y.C. Chen, C.L. Shen, Y.H. Shih, L.H. Tu. Exploring the impact of glyoxal glycation on β-amyloid peptide (Aβ) aggregation in Alzheimer's disease.

The Journal of Physical Chemistry B. 2021 May 21;125(21):5559–71. https://doi.org/10.1021/acs.jpcb.1c02797.

[29] R.R. Watson, V.R. Preedy, editors. *Bioactive nutraceuticals and dietary supplements in neurological and brain disease: prevention and therapy*. Academic Press; 2014 Sep 11.

[30] G.E. Alexander. Biology of Parkinson's disease: Pathogenesis and pathophysiology of a multisystem neurodegenerative disorder. *Dialogues in Clinical Neuroscience*. 2004 Sep 30;6(3):259–80. https://doi.org/10.31887/DCNS.2004.6.3/galexander

[31] X. Han, Y. Liu, Y. Dai, T. Xu, Q. Hu, X. Yi, L. Rui, G. Hu, J. Hu. Neuronal SH2B1 attenuates apoptosis in an MPTP mouse model of Parkinson's disease via promoting PLIN4 degradation. *Redox Biology*. 2022 Jun 1;52:102308. https://doi.org/10.1016/j.redox.2022.102308.

[32] C.M. Deus, J. Teixeira, N. Raimundo, P. Tucci, F. Borges, L. Saso, P.J. Oliveira. Modulation of cellular redox environment as a novel therapeutic strategy for Parkinson's disease. *European Journal of Clinical Investigation*. 2022 Oct;52(10):e13820. https://doi.org/10.1111/eci.13820

[33] A. Rahimmi, S. Tozandehjani, M. Daraei, M. Khademerfan. The neuroprotective roles of dietary micronutrients on Parkinson's disease: A review. *Molecular Biology Reports*. 2022 Aug;49(8):8051–60.

[34] P. Li, C .Song. Potential treatment of Parkinson's disease with omega-3 polyunsaturated fatty acids. *Nutritional Neuroscience*. 2022 Jan 2;25(1):180–91. https://doi.org/10.1080/1028415X.2020.1735143

[35] C.M. Sergi. Epigallocatechin gallate for Parkinson's disease. *Clinical and Experimental Pharmacology and Physiology*. 2022 Oct;49(10):1029–41.https://doi.org/10.1111/1440-1681.13691

[36] B. He, D. Chen, X. Zhang, R. Yang, Y. Yang, .P Chen, Z. Shen. Oxidative stress and ginsenosides: An update on the molecular mechanisms. *Oxidative Medicine and Cellular Longevit*y. 2022 Apr 20;2022, 9299574. https://doi.org/10.1155/2022/9299574.

[37] M.M. Rahman, X. Wang, M.R. Islam, S. Akash, F.A. Supti, M.I. Mitu, M. Harun-Or-Rashid, M. Aktar, K. Kali, M .Sumaiya, F.I. Jahan. Multifunctional role of natural products for the treatment of Parkinson's disease: At a glance. *Frontiers in Pharmacology*. 2022 Oct 6;13:976385. https://doi.org/10.3389/fphar.2022.976385.

[38] M. Zheng, Y. Xin, Y. Li, F. Xu, X. Xi, H. Guo, X. Cui, H. Cao, X. Zhang, C. Han. Ginsenosides: A potential neuroprotective agent. *BioMed Research International*. 2018 Oct;2018, 8174345. https://doi.org/10.1155/2018/8174345.

[39] A. Sheref, Y. Naguib, E. Abouelnour, H. Salem, M. Hassan, H.E. Abdel-Razek. Neuroprotective effect of piracetam and vincamine in a rat model of haloperidol-induced Parkinson's disease. *Bulletin of Egyptian Society for Physiological Sciences*. 2022 Jan 1;42(1):11–26. https://doi.org/10.21608/besps.2021.71203.1099.

[40] E. Ceskova, P. Silhan. Novel treatment options in depression and psychosis. *Neuropsychiatric Disease and Treatment*. 2018 Mar 13;741–7. https://doi.org/10.2147/ndt.s157475.

[41] K.P. Van Der Burg, L. Cribb, J. Firth, D. Karmacoska, J. Sarris. Nutrient and genetic biomarkers of nutraceutical treatment response in mood and psychotic disorders: A systematic review. *Nutritional Neuroscience*. 2021 Apr 3;24(4):279–95. https://doi.org/10.1080/1028415X.2019.1625222.

[42] R. Dobson, G. Giovannoni. Multiple sclerosis–A review. *European Journal of Neurology*. 2019 Jan;26(1):27–40. https://doi.org/10.1111/ene.13819.

[43] S.J. Tabrizi, S. Schobel, E.C. Gantman, A. Mansbach, B. Borowsky, P. Konstantinova, T.A. Mestre, J. Panagoulias, C.A. Ross, M. Zauderer, A.P. Mullin. A biological classification of Huntington's disease: The integrated staging system. *The Lancet Neurology*. 2022 Jul 1;21(7):632–44. https://doi.org/10.1016/S1474-4422(22)00120-X.

44. P. McColgan, S.J. Tabrizi. Huntington's disease: a clinical review. European journal of neurology. 2018 Jan;25(1):24–34.https://doi.org/10.1111/ene.13413.

[45] J. Firth, S.B. Teasdale, K. Allott, D. Siskind, W. Marx. The efficacy and safety of nutrient supplements in the treatment of mental disorders: A meta-review of meta-analyses of randomized controlled trials. *World Psychiatry*. 2019 Oct;18(3):308–24.S. https://doi.org/10.1002/wps.20672.

[46] D. Najarian, P. Sanga, S. Wang, P. Lim, A. Singh, M.J. Robertson, K. Cohen, A. Schotte. A randomized, double-blind, multicenter, noninferiority study comparing paliperidone palmitate 6-month versus the 3-month long-acting injectable in patients with schizophrenia. *International Journal of Neuropsychopharmacology*. 2022 Mar 1;25(3):238–51. https://doi.org/10.1093/ijnp/pyab071.

[47] H.E. Brown, J.L. Roffman. Emerging treatments in schizophrenia: Highlights from recent supplementation and prevention trials. *Harvard Review of Psychiatry*. 2016 Mar 1;24(2):e1–7. https://doi.org/10.1097/HRP.0000000000000101.

[48] J. Sarris, A.C. Logan, T.N. Akbaraly, G.P. Amminger, V. Balanzá-Martínez, M.P. Freeman, J. Hibbeln, Y. Matsuoka, D. Mischoulon, T. Mizoue, A. Nanri. Nutritional medicine as mainstream in psychiatry. *The Lancet Psychiatry*. 2015 Mar 1;2(3):271–4. https://doi.org/10.1016/S2215-0366(14)00051-0.

[49] M. Toricelli, A.A. Pereira, G.S. Abrao, H.N. Malerba, J. Maia, H.S. Buck, T.A. Viel. Mechanisms of neuroplasticity and brain degeneration: Strategies for protection during the aging process. *Neural Regeneration Research*. 2021 Jan 1;16(1):58–67. https://doi.org/10.4103/1673-5374.286952.

[50] S. Mani, M .Sevanan, A. Krishnamoorthy, S. Sekar. A systematic review of molecular approaches that link mitochondrial dysfunction and neuroinflammation in Parkinson's disease. *Neurological Sciences*. 2021 Nov;42(11):4459–69.

[51] K. Kessas, Z. Chouari, I. Ghzaiel, A. Zarrouk, M. Ksila, T. Ghrairi. Role of bioactive compounds in the regulation of mitochondrial dysfunctions in brain and age-related neurodegenerative diseases. *Cells*. 2022 Jan 13;11(2):257. https://doi.org/10.3390/cells11020257.

[52] I. Scala, D. Valenti, D. Scotto' V.Aniello, M. Marino, M.P. Riccio. Epigallocatechin-3-gallate plus omega-3 restores the mitochondrial complex I and F0F1-ATP synthase activities in PBMCs of young children with Down syndrome: a pilot study of safety and efficacy. *Antioxidants*. 2021 Mar 16;10(3):469. https://doi.org/10.3390/antiox10030469.

[53] V. Beldona, A. Patel, K. Patel, N. Abraham. Natural product polyphenol inhibition of amyloid-β aggregation. *Journal of High School Science*. 2022 Jan 24;6(1):1–19. https://shorturl.at/N0LYt

[54] M.H. Rahman, J. Bajgai, A. Fadriquela, S. Sharma, T.T. Trinh, R. Akter, Y.J. Jeong,. Therapeutic potential of natural products in treating neurodegenerative disorders and their future prospects and challenges. *Molecules*. 2021 Sep 2;26(17):5327. https://doi.org/10.3390/molecules26175327.

[55] W. Jing, C. Xiaolan, C .Yu, Q. Feng, Y. Haifeng. Pharmacological effects and mechanisms of tannic acid. *Biomedicine & Pharmacotherapy*. 2022 Oct 1;154:113561. https://doi.org/10.1016/j.biopha.2022.113561.

[56] S. Zhou, X. Dong. Neuroprotective properties of ferulic acid in preclinical models of Alzheimer's disease: A systematic literature review. *Current Medicinal Chemistry*. 2023 Jul 1;30(24):2796–811. https://doi.org/10.2174/0929867329666220906110506.

[57] W. Araki, F. Kametani. Protection against amyloid-β oligomer neurotoxicity by small molecules with antioxidative properties: potential for the prevention of Alzheimer's disease dementia. *Antioxidants*. 2022 Jan 7;11(1):132. https://doi.org/10.3390/antiox11010132.

[58] E. De Lorenzi, D. Franceschini, C. Contardi, R.M. Di Martino, F. Seghetti, M. Serra, F. Bisceglia, A. Pagetta, M. Zusso, F. Belluti. Modulation of amyloid β-induced microglia activation and neuronal cell death by curcumin and analogues. *International Journal of Molecular Sciences*. 2022 Apr 15;23(8):4381. https://doi.org/10.3390/ijms23084381.

[59] F. Han, B. Jiang, M.H. Lue, Z.P. Wang, W. Liu, Y.X. Zhang, J. Xu. Hybrids of polyphenolic acids and xanthone, the potential preventive and therapeutic effects on PD: Design, synthesis, in vitro anti-aggregation of α-Synuclein, and disaggregation against the existed α-Synuclein oligomer and fibril. *Bioorganic & Medicinal Chemistry*. 2022 Jul 15;66:116818. https://doi.org/10.1016/j.bmc.2022.116818.

[60] N.N. Jha, D. Ghosh, S. Das, A. Anoop, R.S. Jacob, P.K. Singh, N. Ayyagari, I.N. Namboothiri, S.K. Maji. Effect of curcumin analogs onα-synuclein aggregation and cytotoxicity. *Scientific Reports*. 2016; 6:1–15.

[61] E. Sadowska-Krępa, B. Kłapcińska, I. Pokora, P. Domaszewski, K. Kempa, T. Podgórski. Effects of six-week Ginkgo biloba supplementation on aerobic performance, blood pro/antioxidant balance, and serum brain-derived neurotrophic factor in physically active men. *Nutrients*. 2017 Jul 26;9(8):803. https://doi.org/10.3390/nu9080803.

[62] M. Naoi, M. Shamoto-Nagai, W. Maruyama. Neuroprotection of multifunctional phytochemicals as novel therapeutic strategy for neurodegenerative disorders: Antiapoptotic and antiamyloidogenic activities by modulation of cellular signal pathways. *Future Neurology*. 2019 Feb 15;14(1):FNL9. https://doi.org/10.2217/fnl-2018-0028.

[63] S. Terracina, C. Petrella, S. Francati, M. Lucarelli, C. Barbato, A. Minni, M. Ralli, A. Greco. Antioxidant intervention to improve cognition in the aging brain: The example of hydroxytyrosol

and resveratrol. *International Journal of Molecular Sciences*. 2022 Dec 10;23(24):15674. https://doi.org/10.3390/ijms232415674.

[64] H. Wang, Y. Guo, Y. Qiao, J. Zhang, P. Jiang. Nobiletin ameliorates NLRP3 inflammasome-mediated inflammation through promoting autophagy via the AMPK pathway. *Molecular Neurobiology*. 2020 Dec;57:5056–68. https://doi.org/10.1007/s12035-020-02071-5.

[65] A. Lama, C. Pirozzi, C. Avagliano, C. Annunziata, M.P. Mollica, A. Calignano, R. Meli, G.M. Raso. Nutraceuticals: An integrative approach to starve Parkinson's disease. *Brain, Behavior, & Immunity-Health*. 2020 Feb 1;2:100037. https://doi.org/10.1016/j.bbih.2020.100037.

[66] G. Schoretsanitis, M. Paulzen, S. Unterecker, M. Schwarz, A. Conca, G. Zernig, G. Gründer, E. Haen, P. Baumann, N. Bergemann, H.W. Clement. TDM in psychiatry and neurology: A comprehensive summary of the consensus guidelines for therapeutic drug monitoring in neuropsychopharmacology, update 2017; a tool for clinicians. *The World Journal of Biological Psychiatry*. 2018 Apr 3;19(3):162–74. https://doi.org/10.1080/15622975.2018.1439595.

19 Impact of Prenatal Diet on the Developing Brain of the Fetus

Kunal Bhattacharya, Biswajit Devanath, Surender Prasad, Tasrina Rahman, and Nongmaithem Randhoni Chanu*
*Corresponding author

19.1 INTRODUCTION

Diet impacts the fetus's and the woman's brain health throughout gestation and the future in life. It is, therefore, very important to study the food routine of pregnant women. The concept that mothers' food patterns may affect children's ability to function psychologically in the future has major consequences for public health practice and policy development, for our understanding of human physiology, for the creation of food products, and for the advancement of the economy (1). Foods and nutrients significantly impact health, especially when pregnant women eat healthily to maximize their child's cognitive and behavioural results. The present evidence has proposed a relationship between food and brain health in the growing fetus (2). For growth and development, the offspring and breastfed child entirely rely on the mother's food and nutrition. In particular, the mother's nutritional state is a significant element that can have both immediate and long-standing effects on an unborn child's health and disease risk. Due to the significance that food and nutrition play in programming the fetal brain's development and subsequent mental processes, the fetus is extremely susceptible to dietary and nutritional changes during pregnancy (3). Numerous questions remain unanswered despite the extensive study on the positive effects of adequate nutrition during pregnancy over the past decades, including the most susceptible times of gestation, the effects of particular foods and nutrients, and the levels of foods and nutrients needed to improve cognitive development. It is important to figure out the ideal foods and nutrient supplements and whether there is an impact on a child's growing brain over the long run (4).

Two of the most frequently studied aspects of fetal programming of neurodevelopment are the mother's diet and nutrition during pregnancy. Enough energy, protein, essential fatty acids, and a number of crucial micronutrients are required during fetal growth to provide the required substrates for producing fetal neurons within the nervous system and as co-factors in physiochemical processes that control normal brain development (5). Extreme cases of inadequate nutrition during gestation, such as those that occurred during starvation, have provided light on the consequences on brain development throughout time. For instance, it was discovered that the cognitive abilities of children of mothers who had nutritional deficiencies during gestation were diminished. Negative neurodevelopmental outcomes from certain nutritional problems or dietary abnormalities may also include neural tube defects, language delays, decreased cognitive capacities, and mental and neurodevelopmental diseases (6).

Around the world, people are becoming more and more concerned about children's cognitive development and the aggravation of mental health issues. According to estimates, 35% of children in Europe have abnormal cognitive development and mental problems like anxiety, mood, and impulse control issues, which negatively affect their life quality and have a significant financial

DOI: 10.1201/9781003300502-19

impact on society. Understanding the causes and effects of mental diseases and applying this information to policies and preventive initiatives is more important than ever because they account for 8.1% of the worldwide disease burden. Studies on both humans and experimental animals show how important food and nutrition are for neurodevelopment (1). The determent and treatment of psychological health issues may be more affordable if an improved diet is provided, particularly in connection to nutrients like iron, zinc, and fatty acids. In healthy children, there is emerging proof that early nutrition and diet might affect later cognitive behaviour and development (2).

Due to its extended period of development and specialization, the brain is extremely susceptible to misprogramming. The effects of early brain misprogramming will have an impact not only on the form and function of the brain but also on other physiological functions. For instance, the brain controls inflammatory and endocrine signals from many 'brain-body axes', controlling every metabolic step necessary for development and growth (1). The effects of a shortfall in a particular food or nutrient depend on the stage of development at which it occurs, how severe the deficiency is, and how long it persists (7).

Due to contemporary non-invasive and computational tools and animal models, fetal brain development has grown in importance over the past twenty years and is now a crucial area of study in neuroscience. The findings allow for a quantitative description of the nutrients acquired necessary for the composition and development of individual nerve cells and entire networks within particular brain areas, as well as for relating the structures to the functions and effect on brain health to complications at both the particular neuron and network levels in fetuses and later-developing children (8). Novel phenotypes are expressed as a result of gene expression variation, which has consequences for understanding evolution and disease risk (9). During embryonic and fetal life, subtle brain structure and function changes can progressively and considerably be exaggerated over time. This causes lasting effects on brain anatomy and physiology (10). Lack of nutrition causes the placental Peg3 gene to be inactive, which alters 56% of the coexpressed genes in the placenta and hypothalamus. This has an impact on embryonic development during gestation (11).

19.2 BRAIN DEVELOPMENT STAGES AND NUTRITIONAL REQUIREMENTS DURING PREGNANCY

The intrauterine environment can profoundly impact the fetal brain's development. To reach its full genetic potential, the fetus must grow in an optimal intrauterine environment free from the harmful effects of illnesses and poisonous substances. It also needs to receive adequate oxygen and nutrients (12). Our roughly 40 weeks within the uterus determine our life extremely outside of it. There is now strong evidence that many neurological conditions that manifest after birth have their roots in fetal development. For instance, epidemiological studies have demonstrated that prenatal rather than perinatal or postnatal factors account for most cerebral palsy cases. It is believed that neurological issues have their origins in utero because minimal cerebral brain dysfunction, which is frequently related to intrauterine growth restriction (IUGR), is defined by general writing, reading, and cognitive impairments in childhood (13). Based on histology and neurochemical research, it has been hypothesized that prenatal injury creates a susceptibility in the growing brain, leading someone with risks such as genetic inheritance to have symptoms of schizophrenia in adolescence or early adulthood. Autism and epilepsy may also be partially caused by neurodevelopmental issues (14).

Over the past few decades, tremendous progress has been made in our knowledge of the intricate series of mechanisms that result in the formation of the entire nervous system from a specialized layer of cells just on the dorsal side of the embryo (15). Different parts of the brain develop at various stages, with the brainstem developing first, the cerebral hemispheres second, and the cerebellum developing last. Therefore, the timing and degree of gestational insults are likely to define the type of neuronal injury, its impact on the person's postnatal function, and the manifestation of specific neurological illnesses (16).

The three prenatal phases of brain development line up with the three trimesters of pregnancy. The neural tube grows in the initial trimester of pregnancy. In the second trimester, the brain grows to its full capacity. For the latter three trimesters of pregnancy, the fetus's brain contains the same number of neurons as an adult brain (17).

The essential stages are:

1. Massive growth of immature brain cells.
2. The movement of the cells in the brain toward a target area and the start of their differentiation into the type of cell best suited to that area.
3. Comparable cell types gathering into discrete locations.
4. Many connections are formed between neurons, both within and between regions.
5. Competition among these connections causes many to be selectively eliminated, with the roughly 100 trillion connections left to stabilize. From roughly five weeks after conception onward, these events do not take place in a strict order but rather overlap in time. No new neurons are added after the age of about 18 months, and the division of different cell types into separate regions is essentially finished. However, the process of eliminating unnecessary connections-clearly a crucial step in determining the structure of the mature brain goes on for years (18).

The number of brain cells in the cerebral cortex increases between 15 and 20 weeks after conception significantly, about the middle of pregnancy. The fetus begins sending its own brain waves around the seventh month, which may be heard via the mother's abdomen. Although it remains essential until birth and for some months following, several lines of evidence indicate that good nutrition is most vital for the development of the brain at this point (19).

There are ten stages in the development of the fetal brain:

1. The creation of the neural preplate and the genesis of the neuroectoderm from the ectoderm.
2. The neural plate is created by the preplate splitting.
3. The neural plate's sides rise up and fold inward to form an arch and a neural tube.
4. The development of neuroepithelium produces juvenile nerve cells called neuroblasts. Neuroblasts migrate, sprout leading appendages, and then assemble in certain, genetically predetermined sites, starting with the area that will eventually become the brainstem and spinal cord.
5. The separation of the neuroepithelium into the subventricular and ventricular zones results in distinct waves of mobile neuroblasts.
6. The neural tube bends, twists, and ascends as a result of ongoing neuroblast production and neuron multiplication, resulting in the formation of three cerebral vesicles initially, followed by another two (medulla, pons, midbrain, diencephalon, telencephalon).
7. The gathering, differentiating, and division of neurons that eventually give rise to the limbic system, striatum, layers I and VII (VIa and VIb), and the nuclei of the medulla, pons, midbrain, and hypothalamus.
8. The development of neocortical layers II through VI as a series of neuroblastic waves produces columns and layers while squeezing between neocortical layers I and VII (VIa and VIb).
9. Dendritic growth, synaptic development, myelination, and neuronal differentiation of active brainstem axons that have made dendritic synaptic connections.
10. The medulla, the pons, the midbrain, the diencephalon, the forebrain, and the neocortex all develop functionally after that (20).

19.3 THE BRAINSTEM

By the seventh month of pregnancy, the majority of the descending spinal-motor fibre tracts have reached their target locations and formed their synaptic connections, and the medulla and pons have nearly completed their cycle of myelination. Because the brainstem continues to develop well after birth, brainstem reflexes occur and disappear at various times after birth (21).

19.4 THE DEVELOPMENT OF THE MEDULLA

The medulla, pons, and midbrain make up the brainstem, which matures in a caudal-to-rostral arc beginning in the sixth or seventh week of pregnancy and lasting far into the first year after birth. As a result, the medulla's nuclei typically arise and myelinate before the pons', which in turn emerge and myelinate before the midbrain. As previously mentioned, the fetus and newborn exhibit this pattern of functional development in behavioural terms (21).

The principal medulla structures are created by the seventh to eighth week of gestation, three weeks after the neural tube is first formed. This group includes the nerves of the reticular activating system along with the glossopharyngeal, hypoglossal, spinal accessory, and vagus cranial nuclei (22). As previously mentioned, these nuclei and activation pathways promote arousal, generalized body movements, head movement, shoulder elevation, heart rate, and respiration.

Around the eighth week of pregnancy, the pons starts to appear following the medulla as the brainstem develops in a caudal to the rostral manner (23). The pons takes longer than the medulla to mature and myelinate, and it does not reach advanced stages of development until about the seventh month of pregnancy. Therefore, many of the pons-related functions arise later in fetal development, around the seventh month of pregnancy. Similar to the medulla, some functions emerge earlier than others and in an erroneous or incomplete manner (24).

19.5 THE DEVELOPMENT OF THE MIDBRAIN

Superior visual colliculus development starts during the eighth week of pregnancy, following the caudal-to-rostral pattern of brainstem maturation that is also typical of the midbrain. The inferior-auditory colliculus starts its cycle of maturation around the sixth week of gestation (20). The superior colliculus establishes itself several weeks after the lower colliculus; therefore, the development cycle is highly drawn out. However, the inferior colliculus does not fully develop until the 17th week. Furthermore, myelination and midbrain development do not reach their peak until soon after birth. The superior colliculus responds to and orients to visual inputs, whereas the inferior colliculus is organized to respond to sound. These midbrain areas are responsible for orienting reflexes that involve turning the head and initiating eye movements when they are activated (25). As a result, the fetus's midbrain matures, making it more sensitive to sound. If adequately stimulated or intensely aroused, the fetus reflexively orients, swivels its head and makes lateral eye movements in response. But when a baby is overstimulated, the brainstem normally reacts by dozing off and causing rapid (lateral) eye movements (REM). It is hypothesized that the late-term fetus will display REM and comparable arousal-induced sleep behaviours (26).

19.6 ROLE OF THYROID HORMONE IN FETAL BRAIN DEVELOPMENT

To achieve optimal neurodevelopment, appropriate fetal concentrations of thyroid hormone must exist from the very beginning of pregnancy. T4 has been shown to be detectable in all fetal fluids before the start of fetal thyroid function, especially by six weeks of pregnancy and in quantities that correspond significantly with circulating maternal T4 levels (27). T3 levels, however, are essentially inconsequential by that point and beyond. Contrarily, T3 has been discovered in purified extracts in much higher concentrations in the human fetal brain as early as nine weeks of pregnancy than one might anticipate based on circulating fetal levels. The latter finding suggests that type II 5-iodothyronine deiodinase produces this iodothyronine locally from T4 (28). In addition, studies demonstrate that by eight weeks of pregnancy, nuclear thyroid hormone receptors are produced in numerous brain areas and are occupied by T3 by ten weeks. Collectively, these findings demonstrate that not only does a transfer of maternal thyroid hormone, specifically T4, to the fetus begin functioning as early as the first trimester, but also that the mechanisms governing the availability of T3 to the fetus's brain are expressed and likely functioning by that time. Because significant fetal thyroid hormone production does not start until midgestation, the development of the fetus's brain depends exclusively on maternal T4 until it is at least 24 to 26 weeks old (29).

Normal thyroid-stimulating hormone (TSH) levels and maternal concentrations of free-thyroxine under the normal range for the pregnancy stages are the hallmarks of the illness known as hypothyroxinemia (30). Primary thyroid failure and autoimmune thyroid disease are anticipated to be substantially less common than maternal hypothyroxinemia brought on by an insufficient iodine intake to fulfil the higher needs imposed by the conceptus. According to a wealth of research, iodine deficiency is the most common cause of cerebral palsy, reduced mental and physical skills, and infertility worldwide. More recent research demonstrates that even mild degrees of iodine deficiency have the potential to negatively impact the outcome of pregnancy. Neurodevelopmental impairments do not just occur in areas of severe iodine deficiency (29). During the embryonic period, there were 23 internationally accepted morphological stages.

19.7 ESSENTIAL FOODS AND NUTRIENTS DURING PREGNANCY AND THEIR EFFECT ON THE BRAIN HEALTH OF THE FETUS

When it comes to cognitive development, there are delicate and crucial stages of growth when outside factors, like food, can have a lasting impact. Early malnutrition affects neurogenesis, which affects cell numbers for any given location. For instance, neonatal starvation can impact the cerebral cortex's size and shape. Variations in neurotransmitter production, uptake and receptor synthesis mechanisms are among the neurochemical changes (31). Modifications in signal transmission and metabolism are reflected in neurophysiology. For example, a food deficit during a crucial period of development can cause long-lasting variations in brain assembly and, consequently, cognitive ability. For instance, findings from the Dutch famine studies showed that pregnant women who experienced malnutrition suffered long-lasting negative consequences on their children's mental health (32).

For proteins involved in brain energy metabolism, iron accretion, dopamine metabolism, and antioxidant action in the fetal and neonatal brain, copper, a divalent cation, is required. The growth of the brain substrates for cognition is significantly influenced by micronutrients (33). Different micronutrient deficiencies may have long-term effects on how individuals grow cognitively. Since vitamin A deficiency is the main contributor to infantile blindness, visual perception is critically dependent on vitamin A. Furthermore, it is essential during periods of rapid growth, such as during pregnancy and early life years (34). The vitamin A derivative retinoic acid, which has been shown to influence numerous molecular signal mechanisms in the growing brain by regulating the expression of numerous genes and promoting cellular growth, is essential for the regulation of neurodevelopment (35). Additionally, it has been demonstrated that retinoic acid transmission is crucial for the formation of striatal functional pathways, which are engaged in dopamine-regulated cognitive and motor activities. A current study discovered that children's skills were selectively impacted by thiamine shortage during the first year of life, producing particular abnormalities in the language domains of grammar and lexical retrieval but not conceptual or general cognitive problems (36).

Due to its contribution, folate is essential for brain development and has a well-documented role in the reduction of neural tube abnormalities during early gestation. Folates appear to affect memory as well as disorders like depression and schizophrenia (37). Low maternal folate levels enhance the likelihood of internalizing and externalizing problems in early childhood during the first trimester of pregnancy. Children of folic acid-supplemented mothers display fewer behavioural problems at

18 months, improved cognitive, psychological adjustment, and attention tests at 4 years, and fewer hyperactivity and peer problems at eight years (38).

All membranes, including those in the grey and white matter, need choline to make them, and this requirement increases with growth. In rodents, research suggests that insufficient choline availability causes deprived brain growth and continuing cognitive and behavioural issues due to the developing central nervous system's sensitivity to choline availability (39). Infants born to pregnant mothers who consume diets low in choline are more likely to have neural tube abnormalities and orofacial clefts. Studies on adults have found that people who consume diets higher in choline have a superior cognitive function, but there are not enough studies with sufficient strength to govern whether choline nutrition during gestation improves the developing brain in newborns, especially memory (40).

Iodine deficiency during the first trimester and the beginning of the second trimester of pregnancy is linked to reduced intellectual capacity and will produce irreversible abnormalities in brain development. Inadequate iodine levels during the initial stage of pregnancy are also negatively correlated with the mental development of children at the age of eight (41). Zinc is also crucial because it is essential for cell growth. It is present in the brain at significant concentrations and is crucial for appropriate cognitive development both before and after birth. It performs both functional and structural roles (42). Early-life iron deficiency can also have a negative impact on how the brain develops. An iron shortage in newborns may result in abnormalities in the metabolism of dopamine and the myelination of neurons that may last for years. The developing hippocampus of children with perinatal iron insufficiency had a different neurochemical signature (43).

Docosahexaenoic acid (DHA), an n-3 fatty acid, and arachidonic acid (AA), an n-6 fatty acid, are the two primary long-chain polyunsaturated fatty acids (LC-PUFAs). DHA starts to accumulate in the developing brain while still in utero, with quantitatively significant build-up taking place in the second half of gestation, timed to correspond with the development spike in the grey matter (44). After delivery, babies receive fatty acids via breast milk, and during pregnancy, the mother's LC-PUFA contributes to the fetus is mediated by maternal-fetal placental transfer. Enhanced maternal plasma levels of DHA lead to increased DHA transfer to the fetus as a result of higher maternal DHA intake. Reduced DHA has been linked to issues with babies' cognitive and behavioural development (45). Improved prenatal AA and DHA status has been linked to better neurodevelopmental outcomes up to at least 18 months of age, and studies looking at the effects of prenatal LC-PUFA status or prenatal LC-PUFA supplementation have also shown that subsequent psychomotor development and cognitive performance in children may also benefit (46).

Additionally, most of the research on zinc, vitamin A, iron and iodine supplements suggests that taking all four together may have a bigger impact on some outcomes than just one component alone. Because of the widespread effects of hypothyroidism on the structure and function of neurons, iodine shortage can impact cognitive function and development (47). Evidence suggests that low iron status negatively affects psychological performance due to the brain's iron-containing enzymes' lower activity and the body's diminished ability to synthesize haemoglobin (48). A zinc deficiency in children has been linked to impaired mental function and an increased risk of depression. Decreased weight of the brain, disrupted formation of the hippocampus, impaired systems of neurotransmitters, and alterations in phosphorylation of protein are just a few of the negative consequences of protein limitation on the brain (49).

19.8 FOOD

The brain has two distinct features. First, it is 60% lipid, and second, it consumes much energy. During its most active phase of fetal growth, it consumes up to 70% of the mother's energy and 60% of the energy delivered postnatally. The fetus does not use fat as a source of energy, but it does

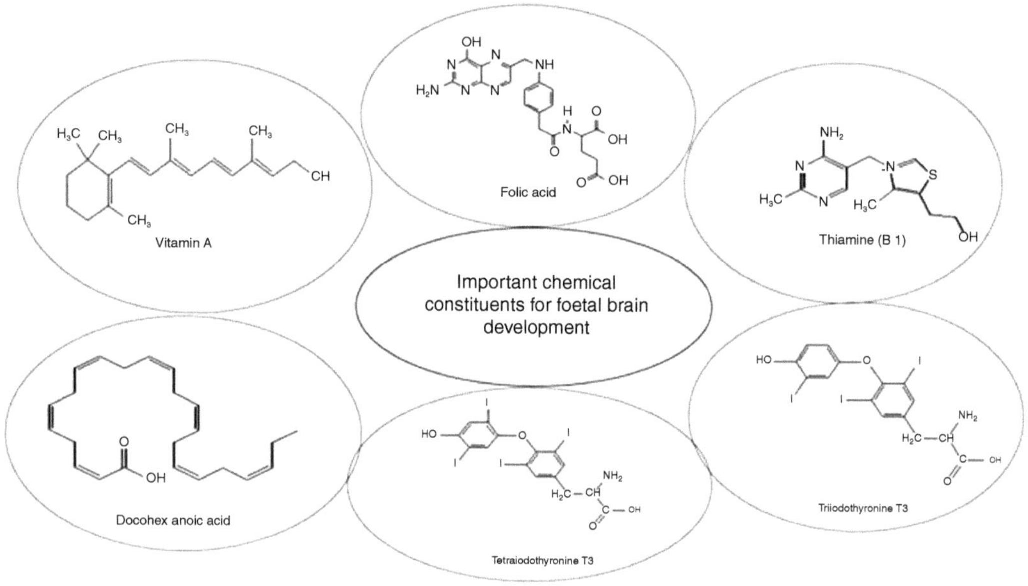

FIGURE 19.1 Structure of important chemical constituent required for fetal brain development.

store fat, which can be a supply of LCPUFAs postnatally while also acting as a vital energy buffer to guarantee the growing brain's demands are met (50).

Every year, approximately 1.4 million kids are born with or acquire severe neurodevelopment abnormalities that have long-term repercussions (51). Since the link between iodine deficiency and cretinism was discovered. Inadequate nutrition has been linked to neurodevelopment abnormalities, which can be reversed with proper nutritional intervention (52).

The mother bears sole responsibility for the development of the brain. Before birth, around 70% of an individual's total number of brain cells split (53). The initial few weeks of embryonic development are the most active for brain cell division (54). A woman is pregnant almost before she realizes it. Because the placenta has not yet developed, the embryo's nutrition is dependent on its mother's health, nutrition, and metabolism (55).

19.8.1 Dietary Guidelines During Pregnancy

The Food Guide Pyramid is the foundation for dietary and nutritional recommendations for women during pregnancy. Due to the requirement for more protein, it is advised to consume more meat, fish, and eggs, with red meat, such as beef, a particularly significant iron source. Eating more fish during pregnancy is encouraged because a sufficient amount of polyunsaturated fatty acids affects how the retina and neuronal system of newborn babies develop. Additionally, it is advised for pregnant mothers to consume more milk products because they are an excellent source of phosphorus, calcium, and riboflavin, all of which are essential for maternal health as well as the development of the fetus's teeth and bone structure (66).

Safe food handling is a crucial component of sound dietary practices to avoid food-related disorders during gestation, including toxoplasmosis and listeriosis. Pregnant women are advised to steer clear of cool, ready-to-eat foods that have not been freshly cooked, raw, or improperly cooked fresh meat to lower their risk of listeriosis and toxoplasmosis. Before eating fresh fruit and vegetables, they should also peel them and wash them thoroughly to get rid of any contaminated

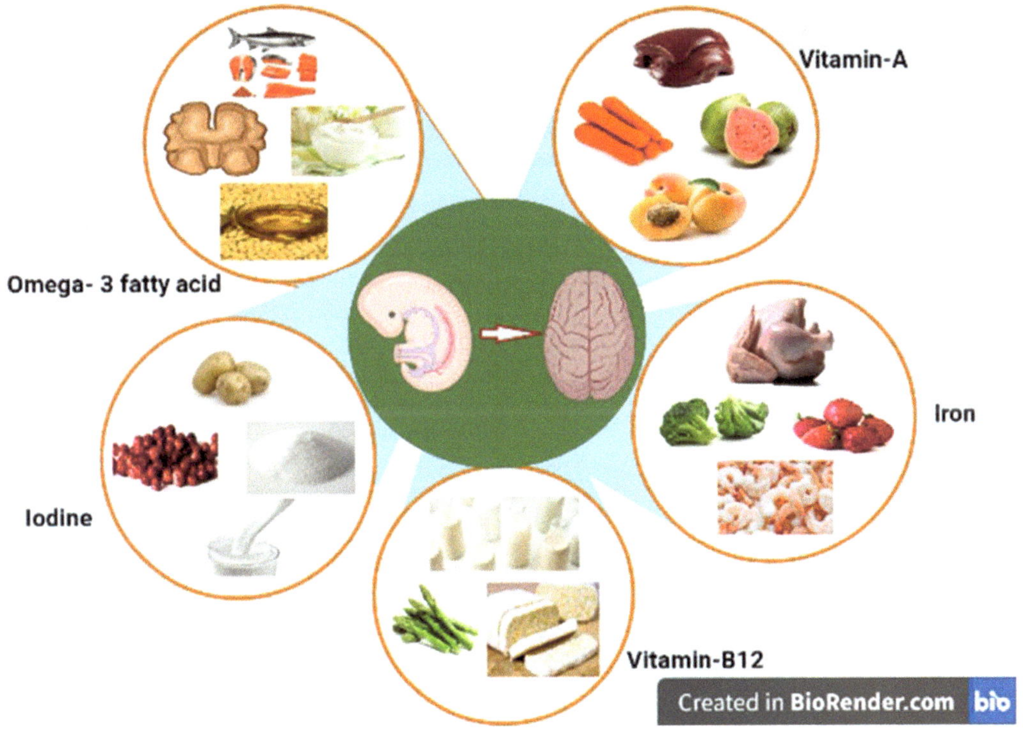

FIGURE 19.2 Nutrients required for fetal brain development.

soil (67). It is advised against pregnant women to consume liver due to high levels of vitamin A, which is related to teratogenic and miscarriage consequences. A healthy lifestyle at the time of gestation should also include reasonable exercise, abstaining from alcohol, avoiding both direct and indirect exposure to tobacco smoke, and avoiding other dangerous substances (68). Preterm birth, gestational diabetes, hypertension, and preeclampsia are all lower risks for pregnant women who exercise. The American College of Obstetricians and Gynaecologists advises pregnant ladies without pregnancy difficulties or pre-existing medical disorders to engage in 30 minutes of exercise each day (69).

Alcohol use is discouraged for women who may become pregnant due to the link between it and fetal alcohol syndrome (FAS). At any point during pregnancy, a harmless amount of alcohol consumption till now has not been identified. Active maternal smoking reduces birth weight and fertility and raises the risk of placental defects, spontaneous abortion, early labour, and neonatal mortality. According to research, pregnant women who smoke are more likely to consume inadequate amounts of antioxidants and other nutrients than those who do not. A recent study found that pregnant women who smoke may negatively impact the offspring's adult cognition (70).

Pregnant women, especially those carrying their first child, are typically driven to change their bad eating habits. Writing advice down can affect knowledge about better eating, but it does not seem to change attitudes or behaviour in the real world. If nutrition counselling for pregnant mothers accounts for both the regular demands of the family and the additional needs that pregnancy imposes, it will have a significant impact. Therefore, it is crucial to educate prospective parents about the nutritional benefits of a healthy diet and make them aware of the potential impact their dietary habits may have on their child's neuronal development (71).

TABLE 19.1
Different nutrients and their sources

		1. Omega 3 fatty acids containing foods (56)
Plant source	Seeds and nuts	Chia seed, walnuts
	Oils	Flax seed oil, soya bean oil, canola oil
Mineral source	Other sea foods and fish	Tuna, salmon, herring
Artificial food	Fortified food	Yogurts, soy beverages
		2. Iron-containing foods (57)
Animal source	Eggs and meat	Chicken, pork, dried beef, Liverwurst, veal, lamp, turkey, ham, beef, liver, eggs (any style)
Mineral source	Seafood	scallops, clams, tuna, mackerel, oysters, shrimp, haddock, sardines
Plant source	Different vegetables	Kale, beet greens, broccoli, chard, spinach, sweet potatoes, collard greens, peas, string beans, and dandelion greens
	Fruit	Figs, dates, raisins, dates, watermelon, prunes, prune juice, and strawberries. Dried peaches and apricots
Artificial foods	Cereals and bread	sliced bread (enriched), wheat-free bread, improved pasta, items made with wheat cereal with bran, corn flour, oat bran, Crusty Wheat, Wheat bread, Refined rice
	Other foods and beans	Beans (garbanzo, kidney, or white, canned), tofu, and tomato-based items (e.g., paste), roasted peas, roasted beans, Lentils, quick breakfast, Molasses, maple syrup, and corn syrup
		3. Iodine-containing foods (58)
Plant source	Potatoes	Potatoes are another excellent source of iodine; however, only organic potatoes should be consumed because conventionally cultivated potatoes have a tendency to absorb a lot of pesticide
	Fruits	A select few fruits, like pineapple, cranberries, and strawberries, are good and abundant sources of iodine
	Beans	Both green beans and navy beans are known for their high iodine content. Some green beans contained 3/mcg of iodine, which is equivalent to 2% of the daily need
	Bananas	Although bananas are not fruits, they are high in potassium and iodine. A single banana has 3/mcg of iodine, which provides 2% of the daily iodine requirement
Animal source	Milk	Although bananas are not fruits, they are high in potassium and iodine. A single banana has 3/mcg of iodine, which provides 2% of the daily iodine requirement
	Eggs	Hard-boiled eggs are popular due to their high iodine content. Eggs are high in Zinc, Vitamin D, Vitamin A and calcium, in addition to iodine
Artificial food	Fortified iodized salt	Using enriched iodized salt instead of conventional salt when cooking is an easy method to integrate iodine into your daily diet. Regular salt is a crystal of sodium and chloride, whereas iodized salt is a mineral enhanced with iodine
	Cheese	Many people do not include cheese and other dairy products, such as pasteurized goods, in their regular diets despite the fact that they are a good source of calcium, protein, and iodine
	Yogurt	Yogurt, particularly organic yogurt, is high in iodine. It contains a lot of iodine in addition to protein and calcium
Mineral source	Seafood	Minerals and nutrients can be found in abundance in seafood. Iodine is one of the minerals that is most frequently present in different kinds of seafood
		4. Vitamin A containing food (59)
Animal-sourced foods	Liver of beef	6,421 mcg (713% DV) per slice: 9,442 mcg (1,049% DV) in 100 grams
	Liver of lamb	2,122 mcg (236% DV) per ounce: 7,491 mcg (832% DV) in 100 games

TABLE 19.1 (Continued)
Different nutrients and their sources

	Feta cheese	35 mcg (4% DV) per ounce: 125 mcg (14% DV) in 100 grams
	Caviar	43 mcg (5% DV) per tablespoon: 271 mcg (30% DV) per 100 grams
	Cream cheese	45 mcg (5% DV) per tablespoon: 308 mcg (34% DV) per 100 grams
	Blue cheese	56 mcg (6 percent DV) per ounce: 198 mcg (22% DV) in 100 grams
	Trout	71 mcg (8% DV) per fillet: 100 mcg (11% DV) in 100 grams
	Hard-boiled egg	74 mcg (8% DV) per 1 big egg: 149 mcg (17% DV) in 100 grams
	Roquefort cheese	83 mcg (9% DV) per ounce: 294 mcg (33% DV) in 100 grams
	Camembert	92 mcg (10% DV) per wedge: 241 mcg (27% DV) in 100 grams
	Cheddar	92 mcg (10% DV) per slice: 330 milligrams (37% DV) in 100 grams
	Limburger cheese	96 mcg (11% DV) per slice: 340 mcg (38% DV) in 100 grams
	Butter	97 mcg (11% DV) per tablespoon: 684 mcg (76% DV) in 100 grams
	Goat cheese	115 mcg (13% DV) per slice: 400 mcg (45% DV) in 100 grams
	Goose liver pate	130 mcg (14% DV) per tablespoon: 1,001 mcg (111% DV) in 100 grams
	Blue-fin tuna	214 mcg (24% DV) per ounce: 757 mcg (84% DV) in 100 grams
	Salmon	229 mcg (25% DV) for a half-fillet: 149 mcg (17% DV) in 100 grams
	King mackerel	388 mcg (43% DV) for a half-fillet: 252 mcg (28% DV) in 100 grams
	Cod liver oil	1,350 mcg (150% DV) per teaspoon: 30,000 mcg (3,333% DV) in 100 grams
	Liver sausage	1,495 mcg (166% DV) per slice: 8,384 mcg (923% DV) in 100 grams
Vegetables source food containing provitamin A	Sweet potato (cooked)	1,836 mcg (204% DV) per cup: 1,043 mcg (116% DV) in 100 grams
	Romaine lettuce (raw)	122 mcg (14% DV) per big leaf: 436 mcg (48% DV) in 100 grams
	Spinach (raw)	141 mcg (16% of DV) per cup: 469 mcg (52% DV) in 100 grams
	Swiss chard (raw)	147 mcg (16% DV) per leaf: 306 mcg (34% DV) in 100 grams
	Sweet red pepper (raw)	257 mcg (29% DV) per big pepper: 157 mcg (17% DV) in 100 grams
	Carrot (cooked)	392 mcg (44% DV) per medium carrot: 852 mcg (95% DV) in 100 grams
	Turnip greens (cooked)	549 mcg (61% DV) per cup: 381 mcg (42% DV) in 100 grams
	Collards (cooked)	722 mcg (80% DV) per cup: 380 mcg (42% DV) in 100 grams
	Kale (cooked)	885 mcg (98% DV) per cup: 681 mcg (76% DV) in 100 grams
	Winter squash (cooked)	1,144 mcg (127% DV) per cup: 558 mcg (62% DV) in 100 grams
Fruits source containing Provitamin-A: Provitamin A is found in higher concentrations in vegetables than in fruits. However, only a few fruits provide adequate amounts.	Cantaloupe.	169 mcg (19%) per large wedge: 172 mcg (19% DV) in 100 grams.
	Mango	181 mcg (20% DV) per medium mango: 54 mcg (6% DV) in 100 grams
	Passion fruit	12 mcg (1% DV) per medium fruit: 64 mcg (7% DV) in 100 grams
	Guava	17 mcg (2% DV) per medium guava: 31 mcg (3% DV) in 100 grams31 mcg
	Nectarine	24 mcg (3% DV) per medium nectarine: 17 mcg (2% DV) in 100 grams17 mcg
	Tangerine	30 mcg (3% DV) per medium tangerine: 34 mcg (4% DV) in 100 grams
	Apricot	34 mcg (4% DV) per medium apricot: 96 mcg (11% DV) in 100 grams
	Papaya	74 mcg (8% DV) per small papaya: 47 mcg (5% DV) in 100 grams
	Watermelon	80 mcg (9% DV) per wedge: 28 mcg (3% DV) in 100 grams
	Pink or red grapefruit	143 mcg (16% DV) per medium grapefruit: 58 mcg (6% DV) in 100 grams
	5. Vitamin- B_{12} containing food (60–65)	
Plant source	Vegetables	Asparagus, Tessa jute, water shield, Japanese butterbur, broccoli, mung bean sprouts, and mushrooms
	leaves of tea	0.1 – 1.2 g of vitamin B12 per 100 g of dry weight
Animal source	Non-vegetables	Fish, shellfish, milk, eggs, and meat
Artificial food	Soybean-based fermented	Vitamin B12 is present in large amounts in Tempe
	Fermented korean vegetables	B12 is present in trace amounts in Kiamichi

19.9 DISCUSSION

Being one of the most prevalent health problems in today's society, mental health illnesses are not only comorbid but also impact how the brains of future generations develop. There is growing evidence that diet-induced parental programming of a child's metabolism, brain function, emotional behaviours, and, ultimately, mental health. Together, these results highlight the significance of epigenetic alterations that might affect how genes are expressed in offspring and the growth of the microbiota-gut-brain axis, which has a real impact on brain health, neurological development, and behaviour.

It is widely acknowledged that the mother's nutritional health during gestation pregnancy can impact the progress of the fetus's brain, which in turn influences the child's behaviour and cognitive abilities. For iron, n-3 fatty acids, and folic acid, the evidence of a relationship between prenatal nutrition and neurological growth appears to be more reliable. Additionally, it is widely accepted that micronutrients positively impact health, particularly when pregnant women eat well to maximize their baby's cognitive development. Recent research has shown that supplementing with a single nutrient is insufficient compared to supplementing with more complicated formulations.

Good nutrition of the mother is a major component inducing the health of both child and mother throughout pregnancy, a crucial time. Maintaining a healthy diet will give you the right kinds and amounts of nutrients to ensure both mother and unborn child as healthy as possible sure both you and your unborn child are as healthy as possible, both physically and mentally. Pregnant women can choose appropriate food to meet their increased nutritional and energy needs while meeting the growing fetus's demands.

19.10 CONCLUSION

The authors concluded that numerous groups would need to work together in sizable national and international consortia to comprehend and address the significant dietary issues that currently negatively impact children's mental health. The information gleaned from this study will strengthen the scientific basis for food suggestions for expectant mothers to enhance the fetus's mental health. Regarding how mental health affects behaviour, education, work potential, and brain diseases in every age group, the discovery and adoption of nutritional recommendations are the ultimate of broad social value for the general community. Therefore, genetic variation caused by nutrition during fetal brain development and its positive and negative effects on future mental health should be considered in future research when assessing the effects of nutrition and making nutritional recommendations.

REFERENCES

(1) Anjos, T., Altmäe, S., Emmett, P., Tiemeier, H., Closa-Monasterolo, R., Luque, V., … & Campoy, C. (2013). *European Journal of Nutrition, 52*(8), 1825–1842.

(2) Nyaradi, A., Li, J., Hickling, S., Foster, J., & Oddy, W. H. (2013). *Frontiers in Human Neuroscience, 7*, 97.

(3) Lindsay, K. L., Buss, C., Wadhwa, P. D., & Entringer, S. (2019). *Biological Psychiatry, 85*(2), 135–149.

(4) Arija, V., & Canals, J. (2021). *Nutrients, 13*(5), 1644.

(5) Lindsay, K. L., Buss, C., Wadhwa, P. D., & Entringer, S. (2019). *Biological Psychiatry, 85*(2), 135–149.

(6) Georgieff, M. K., Ramel, S. E., & Cusick, S. E. (2018). *Acta Paediatrica, 107*(8), 1310–1321.

(7) Benton, D. (2010). *Molecular Nutrition & Food Research, 54*(4), 457–470.

(8) Linderkamp, O., Janus, L., Linder, R., & Skoruppa, D. B. (2009). *International Journal of Prenatal and Perinatal Psychology and Medicine, 21*(1/2), 4–16.

(9) Hanson, M., Godfrey, K. M., Lillycrop, K. A., Burdge, G. C., & Gluckman, P. D. (2011). *Progress in Biophysics and Molecular Biology*, *106*(1), 272–280.
(10) Lindsay, K. L., Buss, C., Wadhwa, P. D., & Entringer, S. (2019). *Biological Psychiatry*, *85*(2), 135–149.
(11) Broad, K. D., & Keverne, E. B. (2011). *Proceedings of the National Academy of Sciences*, *108*(37), 15237–15241.
(12) Heijtz, R. D. (2016, December). *Seminars in fetal and neonatal medicine.* (Vol. 21, No. 6, pp. 410–417). WB Saunders, USA.
(13) Rees, S., & Harding, R. (2004). *Neuroscience Letters*, *361*(1–3), 111–114.
(14) Cohen, E., Baerts, W., & van Bel, F. (2015). *Neonatology*, *108*(4), 269–276.
(15) Stiles, J., & Jernigan, T. L. (2010). *Neuropsychology Review*, *20*(4), 327–348.
(16) Rice, D., & Barone Jr, S. (2000). *Environmental Health Perspectives*, *108*(suppl 3), 511–533.
(17) Dobbing, J. (1990). *Brain, behaviour, and iron in the infant diet* (pp. 1–17). Springer, London.
(18) Linderkamp, O., Janus, L., Linder, R., & Skoruppa, D. B. (2009). *International Journal of Prenatal and Perinatal Psychology and Medicine*, *21*(1/2), 4–16.
(19) De Escobar, G. M., Obregón, M. J., & Del Rey, F. E. (2004). *European Journal of Endocrinology*, *151*(Suppl_3), U25–U37.
(20) Joseph, R. (2000). *Developmental Review*, *20*(1), 81–98.
(21) Herschkowitz, N. (2000). *Brain and Development*, *22*(7), 411–416.
(22) Izac, M. S. M. (2006). *American Journal of Electroneurodiagnostic Technology*, *46*(1), 18–38.
(23) Prayer, D., Kasprian, G., Krampl, E., Ulm, B., Witzani, L., Prayer, L., & Brugger, P. C. (2006). *European Journal of Radiology*, *57*(2), 199–216.
(24) Grady, F. S., Boes, A. D., & Geerling, J. C. (2022). *Frontiers in Neuroscience*, *16*, 1–16. Article no.: 930514.
(25) Sprague, J. M., & Meikle Jr, T. H. (1965). *Experimental Neurology*, *11*(1), 115–146.
(26) MAAKS, D. L. G. (2019). *Burns' Pediatric Primary Care E-Book*, 7th ed., *63–72*.
(27) Miranda, A., & Sousa, N. (2018). *Brain and Behavior*, *8*(2), e00920.
(28) Braun, J. M. (2017). *Nature Reviews Endocrinology*, *13*(3), 161–173.
(29) de Escobar, G. M., Obregón, M. J., & del Rey, F. E. (2004). *Best Practice & Research Clinical Endocrinology & Metabolism*, *18*(2), 225–248.
(30) Skeaff, S. A. (2011). *Nutrients*, *3*(2), 265–273.
(31) Prado, E. L., & Dewey, K. G. (2014). *Nutrition Reviews*, *72*(4), 267–284.
(32) Herlenius, E., & Lagercrantz, H. (2004). *Experimental Neurology*, *190*, 8–21.
(33) Georgieff, M. K., Ramel, S. E., & Cusick, S. E. (2018). *Acta Paediatrica*, *107*(8), 1310–1321.
(34) Thompson, R. A. (2001). *The future of children* (pp. 21–33) Princeton University, Princeton, New Jersey, USA.
(35) Malik, M. A., Blusztajn, J. K., & Greenwood, C. E. (2000). *Journal of Nutritional Biochemistry*, *11*(1), 2–13.
(36) Berk, L. E. (2022). *Infants and Children: Prenatal Through Middle Childhood.* SAGE Publications, New York, USA.
(37) Jain, K. K. (2019). *Medlink-The information resource for clinical neurology.* MedLink Neurology, San Diego, USA.
(38) Naninck, E. F., Stijger, P. C., & Brouwer-Brolsma, E. M. (2019). *Advances in Nutrition*, *10*(3), 502–519.
(39) Derbyshire, E., & Obeid, R. (2020). *Nutrients*, *12*(6), 1731.
(40) Holford, P. (2010). *Optimum nutrition for the mind.* Hachette, UK.
(41) Glinoer, D. (2004). *Best Practice & Research Clinical Endocrinology & Metabolism*, *18*(2), 133–152.
(42) Osredkar, J., & Sustar, N. (2011). *Journal of Clinical Toxicology*, *3*(2161), 0495.
(43) Lozoff, B., & Georgieff, M. K. (2006, September). *Seminars in Pediatric Neurology* (Vol. 13, No. 3, pp. 158–165). WB Saunders, Elsevier USA.
(44) Strickland, A. D. (2014). *Medical Hypotheses*, *82*(5), 522–528.
(45) Bobiński, R., & Mikulska, M. (2015). *Acta Biochimica Polonica*, *62*(3), 499–507.
(46) Escolano-Margarit, M. V., Ramos, R., Beyer, J., Csábi, G., Parrilla-Roure, M., Cruz, F., & Campoy, C. (2011). *The Journal of Nutrition*, *141*(6), 1216–1223.
(47) Black, R. E., Allen, L. H., Bhutta, Z. A., Caulfield, L. E., De Onis, M., Ezzati, M., & Maternal and Child Undernutrition Study Group. (2008). *The Lancet*, *371*(9608), 243–260.

(48) Hercberg, S., & Galan, P. (1989). *Acta Paediatrica, 78*, 63–70.
(49) Cortés-Albornoz, M. C., García-Guáqueta, D. P., Velez-van-Meerbeke, A., & Talero-Gutiérrez, C. (2021). *Nutrients, 13*(10), 3530.
(50) Broadhurst, C. L., Wang, Y., Crawford, M. A., Cunnane, S. C., Parkington, J. E., & Schmidt, W. F. (2002). *Comparative Biochemistry and Physiology Part B: Biochemistry and Molecular Biology, 131*(4), 653–673.
(51) Synnes, A., & Hicks, M. (2018). *Clinics in Perinatology, 45*(3), 393–408.
(52) Kretchmer, N., Beard, J. L., & Carlson, S. (1996). *The American Journal of Clinical Nutrition, 63*(6), 997S–1001S.
(53) Crawford, M. A., Costeloe, K., Ghebremeskel, K., Phylactos, A., Skirvin, L., & Stacey, F. (1997). *The American Journal of Clinical Nutrition, 66*(4), 1032S–1041S.
(54) Bello, B., Reichert, H., & Hirth, F. (2006). *The brain tumor gene negatively regulates neural progenitor cell proliferation in the larval central brain of Drosophila.* Development, 133, 2639–2648.
(55) Williamson, C. S. (2006). *Nutrition Bulletin, 31*(1), 28–59.
(56) Ganesan B., Brothersen C., & McMahon D. J. (2014 Jan 1). *Critical Reviews in Food Science and Nutrition. 54*(1), 98–114.
(57) Monsen E. R., Hallberg L., Layrisse M., Hegsted D. M., Cook J. D., Mertz W., & Finch C. A. *The American Journal of Clinical Nutrition, 31*(1), 134–141.
(58) Haldimann M., Alt A., Blanc A., & Blondeau K. (2005 Sep 1). *Journal of food Composition and Analysis, 18*(6), 461–471.
(59) Arnarson, A. (2024). Healthline. https://www.healthline.com/nutrition/foods-high-in-vitamin-a
(60) Nout, M. J. R., & Rombouts, F. M. (1990). *Journal of Applied Bacteriology, 69*, 609−633.
(61) Okada, N., Hadioetomo, P. S., Nikkuni, S., Katoh, K., & Ohta, T. (1983). *Report of National Food Research Institute (Japan), 43*, 126−129.
(62) Kwak, C. S., Hwang, J. Y., Watanabe, F., & Park, S. C. (2008). *Korean Journal of Nutrition, 41*, 439−447.
(63) Miyamoto, E., Kittaka-Katsura, H., Adachi, S., & Watanabe, F. (2005). *Vitamins 79*, 329–332.
(64) Watanabe, F., Schwarz, J., Takenaka, S., Miyamoto, E., Ohishi, N., Nelle, E., Hochstrasser, R., & Yabuta, Y. (2012). *Journal of Nutritional Science and Vitaminology, 58*, 438−441.
(65) Watanabe, F., Takenaka, S., Katsura, H., Masumder, S. A., Abe, K., Tamura, Y., & Nakano, Y. (1999). *Journal of Agricultural and Food Chemistry, 47*, 2341−2343.
(66) Jackson, A. A., & Robinson, S. M. (2001). *Public Health Nutrition, 4*(2b), 625–630.
(67) Ortega, R. M. (2001). *Public Health Nutrition, 4*(6a), 1343–1346.
(68) Malek, L., Umberger, W., Makrides, M., & Zhou, S. J. (2016). *Public Health Nutrition, 19*(7), 1155–1163.
(69) Caut, C., Leach, M., & Steel, A. (2020). *Maternal & Child Nutrition, 16*(2), e12916.
(70) Forbes, L. E., Graham, J. E., Berglund, C., & Bell, R. C. (2018). *Nutrients, 10*(8), 1032.
(71) Kadawathagedara, M., Ahluwalia, N., Dufourg, M. N., Forhan, A., Charles, M. A., Lioret, S., & de Lauzon-Guillain, B. (2021). *Maternal & Child Nutrition, 17*(3), e13140.